Novel Anti-Corrosion and Anti-Fouling Coatings and Thin Films

Scrivener Publishing
100 Cummings Center, Suite 541J
Beverly, MA 01915-6106

Publishers at Scrivener
Martin Scrivener (martin@scrivenerpublishing.com)
Phillip Carmical (pcarmical@scrivenerpublishing.com)

Novel Anti-Corrosion and Anti-Fouling Coatings and Thin Films

Edited by

Hari Murthy

Dept. of Electronics and Communication Engineering, CHRIST University, Bengaluru, India

Vinay Jha Pillai

Dept. of Electronics and Communication Engineering, CHRIST University, Bengaluru, India

Kukatlapalli Pradeep Kumar

Dept. of Computer Science and Engineering, CHRIST University, Bengaluru, India

and

Matthew Cowan

Dept. of Chemical and Processing Engineering, University of Canterbury, Christchurch, New Zealand

Scrivener Publishing

WILEY

This edition first published 2024 by John Wiley & Sons, Inc., 111 River Street, Hoboken, NJ 07030, USA
and Scrivener Publishing LLC, 100 Cummings Center, Suite 541J, Beverly, MA 01915, USA

For more information about Scrivener publications please visit www.scrivenerpublishing.com.

Wiley Global Headquarters
111 River Street, Hoboken, NJ 07030, USA

For details of our global editorial offices, customer services, and more information about Wiley products visit us at www.wiley.com.

Library of Congress Cataloging-in-Publication Data

ISBN 978-1-394-23428-8

Cover image: Pixabay.Com
Cover design by Russell Richardson

Set in size of 11pt and Minion Pro by Manila Typesetting Company, Makati, Philippines

Printed in the USA

10 9 8 7 6 5 4 3 2 1

Contents

Preface

Corrosion is a highly challenging and damaging phenomenon and therefore various methods of corrosion mitigation have been developed which depend upon the nature of the metal and the environment. Metals such as steel and aluminium are among the most important construction materials, but their safety and aesthetic features are compromised due to corrosion, resulting in economic damage. A major part of the surface engineering of metals is focused on corrosion protection by the use of organic corrosion inhibitors. Organic coatings have been used on many substrates for design purposes, thereby preserving the outward appearance, as well as providing corrosion protection. The primary effect of a coating is to form a physical barrier. Effective corrosion protection by an organic coating requires proper substrate preparation, expert formulation of treatment and coating chemicals, and appropriate application processes, as well as adaptation to different uses and service environments. Recently, nanomaterials and nanocomposite materials have been developed as corrosion inhibitors and are the most noble and effective alternatives to traditional organic corrosion inhibitors. Nanomaterials provide reasonably high anticorrosive activity in both aqueous and solution phases. A unified approach to this task is lacking, however, which highlights the role of all disciplines involved in the creation and use of corrosion protection coatings for metals. Fouling is the process of accumulating unwanted material that is mostly non-living and comprised of detritus and organic or inorganic compounds, or organisms, such as tiny viruses up to giant kelps. This book covers both the processes of biofouling and anti(bio)fouling, and the devices that stop the biofouling process.

The book not only showcases the current state of the technology, but also sparks inspiration for future research. The chapters within this volume span a wide range of topics, reflecting the interdisciplinary nature of organic coatings and material science. This book provides a missing synopsis by providing an understanding of the anticorrosive and antibiofouling effects of nanomaterials and nanocomposites under different environments. It features an up-to-date picture of the quality and chemistry of a substrate surface, its proper preparation by conversion treatment, the function of resins and anticorrosive pigments in paints, and novel concepts for corrosion protection. Overall, this book is written for scholars in academia and industry, working in corrosion engineering, metal surface and coatings engineering, materials science, and applied chemistry. It is intended to provide a better understanding of the mutual roles and responsibilities in corrosion and biofouling protective organic coatings, and to pave the way to a more durable and sustainable preservation of our valuables and resources.

We express our sincere gratitude to the authors who have generously shared their expertise and insights, making this volume possible. Their dedication to advancing the field of material science is evident in the quality and depth of their contributions.

In closing, this book is offered as a resource for those who are eager to explore the frontiers of scientific discovery and technological innovation. We sincerely hope this book will be useful in providing valuable knowledge and inspire new collaborations, spark novel ideas, and contribute to the ongoing dialogue that shapes the future of thin film coatings.

Hari Murthy
Vinay Jha Pillai
Kukatlapalli Pradeep Kumar
Matthew Cowan

Acknowledgements

The authors would like to thank the management at CHRIST Deemed to be University for providing the necessary resources and the motivation to complete the proposed title; Rev. Dr. Fr Jose CC, Vice Chancellor, CHRIST Deemed to be University, and Dr. Anil Pinto, Registrar, CHRIST Deemed to be University, for their unwavering support; Rev. Dr. Fr Sony J. Chundattu, Director, School of Engineering and Technology, CHRIST Deemed to be University, and Dr. Iven Jose, Dean, School of Engineering and Technology, CHRIST Deemed to be University, for ensuring that the resources required for the book are available at all times; and Dr. Inbanila K, Head of Department, Department of Electronics and Communication Engineering, School of Engineering and Technology, CHRIST Deemed to be University, and Dr. Mary Anita, Head of Department, School of Engineering and Technology, CHRIST Deemed to be University, for encouraging us to go ahead with the interdisciplinary topic and providing all the support to ensure that the book comes out in due time. Last but not least, we would like to thank our families who have been the greatest source of inspiration and who have put up with our antics.

1

Antifouling Nanoparticle Coatings for Post-Harvest Food Preservation

Gokuraju Thriveni[1]*, Hari Murthy[1] and CH. Anusha[2]

[1]Department of Electronics and Communication Engineering, School of Engineering and Technology, CHRIST (Deemed to be University), Bengaluru, Karnataka, India
[2]Department of Mechanical Engineering, Vignan Institute of Technology and Science - Deshmukhi, Hyderabad, Telangana, India

Abstract
The reports of the World Food Preservation Center exhibit that the entire number of inhabitants in the globe will reach 9.6 billion by the year 2050. With the increasing population, there is a growing pressure on agricultural produce which is further amplified by the losses between harvest and retail (~17%) and other wastages (~17%). Globally, this is a major concern, and thus, there is no surprise to look into current research developments in food preservation. This chapter provides comprehensive reports on the recent trends in various nanoparticle coatings for the aforementioned application. Silicon dioxide (SiO_2), titanium dioxide (TiO_2), and zinc oxide (ZnO) nano regime coatings on silicon substrate and polymer substrates have been discussed for antifouling applications. This chapter is categorized as follows: Section I delivers the introduction about the significance of food preservation in reference to the reported statistics. In section II, the discussion starts with the materials and methods for post-harvest food saving. Major recent advances in terms of materials or methods to increase the shelf life of cuisine are portrayed in the same section II. Section III entails the appropriate computational methods to envision the interaction of food residuals with coated nanolayers through sensing. The final section (IV) delivers a guideline on feasible research implications to address the shortcomings in food preservation from a broader perspective.

Keywords: Nanoparticles, food preservation, zinc oxide nanoparticles, sensing, nanocoatings, antifouling, food monitoring, food waste

**Corresponding author*: gokuraju.thriveni@christuniversity.in

Hari Murthy, Vinay Jha Pillai, Kukatlapalli Pradeep Kumar and Matthew Cowan (eds.) Novel Anti-Corrosion and Anti-Fouling Coatings and Thin Films, (1–12) © 2024 Scrivener Publishing LLC

1.1 Introduction

Food waste is a highly common occurrence in our day-to-day life which has a significant impact on society, economy, and environment [1]. According to a United Nations (UN) report published in 2021, an estimated 931 million tons of food refuse was produced widely in 2019 [2]. There are almost 690 million people who do not have access to a full meal every day [3] which makes the food waste even more alarming. Wasted food also implies resource misutilization including water and land. Reducing food waste is a paramount requirement under the UN Sustainability Goals to address concerns about food security, water scarcity, global warming, and pollution. After harvesting the produced food is composed of crude resources that are prone to severe warehouse surroundings before they are parceled up and moved to the merchandise [4]. A significant quantity of post-harvest products is bygone during manufacturing, transit, packing, and other arbitrator steps. Moisture content in the produce along with the plant hormones and microbial infestation are major causes of food spoilage.

There is a need to protect post-harvest products from deleterious conditions. The GDP of India is expected to grow 10% annually with food processing covering 32% of the total food market and employing over 1.7 million people [5]. It has been observed that a "one-shoe-fits-all" approach is not feasible due to the wide range of optimal conditions for different crops.

One approach for food preservation is through smart packaging with features incorporated into it such as antifouling, antioxidant, antibacterial, freshness sensors, and pH monitors. Another method to enhance the shelf life is through coatings that repel water and grease, thereby acting as barrier coatings against humidity.

1.2 Materials Support Post-Harvest

Nanomaterial films serve the task of upgrading tastiness and aliment resistance to the extension of pathogens on the exterior of foods [6, 7]. Silicon is a well-known element in the semiconductor industry with atomic number 14 that can be derived from silicon dioxide (SiO_2) which is abundant on Earth [8]. SiO_2 has recorded extensive approaches toward environment-friendly photocatalysis for marine antifouling applications [9]. Silicon is the most common constituent that exists widely in the surface soil layer of the earth, commonly in the form of silica SiO2 in nature. In the present scenario, nano- and microscale SiO2 are employed in the food industry and other fields commonly because of their desired end such as high specific surface area, low toxicity, biocompatibility, optical transparency, and chemical/water/thermal stability [10].

According to the US Food and Drug Administration (FDA), titanium dioxide nanoparticles (TiO2) are safe to utilize in individual meals and manufacturing [11]. TiO2 possesses microbicidal tasks and productive film workability [12]. The white nanosized titanium dioxide particles (nano-TiO2) or transparent are safe and present the best viral sterilization and impediment task [13]. It states that nano-TiO2 and nano-SiO2 composite coatings might possibly sustain post-harvest qualities such as decomposing regard, reduction value, and compression by impeding the extension of bacteria. Furthermore, chitosan consists of nanosized TiO2 film indicating an ethylene degradation task manifests UV light, impeding the ripening process, and persuading changes in the product quality [14, 15]. Nano-TiO2 was widely utilized in the food, pharmaceutical, cosmetic, and chemical industries because of its high resistance, nonpoisonous, germicidal task, and isolated photocatalytic properties [13].

Zinc oxide nanoparticles (ZnO-NPs) also plays a major role in post-harvesting foods by attacking the food pathogens to increase shelf life, increasing food quality that directly influences individual health [16]. ZnO-NPs constructively fight dangerous microorganisms and prevent food contagion along adsorbent-influence laminate impairment and ROS-moderated essential pathogenic that results in being the most efficient disinfectant force resistant to microbes found in food [17]. It is explained that the experimental appeal of a ZnO interruption restrains 0.3% citric acid at a distinct gathering (0, 1, 3, 5, and 8 mM) in *Listeria monocytogenes*, *Escherichia coli*, *Staphylococcus aureus*, and *Bacillus cereus* that markedly reserved a development about tension throughout the way of 12 h. Interlude of ZnO with 5 mM and 8 mM citric acid is more constructive for each tension; 5 mM and 8 mM ZnO are determined to a great extent of development in mango smoothie. Following, it is defined that the ZnO-NPs are able to lower the primary development compute of tension in the mango emulsion [18].

ZnO-NPs possess undeviating contact accompanied by cell exterior bitter cellular membranes that eventually get in and produce free radical damage over the pathogen chamber, which depicts the unease of unit development to unit expiry. Many scientists recommend that ZnO-NPs are a constructive germicide means enclosed by a sustenance safeguarding the area. ZnO-NPs are frequently employed just like a cleansing factor for various implements, vessels used in cuisine manufacturing in expectation of effect along with the decay of food poisonous virulent [19].

ZnO-NPs not only indicate antimicrobic tasks but also prevail steady underneath unfavorable circumstances, like high temperature and high pH values. In recent advancements, ZnO-NPs are considered "generally recognized as safe (GRAS)" by the US FDA for both human and animal consumption [19, 20].

Bio detection systems belonging to ZnO-NPs are employed effectively to identify the various viruses and pollutants in cuisine and aqua in the container [21]. ZnO-NPs are able to control other crucial characteristics of edibles, like vapor amount, immersion, substrate, dispersible, and concept [22]. To ease the possibility of pollutants through viruses prior to or subsequent to extract, it is important to primarily acknowledge the contact of accumulation time and meteorological conditions on a contagious populace among slightly decadent foods [22].

ZnO-NPs are also involved in safeguarding and conserving greenness and nutritiousness in regard to sustenance [23]. It is demonstrated that the application of ZnO-NPs (0, 0.1, 0.5, and 1 g/kg) is conservative for guiding bacterial and physicochemical decay about mayonnaise throughout a frozen repository within the duration of 6 months. Furthermore, ZnO-NPs to mayonnaise concluded at detained development about pathogens, conservation of real and alkaline properties, and enhanced serviceable life in contrast to the regulate (0 g/kg) [24].

After conducting further investigation, it is examined that the antimycotic task of ZnO-NPs is inimical to twofold post-harvest product fungi viruses, *B. cinerea* and *P. expansum*. The implementation in regard to the process of ZnO-NPs upon the progress of mold filaments was also analyzed. It concludes that absorption of ZnO-NPs higher with 3 mmol L^{-1} remarkably hinders through widening either *B. cinerea* or *P. expansum*. Nevertheless, the nanoparticles were established in an effort to occur greater efficiently in comparison to *B. cinerea* than *P. expansum* [25].

The glucose oxidase GOx/ZnONP mist has been put in for the post-harvest remedy toward peach. The restraint and healing clusters were preserved among surrounding temperatures for 15 days, and furthermore, typical quality variables were examined. Inward a divergence from the administer cluster, the GOx/ZnONP mist medicament became exceptionally constructive by preserving the behavioral aspect toward fruits indeed greater than 12 days and additionally revealed a consequential abatement amidst lessen for load, steadiness, TSS, and DPPH clear essential scavenging task of harvest. Hence, GOx/ZnONPs are a suitable way to widen the post-harvest storage life of peaches [26].

It demonstrated that ZnO-NPs are a capable nano-agent to improve Zn content in tomatoes with Zno-Nps at 100 mg L^{-1} applied on the outside layer of tomatoes for post-harvesting of 4 weeks which was usefully proved that ZnO-Nps have enhanced the shelf life of tomatoes [26].

Thus, the above materials and methods act as disinfectants to preserve food and enhance serviceable life.

1.3 Computational Methods to Envision the Interaction of Food Residuals with Coated Nanolayers Through Sensing

Nanotechnology improved the nutritional content of alimentary market merchandise ranging from beginning items to finished goods [27]. Several nano-based approaches, including biochemical tests, microbiological examinations, electrolytic investigations, surface-enhanced Raman scattering (SERS), and the measurement of color, are being established to assist with food inspection as represented in Figure 1.1. These methods rendered it feasible to identify antimicrobial agents, bacteria, herbicides, food sensitivities, toxic substances, and other pollutants throughout consumer goods and the handling of food. In earlier times, there were reports and debates on possibilities for nanotechnology throughout the food and beverage sector [28].

Certainly, nanosensors integrated into food packaging systems serve the purpose of notifying consumers when the expiration of food products occurs. Nanotechnology has the potential to contribute to the creation

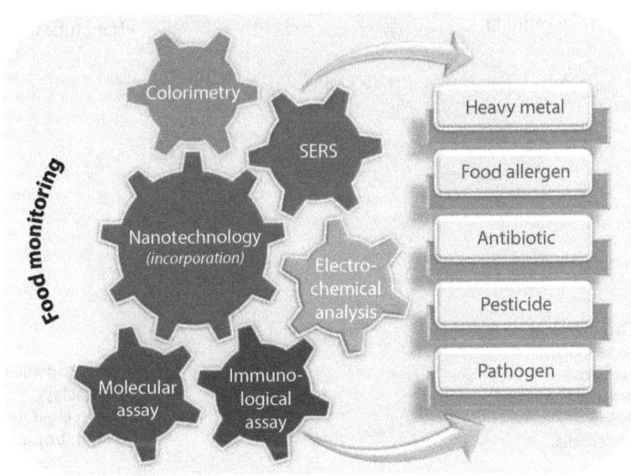

Figure 1.1 Utilizing nanotechnology to track contaminants in food [27].

of more nutritious food options as well [29]. The assortment of versatile nanostructures renders them appropriate for incorporation into both food products and packaging materials, thereby augmenting the nutritional value of foods [30]. Among the pool of 633 accessible nanomaterials, 55 are harnessed within the realms of agriculture and food sciences [31]. As per a contemporary document, it was highlighted that nourishment items associated with nanotechnology are projected to constitute approximately 50% of the overall food products by the year 2020 [32].

Certainly, there is no denying that nanotechnology is fundamentally transforming the landscape in regard to the agro-food sector. Many documented requisitions of nanostructures in food processing encompass the following: (i) enhancing food quality, (ii) fortifying with bioactive, (iii) achieving controlled release of bioactive substances through encapsulation with nanocarriers, (iv) altering food frameworks and tissues, and (v) employing intelligent wrapping systems for detecting and countering biochemical, microbiological, and chemical changes, as depicted in Figure 1.2 [33]. Various types of nanomaterials find extensive uses in creating food items and enhancing their nutritional content. For instance, protein nanoparticles play a crucial role in producing food products due to their ability to enhance protein solubility, resulting during the particular creation of glutenin nanoparticles through desirable operational characteristics in nutrition [34].

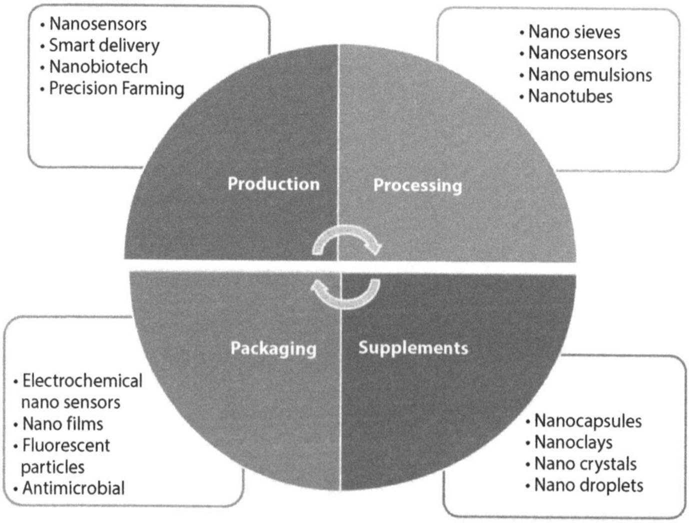

Figure 1.2 An overview of applications of nanomaterials in the food industry [27].

Nanotechnology has found application in the realms of food processing and preservation. This utility involves the creation of a protective barrier to conceal various tastes and flavors. Moreover, nanotechnology facilitates the controlled release and enhanced dispersibility of food ingredients, including water-insoluble supplements [35]. Nanotechnology brings forth several advantages in the field of food processing. These include the creation of a comprehensive texture for food constituents, the encapsulation of food additives, the innovation of novel flavors and sensory experiences, and the regulation of aroma release to heighten the bioavailability of dietary supplements [36].

The food industry has been transformed by nanotechnology, introducing groundbreaking delivery methods for producing agrochemicals in nano-form, enhancing nutritional aspects, and fostering novel product creation via bioactive encapsulation. Notably, nanoencapsulation distinctly enhances the nutritional profile of the product [37].

In recent times, the industrial utilization of food nanotechnology has spanned from intelligent packaging to the development of interactive foods. A contemporary term, "active packaging," describes the practice of packaging food items in a way that transforms the state of the packaged product to enhance sensory attributes and ensure food safety. This approach extends shelf life by preserving the quality of food products [38].

Nanotechnology's integration into the food industry fosters innovation across various aspects of food items. This encompasses enhancements in texture, taste, sensory characteristics, and product color intensity, along with the steadiness and storage life of products [39].

1.4 Feasible Research Implications to Address the Shortcomings in Food Preservation

Food preservation encompasses a series of diverse food processing stages aimed at upholding food quality to attain optimal benefits and nutritional value. This process involves growing, harvesting, processing, packaging, and distributing foods. The primary goals of food preservation include mitigating inadequate agricultural planning, creating value-added products, and diversifying dietary options. Food spoilage can stem from an array of chemical and biochemical reactions [40].

Traditional methods of food preservation, including withering, chilling, hygiene, and synthetical conservation, are widely employed on a global scale. Ongoing technological improvements and progress are driving the advancement of present techniques and the emergence of novel approaches, for instance, illumination, high-pressure technology, and barrier science [41]. The realm of food preservation has evolved into a highly collaborative field, encompassing stages tied to the cultivation, harvesting, processing, packaging, and distribution of foods. Consequently, adopting an integrated approach proves beneficial for effectively preserving food items throughout the entire production and processing journey.

Currently, the worldwide processed food market stands at approximately $7 trillion and is experiencing steady growth [42]. The rapid pace of globalization and industrialization plays a pivotal role in driving the advancement of food processing industries across various nations. According to the UNIDO Industrial Statistics Database (2005), an examination reveals that food processing holds promising prospects within the manufacturing sector of developing countries, and the contribution of these industries to a nation's GDP rises in proportion to its national income [43].

References

1. Ruzi, M., Celik, N., Onses, M.S., Superhydrophobic coatings for food packaging applications: A review. *Food Packag. Shelf Life*, 32, 100823, Jun. 2022, doi: 10.1016/j.fpsl.2022.100823.
2. Prokic, D., Stepanov, J., Curcic, L., Stojic, N., Pucarevic, M., The role of circular economy in food waste management in fulfilling the United Nations' sustainable development goals. *Acta Univ. Sapientiae Aliment.*, 15, 1, 51–66, Dec. 2022, doi: 10.2478/ausal-2022-0005.
3. *The State of Food Security and Nutrition in the World 2020. FAO, IFAD, UNICEF, WFP and WHO*, 2020, doi: 10.4060/ca9692en.
4. Peerzada Gh, J. *et al.*, An overview on smart and active edible coatings: safety and regulations. *Eur. Food Res. Technol.*, May 2023, doi: 10.1007/s00217-023-04273-2.
5. Kumar, K.P., Murthy, H., Pillai, V.J., Prathap, B.R., Blockchain-Enabled Model for Minimizing Post Harvest Losses. *ECS Trans.*, 107, 1, 17475–17482, Apr. 2022, doi: 10.1149/10701.17475ecst.
6. Hafez, E.M., Osman, H.S., Gowayed, S.M., Okasha, S.A., Omara, A.E.D., Sami, R., Abd El-Razek, U.A., Minimizing the adversely impacts of water deficit and soil salinity on maize growth and productivity in response to the application of plant growth-promoting rhizobacteria and silica nanoparticles. *Agronomy*, *11*, 4, 676, 2021.
7. Eldib, R., Khojah, E., Elhakem, A., Benajiba, N., Helal, M., Chitosan, nisin, silicon dioxide nanoparticles coating films effects on blueberry (Vaccinium myrtillus) quality. *Coatings*, *10*, 10, 962, 2020.

8. Putwa, S., Curtis, I.S., Dasog, M., Nanostructured silicon photocatalysts for solar-driven fuel production. *Iscience*, *26*, 4, 2023.
9. Wang, Y., Lu, Q., Han, S., Zhang, D., Cong, H., Chen, C., Design and synthesis of novel Pt/Fe3O4@ SiO2@ WO3 magnetic core/shell microspheres for solar light driven photocatalytic production of hypochlorous acid. *Mater. Res. Bull.*, 112347, 2023.
10. Zhang, W., Ahari, H., Zhang, Z., Mahdi Jafari, S., Role of silica (SiO2) nano/micro-particles in the functionality of degradable packaging films/coatings and their application in food preservation. *Trends Food Sci. Technol.*, 2023.
11. https://www.ecfr.gov/current/title-21/chapter-I/subchapter-A/part-73#73.575
12. Rokayya, S., Khojah, E., Elhakem, A., Benajiba, N., Chavali, M., Vivek, K., Helal, M., Investigating the nano-films effect on physical, mechanical properties, chemical changes, and microbial load contamination of white button mushrooms during storage. *Coatings*, *11*, 1, 44, 2021.
13. Shanmugam, B.K., Rangaraj, S., Subramani, K., Srinivasan, S., Aicher, W.K., Venkatachalam, R., Biomimetic TiO2-chitosan/sodium alginate blended nanocomposite scaffolds for tissue engineering applications. *Mater. Sci. Engi. C*, *110*, 110710, 2020.
14. Yuan, Z., Wang, M.L., Li, X., Effects of chitosan/TiO2 composite coating on keeping-fresh of stauntonvine. *Adv. Mater. Res.*, *530*, 68–73, 2012.
15. Jayakumar, R., Ramachandran, R., Kumar, P.S., Divyarani, V.V., Srinivasan, S., Chennazhi, K.P., Nair, S.V., Fabrication of chitin–chitosan/nano ZrO2 composite scaffolds for tissue engineering applications. *Int. J. Biol. Macromol.*, *49*, 3, 274–280, 2011.
16. https://encyclopedia.pub/entry/43221#:~:text=ZnO%2DNPs%20are%20an%20important,safety%20insurance%20of%20food%20products.
17. Mirhosseini, M. and Firouzabadi, F.B., Antibacterial Activity of Zinc Oxide Nanoparticle Suspensions on Food-Borne Pathogens. *Int. J. Dairy Technol.*, 66, 291–295, 2013.
18. Firouzabadi, F.B., Noori, M., Edalatpanah, Y., Mirhosseini, M., ZnO Nanoparticle Suspensions Containing Citric Acid as Antimicrobial to Control Listeria Monocytogenes, Escherichia Coli, Staphylococcus Aureus and Bacillus Cereus in Mango Juice. *Food Control*, 42, 310–314, 2014.
19. Joshi, H., Dave, R., Venugopalan, V.P., Pumping Iron to Keep Fit: Modulation of Siderophore Secretion Helps Efficient Aromatic Utilization in Pseudomonas Putida Kt2440. *Microbiology*, 160, 1393–1400, 2014.
20. Duncan, T.V., Applications of Nanotechnology in Food Packaging and Food Safety: Barrier Materials, Antimicrobials and Sensors. *J. Colloid Interf. Sci.*, 363, 1–24, 2011.
22. Paul, S.K., Dutta, H., Sarkar, S., Sethi, L.N., Ghosh, S.K., Nanosized Zinc Oxide: Super-Functionalities, Present Scenario of Application, Safety Issues, and Future Prospects in Food Processing and Allied Industries. *Food Rev. Int.*, 35, 505–535, 2019.

23. McClements, D.J. and Xiao, H., Is Nano Safe in Foods? Establishing the Factors Impacting the Gastrointestinal Fate and Toxicity of Organic and Inorganic Food-Grade Nanoparticles. *NPJ Sci. Food*, 1, 6, 2017.
24. Hakimian, F., Emamifar, A., Karami, M., Evaluation of Microbial and Physicochemical Properties of Mayonnaise Containing Zinc Oxide Nanoparticles. *LWT-Food Sci. Technol.*, 163, 113517, 2022.
25. He, L., Liu, Y., Mustapha, A., Lin, M., Antifungal Activity of Zinc Oxide Nanoparticles against Botrytis cinerea and Penicillium expansum. *Microbiol. Res.*, 166, 207–215, 2011.
26. Batool, R., Kazmi, S.A.R., Khurshid, S., Saeed, M., Ali, S., Adnan, A., Fatima, N., Postharvest shelf life enhancement of peach fruit treated with glucose oxidase immobilized on ZnO nanoparticles. *Food Chem.*, *366*, 130591, 2022.
27. Shafiq, M., Anjum, S., Hano, C., Anjum, I., Abbasi, B.H., An Overview of the Applications of Nanomaterials and Nanodevices in the Food Industry. *Foods*, *9*, 148, 2020.
28. Kiss, É., Nanotechnology in Food Systems: A Review. *Acta Aliment.*, *49*, 460–474, 2020.
29. Gokularaman, S., Cruz, S.A., Pragalyaashree, M., Nishadh, A., Nanotechnology approach in food packaging-review. *J. Pharm. Sci. Res.*, *9*, 1743–1749, 2017.
30. Aditya, A., Chattopadhyay, S., Gupta, N., Alam, S., Veedu, A.P., Pal, M., Singh, A., Santhiya, D., Ansari, K.M., Ganguli, M., ZnO nanoparticles modified with an amphipathic peptide show improved photoprotection in skin. *ACS Appl. Mater. Interfaces*, *11*, 56–72, 2018.
31. Belluco, S., Gallocchio, F., Losasso, C., Ricci, A., State of art of nanotechnology applications in the meat chain: A qualitative synthesis. *Crit. Rev. Food Sci. Nutr.*, *58*, 1084–1096, 2018.
32. Jiménez, D., de Miguel-Díez, J., Guijarro, R., Trujillo-Santos, J., Otero, R., Barba, R., Muriel, A., Meyer, G., Yusen, R.D., Monreal, M., Trends in the management and outcomes of acute pulmonary embolism: Analysis from the RIETE registry. *J. Am. Coll. Cardiol.*, *67*, 162–170, 2016.
33. Steinvil, A., Zhang, Y.-J., Lee, S.Y., Pang, S., Waksman, R., Chen, S.-L., Garcia-Garcia, H.M., Intravascular ultrasound-guided drug-eluting stent implantation: An updated meta-analysis of randomized control trials and observational studies. *Int. J. Cardiol.*, *216*, 133–139, 2016.
34. Xu, W., *Three-Dimensional Printing of Wood Derived Biopolymers Towards Biomedical Applications*, Ph.D. Thesis, Åbo Akademi University, Turku, Finland, 2019.
35. McClements, D.J., DeLoid, G., Pyrgiotakis, G., Shatkin, J.A., Xiao, H., Demokritou, P., The role of the food matrix and gastrointestinal tract in the assessment of biological properties of ingested engineered nanomaterials (iENMs): State of the science and knowledge gaps. *NanoImpact*, *3*, 47–57, 2016.
36. Cho, Y.-H. and Jones, O.G., Assembled protein nanoparticles in food or nutrition applications. *Adv. Food Nutr. Res.*, *88*, 47–84, 2019.

37. Yoo, S.M., Keum, K.C., Yoo, S.Y., Choi, J.Y., Chang, K.H., Yoo, N.C., Yoo, W.M., Kim, J.M., Lee, D., Lee, S.Y., Development of DNA microarray for pathogen detection. *Biotechnol. Bioprocess Eng.*, 9, 93–99, 2004.
38. Singh, T., Shukla, S., Kumar, P., Wahla, V., Bajpai, V.K., Rather, I.A., Application of nanotechnology in food science: Perception and overview. *Front. Microbiol.*, 8, 1501, 2017.
39. Salgado, P.R., Di Giorgio, L., Musso, Y.S., Mauri, A.N., Bioactive packaging: combining nanotechnologies with packaging for improved food functionality, in: *Nanomaterials for Food Applications*, vol. pp, pp. 233–270, Elsevier, Amsterdam, The Netherlands, 2019.
40. Amit, S.K., Uddin, M.M., Rahman, R., Islam, S.M., Khan, M.S., A review on mechanisms and commercial aspects of food preservation and processing. *Agric. Food Secur.*, 6, 1, 1–22, 2017.
41. Blum, D., Food that lasts forever. *TIME Mag.*, 2012.
42. *World Food Market Overview Marketing Essay*, 2013, http://www.ukessays.com/essays/marketing/world-food-market-overview-marketing-essay.php?cref=1.
43. Wilkinson, J. and Rocha, R., *Agri-processing and developing countries*, World Bank, Washington, DC, 2008.

2

Toward Stable Electrochemical Water Splitting: Anticorrosive Properties of Transition Metal–Based Electrocatalytic Coatings

Kaushal Gavankar, Suchitra Sapakal and Anamika V. Kadam*

Department of Physics, The Institute of Science, Dr. Homi Bhabha State University, Mumbai, India

Abstract

The electrocatalytic route of water splitting has emerged promising and rational method for hydrogen production, offering an eco-friendly alternative to traditional fossil fuel–based sources. However, maintaining the stability of electrodes long-term in harsh conditions such as strongly acidic or basic electrolytes is crucial. The present study addresses this challenge by exploring cost-effective solutions that improve stability while sustaining high efficiency, fast kinetics, and low overpotential. The primary focus is to study anticorrosive catalyst coatings of transition metal–based compounds such as FeO, NiFeN, and NiO catalysts. Also, it includes some binary-ternary metal complexes, as well as carbon- and graphene-based coatings. It is being approached first as a cheap, abundant alternative to conventional noble metal–based catalysts and electrodes for next-generation hydrogen-oxygen evolution coatings. This review includes a detailed description of the synthesis methods used for these materials and their performance in harsh conditions. This study also highlights the practical implementation of this work for stable hydrogen and oxygen evolution.

Keywords: HER, OER, anticorrosive coatings, pH, electrode degradation, seawater electrolysis, corrosion, fouling

**Corresponding author*: anamika@iscm.ac.in

Hari Murthy, Vinay Jha Pillai, Kukatlapalli Pradeep Kumar and Matthew Cowan (eds.) Novel Anti-Corrosion and Anti-Fouling Coatings and Thin Films, (13–42) © 2024 Scrivener Publishing LLC

2.1 Introduction

Decarbonizing highly carbon-dependent industries such as steel manufacturing can require colossal amounts of renewable energy. Hydrogen is considered a key solution in this regard. Electrocatalysis of water using green energy is the cleanest way to produce hydrogen. But major challenges in the production of green hydrogen are the efficiency and durability of the electrolyzers [1]. Alkaline electrolysis, solid oxide electrolysis, and proton exchange membrane electrolysis are currently proven technologies in the market. Yet, these technologies are also prone to issues such as high cost, low efficiency, and limited durability. For the implementation of green hydrogen production on a global level, it will be mandated to facilitate seawater electrolysis, which itself is a challenging task due to harsh conditions causing corrosion and fouling of the critical components of electrolyzers. Noble metals like platinum group metal (PGM)–based compounds are widely used as primary electrocatalysts due to their robustness, high catalytic activity, and corrosion resistance, but they end up making the device very expensive. For large-scale manufacturing of electrolyzers, developing cheap, abundant, and stable electrocatalysts is essential. According to a recent comparison, modern alkaline electrolyzers with nickel-based electrocatalysts produce comparable or greater efficiency than acidic polymer/proton electrolyte membrane (PEM) water electrolysis with PGM-based electrocatalysts. Yet, non-noble metals exhibit low stability and corrosion resistance, which makes alkaline electrolyzers a high-maintenance technology. PEM electrolyzers also exhibit a quick start time, unlike alkaline electrolysis, which may require hours to reach its peak performance. However, their high cost is a significant disadvantage [2]. Due to the association of multiple factors, it is very difficult to address the superiority of a single technology over others. Instead, detailed knowledge of the water-splitting process and the particular parameters under which the reaction will be undergoing is necessary when selecting a specific technology [3]. During the electrocatalytic process of water splitting, hydrogen is evolved at the cathode while oxygen is evolved at the anode. This causes Spontaneous oxidation-reduction reactions in different microregions on the surface of metals, producing rust layers [4]. Corrosion causes changes in the structure and composition of the material, which lowers the system's efficiency and longevity, increases maintenance costs, and reduces the overall economic viability of the process. This has been a common issue in all types of electrolyzers, primarily due to harsh pH conditions, highly reactive electrolytes, contaminants, fluctuations in potential, and high temperatures, with the only feasible solution being the use of inert noble metals. Enhancing the stability of non-noble metal electrocatalysts will revolutionize electrocatalytic water splitting, making it more sustainable and economical. Hence, catalyst corrosion is the focal point of the present work. To address and mitigate corrosion in such technologies, it is necessary to understand the corrosion mechanism in each case. Here, we will first understand the operation of various technologies and their operating parameters, along with their correlation to corrosion. Then, different strategies to prevent corrosion will be summarized, with an emphasis on anticorrosive catalytic coatings and thin films.

Harsh pH conditions provide excess ions, which increase the conductivity of the medium. In their works, researchers have mainly focused on electrocatalytic water splitting where the pH ranges from 1 to 4 or 11 to 14 [5] and elucidated hydrogen or oxygen evolution mechanisms. Two electrons are transported throughout the hydrogen evolution reaction (HER) process. Strong acids are recommended due to their ability to provide excess protons in acidic electrolytes, as the reaction begins with the H^+ ions segregation from electrolytes onto the electrode surface (Volmer reaction) [6]. Following this, the adsorbed H species keep on combining with one another to form an H-H bond [7], and a hydrogen molecule is then released through either the process of Tafel or the Heyrovsky, depending on the case [8, 9]. The proton source in the Volmer processing step is where alkaline electrolytes differ most significantly from acidic electrolytes, the formation of the adsorbed H species after the breaking of the H-O link, which typically occurs after the adsorption of water molecules. In order to produce hydrogen, adsorbed H species then interact with one another or adsorb on other water molecules. HER activity in acidic media is two to three times greater than that in alkaline media [10]. Evidently, the reaction route is affected by the pH of the electrolytes and decides whether it begins with direct H^+ adsorption or water dissociation. In an alkaline solution, it is critical to break the strong covalent H-O-H bond despite the H_3O^+ weak covalent bond [10, 11]. As a result, it is thought that the aq OH^- concentration on the active sites, metal-hydrogen binding energy, water adsorption, and water dissociation capability are the key barriers to alkaline HER. Therefore, providing active sites with a superior capacity to dissociate water molecules in an alkaline environment is a key component for the analytical design of HER catalysts [12]. Platinum generally shows the best catalytic response for HER in both acidic and alkaline media [7], having its free energy of hydrogen adsorption value, or Gibbs free energy (ΔGH^*) close to zero, showing the lowest overpotential, a low Tafel slope, and superior long-term durability. Therefore, it is the most commonly used cathode material in water electrolysis. Other materials may have higher hydrogen adsorption energies, leading to less favorable Gibbs free energy (ΔGH^*) for the HER, which results in slower reaction kinetics and reduced hydrogen production rates. Also, they possess higher chemical reactivity, which reduces stability during redox reactions [13]. Many electrocatalytic systems usually operate in high-purity water with acidic, alkaline, or buffer systems [7]. Due to the limited availability of highly filtered water, which may impede scalability [7], seawater electrolysis offers a potentially abundant and sustainable source of hydrogen. Seawater is a plentiful resource as it makes up over 96.5% of the Earth's surface water [14]. However, dissolved ions including Na^+, Cl^-, Ca^{2+}, and Mg^{2+} make up more than 3.5% of the salts in salt water and are involved in a number of competitive electrochemical processes. By creating insoluble compounds at the cathode and the anode, they can react and poison the water-splitting catalysts, causing degradation and deactivation of the catalysts as well as a reduction in electrolysis efficiency [15]. For instance, on the electrode surface, magnesium hydroxide precipitates are insoluble. The clogging of MgOH on the electrode surface might hinder mass transfer between electrocatalysts and electrolytes [16]. Prevention of fouling and corrosion on the catalyst surface has been a very difficult challenge and will be a very important one as the industry makes progress toward commercialization. It is critical to produce active, sustainable, and selective anode materials in order to minimize the formation of poisonous and corrosive layers of chloride complexes during seawater electrolysis [17]. Major concerns with HER and oxygen evolution reaction (OER) in seawater splitting are connected with small pH changes at the electrode [18]. The pH at the anode continuously drops during electrolysis due to OH oxidation, whereas the pH at the cathode rises as a result of H reduction [18]. Even at modest geometric activity, this pH volatility, which ranges from 5 to 9 pH units, significantly reduces the effectiveness of the catalyst [18, 19]. The anodic chamber's constant pH decrease raises acidity near the electrode surface, causing severe anode corrosion. Therefore, addressing corrosion pathways in seawater electrolysis is imminent. Here, we should first understand the corrosion mechanism in detail before going toward the methods to enhance stability and prevent corrosion.

2.2 Corrosion Mechanisms

Among many responsible factors for corrosion, such as temperature, potential, and contaminants (depicted in Figure 2.1), the pH of the reaction medium or electrolyte influences the corrosion reaction significantly.

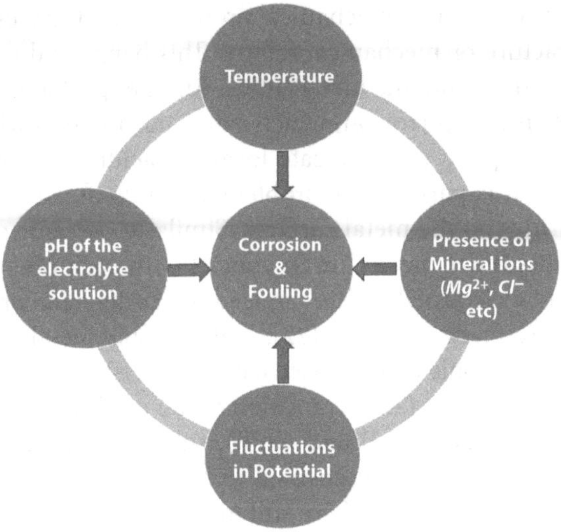

Figure 2.1 Parameters inflicting corrosion and fouling.

Thus, depending on the pH conditions, there are two types of corrosion mechanisms [20]: hydrogen evolution corrosion, which occurs under strong acidic conditions, and oxygen evolution corrosion, which occurs under mildly acidic or neutral conditions. Hydrogen evolution corrosion can be seen in the form of bubbles appearing on the surface. Thus, the type of corrosion becomes a major issue in acidic technologies such as PEM electrolyzers. This technology is effective and suited for on-demand hydrogen production because it uses a solid polymer electrolyte to assist proton transfer. However, at low pH levels, the acidic environment can cause the PEM and catalyst layers to corrode, shortening their lifespans. The overall reaction consists of simultaneous oxidation and reduction reactions with the participation of protons, as given by Equations 2.1 and 2.2 [21].

$$\text{Oxidation reaction: } M \rightarrow M^{x+} + xe^- \tag{2.1}$$

$$\text{Reduction reaction: } 2H^+ + 2e^- \rightarrow H_2 \uparrow \tag{2.2}$$

Here, the dissolution of metal occurs rather than producing a rust layer on the surface. Hydrogen evolution corrosion involves the formation of hydrogen gas at the metal surface as a result of electrochemical reactions, leading to the risk of hydrogen embrittlement when the metal is under stress, weakens the metal's mechanical properties, and makes it more susceptible to fracture or mechanical failure. This happens due to the diffusion of hydrogen gas into the metal lattice. Another problem seen in acidic electrolysis is crevice corrosion, which can occur in areas where stagnant electrolytes or trapped gasses create localized acidic environments. This type of corrosion is particularly problematic in confined spaces or areas with irregularities on the metal surfaces. Similarly, galvanic corrosion can occur when dissimilar metals are in contact with each other in the presence of an acidic electrolyte. This can lead to accelerated corrosion of the less noble (more reactive) metal [22]. Unlike noble metals such as Pt, Ir, and Ru, most non-noble metal–based electrocatalysts are unable to endure strong acidic conditions [23]. Transition metal oxides (FeO, MoO_2, NiO, and CoO), phosphides (Ni_2P and Co_2P), sulfides (WS_2 and MoS_2), and carbides (Mo_2C) have been discovered to have good activity due to their unique electronic structure and are still undergoing research for enhancement in stability before being considered as promising materials for water splitting [23–27]. Oxides, in particular, are simpler to synthesize and can withstand harsh pH conditions with suitable modifications. Transition metal oxides like MnO, FeO, and CoO have shown corrosion-resistant properties and good catalytic activity, usually in low-pH media. MnO, due to its strong metal oxide bond strength, shows good stability in acidic media but projects lower catalytic activity than others [28]. The majority of transition metals have either 'too strong' or 'too weak' M-H bonds [23], resulting in low HER activity. Similarly, the Sabatier principle states that 'too strong' or 'too weak' M-O interactions result in decreased OER activity [29, 30]. Much research is therefore focused on the development of high-performance catalysts that function under mild circumstances, which can make it more realistic to design cost-effective electrolyzers with more minor corrosion difficulties [7]. Oxygen evolution corrosion occurs due to the participation of oxygen atoms, generating a hydroxide layer on the surface. This can be explained by Equations 2.3 to 2.5 [21].

$$\text{Oxidation reaction: } M \rightarrow M^{x+} + xe^- \quad (2.3)$$

$$\text{Reduction reaction: } O_2 + 2H_2O + 4e^- \rightarrow 4OH^- \quad (2.4)$$

$$\text{Precipitation reaction: } M^x + xOH^- \rightarrow M(OH)_x \downarrow \quad (2.5)$$

Lack of H^+ or OH^- ions interferes with mass transfer and increases the amount of energy required to separate water molecules even in low pH settings [31]. Harsh pH conditions also increase the conductivity of the electrolyte, which endorses its necessity for an efficient water-splitting reaction. Alkaline electrolysis or, to be precise, alkaline electrolyzers, which are renowned for their cost-effectiveness, work with a highly conductive alkaline electrolyte (often potassium hydroxide, KOH, or sodium hydroxide, NaOH). These electrolyzers' durability and effectiveness can be significantly reduced at higher pH levels due to substantial corrosion of the electrode materials and current collectors. Nickel-based materials are primarily employed as catalysts and electrode components in several alkaline electrolyzers. In an alkaline atmosphere, nickel may be susceptible to corrosion, which may reduce the electrolyzer's longevity and effectiveness. For example, nickel can undergo corrosion in the following pathways [31, 32]:

In the alkaline electrolyte, nickel (Ni) at the anode can undergo an oxidation reaction with hydroxide ions (OH^-) to form nickel hydroxide ($Ni(OH)_2$):

$$Ni + 2OH^- \rightarrow Ni(OH)_2 + 2e^- \tag{2.6}$$

At higher anodic potentials, nickel hydroxide can further react with hydroxide ions and water to form nickel oxyhydroxide (NiOOH):

$$Ni(OH)_2 + OH^- \rightarrow NiOOH + H_2O + e^- \tag{2.7}$$

Under the anodic polarization, nickel atoms lose electrons and oxidize to form nickel ions (Ni^{2+}) that dissolve into the electrolyte:

$$Ni \rightarrow Ni^{2+} + 2e^- \tag{2.8}$$

The dissolved nickel ions in the electrolyte can lead to the corrosion of the nickel electrode. The corrosive attack involves the continuous dissolution of nickel atoms from the surface into the electrolyte, which may form soluble nickel complexes in the alkaline solution. This corrosion can be further accelerated under other operating conditions such as high current density, the presence of impurities, and high temperatures.

Seawater contains chloride ions, which can lead to accelerated corrosion of metallic components in the electrolyzer. Corrosion can degrade the system's efficiency and longevity, increasing maintenance costs and reducing the overall economic viability of the process. One of the major challenges is the higher requirement for anode material, which mediates the OER [17]. For the OER in seawater electrolysis over a wide range of potentials, the anode material must be resistant to chloride corrosion and have good catalytic activity [33–36]. The chloride ions present in the acidic electrolyte form complexes with metal ions and can exacerbate the corrosion process, accelerating the degradation of metallic components. Even though stainless steel is often quite resistant to corrosion, it can nevertheless corrode when there are chloride ions present, like in seawater. Chloride ions can break down the preventive oxide layer on the stainless steel's surface, leading to pitting or crevice corrosion [37].

2.3 Methods to Prevent Corrosion

Numerous tactics can be used to tackle corrosion and the consumption of catalysts based on non-noble metals during the water-splitting process. However, it is important to acknowledge that the surface of the catalyst or electrode material cannot be coated with any conventional anticorrosive material. The surface of the catalyst plays a key role in water electrolysis, and thus, the protective coating must also show catalytic properties at the same time [31]. Hence, protective methods other than coatings are also popular in these devices.

2.3.1 To Avoid pH Attack

Adjusting the pH of the electrolyte solution to a range where the electrocatalyst is less prone to corrosion can prevent pH attacks, most commonly acid attacks. This can be achieved using buffer solutions or selecting suitable acids or bases. Also, an electrolyte composition that minimizes the corrosive effects can be chosen. For example, alkaline pH media are less corrosive for the electrocatalysts than the acidic media but it is seen that alkaline media offers a lower catalytic response of HER than acidic media [39]. Therefore, for significant results, studying catalysis in all different media is necessary along with emphasis on material selection.

2.3.2 Material Selection

Material properties are the main deciding factor for the stability of the catalyst. Selecting materials that have good stability specific to acidic or alkaline environments is the primary strategy. Using metal oxides, nitrides, and phosphides as electrocatalysts has been a central part of research interest. Oxides, in particular, are simpler to synthesize and can withstand harsh pH conditions with suitable modifications.

2.3.3 Surface Functionalization

Techniques like surface functionalization, alloying, or doping can improve stability and resistance to corrosion. Using nanostructured transition metal oxides as electrocatalysts or coating them on other catalytic materials has been done over the years for various combinations of materials, substrates, and electrolytes, as well as synthesis and deposition methods.

2.3.4 Protective Coating Deposition

Protective coatings are used to increase the stability and durability of the catalysts. PVD methods, such as sputtering or evaporation, are commonly used methods to deposit thin films and coatings [40]. A thin film is made *via* sputtering, which involves the ejection of atoms from a target material onto a substrate by a high-energy plasma [31]. While in evaporation the catalytic material is heated to a high temperature, causing it to vaporize. The vapor is then condensed onto a substrate to create a thin layer. PVD techniques enable fine control of layer thickness and composition. Using an aqueous solution as the reaction system in a unique closed vessel known as an autoclave is called a hydrothermal or solvothermal approach. It is used to generate a high-temperature, high-pressure reaction environment, wherein the process dissolves and recrystallizes a substance that is poorly soluble or insoluble under normal conditions, which is typically used to synthesize the catalytic materials. After synthesis, these materials generally in the form of powders are dried and coated on the substrate [42, 43]. Electrodeposition or electroplating can also be used to coat substrates with the catalytic materials in the least amount of time with significantly lower efforts, making it a very efficient method. This method, also sometimes referred to as anodizing, can also be utilized for oxidating the electrode. The oxide layer serves as a barrier to protect the inner metallic surface from the corrosion caused by the electrolysis medium [40]. Another similar method that is generally used for corrosion prevention on stainless steel is passivation. This can also imply the formation of a nitride, phosphide, or sulfide layer, and the method is called nitriding, phosphating, or sulfating. Such coatings not only prevent corrosion but can also enhance the efficiency of the reaction [44–46].

2.3.5 Corrosion Engineering

Artificially developing such layers on the substrate or catalyst surface using strong corrodents is referred to as corrosion engineering, where the surface is corroded to prevent further corrosion and protect the bulk of the material. In general, autogenous corrosion processes of metallic materials can be used in corrosion engineering to build effective and stable electrocatalysts. [40]. Since the majority of substrates have low catalytic activity, it can be a good idea to continuously erode them using strong corrodents like O_2, Fe^{3+}, H_2O_2, or $S_2O_8^{2}$ to produce metal ions in order to stop further deterioration [47]. Corrosion engineering is always used to create powder hydroxides or oxides, which have not only significant OER characteristics but only moderate HER activity [48]. Fortunately, significant HER activity can be obtained by converting hydroxides or oxides into phosphides or selenides. Self-supporting electrocatalysts have also been created using foams or plates consisting of nickel, CoNi, NiAl, Fe, NiFe, and stainless steel through the use of corrosion engineering (Figure 2.2). The simple corrosion of metal substrates near the interfaces results in the production of numerous metal ions, including Ni^{2+}, Fe^{3+}, and Co^{2+}, which are essential for the development of electrochemically active corrosion layers [49, 50]. Additionally, some electrocatalysts created through corrosion engineering are extraordinarily durable and can withstand elevated current densities of up to 1000 mA/cm^2 and extremely alkaline environments of 10 M KOH. Therefore, corrosion engineering can be a feasible method for the mass production of powder electrocatalysts, demonstrating outstanding superiority over the traditional expensive and tedious synthesis methods [48].

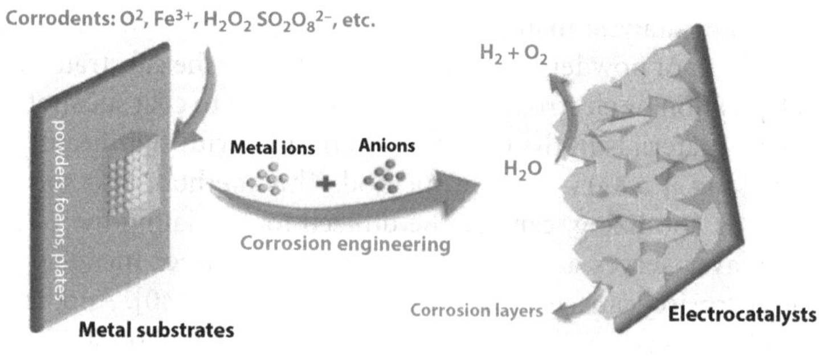

Figure 2.2 Illustration of corrosion engineering technique for synthesis of robust anticorrosive catalytic coatings.

The substrate plays a key role in any electrochemical system. The current collectors, such as carbon paper and copper foam, become inactivated during stability tests on large densities over the long term but substrates like nickel foam or AISI 304 stainless steel are known to be very stable for water splitting, even at high potentials. Stainless steel with Cl_2 treatment also shows moderately achievable success in terms of efficiency [37]. Due to its porous structure, significant surface area, high intrinsic strength, and electrical conductivity, Ni foam (NF) has been extensively used to create self-supporting electrocatalysts [50–52]. Catalytic layers and anticorrosive coatings can either be deposited on the substrates using the different methods mentioned above or can be developed on the substrate itself through corrosion engineering. Multi-metallic alloys (such as NiFe, CoNi, and NiAl) have also been used as substrates and to produce very effective electrocatalysts by corrosion engineering [52]. Because of the synergistic action of several metal sites, alloys always produce better stability and higher electrocatalytic activity than Ni or Fe [53–57].

2.4 Anticorrosive Catalytic Coating

Transition Metal Oxides (TMOs)

The strong crystal structure and chemical composition of transition metal oxides contribute to their increased stability. These oxides' metal-oxygen interaction gives greater corrosion resistance than pure metals, making them less vulnerable to chemical attack and surface degradation. Transition metal oxides can operate as efficient and long-lasting anticorrosive catalytic coatings when coated on electrodes or another catalyst material. Transition metal oxides in their most stable oxidation states can be suitable for such purposes. For example, Fe shows two most stable oxidation states 2+ and 3+, which can be seen in compounds like FeO and naturally occurring mineral Fe_2O_3 (hematite) [58].

2.4.1 FeO

$CuCo_2S_4$ (CCS) has been regarded as an excellent non-precious metal-based HER electrocatalyst as compared to their counterparts (the comparison of Tafel slopes was shown in Figure 2.3a) due to their superior electrochemical qualities and higher electrical conductivity (low resistivity of 10^{-4} Ω) [57–60]. Irrespective of its remarkable performance qualities, it

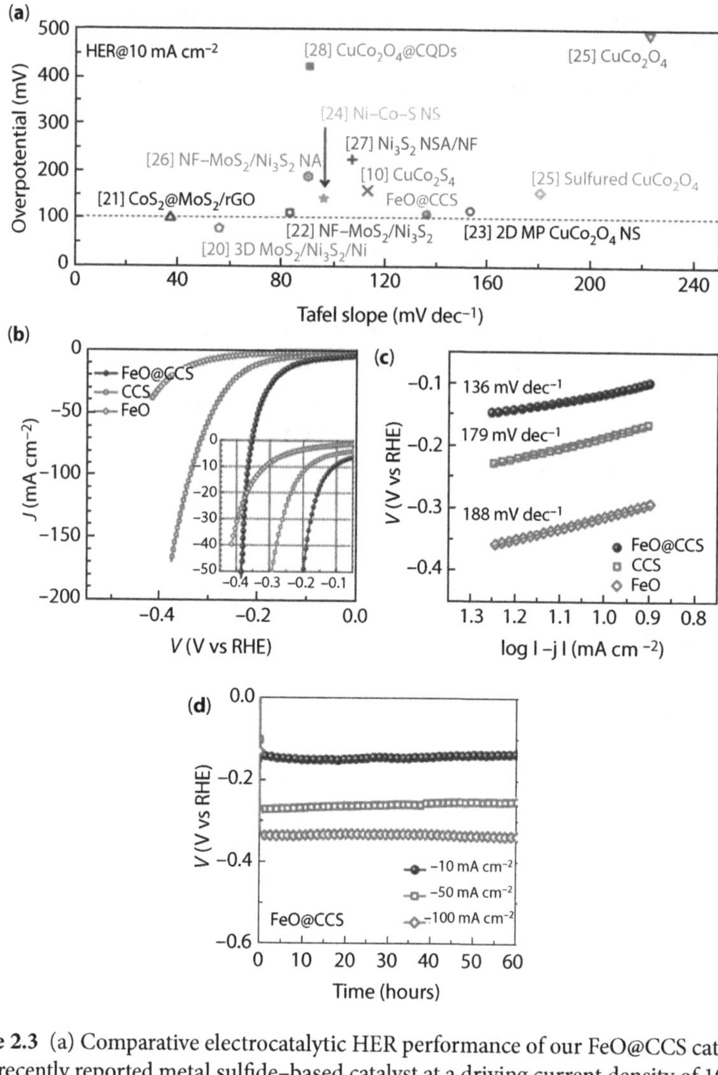

Figure 2.3 (a) Comparative electrocatalytic HER performance of our FeO@CCS catalyst and a recently reported metal sulfide-based catalyst at a driving current density of 10 mA cm^{-2} in a 1 M KOH electrolyte. (b) Polarization curve of FeO@CCS. (c) Tafel plots for FeO@CCS, CCS, and FeO HER electrocatalysts. (d) Chronopotentiometric stability curves for HER of the optimized FeO@CCS electrocatalyst at 10, 50, and 100 mA cm^{-2} [41]. Copyright (2021) Elsevier.

undergoes corrosion and surface oxidation after a large number of cycles in alkaline electrolytes [61]. Coating such sulfide-based catalyst materials with highly stable and low-cost metal oxides like FeO can endorse efficient water splitting in an alkaline medium [58, 62]. Ahmed *et al.* [41] deposited an anticorrosive FeO coating by magnetron sputtering on hydrothermally synthesized CCS to study the synergistic effects between the two and concluded that FeO-coated CCS shows significantly greater corrosion resistance. It requires a lower overpotential value (107 mV at 10 mA/cm ^{2}) depicted in Figure 2.3b compared to known binary/ternary catalysts and has a somewhat higher Tafel slope value (136 mV/dec) compared to multilayer binary combination heterostructures (Figure 2.3c). The stability of the FeO@CCS electrocatalyst at various current densities is shown by the chronopotentiometric curve (Figure 2.3d). At every applied current rate, the potentials remain constant for more than 60 hours, demonstrating increased catalyst stability. Instead of a curve that is more inclined toward the *Y*-axis, or the low-frequency zone, the EIS curves display a steeper line (high-frequency region) [41]. Due to the effective diffusion at the electrode/electrolyte interface, FeO@CCS has outstanding alkaline medium endurance [41]. Considering the large availability, low cost, and nontoxicity of Fe-based compounds, it will be a milestone to obtain stable and efficient electrocatalysts.

2.4.2 MoO_2-Based Coating

Similarly in the case of molybdenum, Mo_3O and MoO_2 are some of the most stable oxides. Combining these materials with other suitable catalytic components can result in the construction of robust highly active electrocatalysts [63]. For HER catalysts, MoS_2 $MoSe_2$ [63–65], Mo_2C [66], and MoO_2 [67] have undergone significant research. However, inadequate efficiency and stability restricted their application. Molybdenum oxides (MoOx), which consist of the simple binary oxides MoO2 and MoO3, are transition metal oxides with unique structural properties [42, 63]. Sapakal *et al.* [42] synthesized a MoO_2 catalyst with a binder-free hydrothermal method for the HER electrode. This film deposited on stainless steel plate showed good HER activity and stability for over 9 hours. This study can be extended to study synergistic effects as done by Ahmed *et al.* [41], with other catalyst materials for their utilization as anticorrosive coatings.

2.4.3 NiO-Based Coating

Ashwini *et al.* [43], who performed a similar study on NiO films coated on 304-grade stainless steel substrates by hydrothermal method, showed that the Ni content of stainless steel also gets converted to NiO and this adds up to the overall performance and stability of the electrode. This film acts as an excellent catalyst for OER with a low overpotential of 133 mV and retains its peak performance for more than 12 hours [43].

2.4.4 Ni-Fe on Iron Foam (IF)

Iron foam (IF) is widely used as an electrode as well as a substrate material for various catalyst materials. IF can be alloyed with other metals by corrosion engineering to make robust catalyst layers on the surface. Using corrosion engineering, Liu *et al.* [47] created the NiFe layered double hydroxide (LDH) catalyst by drying the cleaned iron plate or foam in a 1 × 10^{-3} M $NiSO_4$ solution for 12 hours [40]. The O_2 corrosion of Fe served as the primary catalyst for the creation of NiFe LDH [40]. The NiFe LDH catalyst could also run at a high current density of 1000 mAcm^{-2} for more than 5000 hours in 1 M KOH [36] and 1050 hours in 10 M KOH without seeing a noticeable voltage increase (Figure 2.4).

As substantial OER catalysts, Zou *et al.* [68] produced amorphous Ni-Fe hydroxides on Ni_3S_2/NF surfaces. The Ni_3S_2/NF substrates were hydrothermally synthesized before being immersed in a $FeCl_3$/$NaNO_3$ aqueous solution at 100°C for 5 s to create Ni-Fe-OH thin films [40]. It was discovered that Ni-, Ni-Co–, Ni-Mn–, and Ni-Cr–based amorphous hydroxide films may also develop on Ni_3S_2/NF by substituting Ni^{2+}, Co^{2+}, Mn^{2+}, and Cr^{3+}, for Fe^{3+} in the reaction solutions [40]. The Ni-Fe-OH@ Ni_3S_2/NF displayed the highest OER activity, with current densities of 10 and 100 mA/cm^2 at overpotentials of 165 and 240 mV in 1 M KOH electrolyte, respectively. The OER performances in 30 wt% KOH (7.64 M) industrial alkaline electrolyte were also determined, and it was found that current densities of 500, 1000, and 1500 mA/cm^2 required low overpotentials of 370, 469, and 565 mV, respectively. The increased OER activity was due to the much higher conductivity and higher concentration of OH

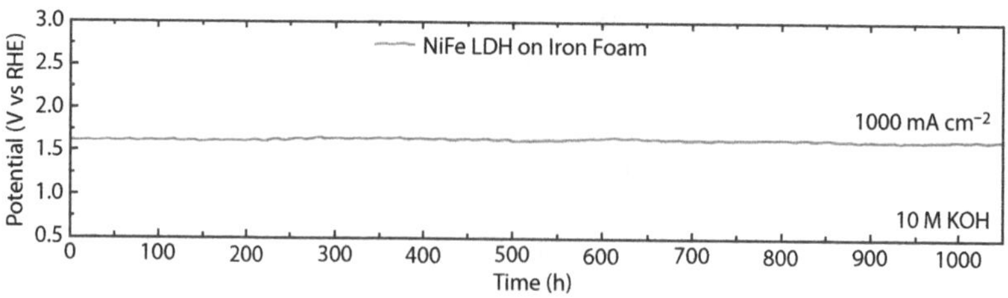

Figure 2.4 Chronopotentiometric curve of NiFe LDH layer coated on iron foam in 10 M KOH at a current density of 1000 mA cm^{-2} [47]. Copyright 2018 under creative commons license, Nature Communications.

groups. The Ni-Fe-OH@Ni_3S_2/NF catalyst could function at a high current density of 1000 mA/cm^2 for 50 hours in 30% KOH [38, 69].

2.4.5 NiFeN

Luo [70] manufactured a 3D core-shell metal-nitride catalyst in which NiFeN nanoparticles atop NiMoN nanorods were evenly decorated. In powerful alkaline salt water electrolysis, this catalyst was able to selectively produce oxygen. The cell voltages are 1.608 and 1.709 V, respectively, when NiMoN nanorods are used as the cathode, in order to attain the 500 and 1000 mA/cm^2 at 60°C current densities required by the industry. Since amorphous NiFe oxide and NiFe oxy(hydroxide) layers formed as active sites during the OER process and decreased chloride anion corrosion, the electrochemical performance was improved [71].

2.4.6 NiFeBx

In seawater electrolysis, Ni-Fe alloy substrates are exposed to an excessive chloride ion environment and corrode [71]. Jihong *et al.* [72] used boron powder to direct heat boronize a commercially available NiFe alloy plate before electrochemically oxidizing it. The alloy interlayer of $NiFeB_x$, the NiFe alloy substrate, and the surface-oxidized $NiFeB_x$ alloy layer were used to create this electrode (see Figure 2.5, a and b) [72]. The outermost oxidized $NiFeB_x$ layer contains boron species as metaborate, and their presence aids in the development and stability of the catalytic active phase (NiFe)-OOH. The application of a $NiFeB_x$ interlayer efficiently avoids severe oxidative corrosion of the anode material in a chloride environment [72]. With an almost unchanging electrode surface, the metal oxygen evolution electrode (MOEE) can facilitate the OER for more than 70 hours. The concentration of chloride ions in the electrolyte stays almost constant during electrolysis, and the MOEE has a Faradaic efficiency of 100% for oxygen evolution [72]. The MOEE is an oxygen evolution electrode that is extremely active and corrosion-resistant even in the presence of chloride ions [70, 72]. The presence of a $NiFeB_x$ interlayer effectively prevents severe oxidative corrosion of the MOEE [72]. Open circuit potential (OCP) studies, electrochemical impedance spectroscopy (EIS), and typical potentiodynamic curves (shown in Figure 2.5b) were also used to evaluate the corrosion behavior of MOEE and NiFe in a chloride-rich solution. The corrosion potential of MOEE (1.46 V vs. RHE) was found to be higher than that of NiFe (1.41 V vs. RHE), while MOEE has a lower corrosion current ($5.30*10^{-7}$ A) than NiFe ($9.59*10^{-6}$ A) [72]. The results suggest that

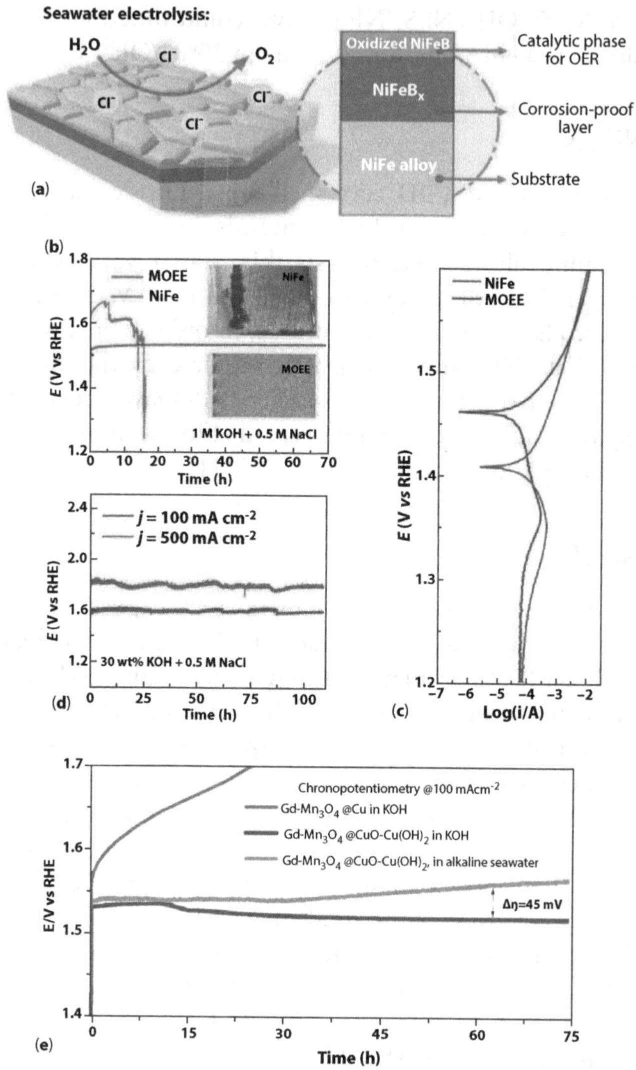

Figure 2.5 (a) Illustration of the multilayered anode composed of the Ni-Fe alloy substrate for structural support, NiFe-B alloys as a corrosion-proof layer, and surface-oxidized Ni-Fe-B as OER catalyst indicates higher selectivity and corrosion resistance in seawater electrolysis. (b) Chronopotentiometric curves with NiFe and MOEE as the electrodes in 1 M KOH solution containing 0.5 M NaCl. The photographs of the electrodes after OER electrocatalysis are provided in the insets. (c) Potentiodynamic polarization curves. (d) Chronopotentiometric curves of MOEE as an electrode in 30 wt% KOH electrolyte with 0.5 M NaCl at current densities of 100 and 500 mA cm^{-2} for 100 hours [72]. Copyright 2021, Wiley-VCH. (e) Durability by chronoamperometry test at a constant current density of 100 mA cm^{-2} in alkaline seawater [74]. Copyright 2022, American Chemical Society.

MOEE has greater corrosion resistance than NiFe in the chloride environment (Figure 2.5, b and d) [72]. Furthermore, it was shown that MOEE has a higher optical circuit potential than NiFe (1.39 V against RHE) (Figure 2.5c). A more condensed and steady passive film, according to earlier studies, may explain why an OCP value is higher [49, 50]. When compared to NiFe, a positive OCP value for MOEE highlights the critical function of the $NiFeB_x$ interlayer as a true insulator [72, 73].

2.4.7 Gd-Mn3O4@CuO-Cu (OH)2 Nanostructure

Tanveer *et al.* [74] also synthesized another novel gadolinium (Gd)-doped Mn_3O_4 nanosheets in an array of CuO-$Cu(OH)_2$ nanostructure array. First, these $CuOCu(OH)_2$ nanostructures were prepared by anodizing a Cu foil, and then gadolinium-doped Mn_3O_4 nanosheets were catholically deposited on it by electrodeposition using a standard three-electrode setup. The hierarchical structures on the surface exhibited large specific surface area, good conductivity, and ionic mobility, therefore possessing excellent catalytic activity and corrosion resistance [31] (durability showed by CA curves shown in Figure 2.5e).

2.4.8 Cobalt-Based Coatings

Cobalt is another promising transition metal with wide-ranging applications in catalysis. Yang *et al.* [75] concluded using density functional theory that single transition metal atoms anchored by vacancies on the copper-sulfide (CuS) surface exhibit extraordinary stability and serve as active centers for water splitting. Co@CuS exhibits good HER performance, with a hydrogen adsorption-free energy of 0.03 eV. Co@CuS exhibits high catalytic activity toward OER and has a comparatively low OER overpotential of 0.41 V. Co@CuS appears as a viable bifunctional HER/OER electrocatalyst candidate, supporting the use of Co layers as feasible corrosion-resistant catalytic coatings. Similarly, cobalt alloys with other elements such as nickel perform well as electrode materials and can be used to produce powerful anticorrosive catalysts *via* corrosion engineering approaches. CoNi-based OER catalysts were explored by Huang *et al.* [76] using comparable production techniques. On CoNi foams, extremely thin metal-organic structure (MOF) nanosheet arrays were created using an *in situ* self-dissociation assembly approach [40]. In order to start corrosion on the CoNi foams, benzenedicarboxylic acid (BDC) was used. This chemical released Co^{2+} and Ni^{2+} ions, which interacted with BDC to create

Co_9Ni-MOF nanosheet arrays with an 8.2-nm thickness [76]. Because of the mutually beneficial interactions between unsaturated metal active sites and bimetallic electron coupling [40], this as-prepared Co_9Ni-MOFs demonstrated significant OER activity, with a low overpotential of 215 mV to achieve a current density of 10 mA/cm^2 with negligible decay even after 300 hours of continuous long-term operation, confirming its anticorrosive nature. The dilemma of limited mass transport and metal site exposure for traditional bulk MOFs is addressed in this work by synthesizing ultrathin MOF nanosheets on conductive substrates *via* corrosion engineering.

2.4.9 Cr_2O_3@CoO_x

Jiaxin *et al.* [81] performed direct electrolysis of natural salt water that has not been alkalized or acidified, with long-term stability for more than 100 hours at 500 mAcm^{-2} and performance comparable to a typical PEM electrolyzer operating in high-purity water. This is accomplished by coating a transition metal oxide catalyst (CoO_x) with a Lewis acid layer (for example, Cr_2O_3), which allows water molecules to split dynamically and hydroxyl anions to be captured. This Lewis acid coating avoided chloride attack corrosion and fouling on the electrodes.

2.5 Carbon-Based Anticorrosive Catalytic Coatings

Carbon coatings are particularly useful for corrosion protection because of carbon's unique properties, such as chemical inertness, high strength, and low reactivity with most chemicals. Carbon has been widely used as an electrode material widely due to its superior qualities. There have been attempts to use graphene and other carbon-based materials either as dopants or coatings for electrocatalysts [77, 78].

2.5.1 Carbon-Based Coatings

Single-layer graphene has been investigated as a potential anticorrosive coating for electrocatalysts due to its remarkable properties. Single-layer graphene is chemically inert, which means it is less likely to react with corrosive species in the environment. This inertness can help protect the underlying electrocatalyst. Graphene is also an excellent conductor of electricity [78]. Graphene oxide (GO) has also gained significant attention as an anticorrosive coating on electrocatalysts. GO can form a passivation layer that inhibits the direct contact between the electrocatalyst and corrosive

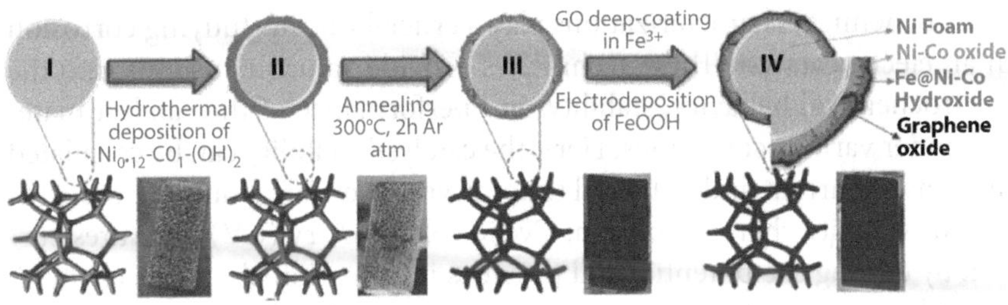

Figure 2.6 Composition and structural visualization of GO@Fe@Ni-Co@ Ni foam electrode [79]. Copyright 2018, American Chemical Society.

agents. By synthesizing multilayered GO on FeOOH/-Ni-Co, Lee *et al.* [79] (Figure 2.6) were able to produce current densities of 500 and 1000 mA/cm^2 in alkaline seawater with 303- and 345-mV overpotentials, respectively. The active site would be shielded from chloride corrosion if GO was layered over the catalyst to prevent direct contact between the electrode and seawater. This GO@Fe@Ni-Co@Ni foam showed excellent stability for overall alkaline seawater splitting with negligible degradation over 378 hours at 1 Acm^{-2}. These results also indicate the superiority of this GO coating over the corrosion engineering-enabled Co_9Ni-MOF synthesized by Huang *et al.* [76], Therefore, it has been regarded as a potential anticorrosive coating not just for alkaline but for alkaline seawater electrolysis as well.

The carbon-coated sodium cobalt-iron pyrophosphate nanoparticle catalyst showed insignificant variations in OER activity in alkaline saline water compared to electrocatalytic performance in alkaline freshwater [80]. With a 100% faradic efficiency and a stable curve under 100 hours at 50 $mAcm^{-2}$, this catalyst displayed not only excellent selectivity for OER but also remarkable corrosion resistance [80]. These findings demonstrated the broad potential of carbon- and graphene-based coatings in anticorrosive catalytic applications for water splitting. These substances can make the electrolyzers more reliable and cost-effective because they are less expensive and simpler to synthesize.

2.6 Discussion

2.6.1 Stability Characterization

There are a number of ways to study the corrosion mechanism, and the obtained data can be used to compare various tactics and their efficacy.

Electrochemical characterizations play a crucial role in studying corrosion in an electrocatalyst. These techniques provide valuable insights into the electrochemical behavior, stability, and performance of the catalyst material under various conditions. Here, the catalyst's stability can be calculated in terms of current and potential time curves. There are a number of methods for stability characterization. Cyclic voltammetry (CV) measures current by cycling the potential and provides information about the catalyst's redox behavior and the presence of any electrochemically active species on the surface. This involves sweeping the applied potential between two limits at a controlled scan rate. More than 5000 cycles are usually required at a specific scan rate, usually 100 mV s^{-1} [13]. Changes in the CV curves over time can indicate corrosion and degradation processes. The shift in overpotential before and after the CV cycles at a specific rate is examined by linear sweep voltammetry (LSV). Less change in overpotential is associated with better stability. EIS measures the response of the electrocatalyst to small sinusoidal perturbations in potential or current over a range of frequencies. This technique is particularly useful for probing the catalyst-electrolyte interface and evaluating the corrosion resistance of the catalyst. Another method to characterize the stability of catalysts is chromatoamperometry (I-t curve) at constant potential or chromatopotentiometry (E-t curve) at constant current. In chronoamperometry, a constant potential step is applied to the working electrode (catalyst) by the potentiostat. The potential is set at a value within the region of interest for the catalyst's corrosion. This potential is usually determined based on the catalyst's working conditions and expected corrosion potential. Once the potential step is applied, the potentiostat measures the current response from the working electrode over time. The current recorded during the experiment provides information about the catalyst's corrosion rate. As corrosion progresses, the current response may change due to the degradation of the catalyst material. Chronoamperometry can be conducted over extended periods to simulate real-world operating conditions, and mathematical models or kinetic equations can be used to interpret the current-time data and extract corrosion rates, degradation kinetics, and other relevant parameters [13, 38].

2.6.2 Future Prospects and Challenges

This comprehensive study on corrosion in electrocatalytic water splitting using transition metal-based electrocatalysts across various pH media, including the challenging environment of seawater electrolysis, has shed light on critical insights. The potential of techniques such as corrosion engineering as cheap and simple pathways for synthesizing anticorrosive layers on catalysts was elaborated, in comparison with other deposition methods. Distinct behaviors were observed among the catalysts under investigation, highlighting both strengths and limitations. FeO-based catalytic coatings prepared by Ahmed *et al.* [41], who enhanced the corrosion resistance of electrocatalysts in an alkaline medium, are far from attaining the endurance and efficiency of benchmark materials. The low cost of these coatings still projects an advantage and need for further research to improve the overall performance of this material.

While nickel-based catalysts exhibited remarkable stability and efficiency, like all other catalysts, their application was limited to mild alkaline electrolytes and faced severe corrosion issues in aggressive environments, especially seawater. Catalytic systems designed with anticorrosive layers like $NiFeB_x$ [72] showed impressive water-splitting performance in alkaline seawater, while Cr_2O_3@CoO_x synthesized by Jiaxin *et al.* [81] showed sustainable performance in neutral seawater which is a significant achievement. Acidic media still propose a major hurdle for the application of TM-based catalysts in water-splitting and replacing benchmark precious metal–based materials. Considering all the observations and challenges, the implementation of electrocatalytic water splitting using non-noble metal–based electrocatalysts is yet to become commercially feasible.

To address these challenges, we propose several potential solutions. First, further research into advanced protective coatings or surface modifications could enhance the durability of susceptible catalysts, making them more resilient in corrosive conditions. Second, exploring alternative corrosion-resistant catalyst coatings may offer promising avenues for improved performance. Exploring carbon-based coatings in combination with TM electrocatalysts has been a new advancement in the field which requires further research.

In the future, large-scale practical implementation of water electrolysis technologies may face a greater number of challenges. Although many TMOs have been explored, electrode stability, proper hydrogen evolution rate, and achieving high HER and OER at low potential are still major questions. Also, the use of direct seawater may cause biofouling in the devices, a hurdle yet to be studied. To solve such issues, more advancements in the protective layers may be necessary. As we move toward a sustainable future, water splitting remains a critical technology in the energy landscape. The breakthroughs in anticorrosive catalytic coatings presented in this chapter contribute significantly to unlocking the full potential of water electrolysis as a clean and efficient means of producing hydrogen. This, in turn, can accelerate the adoption of renewable energy sources and drive the transition to a carbon-neutral society.

2.7 Conclusion

In conclusion, this chapter has explored the remarkable advancements in the development of anticorrosive catalytic coatings for stable and efficient water splitting, which play a pivotal role in the realization of sustainable energy systems. By delving into the fundamental principles of water electrolysis and the challenges posed by corrosion in electrocatalysis, we have gained a deeper understanding of the significance of this research area. Throughout the chapter, we have highlighted various strategies employed to design and fabricate anticorrosive catalytic coatings, ranging from advanced materials selection to innovative surface engineering techniques. The integration of state-of-the-art nanomaterials, protective layers, and composite structures has demonstrated promising results in enhancing the durability and longevity of electrocatalysts under harsh operational conditions. Moreover, we have explored the impressive catalytic efficiencies achieved by these coatings, leading to significant improvements in hydrogen and oxygen evolution reactions. The reduced overpotential and enhanced reaction kinetics are clear indicators of the potential these coatings hold for enabling efficient water-splitting processes. Here, recent progress in transition metal-based systems was summarized along with future prospects and research requirements. Last, a brief overview of carbon-based coatings was also included due to their potential as next-generation anticorrosive catalytic materials.

Acknowledgement

This research work is fully funded by DST-SERB, New Delhi under the project grant number (EEQ/2021/000633).

References

1. Suen, N.T., Hung, S.F., Quan, Q., Zhang, N., Xu, Y.J., H.M., Electrocatalysis for the oxygen evolution reaction: recent development and future perspectives. *Chem. Soc. Rev.*, 46pp, 2, 337–365, 2017, doi: 10.1039/c6cs00328a.

2. d, M., Rashid, M., Al Mesfer, M.K., Naseem, H., Danish, M., Hydrogen Production by Water Electrolysis: A Review of Alkaline Water Electrolysis, PEM Water Electrolysis and High Temperature Water Electrolysis (2015). *Int. J. Eng. Adv. Technol. (IJEAT)*, 4, 3, 2249–8958.
3. Schalenbach, M., Tjarks, G., Carmo, M., Lueke, W., Mueller, M., Stolten, D., Acidic or Alkaline? Towards a New Perspective on the Efficiency of Water Electrolysis. *J. Electrochem. Soc.*, 163, 11, F3197–F3208, 2016, doi: 10.1149/2.0271611jes.
4. Le Bozec, N., Compere, C., L'Her, M., Laouenan, A., Costa, D., Marcus, P., Influence of stainless steel surface treatment on the oxygen reduction reaction in seawater. 43, 4, 765–786, 2001, https://doi.org/10.1016/S0010-938X(00)00113-X.
5. Strmcnik, D., Uchimura, M., Wang, C., Subbaraman, R., Danilovic, N., van der Vliet, D., Markovic, N.M., Improving the hydrogen oxidation reaction rate by promotion of hydroxyl adsorption. *Nat. Chem.*, 5, 4, 300–306, 2013, doi: 10.1038/nchem.1574.
6. Fang, Y.-H., Wei, G.-F., Liu, Z.-P., Catalytic Role of Minority Species and Minority Sites for Electrochemical Hydrogen Evolution on Metals: Surface Charging, Coverage, and Tafel Kinetics. *J. Phys. Chem. C*, 117, 7669, 2013, https://doi.org/10.1021/jp400608p.
7. Xiao, X., Yang, L., Sun, W., Chen, Y., Yu, H., Li, K., Jia, B., Zhang, L., Ma, T., Electrocatalytic Water Splitting: From Harsh and Mild Conditions to Natural Seawater. *Nano-Micro Small*, 2021, https://doi.org/10.1002/smll.202105830.
8. Murthy, A.P., Madhavan, J., Murugan, K., Recent advances in hydrogen evolution reaction catalysts on carbon/carbon-based supports in acid media. *J. Power Sources*, 398, 9, 26, 2018, doi: 10.1016/j.jpowsour.2018.07.040.
9. Luo, W., Wang, Y., Cheng, C., Ru-based electrocatalysts for hydrogen evolution reaction. *Mater. Today Phys.*, 15, 100274, 2020, doi: 10.1016/j.mtphys.2020.100274.
10. Ma, H., Yang, X., Wang, Z., Jiang, Q., Engineering the interface of porous CoMoO3 nanosheets with Co3Mo nanoparticles for high-performance electrochemical overall water splitting. *J. Mater. Sci. & Technol.*, 137, 184–192, 2022, https://doi.org/10.1016/j.jmst.2022.07.044.
11. Li, P., Zhao, G., Cui, P., Cheng, N., Lao, M., Xu, X., Dou, S.X., Sun, W., Nickel single atom-decorated carbon nanosheets as multifunctional electrocatalyst supports toward efficient alkaline hydrogen evolution. *Nano Energy*, 83, 105850, 2021, https://doi.org/10.1016/j.nanoen.2021.105850.
12. Hu, C., Zhang, L., Gong, J., Recent Progress of Mechanism Comprehension and Design of Electrocatalysts for Alkaline Water Splitting. *Energy Environ. Sci.*, 2019, doi: 10.1039/c9ee01202h.
13. Wang, S., Lu, A., Zhong, C.J., Hydrogen production from water electrolysis: role of catalysts. *Nano Converg*, 11, 8, 2021, doi: 10.1186/s40580-021-00254-x.

14. Dresp, S., Dionigi, F., Klingenhof, M., Strasser, P., Direct electrolytic splitting of seawater: opportunities and challenges. *ACS Energy Lett.*, 4, 933–942, 2019.
15. Kirk, D.W. and Ledas, A.E., Precipitate formation during sea water electrolysis. *Int. J. Hydrog. Energy*, 7, 12, 925–932, 1982. https://doi.org/10.1016/0360-3199(82)90160-4.
16. Dionigi, F., Reier, T., Pawolek, Z., Gliech, M., Strasser, P., Design criteria, operating conditions, and nickel–iron hydroxide catalyst materials for selective seawater electrolysis. *ChemSusChem*, 9, 9, 962–972, 2016. https://doi.org/10.1002/cssc.201501581.
17. Li, J., Liu, Y., Chen, H., Zhang, Z., Zou, X., Design of a Multilayered Oxygen-Evolution Electrode with High Catalytic Activity and Corrosion Resistance for Saline Water Splitting. *Adv. Funct. Mater.*, 31, 2021, doi: 10.1002/adfm.202101820.
18. Peng, M., Huang, J., Zhu, Y., Zhou, H., Hu, Z., Liao, Y.-K., Lai, Y.-H., Chen, C.-T., Chu, Y.-H., Zhang, K.H.L., Fu, X., Zuo, F., Li, J., Sun, Y., Structural anisotropy determining the oxygen evolution mechanism of strongly correlated perovskite nickelate electrocatalyst. *ACS Sustain. Chem. & Eng.*, 9, 11, 4262–4270, 2021, doi: 10.1021/acssuschemeng.1c00596.
19. Gayen, P., Saha, S., Ramani, V.K., Selective Seawater Splitting Using Pyrochlore Electrocatalyst. *ACS Appl. Energy Mater.*, 3, 3978–3983, 2020, doi: 10.1021/acsaem.0c00383.
20. Vu, A.Q., Vuillemin, B., Oltra, R., Allely b, C., Cut-edge corrosion of a Zn–55Al-coated steel: a comparison between sulphate and chloride solutions. *Corrosion Sci.*, 53, 9, 3016–3025, 2011, doi: 10.1016/j.corsci.2011.05.048.
21. Arukalam, I.O., Durability and synergistic effects of KI on the acid corrosion inhibition of mild steel by hydroxypropyl methylcellulose. *Carbohydr. Polym.*, 112, 291–299, 2014. https://doi.org/10.1016/j.carbpol.2014.05.071.
22. Gras, J.M. and Spiteri, P., Corrosion of stainless steels and nickel-based alloys for alkaline water electrolysis. *Int. J. Hydrog. Energy*, 18, 7, 561–566, 1993. https://doi.org/10.1016/0360-3199(93)90175-A.
23. Cavaliere, P., *Water Electrolysis for Hydrogen Production*, Springer Science and Business Media LLC, 2023.
24. Liu, P., Rodriguez, J.A., Muckerman, J.T., Desulfurization of SO2 and thiophene on surfaces and nanoparticles of molybdenum carbide: unexpected ligand and steric effects. *J. Phys. Chem. B*, 108, 40, 15662–15670, 2014, doi: 10.1021/jp040267a.
25. Walter, M.G., Warren, E.L., Mckone, J.R., Boettcher, S.W., Mi, Q., Santori, E.A., Lewis, N.S., Solar Water splitting Cells. *Chem. Rev.*, 110, 11, 6446–6473, 2010, doi: 10.1021/cr1002326.
26. Vrubel, H. and Hu, X., Molybdenum boride and carbide catalyze hydrogen evolution in both acidic and basic solutions. *Angew. Chem. Int. Ed.*, 51, 51, 12703–12706, 2012, doi: 10.1002/anie.201207111.

27. Huang, J., Li, Y., Xia, Y., Zhu, J., Yi, Q., Wang, H., Xiong, J., Sun, Y., Zou, G., Flexible cobalt phosphide network electrocatalyst for hydrogen evolution at all pH values. *Nano Res.*, 10, 3, 1010–1020, 2017, doi: 10.1007/s12274-016-1360-y.
28. Huang, Z.F., Song, J., Li, K., Tahir, M., Wang, Y.T., Pan, L., Wang, L., Zhang, X., Zou, J.J., Hollow cobalt-based bimetallic sulfide polyhedra for efficient all-pH-value electrochemical and photocatalytic hydrogen evolution. *J. Am. Chem. Soc.*, 138, 4, 1359–1365, 2016, doi: 0.1021/jacs.5b11986.
29. Huynh, M., Shi, C., Billinge, S.J.L., Nocera, D.G., Nature of Activated Manganese Oxide for Oxygen Evolution. *J. Am. Chem. Soc.*, 137, 47, 14887–14904, 2015, doi: 10.1021/jacs.5b06382.
30. Laursen, A.B., Varela, A.S., Dionigi, F., Fanchiu, H., Miller, C., Trinhammer, O.L., Rossmeisl, J., Dahl, S., Electrochemical Hydrogen Evolution: Sabatier's Principle and the Volcano Plot. *Chem. Educ.*, 89, 12, 1595–1599, 2012, doi: 10.1021/ed200818t.
31. Tahir, A., Arshad, F., ul Haq, T., Hussain, I., Hussain, S.Z., ur Rehman, H., Roles of Metal Oxide Nanostructure-Based Substrates in Sustainable Electrochemical Water Splitting: Recent Development and Future Perspective. *ACS Appl. Nano Mater.*, 2023.
32. Bu, X., Wei, R., Cai, Z., Quan, Q., Zhang, H., Wang, W., Li, F., Po, Y.S., Meng, Y., Chan, K.S., Wang, X., Hoa, J.C., More than physical support: The effect of nickel foam corrosion on electrocatalytic performance. *Appl. Surf. Sci.*, 538, 147977, 2021, doi: 10.1016/j.apsusc.2020.147977.
33. Vos, J.G., Tim, A., Wezendonk, A.J., Koper, M.T.M., MnOx/IrOx as selective oxygen evolution electrocatalyst in acidic chloride solution. *J. Am. Chem. Soc.*, 140, 32, 10270–10281, 2018, doi: 10.1021/jacs.8b05382.
34. Kuanga, Y., Kenneya, M.J., Menga, Y., Hunga, W.-H., Liuf, Y., Huanga, J.E., Prasannag, R., Lib, P., Lib, Y., Wanglı, L., Lind, M.-C., McGeheeg, M.D., Sunb, X., Daia, H., Solar-driven, highly sustained splitting of seawater into hydrogen and oxygen fuels. *Proceedings of the National Academy of Sciences*, vol. 116, pp. 6624–6629, 2019, doi: 10.1073/pnas.1900556116.
35. Jamesh, M.-I., Moni, P., Prakash, A.S., Harb, M., ORR/OER activity and zinc-air battery performance of various kinds of graphene-based air catalysts. *Mater. Sci. Energy Technol.*, 4, 1–22, 2021, doi: 10.1016/j.mset.2020.12.001.
36. Tong, W. *et al.*, Electrolysis of low-grade and saline surface water. *Nat. Energy*, 5, 5, 367–377, 2020, doi: 10.1038/s41560-020-0550-8.
37. Schafer, H., Beladi-Mousavi, S.M., Walder, L., Wollschlager, J., Kuschel, O., Ichilmann, S., Sadaf, S., Steinhart, M., Kupper, K., Schneider, L., Surface Oxidation of Stainless Steel: Oxygen Evolution Electrocatalysts with High Catalytic Activity. *ACS Catal.*, 5, 4, 2671–2680, 2015, doi: 10.1038/s41560-020-0550-8.
38. Gueshi, T., Tokuda, K., Matsuda, H., Voltammetry at partially covered electrodes: Part I. Chronopotentiometry and chronoamperometry at model

electrodes. *J. Electroanal. Chem. Interfacial Electrochem.*, 892, 247–260, 1978, https://doi.org/10.1016/S0022-0728(78)80188-0.

39. Strmcnik, D., Uchimura, M., Wang, C., Subbaraman, R., Danilovic, N., van der Vliet, D., Paulikas, A.P., Stamenkovic, V.R., Markovic, N.M., Improving the hydrogen oxidation reaction rate by promotion of hydroxyl adsorption. *Nat. Chem.*, 5, 300–306, 2013, https://doi.org/10.1038/nchem.1574.
40. Liu, X., Gong, M., Deng, S., Zhao, T., Shen, T., Zhang, J., Transforming damage into benefit: corrosion engineering enabled electrocatalysts for water splitting. *Adv. Funct. Mater.*, 31, 11, 2009032, 2021, doi: 10.1002/adfm.202009032.
41. Ahmed, A.T.A., Ansari, A.S., Pawar, S.M., Shong, B., Kim, H., Im, H., Anti–corrosive FeO decorated CuCo2S4 as an efficient and durable electrocatalyst for hydrogen evolution reaction. *Appl. Surface*, 539, 148229, 2021, doi: 10.1016/j.apsusc.2020.148229.
42. Sapakal, S.N., Khan, A., Khobragade1, J.M., Kadam, A.V., Synthesis of MoO2 on steel substrate for hydrogen evolution reaction in water splitting: a new approach. *J. Mater. Sci.*, 58, 1403–1414, 2023, https://doi.org/10.1007/s10853-022-08079-7.
43. Ashwini, P., Sapakal, S., Khan, A., Kadam, A.V., Synthesis of NiO thin film on 304-grade stainless steel substrate for oxygen evolution reaction. *Surf. Interfaces*, 37, 102706, 2023, https://doi.org/10.1016/j.surfin.2023.102706.
44. Theerthagiri, J., Lee, S.J., Murthy, A.P., Madhavan, J., Choi, M.Y., Fundamental aspects and recent advances in transition metal nitrides as electrocatalysts for hydrogen evolution reaction: A review. *Curr. Opin. Solid State Mater. Sci.*, 24, 1, 100805, 2020, https://doi.org/10.1016/j.cossms.2020.100805.
45. Liu, D., Xu, G., Yang, H., Wang, H., Xia, B.Y., Rational Design of Transition Metal Phosphide-Based Electrocatalysts for Hydrogen Evolution. *Adv. Funct. Mater.*, 33, 7, 2208358, 2023, https://doi.org/10.1002/adfm.202208358.
46. Zahra, R., Pervaiz, E., Yang, M., Rabi, O., Saleem, Z., Ali, M., Farrukh, S., A review on nickel cobalt sulphide and their hybrids: Earth abundant, pH stable electro-catalyst for hydrogen evolution reaction. *Int. J. Hydrog. Energy*, 45, 46, 24518–24543, 2020, https://doi.org/10.1016/j.ijhydene.2020.06.236.
47. Liu, Y., Liang, X., Gu, L., Zhang, Y., Li, G.D., Zou, X., Chen, J.S., Corrosion engineering towards efficient oxygen evolution electrodes with stable catalytic activity for over 6000 hours. *Nat. Commun.*, 9, 1, 2609, 2018, https://doi.org/10.1038/s41467-018-05019-5.
48. Roberge, P.R., *Handbook of corrosion engineering*, vol. 1128, Mcgraw-hill, New York, 2000.
49. Liu, W.-J., Xu, Z., Zhao, D., Pan, X.-Q., Li, H.-C., Hu, X., Fan, Z.-Y., Efficient electrochemical production of glucaric acid and H_2 via glucose electrolysis. *Nat. Commun.*, 11, 1, 265, 2020, https://doi.org/10.1038/s41467-019-14157-3.
50. Zhong, D., Zhang, L., Li, C., Li, D., Wei, C., Zhao, Q., Li, J., Gong, J., Nanostructured NiFe (oxy) hydroxide with easily oxidized Ni towards

efficient oxygen evolution reactions. *J. Mater. Chem. A*, 6, 35, 16810–16817, 2018, https://doi.org/10.1039/C8TA04721A.

51. Zhong, D., Liu, L., Li, D., Wei, C., Wang, Q., Hao, G., Zhao, Q., Li, J., Facile and fast fabrication of iron-phosphate supported on nickel foam as a highly efficient and stable oxygen evolution catalyst. *J. Mater. Chem. A*, 5, 35, 18627–18633, 2017, https://doi.org/10.1039/C7TA05580C.
52. Kang, Y., Xie, H., Liu, D., Gao, M., Chu, P.K., Ramakrishna, S., Yu, X.F., Facile mass production of self-supported two-dimensional transition metal oxides for catalytic applications. *Chem. Commun.*, 55, 76, 11406–11409, 2019, https://doi.org/10.1039/C9CC06261K.
53. Rosalbino, F., Scavino, G., Actis Grande, M., Electrocatalytic activity of Ni–Fe–M (M= Cr, Mn, Cu) sintered electrodes for hydrogen evolution reaction in alkaline solution. *J. Electroanal. Chem.*, 694, 114–121, 2013, https://doi.org/10.1016/j.jelechem.2013.01.039.
54. Ma, S., Yuan, H., Cai, L., Wang, X., Long, H., Chai, Y., Tsang, Y.H., One step synthesis of Fe4. 4Ni17. 6Se16 coupled NiSe foam as self-supported, highly efficient and durable oxygen evolution electrode. *Mater. Today Chem.*, 9, 133–139, 2018, https://doi.org/10.1016/j.mtchem.2018.05.003.
55. Fan, J., Chen, Z., Shi, H., Zhao, G., *In situ* grown, self-supported iron–cobalt–nickel alloy amorphous oxide nanosheets with low overpotential toward water oxidation. *Chem. Commun.*, 52, 23, 4290–4293, 2016, https://doi.org/10.1039/C5CC09699E.
56. Lili, Z., Yang, L., Lu, J., Jia, J., Yu, J., Deng, Y., Shao, M., Zhou, W., One-step synthesis of Fe-Ni hydroxide nanosheets derived from bimetallic foam for efficient electrocatalytic oxygen evolution and overall water splitting. *Chin. Chem. Lett.*, 29, 12, 1875–1878, 2018, https://doi.org/10.1016/j.cclet.2018.10.026.
57. Martindale, B.C.M. and Reisner, E., Bi–functional Iron-only electrodes for efficient water splitting with enhanced stability through *in situ* electrochemical regeneration. *Adv. Energy Mater.*, 6, 1502095–1502103, 2016, https://doi.org/10.1002/aenm.201502095.
58. Guo, Y., Park, T., Yi, J.W., Henzie, J., Kim, J., Wang, Z., Jiang, B., Bando, Y., Sugahara, Y., Tang, J., Yamauchi, Y., Nano-architectonics for transition-metal-sulphide-based electrocatalysts for water splitting. *Adv. Mater.*, *31*, 1807134–1807167, 2019, https://doi.org/10.1002/adma.201807134.
59. Wu, X., Li, S., Liu, J., Yu, M., Mesoporous hollow nested nanospheres of Ni, Cu, Co–based mixed sulphides for electrocatalytic oxygen reduction and evolution. *ACS Appl. Nano Mater.*, 2, 4921–4932, 2019, doi: 10.1021/acsanm.9b00891.
60. Du, X., Fu, J., Zhang, X., Controlled synthesis of CuCo2S4@ Ni (OH) 2 hybrid nanorod arrays for water splitting at an ultralow cell voltage of 1.47 V. *Chem. Asian J.*, 14, 19, 3386–3396, 2019, https://doi.org/10.1002/asia.201901137.

61. Jagadeesh, R.V., Natte, K., Junge, H., Beller, M., Nitrogen-doped graphene-activated iron–oxide–based nanocatalysts for selective transfer hydrogenation of nitroarenes. *ACS Catal.*, 5, 1526–1529, 2015, doi: 10.1021/cs501916p.
62. Zhang, H., Xi, B., Gu, Y., Chen, W., Xiong, S., Interface engineering and heterometal doping Mo-NiS/Ni (OH) 2 for overall water splitting. *Nano Res.*, 14, 3466–3473, https://doi.org/10.1007/s12274-021-3557-y.
63. Bento, A., Sanches, A., Medina, E., Nunes, C.D., Vaz, P.D., MoO2 nanoparticles as highly efficient olefin epoxidation catalysts. *Appl. Catal. A*, 504, 399-407, 2015, https://doi.org/10.10 16/j.apcata.2015.03.024.
64. Lu, C.P., Li, G., Mao, J., Wang, L.M., Andrei, E.Y., Bandgap, mid-gap states, and gating effects in MoS2. *Nano Lett.*, 14, 4628–4633, 2014, https://doi.org/10.1021/nl501659n.
65. Gu, Y., Xi, B., Tian, W., Zhang, H., Fu, Q., Xiong, S., Boosting selective nitrogen reduction via geometric coordination engineering on single-tungsten-atom catalysts. *Adv. Mater.*, 33, 25, 2100429, 2021, https://doi.org/10.1002/adma.202100429.
66. Ramakrishnan C, V., Alex, A.N., Nair, A.N., John, N.S., Designing metallic MoO2 nanostructures on rigid substrates for electrochemical water activation. *Chem. A Eur. J.*, 24, 18003–18011, 2018. https://doi.org/10.1002/chem.201803570.
67. Jin, Y., Wang, H., Li, J., Yue, X., Han, Y., Shen, P.K., Cui, Y., Porous MoO2 nanosheets as non-noble bifunctional electrocatalysts for overall water splitting. *Adv. Mater.*, 28, 19, 3785–3790, 2016, https://doi.org/10.1002/adma.201506314.
68. Xu, Z., Liu, Y., Li, G.-D., Wu, Y., Liu, D.-P., Li, W., Li, H.-W., Wang, D., Zhang, Y., Zou, X., Ultrafast formation of amorphous bimetallic hydroxide films on 3D conductive sulfide nanoarrays for large-current-density oxygen evolution electrocatalysis. *Adv. Mater.*, 29, 22, 1700404, 2017, https://doi.org/10.1002/adma.201700404.
69. Zou, X., Liu, Y., Li, G.-D., Wu, Y., Liu, D.-P., Li, W., Li, H.-W., Wang, D., Zhang, Y., Zou, X., Ultrafast Formation of Amorphous Bimetallic Hydroxide Films on 3D Conductive Sulfide Nanoarrays for Large-Current-Density Oxygen Evolution Electrocatalysis. *Adv. Mater.*, 2017.
70. Luo, Y., non-noble metal-nitride based electrocatalysts for high-performance alkaline seawater electrolysis. *Nat. Commun.*, 10, 1, 5106, 2019, https://doi.org/10.1038/s41467-019-13092-7.
71. ul Haq, T. and Haik, Y., Strategies of Anode Design for Seawater Electrolysis: Recent Development and Future Perspective. *Small Sci.*, 2, 2200030, 2022, https://doi.org/10.1002/smsc.202200030.
72. Li, J., Design of a multilayered oxygen-evolution electrode with high catalytic activity and corrosion resistance for saline water splitting. *Adv. Funct. Mater.*, 31, 27, 2101820, 2021.
73. Shin, H., Xiao, H., Goddard III, W.A., In silico discovery of new dopants for fe-doped ni oxyhydroxide ($Ni_{1-x}Fe_xOOH$) catalysts for oxygen

evolution reaction. *J. Am. Chem. Soc.*, 140, 22, 6745–6748, 2018, DOI: 10.1021/jacs.8b02225.

74. ul Haq, T., Mansour, S., Haik, Y., Electronic and structural modification of Mn3O4 nanosheets for selective and sustained seawater oxidation. *ACS Appl. Mater. Interfaces*, 14, 18, 20443–20454, 2022, doi: 10.1021/acsami.1c24304.
75. Yang, Y., Liu, J., Xiong., B., Vacancy-mediated transition metals as efficient electrocatalysts for water splitting. *Nanoscale*, 14, 19, 7181–7188, 2022, https://doi.org/10.1039/D2NR01259F.
76. Huang, L., Gao, G., Zhang, H., Chen, J., Fang, Y., Dong, S., Self-dissociation-assembly of ultrathin metal-organic framework nanosheet arrays for efficient oxygen evolution. *Nano Energy*, 68, 104296, 2020, https://doi.org/10.1016/j.nanoen.2019.104296.
77. Hao, Q., Li, S., Liu, H., Mao, J., Li, Y., Liu, C., Zhang, J., Tang, C., Dual tuning of nickel sulfide nanoflake array electrocatalyst through nitrogen doping and carbon coating for efficient and stable water splitting. *Catal. Sci. Techno.*, 9, 3099–3108, 2019, https://doi.org/10.1039/C9CY00688E.
78. Yang, X., Cui, J., Liu, X., Zhang, Q., Wang, D., Ye, J., Liu, L., Boosting photocatalytic overall water splitting over single-layer graphene coated metal cocatalyst. *Appl. Catal., B*, 325, 122369, 2023. ISSN 0926-3373, https://doi.org/10.1016/j.apcatb.2023.122369.
79. Jadhav, A.R., Kumar, A., Lee, J., Yang, T., Na, S., Lee, J., Luo., Y., Stable complete seawater electrolysis by using interfacial chloride ion blocking layer on catalyst surface. *J. Mater. Chem. A*, 46, 8, 24501–24514, 2020, https://doi.org/10.1039/D0TA08543J.
80. Song, H.J., Yoon, H., Ju, B., Lee, D.Y., Kim, D.W., Electrocatalytic Selective Oxygen Evolution of Carbon-Coated Na2Co1−x fe x P2O7 nanoparticles for alkaline seawater Electrolysis. *ACS Catal.*, 10, 1, 702–709, 2019, doi: 10.1021/acscatal.9b04231.
81. Jiaxin, G., Zheng, Y., Hu, Z., Zheng, C., Mao, J., Du, K., Jaroniec, M., Qiao, S.-Z., Ling, T., Direct seawater electrolysis by adjusting the local reaction environment of a catalyst. *Nat. Energy*, 8, 3, 264–272, 2023, https://doi.org/10.1038/s41560-023-01195-x.

3

Ionic Liquids in Marine Anti-Fouling Coatings

Samantha Taylor[1,2,3], Sajith K. Baiju[1,2,3], Madison K. McQuinlan[1,2,3] and Matthew G. Cowan[1,2,3]*

[1]*Department of Chemical and Process Engineering, University of Canterbury, Christchurch, New Zealand*
[2]*New Zealand Product Accelerator, University of Canterbury, Christchurch, New Zealand*
[3]*MacDiarmid Institute for Advanced Materials and Nanotechnology, University of Canterbury, Christchurch, New Zealand*

Abstract

Biofouling is the accumulation of marine organism colonies on submerged surfaces. Unchecked, this phenomenon is crippling to the marine industry, affecting both static structures and mobile vessels. The two major problems are the reduced efficiency of vessels and the transport of invasive species across marine environments. While biofouling is solvable through coatings containing heavy metal biocides, the accumulation of heavy metals makes the environmental impact unacceptable.

This chapter reviews the strategies being pursued to mitigate biofouling while eliminating the environmental concerns associated with heavy metal biocides. This chapter briefly summarizes the leading reviews and current state-of-the-art technology in marine anti-fouling coatings, covering fundamental principles, rational design strategies, and emerging technologies, before focusing on ionic liquid formulations being investigated for this application.

Keywords: Ionic liquid, poly(ionic liquid), anti-fouling, marine, foul release, foul degrading, foul resistant

**Corresponding author*: Matthew.cowan@canterbury.ac.nz

Hari Murthy, Vinay Jha Pillai, Kukatlapalli Pradeep Kumar and Matthew Cowan (eds.) Novel Anti-Corrosion and Anti-Fouling Coatings and Thin Films, (43–76) © 2024 Scrivener Publishing LLC

3.1 Introduction

Biofouling is the undesirable accumulation of marine bacteria, flora, and fauna on submerged surfaces. Biofouling is a natural phenomenon that leads to the formation of complex flora and fauna colonies on both static marine structures and mobile vessels. The exact mechanism for the formation of a biofouling system is highly variable, influenced by the water conditions and nutrient availability. The consequences of biofouling severely affect multiple industries.

3.1.1 Societal, Environmental, and Economic Context to Marine Biofouling

Biofouling influences the financial stability and environmental impact of industries dependent on the marine and coastal environment, such as maritime trade, aquaculture, fisheries, and tourism.

In maritime trade, biofouling increases the fuel consumption of ocean-going vessels, increasing the cost due to slower travel and worse fuel efficiency—which also results in increasing CO_2 emissions. The marine industry is currently responsible for annual emissions of approximately 900 million tonnes of CO_2 and it is reported that biofouling of a 5-mm calcareous coverage can trigger an increase in GHG emissions by up to 50% (Figure 3.1) [1]. In 2011, the International Maritime Organisation (IMO) pledged to reduce the total greenhouse gas (GHG) emissions from

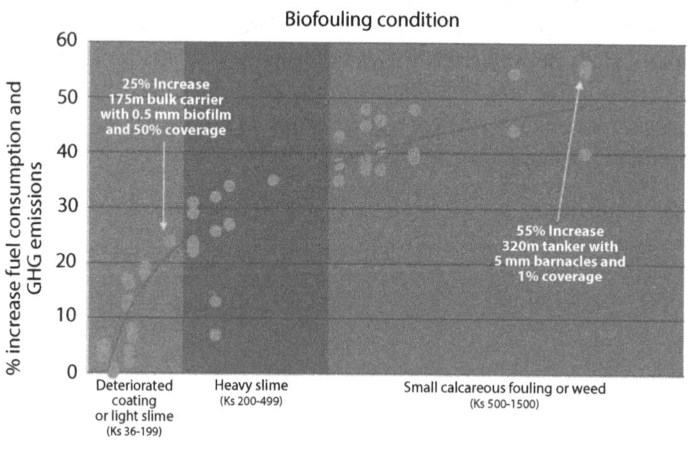

Figure 3.1 Compendium of results from published research studies for the impact of ship hull biofouling on GHG emissions [1].

shipping. This requires a switch to greener fuels or reduced fuel consumption, and biofouling prevention is the best path toward reduced fuel consumption [2].

Biofouling poses significant environmental threats beyond CO_2 emissions, impacting biosecurity and commercial fisheries. The local environment provided by accumulated biofouling provides a refuge for macrofouling species that can then be transported across large areas of the ocean and establish new invasive populations in formerly pristine waters, potentially causing the collapse and extinction of native populations [3]. Biofouling has been identified as the most significant vector for the introduction of invading species within a marine environment, at global and local scales [4]. Similarly, fish species can be impacted by the growth of invasive species and the accumulation of biofouling that competes for nutrients, reducing the productivity of commercial seafood production [4–6].

Both maritime and biosecurity industries therefore agree that the prime focus of biofouling research is the development of a technology that prevents biofouling at the initial biofilm formation stage. The marine industry seeks a preventative solution because the initial biofilm forms within minutes to hours of the vessel being in the water, thus imposing an almost immediate increase in the hydrodynamic drag [7]. The biosecurity industry seeks a preventative solution because intervention at the transfer stage of invasion, where the foulants are released from the vessel, is very difficult. In the time required to implement such solutions, the negative effects of the invader have begun [4].

For these reasons, there is a constant demand for effective anti-fouling coatings for static and dynamic applications. The majority of solutions currently in use and under development target the initial conditioning stage of biofouling in order to have the greatest improvement in vessel efficiency. While heavy metal coatings, acting as chemical biocides, are extremely effective at inhibiting biofilm formation, they accumulate in the food chain and therefore impose an ongoing environmental concern due to off-target effects [8]. This has led to multiple local and national governments introducing legislation to limit the use of anti-fouling coatings containing heavy metals, particularly tin. While copper- and zinc-containing coatings are currently accepted internationally, they too pose accumulation hazards. The leading strategy for avoiding heavy metals is slippery liquid-infused porous surfaces (SLIPS); however, they provide only a physical barrier to adhesion and any abrasion allows fouling species to colonize the underlying substrate [9].

In summary, the anti-fouling coating market in the global maritime industry is large, and it is estimated that anti-fouling coatings could save the marine industry 150 billion USD annually [10] by limiting corrosion, weight, and drag to increase fuel efficiency, time intervals between dry-docking, and engine lifetime [11]. A gap in the current market exists for a robust anti-fouling coating that is effective at preventing the adhesion of marine organisms without posing long-lasting environmental concerns.

3.1.2 Mechanism of Biofouling

The ocean contains a diverse range of nutrients, bacteria, flora spores, and fauna larvae competing for solid space to adhere to and commence their life cycle, so when a new surface enters the ocean, it is immediately colonized [12]. Intuitively, this fouling is visually obvious through large predatory macrofoulers such as mussels, oysters, and shellfish. However, the biofouling process commences *via* four steps starting with adherence of extracellular polymeric substances (EPS), bacteria, larvae, and finally, the arrival of larger predators (Figure 3.2).

The four stages of biofouling described below are complex and occur through physical, covalent, and biochemical interactions and reactions [13]. An ideal anti-fouling coating would combat all fouling stages, effectively combatting over an estimated 4000 different biofouling species, all using a range of mechanisms honed by evolution to adhere to a surface in the ocean environment under a wide variety of conditions [13].

3.1.2.1 Stage 1: Conditioning Film Formation

The first stage of anti-fouling is the physical adhesion of organic molecules to the surface, such as proteins, polysaccharides, and glycoproteins [14]. The organic molecules that adhere are EPS, which are secreted by bacteria

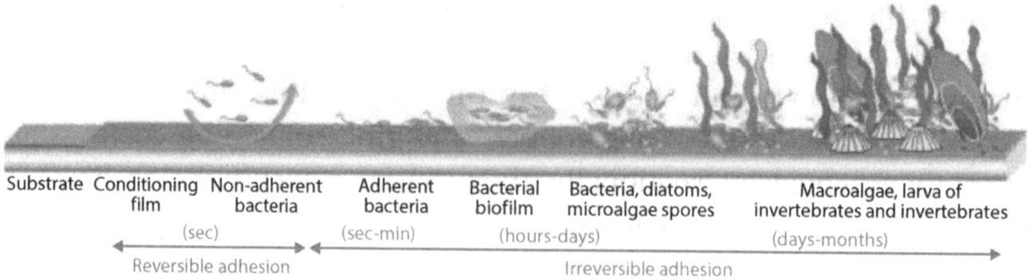

Figure 3.2 Illustration of the stages of biofouling on submerged surfaces in the ocean. The process starts with extracellular polymeric substances forming a conditioning film, followed by bacterial colonization, larvae and algae adhesion, and finally, the attachment of larger predatory species [7].

and other colonizing marine organisms. EPS create a film on the surface, attaching *via* physical forces such as Van der Walls forces and electrostatic interaction [15, 16]. The EPS layer creates a nutrient-dense medium that allows for intercellular communication between simple organisms and increases the stickiness of the surface, which allows the next stage of fouling to occur [16, 17]. Conditioning film formation occurs within minutes of the surface being submerged [12].

3.1.2.2 Stage 2: Primary Colonization

The second stage of fouling occurs within 24 hours of submersion and consists of bacteria and unicellular algae adsorbing to the surface to make use of the conditioning film. At first, these species are reversibly adsorbed onto the surface *via* weak noncovalent bonds. Irreversible adhesion follows, as the bacteria anchor to the surface *via* the secretion of exopolymers [7, 18]. This rough and nutrient-rich film provides lower-energy attachment points and sufficient food for the third stage of fouling to occur [14, 16].

3.1.2.3 Stage 3: Secondary Colonization

The third stage of fouling occurs in the days after submersion and reflects a diversification of the biofilm microorganisms, growing to include algal spores, diatoms, spores of microalgae, and barnacle larvae [9]. The microorganisms adhere by secretion of mucilaginous extracellular polymers and polysaccharides that contain carboxylate, monoester-sulfate, and phosphate groups to covalently attach to the surface [19, 20]. Once adhered to the surface, the organisms grow and divide, giving rise to colonies. The formation of this more complex community is considered microfouling and is irreversible [7].

3.1.2.4 Stage 4. Secondary Colonization

The fourth stage of fouling is the irreversible attachment of macroorganisms, which typically involves macroalgae (seaweed) and marine invertebrates (mussels, oysters, shellfish, etc.). Macrofoulers adhere reversibly *via* the secretion of weakly acidic mucopolysaccharides and irreversibly *via* bioadhesive polyamides [21]. This stage typically occurs over a period of days to months and the layer evolves into a complex biological community that can grow several meters thick under idealized conditions [7, 8, 13, 14].

3.1.3 Factors Affecting Speed and Extent of Biofouling

Temperature, solar radiation, pollution, and the presence of other organisms all affect biofouling. These parameters are often associated with the local water conditions and thus cannot be easily modified to control potential fouling [8]. However, it is important to understand how these factors influence biofouling and therefore experimental design of materials testing.

3.1.3.1 Seasonal Factors: Temperature and Sunlight

Biofouler growth tends to be fastest and greatest at warmer temperatures [22]. Similarly, solar radiation provides light energy for photosynthesis and thus influences nutrient availability. Regions with marked seasonal variations in temperature and sunlight hours also observe associated variations in marine growth and hence biofouling. In regions where there is little seasonal variation in temperature, such as tropical climates close to the equator, fouling occurs at a relatively constant rate throughout the year [8]. This is important when designing studies to compare the effectiveness of anti-fouling coatings because a second-generation material tested over winter months may show improved performance when, in reality, the performance variation is a seasonal effect.

3.1.3.2 Pollution

Pollution can block solar radiation and slow growth, as well as deplete nutrients and oxygen causing starvation and asphyxiation, slowing biofouling. Alternatively, pollutants can alter selection pressures, reducing biodiversity and allowing invasive species to dominate marine communities, thereby accelerating biofouling [8, 23]. Furthermore, pollutants can interfere with the mechanism of action of anti-fouling coatings, either by accelerating the depletion of biocide molecules through reaction or catalysis or by interacting with and obscuring surface functionality. Therefore, it is important to consider the potential impact on pollutants when completing 'real-world' immersion testing, in comparison to synthetic seawater prepared in a laboratory.

3.1.4 Background of Anti-Fouling Coatings

The earliest recorded anti-fouling coatings include wax, tar, asphalt, and pitch for coating wooden boat hulls [13, 14, 16]. Reports from the early 15th century diversify into the use of arsenic, mercury, copper, and lead incorporated into tar and wax coatings [7, 24]. The heavy metals act as potent biocides, disrupting cellular activity and thus inhibiting growth and colonization.

The drawbacks of these coatings, which still challenge the industry today, include short life cycles (heavy metals leach into seawater), coating degradation (requiring frequent reapplication), off-target toxicity, and corrosion of iron components of the boats [16]. Interestingly, for the context of this book, heavy metals were consequently discontinued primarily due to their corrosive nature [8].

3.1.4.1 Heavy Metal Biocide Coatings

The direction of technology improvement led to less-corrosive heavy metals used as biocides and the ability to disperse the biocides within a slow-release matrix. The mid-19th century saw the first anti-fouling paints, with biocides such as copper, arsenic, or mercury, dispersed in a coating such as linseed oil, shellac, or rosin [14, 16].

The first organometallic paints were developed in 1950 and marked a major improvement in anti-fouling coating efficacy [14]. The organometallic compounds are highly toxic, reducing the concentrations required. Furthermore, the development of self-polishing coatings, which constantly exposed a fresh biocide surface, prolonged coating lifetimes. These organometallics are RMeX-type compounds, consisting of R, an organic radical; Me, a metal; and X, a halide or organic acid residue [14].

The premier example of organometallic coatings is tributyltin (TBT), a liquid-soluble compound that can easily penetrate and rupture cells to stop respiration and photosynthesis. However, the persistence in the food chain and accumulation in marine mammals makes TBT harmful to the environment. As a result, TBT and other organometallic compounds were banned by the International Maritime Organisation (IMO) first by the Convention on the Control of Harmful Anti-fouling Systems on Ships in 2001, with the ban coming into effect in 2008 [25].

3.1.4.2 Heavy Metal Substitutes

Without TBT, the marine industry returned to the use of copper-containing compounds. Copper metal was incorporated into similar self-polishing polymers, providing control over the release rate. Copper, like tin, can bioaccumulate. However, the justification for copper use is that copper is naturally present in the marine environment at relatively high concentrations [8].

In response to the bioaccumulation of copper, some national and local legislative bodies have limited the use of copper anti-fouling coatings. For example, in July 2018, the California Pesticides Regulatory Authority established an extreme permissible copper leaching rate of 9.5 μg/cm^2/day for copper-based anti-fouling products used on recreational vessels [28]. Similarly, Washington State (USA) has mandated that by 2021 the use of copper-based paints on commercial boats be phased out to protect salmon, and the European Commission is encouraging member states to investigate and finance alternative anti-fouling coatings.

Efforts to replace heavy metals in anti-fouling coatings have focused on the use of zinc-based and organic pesticides and herbicides such as dichlofluanid, chlorothalonil, chlorine dioxide DCOIT, zinc pyrithione, diuron, TCMS pyridine, irgarol 1051, and zineb are limited in anti-fouling products [26]. While these compounds improve the effectiveness of copper self-polishing systems to match the performance of TBT [16], some of the booster biocides are persistent in the environment and continuously leach from anti-fouling paints. This poses similar concerns, around buildup in aquatic products and increasing antibiotic resilience in marine bacteria.

Overall, the commercial importance of anti-fouling coatings and the environmental persistence and off-target effects of light metal biocides such as copper and zinc reinforce the need for a new generation of coatings to prevent marine biofouling.

3.2 Current Anti-Fouling Design Strategies

3.2.1 Requirements of Novel Anti-Fouling Coating

Before embarking on understanding potential solutions to the biofouling problem, it is useful to clearly define the requirements of an anti-fouling coating:

3.2.1.1 Performance Specifications

1. Prevents all stages of fouling, particularly the formation of the initial biofilm.
2. Cost-effective, balancing the durability and lifetime against the cost of manufacture and application (ca. 90 months for dynamic moving systems and 120 days for static objects).
3. Ability to be coated through a simple and efficient method, e.g., spray coating.

3.2.1.2 *Necessary Considerations*

4. Legally compliant, not releasing an environmentally accumulating biocide.
5. Noncorrosive or compatible with anticorrosive primer, to prevent corrosion of the boat hull or static object surface [14].

3.2.2 Types of Anti-Fouling Coatings

Anti-fouling coatings can be classified into three mechanisms of action (Figures 3.3, 3.4, and 3.5) [10]:

1. Foul-degrading: kills fouling species.
2. Foul-resistant: prevents adhesion of microfouling organisms.
3. Foul-release: allows adhesion of microfouling organisms, but with low release energies.

Fouling-resistant coatings work by preventing the adhesion of microfoulers to the surface, (proteins, algae, and bacteria), which is commonly achieved with a strongly hydrated surface, creating a physical and energy barrier for the foulants. Fouling-release coatings allow for only weak adhesion of foulants to the surface, which means that they can easily be removed from the surface by mechanical or shear force. These forces can be imposed by a water jet or by the movement of the ocean when the vessel moves above a certain speed. Fouling-degrading coatings work by degrading or

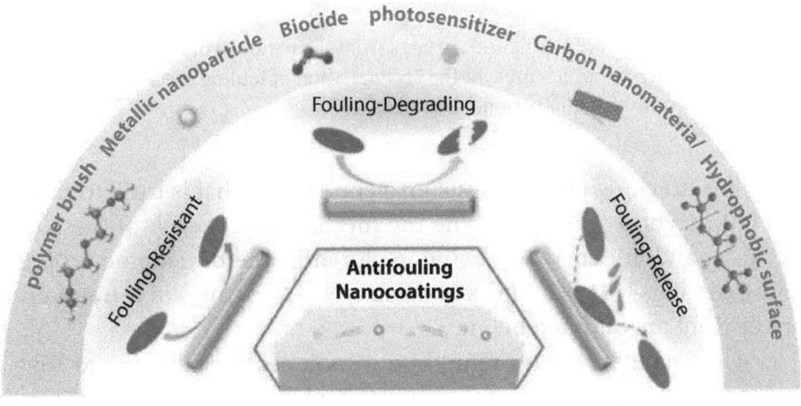

Figure 3.3 Illustration of the three classes of anti-fouling coating mechanisms, along with the most common design strategies for achieving these effects [27].

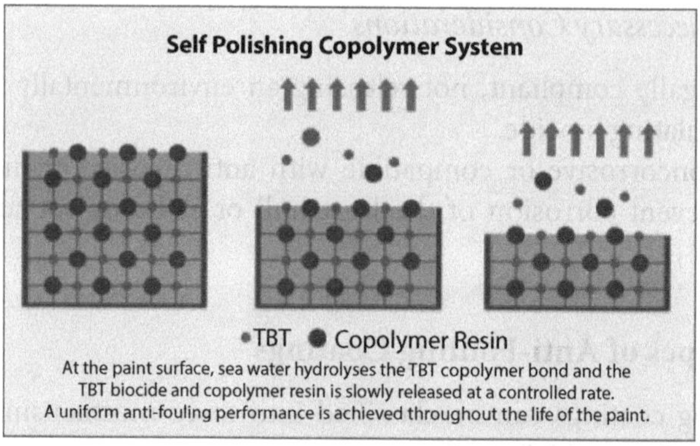

Figure 3.4 Self-polishing copolymer system mechanism for controlled TBT biocide release [28].

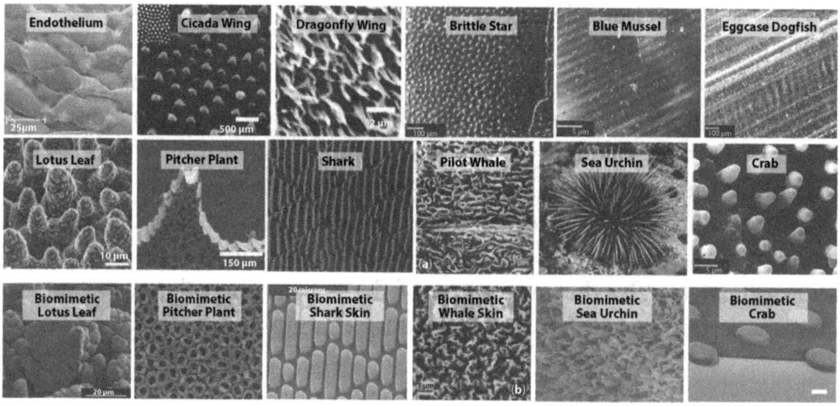

Figure 3.5 Nanostructured skins and surfaces from the natural world and created in laboratories (third row) which prevent the accumulation of fouling species. [Reproduced with permission Advanced Functional Materials] [37].

killing the attaching organisms using a component that is toxic to the foulants [10]. Hybrid paints combine the mechanisms of each of these types of coating and are currently being explored in the development of anti-fouling coatings [14]. Examples of these types of coatings in use and under development are outlined below.

As will become evident, ionic liquid moieties can be tailored toward each mechanism of action, positioning them as a useful material for the design of next-generation anti-fouling coatings.

3.2.2.1 *Foul Degrading Coatings*

Typical foul-degrading coatings rely on toxic components that can kill or inhibit the growth of fouling species. The biocide activity of these coatings is highly effective at preventing the formation of Stage 1 fouling layers. Historically, foul-degrading coatings relied on heavy metals and research focused on self-polishing paint strategies to continually expose 'fresh' surfaces with high concentrations of heavy metals. In response to the environmental impacts of heavy metals, recent research has examined the effectiveness of organic biocides and enzymes for breaking down the biofilm layers.

3.2.2.1.1 Self-Polishing Paints

TBT self-polishing paints were used by an estimated 70% of commercial shipping vehicles in 1999 due to their high effectiveness against fouling. Self-polishing paints are a highly effective design for continually exposing fouling species to high concentrations of biocide and preventing the accumulation of organic material. Furthermore, formulations could be tuned to suit the activity of the vessel. For example, sedentary ships that spend a lot of time in port require a faster rate of release, whereas faster vessels require slower rates of release. This means that self-polishing coatings can be optimized for maximum efficiency depending on the type of vessel.

The archetypal self-polishing paint is an acrylic polymer with TBT groups bonded to the main chain *via* ester linkages. The ester linkage is hydrolytically unstable, which allows constant release of TBT into the surrounding seawater [14]. Elegantly, the removal of TBT embrittles the polyacrylate, which is then worn away by shear stress from tides (fixed objects) or movement (boats). This creates a self-polishing effect (Figure 3.4), where fresh surfaces rich in TBT are constantly exposed, and the release rate can be controlled through either modification of the strength of the chemical bonding to TBT or by altering the polymer matrix. The rate of release of TBT is constant regardless of the ship's velocity.

3.2.2.1.2 Tin-Free Self-Polishing Paints

The ban on TBT in self-polishing coatings resulted in the design of tin-free self-polishing coatings (TF-SPCs). TF-SPCs use three mechanisms in an attempt to mimic TBT coatings [29]:

1. Self-polishing coatings, where the polymer surface is worn away, constantly exposing fresh biocide at the surface. Biocide depletion occurs at a constant rate.

2. Controlled depletion coatings, where the polymer allows infiltration of seawater to solubilize biocides and accelerate diffusion to the surface. The polymer is also slowly worn away, but initial biocide release occurs rapidly and decreases steadily over time.
3. Contact leaching coatings, where there is no decomposition of the polymer matrix over time. This leads to the formation of a boundary layer, limiting the rate of biocide release to the point toxic concentrations cannot be maintained and the film becomes ineffective (≤24 months).

An excellent review explaining the idealized design of TF-SPCs was published in 2009 by Bressy *et al.* [29]. Their review explains nuances vital to the rational development of anti-fouling coatings mimicking the effectiveness of TBT, such as standards and experimental procedures for quantifying biocide release, the link of hydration/decomposition kinetics and polymer degradation, and the specific biocide and polymer and chemistries used in TF-SPCs.

3.2.2.1.3 Enzymes as Biocide Mimics or Biocide Producers

Enzymes have been included in anti-fouling coatings to act either directly, where enzymes degrade the adhesives of the foulants or directly harm them, or indirectly, where enzymes catalyze reactions that produce substances that prevent foulers from adhering [30].

Examples, where enzymes act directly, include the disruption of the conditioning biofilm created from EPS. Studies have reported the use of hydrolases to hydrolyze the EPS and induce cell lysis, specifically using mixtures of lipase, xylanase, and viscozyme to target a variety of biofilm-forming bacteria [17], and tethering to immobilize xylanase on surfaces to provide long-lasting hydrolysis of polysaccharides [31].

Future design challenges facing enzyme-based coatings include the ability to maintain activity in higher temperature waters and evaluating the effective lifetime of the films.

3.2.2.1.4 Biomimetic Biocides[i]

Biomimetics is the mimicry of biological solutions and has been investigated, or perhaps inspired, by each class of anti-fouling material design. In the context of foul-degrading coatings, research has focused on replicating the biocidal compounds produced by marine organisms [30]. Studies have included macroalgae, marine bacteria, corals, and microbes [16, 32].

In contrast to heavy metal biocides, biomimetic biocides are organic molecules that biodegrade and therefore do not accumulate in the environment. A leading example is butenolide and its derivatives, which have shown good anti-fouling performance during field tests in simple polymer matrices [33].

3.2.2.2 Foul-Resistant Coatings

Foul-resistant coatings rely on preventing the adhesion of foulers to the surface. The materials utilize surface morphologies and/or charge distribution to produce an inhospitable surface for bacteria and the initial conditioning film.

3.2.2.2.1 Polymer Brush Coatings

Polymer brush coatings consist of a densely packed array of polymer chains. These are attached at their end to an interface and stretched out into the solution creating a coating that is just a few nanometers thick [10]. Polymer brush coatings operate *via* two mechanisms.

Hydration layers are the first mechanism, wherein the polymer uses hydrogen bonding and electrostatic interactions to create highly ordered layers of water molecules above the surface. This presents a thermodynamic challenge to EPS and organisms approaching the surface because the displacement of ordered water layers requires overcoming a high activation energy.

Similarly, polymer brushes can work by compressive entanglement, where an approaching foulant causes compression of the polymer brushes, reducing the number of possible system confirmations and making fouling adhesion entropically unfavorable.

Typical challenges of polymer brush systems include the synthetic expense and the durability of the coatings. The same polymers suitable for creating ordered water layers are susceptible to swelling, which severely reduces mechanical strength and lifetime. Similarly, polarized bonds (again useful for forming ordered water layers) are susceptible to enzymatic degradation.

Polymers explored for this strategy include poly(ethylene glycol) (PEG) [34], polyoxazolines, polysaccharides, polyglycerol dendrons, polypeptoids, polyacrylamides [35], and poly(oxazolines) [36]. As we later describe, zwitterionic poly(ionic liquid) (PIL) materials also feature heavily in this class of compounds.

3.2.2.2.2 Nanostructured Surfaces and Biomimetics

Marine organisms such as sharks, dolphins, and whales survive for decades without becoming colonized by marine fouling. This can be attributed partly to the nanostructure of their skin (Figure 3.5). A wide variety of materials have been used to mimic these nanostructures, covering polymers to metal oxides, and have been reviewed by Maan *et al.* [37].

Generally, the nanostructured topographies are chemically inert and rely on surface irregularities to generate localized turbulent flow that disrupts the binding of other organisms [16, 32]. Another approach relies on microorganisms' preference to adhere to surfaces slightly larger than themselves to maximize surface contact and growth potential. Therefore, creating textured surfaces with nanostructures smaller than the microorganisms reduces the prevalence and favorability of adhesion [10].

3.2.2.2.3 Electrostatic Fibers

Similar to forming an ordered water surface using surface charge, electrostatically charged fibers have been combined with epoxy additives to produce fouling-resistant coatings [14]. The constant movement of fine fibers, ranging between 50 and 100 μm in length, acts similar to a sea anemone by preventing fouling due to the constant movement of the fibers *via* ocean and tidal currents. Although the fibers create a rough surface that increases drag force, they are effective and can be found in many commercial anti-fouling coatings [14].

3.2.2.3 Fouling Release Coatings

Fouling release coatings operate by allowing weak adhesion of fouling species to the surface. Due to the weakness of this interaction, shear stress (e.g., from a vessel sailing) is sufficient to displace accumulated fouling layers—effectively cleaning the hull. While biocide-free, the disadvantage of fouling release coatings is that any surface abrasion or damage results in exposed areas that become accessible as adhesion sites.

The mechanism of operation relies on controlling and limiting the possible intermolecular interactions between the coating surface and the fouling species. By designing surfaces optimized to have surface energies of ca. 23 N mNm^{-1} (20 to 30 mNm^{-1}), any fouling species that are attached can be removed by the application of minimal shear stress.

3.2.2.3.1 Controlled Surface Tension Designs

Slippery surfaces with surface tension in the optimal range, such as poly(-silicones) have been commercially produced as coatings such as Intersleek 900, a fluoropolymer-modified silicone-based coating. Intersleek 900 combines hydrophobic, ultrasmooth surfaces with hydrophilic regions that prevent strong adhesion of macrofoulers that rely predominantly on either hydrophilic (diatoms) or hydrophobic (barnacles) adhesives. However, one limitation of this coatings class is that they develop a slime layer that cannot be removed even under high shear conditions, causing significant drag compared to a fouling degrading coating [17].

Commonly, this mixture of hydrophilic and hydrophobic sites is combined with low elastic modulus, which makes the surface dynamic and deformable, decreasing the probability that microorganisms that land on the surface can settle and bind [39]. This strategy is taken further with slippery liquid-infused polymer (SLIP) designs.

3.2.2.3.2 Slippery Liquid-Infused Polymers

Biomimetics has provided insight into the design of foul-release coatings. For example, some fish and pitcher plants produce mucus that acts as a protective layer against the adhesion of fouling species. Constant production and renewal of the mucus layer are similar to the self-polishing strategy of TBT polymers, however without the heavy metal components. This approach has been somewhat mimicked with the creation of SLIPs, which actually operate *via* a slightly different mechanism.

SLIPS are created by infusing microporous membranes with volatile fluorinated oil lubricants [40]. A liquid surface is inherently smooth on a molecular scale and, hence, does not provide any adhesion sites [9]. Because of the weak adhesion of the, effectively liquid, surface, SLIPs display excellent anti-fouling properties under dynamic conditions, i.e., periodically exposed to shear stress (Figure 3.6), although they appear out-performed by biocide coatings under static water conditions [41].

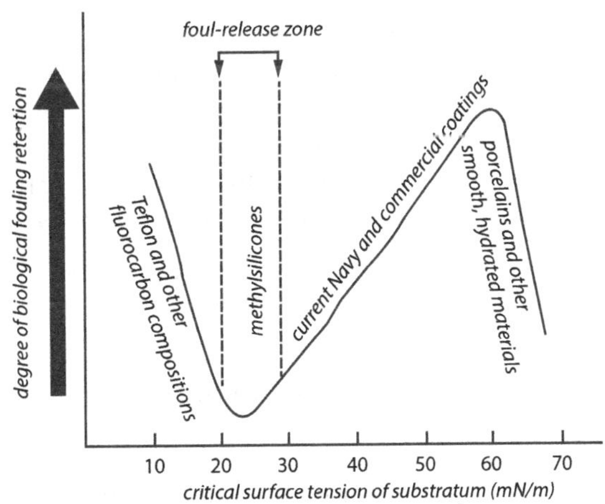

Figure 3.6 Relating the degree of biological fouling to the surface tension of a solid surface. *substrate* [38].

3.2.3 Summary of Anti-Fouling Strategies and Opportunity for Ionic Liquids

The above sections provide a very brief overview of an incredibly large field and readers are pointed toward the many review articles referenced on this topic.

Setting the context by listing and illustrating the three design strategies behind anti-fouling coatings (foul degradation, foul resistance, and foul release), we are now positioned to evaluate the research reporting experimentation of ionic liquids being used for marine anti-fouling. We see that all strategies are viable, with their own strengths and weaknesses, able to summarize that the goal of a new anti-fouling coating is to prevent all four stages of fouling from the formation of conditioning films to the arrival of macrofoulers.

3.3 Poly(Ionic Liquids) as Anti-Fouling Coatings

PILs are polymers that contain ionic groups. PILs can be combined with ionic liquids to create gels. Both PILs and their gels have been explored for use in many fields, from solvents to catalysts, to gas separation due to the range of physical properties that can be obtained from liquid salts with high stability and low vapor pressure. In particular, the surface properties of PILs make them interesting for anti-fouling applications.

PILs can be designed for all three leading anti-fouling strategies. Long-chain cationic ionic liquids exhibit the ability to penetrate cell walls and membranes, facilitating leakage of low-molecular-weight material, and ionic liquids degrade cell machinery directly [43]—making both potent biocides suitable for foul-degrading coatings [44–46]. Zwitterionic ionic liquids spread intense electrostatic charge across surfaces, making adhesion difficult and suiting them for use as foul-resistant coatings. Similarly, strongly hydrophobic ionic liquids can act as foul-release coatings. There is a small number of encouraging studies that suggest PILs will make effective anti-fouling coatings in the marine environment.

3.3.1 Experimental Methods for Evaluating Poly(Ionic Liquid) Coatings

Multiple stages of testing can be used to evaluate the potential and actual performance of PILs for anti-fouling applications. The most commonly performed evaluation is bacterial adhesion or survival studies, which provide insight into Stage 2, i.e., the formation of a viable biofilm. A more realistic proxy method is the deposition of microalgae and diatoms in solution, which provides insight into Stages 2 and 3 of the fouling process, i.e., the formation and diversification of the biofilm. The ultimate test for PIL coatings is immersion testing in marine environments, which provides direct evidence of the coating's effectiveness.

3.3.1.1 *Anti-Bacterial Activity*

Bacterial adhesion or survival tests are performed by the growth of a cultured bacterial strain in the presence of the toxic component, or on the coating surface. *E. coli* and *S. aureus* are most commonly used for these tests presumably because they are readily available and provide contrast between Gram-negative and -positive species. Bacterial survival rates can then be evaluated by optical density measurements [47], fluorescence measurements [48], or zone-of-inhibition measurements [49]. The efficacy of the biocide or surface is then evaluated by comparing the results to a blank and, ideally, by comparing the results to a common biocide [such as known concentrations of copper(I) oxide as a standard]. These results provide a measure of anti-bacterial activity, which can be related to the ability of a coating to resist Stage 2 biofilm formation. Logically, some correlation is expected between anti-bacterial efficacy and marine anti-fouling performance. However, bacterial tests ignore the complications of the liquid phase, let alone the complexity of the marine environment, and we have been unable to find extensive studies showing a clear correlation between anti-bacterial and marine anti-fouling performance.

3.3.1.2 *Micro-Algae Inhibition*

Micro-algae inhibition provides more realistic marine anti-fouling tests by taking place in solution and using microorganisms more likely to be found in the marine environment. The procedure involves cultivating a solution of the desired algae, then immersing the substrate in the culture medium and incubating for the desired growth/adhesion time. Samples are then evaluated using optical microscopy and image analysis software to compare colonization of the coated substrate against blanks [11, 50–53]. While providing a more realistic test and insight into Stage 2 and 3 colonization, we found no evidence directly correlating micro-algae inhibition to immersion anti-fouling results.

3.3.1.3 Immersion Testing (Macrofoulers)

Full immersion in an uncontrolled marine environment provides the true test of whether the material performs well as a marine anti-fouling coating. A major advantage is visible evidence of macrofouler presence, e.g., tubeworm growth. However, due to the complexity of the environment, there is usually a loss of resolution in mechanistic insights. To carry out the experiment, the coated substrate is immersed in a container suitable for mimicking the intended application (e.g., control of sunlight and current exposure). The substrate is then removed at suitable intervals, e.g., weekly, and visually inspected. The Naval Ships' Technical Manual Chapter 081 is used to quantify anti-fouling performance by assigning a fouling rating to the species present [54]. Foul ratings can then be converted to a total foul rating, which takes into account the coverage of each level of foulant species. Generally, to avoid the risk of translocating invasive species, hull cleaning would be undertaken when the foul rating reaches 40.

The time, sample size, and infrastructure investment required for immersion testing, alongside lack of insight into the underlying anti-fouling mechanism, mean that the majority of publications regarding the potential of PILs for marine anti-fouling only experimentally study the materials for anti-bacterial or microalgae inhibition.

3.3.2 Anti-Fouling Performance of Poly(Ionic Liquids)

A summary of PIL materials studied for marine anti-fouling is presented in Table 3.1. Studies covering immersion testing are the most important, highlighting the actual effectiveness of PIL materials for marine anti-fouling. Bacterial studies are summarized, but not included in this review.

3.3.2.1 Immersion Testing (Macrofoulers)

Four studies were identified that examined the performance of PIL coatings during full immersion studies. The most impressive study is the testing of a poly(acrylate) with pendant zwitterionic ammonium sulfonate groups to produce a polymer brush [55], along with two studies on the incorporation of imidazolium, pyridinium, and ammonium-functionalized chitosan and paint mixtures [49, 56].

The zwitterionic polyacrylate reported by Liu *et al.* was tested in the Nan'ao Sea, Shenzhen Province for a period of 6 weeks [55]. Impressively, substrates coated with the zwitterionic PILs showed only minor colonization by barnacle larvae after 6 weeks (Figure 3.7). Although the study was

Table 3.1 The types of poly(ionic liquids) and their performance in anti-fouling testing are summarized, showing which level of experimental work has been completed and what control materials were used in the study.

Type of IL coating	Control material	Antibacterial activity	Microalgae/diatom testing	Macrofouler/ field testing
Zwitterionic polymer brushes [55]	resin on steel plates	-	Effective against algae (*Porphyridium* sp.) and diatoms (*Navicula* sp.)	Minor barnacle larvae attachment after 6 weeks
Small molecules containing ionic liquid moieties [56]	Diuron®	Effective against *A. hydrophilia*	-	6–10% tubeworm coverage and <3% barnacle coverage
Imidazolium/ammonium/ pyridinium IL-modified chitosan [49]	Diuron®	Effective against all bacteria (ammonium>imidazolium> pyridinium)	-	0% macroalgae coverage. Colonized by macrofoulers over 147 days
Phosphonium PIL	Epoxy-coating	Not effective against bacteria		Colonized by macrofoulers over 92 days

(*Continued*)

Table 3.1 The types of poly(ionic liquids) and their performance in anti-fouling testing are summarized, showing which level of experimental work has been completed and what control materials were used in the study. (*Continued*)

Type of IL coating	Control material	Antibacterial activity	Microalgae/diatom testing	Macrofouler/ field testing
C_{16}-methylimidazolium and other alkylimidazolium and ammonium ionic liquids [53]	0 μM IL	-	5 μM IL concentration inhibited *Navicula* Sp. biofilms and 10 μM concentration prevented adhesion	
Imidazolium-based IL polystyrene nanospheres [11]	Pure PDMS coating	-	Approximately 85% decrease in adhesion	
Zwitterionic-functionalized liquid metal-based PDMS (polydimethylsiloxane) GLM-TISM [50]	Pure PDMS coating	99.7% bactericidal efficacy for *E. coli* and 99% *S. aureus*	71% decrease in *Dunaliella* adhesion, and 77% decrease in *Porphyridium* adhesion	
Poly(ionic liquid) coatings with polymer brushes grafted to TiO2 nanowires [51]	Polymer brushes grafted to TiO_2 nanowires	-	Coatings with PIL had an ~40% decrease in *Navicula* adhesion and an ~50% decrease in *Dunaliella* adhesion	

(*Continued*)

Table 3.1 The types of poly(ionic liquids) and their performance in anti-fouling testing are summarized, showing which level of experimental work has been completed and what control materials were used in the study. (*Continued*)

Type of IL coating	Control material	Antibacterial activity	Microalgae/diatom testing	Macrofouler/ field testing
poly(ionic liquid) brushes on biomimetic sharklet resin surface [52]	Smooth surface and blank sharklet surface	Much lower attachment density of bacteria (Figure 3.5)	60.1% removal of *Dunaliella* and 55.9% removal of *Porphyridium*	-
Zwitterionic polymer-functionalized MXene nanosheets [48]	N/A	Antibacterial rates of 97% for *E. coli*, and 84% for *S. aureus*	-	-
Polymer brush coating with ionic liquid [47]	Polymer brush coating without IL	IL brushes attracted more biofilm but better removal	-	-
Pyrrolidinium poly(ionic liquid) membrane [57]	Sterile PET	50% inhibition of *S. aureus* and *E. coli* after 24 hours. Antibacterial activities decreased with chain length	-	-
Amino acid-based zwitterionic poly(serine methacrylate) (pSerMA) [58]	Gold and glass	No viable cells were observed on pSerMA-grafted surfaces after 1 week		

(*Continued*)

Table 3.1 The types of poly(ionic liquids) and their performance in anti-fouling testing are summarized, showing which level of experimental work has been completed and what control materials were used in the study. (*Continued*)

Type of IL coating	Control material	Antibacterial activity	Microalgae/diatom testing	Macrofouler/ field testing
Epoxy-amine networks from ionic liquid monomers [59]	Non-ionic epoxy-amine network	Ionic liquid performance superior. 87—92% inhibition of *E. coli*	-	-
Polymer brushes containing quaternized ammonium and fluorinated monomers [60]	Cotton	Fluorinated copolymers showed the best antibacterial properties	-	-
1-Alkyl-3-methylimidazolium chloride ionic liquids [61]. [C*n*mim][Cl], n=8-14	None	$n = 14$, exhibited the greatest antibiofilm activity	-	-
Phosphonium-based ionic liquids (generic formula [PR3R′]X (R and R′ are alkyl groups; X is halide) [62]		Strongly effective against cocci		

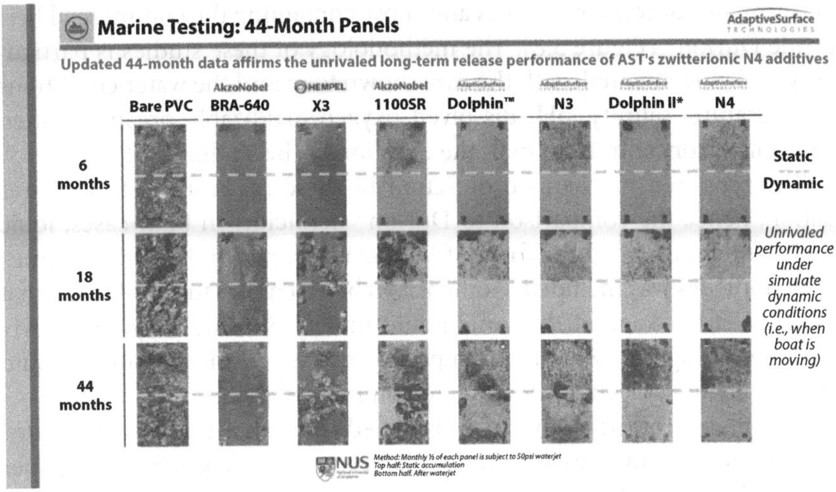

Figure 3.7 Performance of SLIPs Dolphin™ and Dolphin II under static (above yellow line) and dynamic (below yellow line) testing conditions over a period of 44 months. Figure reproduced from Adaptive Surface Technologies [42].

performed over winter months, the control substrate used by the authors showed contrastingly rapid colonization by barnacle larvae.

Importantly, the authors were able to correlate the immersion testing results with microalgae inhibition testing to gain mechanistic insights. The key finding was a difference in the lifetime performance of the zwitterionic PIL compared to the anionic sulfonate polymer. The zwitterionic PIL coating maintained anti-fouling performance against algae (*Porphyridium* sp.) and diatoms (*Navicula* sp.) despite immersion in simulated seawater (1 M Ca^{2+}). While the authors attribute this to the non-zwitterionic sulfonate being better able to maintain a low contact angle, with the argument that this provides a highly ordered water shell that acts as a barrier to adhesion, examination of the supporting information reveals the change in contact angle of the zwitterionic PIL to be even higher, detracting from that explanation of the mechanism. Regardless of the actual mechanistic explanation, the performance against microalgae adhesion is superior for the zwitterionic PIL and suggests this design should be explored more rigorously in the future.

Two immersion studies have been performed by the Elshaarawy research group in Egypt, including ionic liquid moieties into commercial paint [56] and onto a chitosan polymer [49]. The performance of both coatings was tested using static immersion tests lasting 147 days over summer months in the Alexandria Eastern Harbor and compared to the performance of control samples and those containing the commercial herbicide Durion® (Figure 3.8). The methodology of these studies is particularly well-documented, with the care shown to record the water conditions (temperature, salinity, pH, dissolved oxygen, oxidizable organic matter, and sulfate concentration) over the duration of the studies [56].

The proposed mechanism of action in these materials is foul degradation, hence the comparison to Durion® herbicide. In both cases, ionic liquid functionality in the form of imidazolium, pyridinium, and ammonium moieties was included. Comparison between the studies showed that the covalent linkage of the ionic liquid moiety into the polymer matrix (chitosan) (Figure 3.8) was far superior to the inclusion of ionic liquid moiety as free molecules (doped into commercial paint).

A limitation of the study on IL-modified chitosan is that visual data for the unmodified chitosan was not supplied, and a later figure from the paper (Figure 3.8) suggests that unmodified chitosan has comparatively superior anti-fouling performance, particularly to adhesion of bryozoa.

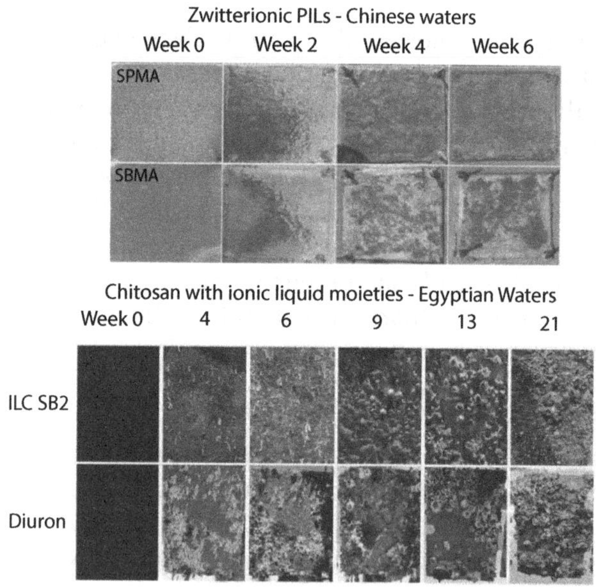

Figure 3.8 Top: Field testing results from zwitterionic PILs (SPMA and SBMA) showing promising anti-fouling performance over a 6-week period over winter in coastal Chinese waters. Modified from reference [55]. Bottom: Field testing results of chitosan modified with ionic liquid moieties. These materials showed promising anti-fouling performance over a 147-day (21 weeks) period during summer in coastal Egyptian waters. Modified from reference [49].

Interestingly, the immersion testing performance of these materials was compared to anti-bacterial testing and examined for correlations of efficacy, with the rationale that the bacterial biofilm increases oxygen generation and provides sites for further fouling to occur. Tests were performed using *E. coli*, *S. aureus*, *A. hydrophilia*, and *Vibrio*. The ionic liquid–modified small molecules showed good activity against *A. hydrophilia*, but little activity against three bacteria. In contrast, the proposed biocides without ionic liquid groups showed superior performance. This suggests that the presence of an ionic liquid moiety made little positive difference to the anti-bacterial performance.

In contrast to the small-molecule study, chitosan modified with ionic liquid showed significantly better inhibition of all bacteria studied than both the control chitosan and Diuron®-infused chitosan. Furthermore, microscopy evidence showing cell membrane damage of *E. coli* was obtained and supported the classification of ionic-liquid functionalized polymer brushes as potentially effective *via* the foul-degrading mechanism.

These immersion testing results reflect the anti-microbial effectiveness of the ionic liquid–modified chitosan. For example, the test panel showed no red, brown, or green algae growth after 21 weeks, in comparison to the control and blank, which showed coverages up to 80% over the course of the immersion testing. However, because only one formulation of the ionic liquid–modified chitosan was tested there is not enough data to link anti-bacterial efficacy to immersion testing success. Furthermore, the inhibition of the agal biofilm did not prevent colonization by macrofoulers.

A 2023 study by Baiju *et al.* reported immersion testing of PILs ion gels composed of phosphonium ionic liquids, with some formulations containing copper oxide as a biocide [64]. Immersion testing results indicated that coatings with higher hydrophobicity, measured using contact angle studies, were superior for retarding macrofoulers. Interestingly, the addition of copper oxide as a biocide led to worse anti-fouling performance and was linked to a decrease in the hydrophobicity of the coatings (Figure 3.9).

A major weakness of the immersion testing examples reported to date for PILs is that no comparison was made to leading commercial anti-fouling paint performance, for example, copper-containing paints such as Hempaguard [63]. Hempaguard products are rated up to 90-month or, more relevant to these comparisons, 120-day idle periods. If a new environmentally benign anti-fouling coating is to be adopted, then it must be able to match or exceed the performance of established candidates so direct comparison under identical test conditions should be made.

Regardless of limitations to the current studies, ionic liquid–modified polymers show effective prevention of algal biofilm growth. Preventing

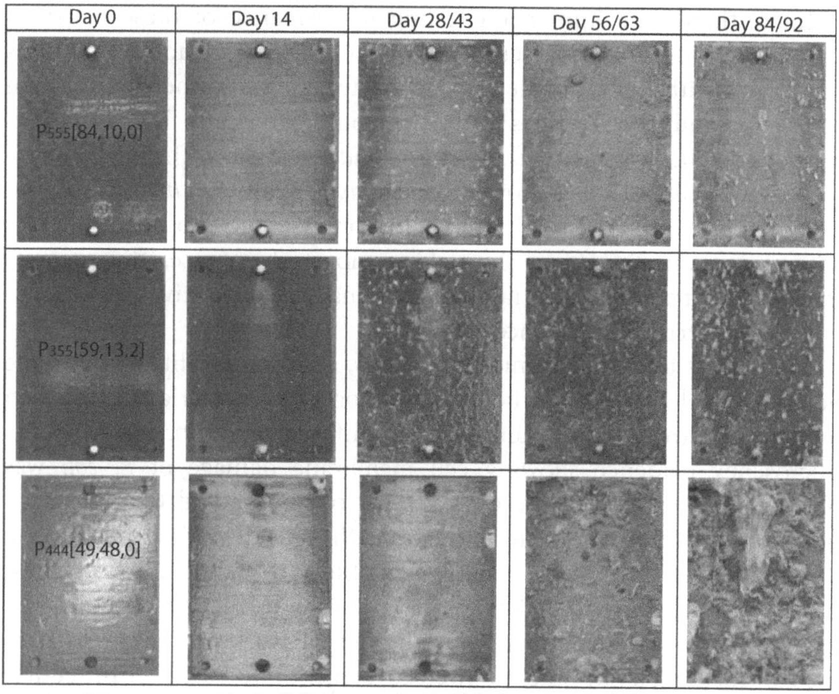

Figure 3.9 Performance of phosphonium-based ion gels containing phosphonium cations with tributyl, trioctyl, and trioctyl carbon chains and copper oxide through immersion testing in New Zealand coastal waters. Names denote P_{xxx}[AA, BB, CC] where x = carbon chain length, AA = weight % [$P_{XXXvinylbenzene}$][Docusate], BB = weight % [P_{XXX8}][Docusate], and CC = weight % copper oxide. Reproduced with permission from *Polymers* [64].

biofilm growth is an integral part of overall anti-fouling performance. Therefore, these results are encouraging and suggest that additional work on ionic liquids and PILs to prevent macrofouling attraction could position them as a viable anti-fouling option. Meanwhile, multiple studies have addressed algal and bacterial biofilm inhibition under controlled laboratory conditions.

3.3.2.2 Micro-Algae Inhibition

We have identified five studies that examined the effectiveness of ionic liquids and ionic liquid–modified surfaces for inhibition of micro-algae structures. These exemplars reveal insights into the effective design of ionic liquids for algal inhibition and demonstrate their use on a variety of surfaces.

We identified a solitary example of a small-molecule ionic liquid studied for inhibition of micro-algae, particularly *Navicula* sp. [53]. The study is particularly revealing, illuminating both the best ionic liquid structures and exploring their mechanisms of action.

A series of alkylimidazolium and trimethyl-alkyl-ammonium salts were explored for algae inhibition using a chlorophyll assay. The results of the assay were impressive and revealed a trend of increasing effectiveness with alkyl chain length. In particular, C_{16}-methylimidazolium and C_{16}trimiethylammonium salts showed the inhibition of algal biofilms at concentrations of only 5 μM and prevention of algal adhesion at concentrations of only 10 μM. Furthermore, the authors demonstrated the ability of 5 μM C_{16}-methylimidazolium solutions to prevent biofilm formation on plastic composites in both seawater and freshwater.

Mechanism of action was determined *via* a SYTOX® Green assay and fluorescence imaging, which indicated that exposure of *Navicula* sp. to 10 μM C_{16}-methylimidazolium caused catastrophic damage to cell membranes, increasing permeability resulting in the release of both genetic material and chlorophyll into the solution. The authors note the limitations of the study and note that the mechanism of action has not been definitively proven.

The remaining four studies involve the attachment of polymer brushes to a variety of polymeric and metal surfaces. In all cases, the mechanism proposed is foul-resistant, i.e., increasing the energetic barrier needing to be overcome for surface adhesion to take place.

An innovative study from He *et al.* [52] reported the use of 3D printing to create a biomimetic sharklet surface and then functionalize that surface with polymer chains of imidazolium ionic liquids. The authors studied the adhesion of proteins (BSA), bacteria (*S. aureus* and *E. coli*), and algal spores (*Porphyridium* and *Dunaliella*) (Figure 3.10). Interestingly, the sharklet topology caused a minor reduction in algae adhesion and functionalization with ionic liquid brushes led to a >3× drop in adhesion. This result was consistent across static and dynamic shear tests. Possible explanations included the significant increase in surface hydrophilicity from a water contact angle >75° to <15°. However, the performance was attributed to the ionic liquid brushes permeabilizing the algae cell membranes. This conclusion was supported by the bacterial testing which showed a large increase in the number of dead bacterial cells on surfaces with ionic liquid modification.

Similar studies have been performed on poly(styrene) spheres [11] and titanium dioxide (TiO_2) [51] surfaces. The coated surfaces were tested for adhesion and viability of *Porphyridium, Dunaliella*, and *Navicula* spores. In both cases, the addition of long-chain cationic ionic liquid functionality resulted in the largest decreases in algal spore viability but not complete inhibition of agal viability.

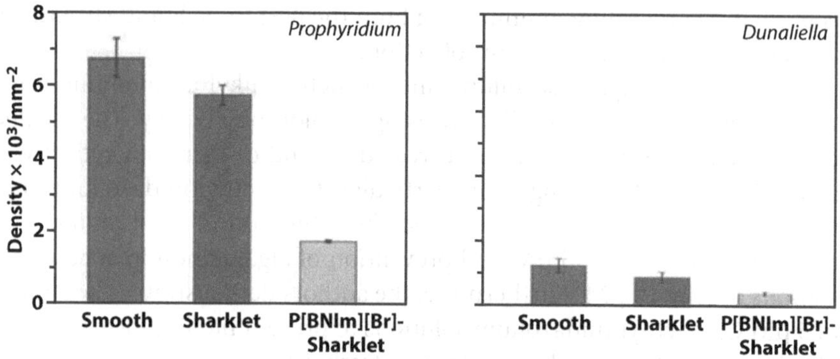

Figure 3.10 Attachment of *Porphyridium* (left) and *Dunaliella* (right) to acrylic resin patterned into a sharklet topology and surface-functionalized with long-chain imidazolium groups under static conditions. Modified from reference [52].

To make an ionic liquid–based self-polishing foul-degradation surface, Wang *et al.* [11] used hydrolyzable methacrylate linkages to promote the release and 'polishing' of the ionic liquid functionality on the surface. The results suggested the self-polishing coatings showed much more effective inhibition of algae adhesion, which they hypothesized was due to the release of fluorinated anions associated with the ionic liquid head groups. As the study was limited to 24 hours, the longevity of this self-polishing effect remains an open question.

To combat the longevity and susceptibility to damage of anti-fouling coatings, Wang *et al.* [50] introduced the concept of self-healing coating based on gallium metal nanodroplets functionalized with zwitterionic brushes to prevent algae adhesion. Blending the gallium nanodroplets into the PDMS polymer resulted in a >70% reduction in *Dunaliella* and *Porphyridium* adhesion. Furthermore, scratch tests showed that PDMS containing gallium nanodroplets would heal. The mechanism for this healing was proposed as a combination of the elasticity of the PDMS and the ability of gallium nanodroplets to catalyze the polymerization of carbon-carbon double bonds at the PDMS interfaces.

3.3.3 Ionic Liquid Properties for Anti-Fouling

Several studies have been carried out investigating the antimicrobial properties of ionic liquids. The exact mechanism for the antimicrobial activity of ionic liquids remains unknown. It is hypothesized that the antimicrobial mechanism of ionic liquids is similar to that of cationic biocides and surfactants due to the structural similarities [44]. Both surfactants and ionic liquids contain a charged hydrophilic head group and one or more hydrophobic tails, suggesting a common mechanism of action [45]. Cationic surfactants exhibit antimicrobial activity *via* the destruction of cellular membranes. At alkaline or neutral pH, cationic surfactants show a high affinity for the negatively charged interfaces of bacteria, viruses, and fungi. The surfactants first adsorb to and penetrate cellular walls and membranes. This causes low-molecular-weight material to leak from the cell. Proteins and nucleic acids are then degraded, ultimately leading to cell lysis and death. Anionic surfactants function *via* the denaturation of proteins [43]. It is predicted that ionic liquids function in the same way to suppress biological activity [46].

Ionic liquids containing imidazolium, pyridinium, ammonia, and phosphonium cationic constituents have been identified to possess potent antimicrobial efficacy [40]. The antimicrobial properties of ionic liquids can be tuned by the selection of an antimicrobial anion. A 2020 study by Wylie *et al.* investigated the antimicrobial effects of SLIPS infused with phosphonium ionic liquids with different counterions. Bis(trifluoromethylsulfonyl) imide ([Tf2N]-) is a common anion used in ionic liquids (Figure 3.8). This was compared to docusate ([AOT]-), a known anionic surfactant (Figure 3.7). Phosphonium ionic liquids containing the AOT counterion displayed significant improvement in antimicrobial activity over the Tf2N counterion. As a surfactant, AOT lowers the surface tension between water and a hydrophobic surface, improving the wettability and hence the coverage of the anti-fouling coating. The AOT-infused SLIPS achieved very low super-hydrophilic contact angles with water. It was proposed that this facilitates the formation of a water layer on top of the SLIPS coating. This inhibits the interaction of bacterial cells with the solid surface consequently reducing cell adhesion [40].

References

1. Analysing the Impact of Marine Biofouling on the Energy Efficiency of Ships and the GHG Abatement Potential of Biofouling Management Measures. *Int. Marit. Organ.*, 2022.
2. Korberg, A.D., Brynolf, S., Grahn, M., Skov, I.R., Techno-economic assessment of advanced fuels and propulsion systems in future fossil-free ships. *Renew. Sust. Energ. Rev.*, 142, 110861, 2021.
3. Butler, A.J. *et al.*, Techniques for the Quantification of Biofouling, in: *Biofouling*, S. Dürr, and J. Thomason, (Eds.), pp. 319–332, Wiley-Blackwell, Chichester, UK, 2010.

4. Davidson, I. *et al.*, Mini-review: Assessing the drivers of ship biofouling management–aligning industry and biosecurity goals. *Biofouling*, 32, 411–428, 2016.
5. Hopkins, G. *et al.*, Managing biofouling on submerged static artificial structures in the marine environment–assessment of current and emerging approaches. *Front. Mar. Sci.*, 8, 759194, 2021.
6. Dürr, S. and Thomason, J.C., *Biofouling*, 2010.
7. Lejars, M., Margaillan, A., Bressy, C., Fouling release coatings: a nontoxic alternative to biocidal antifouling coatings. *Chem. Rev.*, 112, 4347–4390, 2012.
8. Yebra, D.M., Kiil, S., Dam-Johansen, K., Antifouling technology—past, present and future steps towards efficient and environmentally friendly antifouling coatings. *Prog. Org. Coat.*, 50, 75–104, 2004.
9. Wong, T.-S. *et al.*, Bioinspired self-repairing slippery surfaces with pressure-stable omniphobicity. *Nature*, 477, 443–447, 2011.
10. Maan, A.M.C., Hofman, A.H., Vos, W.M., Kamperman, M., Recent Developments and Practical Feasibility of Polymer-Based Antifouling Coatings. *Adv. Funct. Mater.*, 30, 2020.
11. Wang, B. *et al.*, Fabrication of ionic liquid-functionalized polystyrene nanospheres via subsurface-initiated atom transfer radical polymerization for anti-fouling application. *Prog. Org. Coat.*, 171, 107044, 2022.
12. Dahlström, M., Eklund, B., Wiklund, A.-K.E., Lagerström, M., Wrange, A.-L., Biofouling and antifouling–why should we care?, in: *Changing leisure boat antifouling practices in the Baltic Sea*, pp. 20–24, 2018.
13. Cao, S., Wang, J., Chen, H., Chen, D., Progress of marine biofouling and antifouling technologies. *Chin. Sci. Bull.*, 56, 598–612, 2010.
14. Almeida, E., Diamantino, T.C., de Sousa, O., Marine paints: The particular case of antifouling paints. *Prog. Org. Coat.*, 59, 2–20, 2007.
15. Ciriminna, R., Bright, F.V., Pagliaro, M., Ecofriendly Antifouling Marine Coatings. *ACS Sustain. Chem. Eng.*, 3, 559–565, 2015.
16. Chambers, L.D., Stokes, K.R., Walsh, F.C., Wood, R.J.K., Modern approaches to marine antifouling coatings. *Surf. Coat. Technol.*, 201, 3642–3652, 2006.
17. Aykin, E., Omuzbuken, B., Kacar, A., Microfouling bacteria and the use of enzymes in eco-friendly antifouling technology. *J. Coat. Technol. Res.*, 16, 847–856, 2019.
18. Donlan, R.M., Biofilm formation: a clinically relevant microbiological process. *Clin. Infect. Dis.*, 33, 1387–1392, 2001.
19. Petrone, L., Molecular surface chemistry in marine bioadhesion. *Adv. Colloid Interface Sci.*, 195-196, 1–18, 2013.
20. McWhirter, M.J., Bremer, P.J., Lamont, I.L., McQuillan, A.J., Siderophore-Mediated Covalent Bonding to Metal (Oxide) Surfaces during Biofilm Initiation by Pseudomonas aeruginosa Bacteria. *Langmuir*, 19, 3575–3577, 2003.

21. Gao, Z., Bremer, P.J., Barker, M.F., Tan, E.W., McQuillan, A.J., Adhesive Secretions of Live Mussels Observed *in Situ* by Attenuated Total Reflection–Infrared Spectroscopy. *Appl. Spectrosc.*, 61, 55–59, 2007.
22. Awad, M.M., *Fouling of heat transfer surfaces*, INTECH Open Access Publisher, London, 2011.
23. Kucera, J., Biofouling of polyamide membranes: Fouling mechanisms, current mitigation and cleaning strategies, and future prospects. *Membranes*, 9, 111, 2019.
24. Dafforn, K.A., Lewis, J.A., Johnston, E.L., Antifouling strategies: history and regulation, ecological impacts and mitigation. *Mar. Pollut. Bull.*, 62, 453–465, 2011.
25. Gipperth, L., The legal design of the international and European Union ban on tributyltin antifouling paint: Direct and indirect effects. *J. Environ. Manage.*, 90, S86–S95, 2009.
26. Venkatnarayanan, S., Murthy, P.S., Kirubagaran, R., Venugopalan, V.P., Chlorine dioxide as an alternative antifouling biocide for cooling water systems: Toxicity to larval barnacle Amphibalanus reticulatus (Utinomi). *Mar. Pollut. Bull.*, 124, 803–810, 2017.
27. Kumar, S., Ye, F., Dobretsov, S., Dutta, J., Nanocoating is a new way for biofouling prevention. *Front. Nanotechnol.*, 3, 771098, 2021.
28. Miller, J., Tributyltin: Case Study of an Environmental Contaminant. *J. Environ. Qual.*, 26, 1193, 1997.
29. Bressy, C., Margaillan, A., Faÿ, F., Linossier, I., Réhel, K., Hellio, C., Yebra, D., *Advances in Marine Antifouling Coatings and Technologies*, pp. 445–491, Woodhead Publishing, 2009.
30. Chen, L. *et al.*, Biomimetic surface coatings for marine antifouling: Natural antifoulants, synthetic polymers and surface microtopography. *Sci. Total Environ.*, 766, 144469, 2021.
31. Baker-Branstetter, R.W., Bartlett, M.E., Shuler, S.A., Messersmith, R.E., Covalent immobilization of xylanase and lysing complex into polymer scaffolds with long-term activity retention. *J. Coat. Technol. Res.*, 20973-978, 2023.
32. Myan, F.W.Y., Walker, J., Paramor, O., The interaction of marine fouling organisms with topography of varied scale and geometry: a review. *Biointerphases*, 8, 2013.
33. Chiang, H.Y., Cheng, J., Liu, X., Ma, C., Qian, P.-Y., Synthetic Analogue of Butenolide as an Antifouling Agent. *Mar. Drugs*, 19, 481, 2021.
34. Xie, Q., Pan, J., Ma, C., Zhang, G., Dynamic surface antifouling: mechanism and systems. *Soft Matter.*, 15, 1087–1107, 2019.
35. Knowles, B.R., Wagner, P., Maclaughlin, S., Higgins, M.J., Molino, P.J., Silica nanoparticles functionalized with zwitterionic sulfobetaine siloxane for application as a versatile antifouling coating system. *ACS Appl. Mater. Interfaces*, 9, 18584–18594, 2017.

36. Selim, M.S. *et al.*, Silicone/ZnO nanorod composite coating as a marine antifouling surface. *Appl. Surface Sci.*, 466, 40–50, 2019.
37. Maan, A.M.C., Hofman, A.H., de Vos, W.M., Kamperman, M., Recent Developments and Practical Feasibility of Polymer-Based Antifouling Coatings. *Adv. Funct. Mater.*, 30, 2000936, 2020.
38. Baier, R.E., Surface behaviour of biomaterials: The theta surface for biocompatibility. *J. Mater. Science: Mater. Med.*, 17, 1057–1062, 2006.
39. Peng Hu, Q.X., Ma, C., Zhang, G., Silicone-Based Fouling-Release Coatings for Marine Antifouling. *Langmuir*, 2020.
40. Wylie, M.P., Bell, S.E., Nockemann, P., Bell, R., McCoy, C.P., Phosphonium ionic liquid-infused poly (vinyl chloride) surfaces possessing potent antifouling properties. *ACS Omega*, 5, 7771–7781, 2020.
41. Kratochvil, M.J. *et al.*, Slippery liquid-infused porous surfaces that prevent bacterial surface fouling and inhibit virulence phenotypes in surrounding planktonic cells. *ACS Infect. Dis.*, 2, 509–517, 2016.
42. *44 Months In, AST Marine Coatings Lead the Pack – Adaptive Surface Technologies*, 2023, https://www.adaptivesurface.tech/2023/05/09/44-months-in-ast-marine-coatings-lead-the-pack/.
43. Falk, N.A., Surfactants as antimicrobials: a brief overview of microbial interfacial chemistry and surfactant antimicrobial activity. *J. Surfactants Deterg.*, 22, 1119–1127, 2019.
44. Pendleton, J.N. and Gilmore, B.F., The antimicrobial potential of ionic liquids: A source of chemical diversity for infection and biofilm control. *Int. J. Antimicrob. Agents*, 46, 131–139, 2015.
45. Blesic, M. *et al.*, Self-aggregation of ionic liquids: micelle formation in aqueous solution. *Green Chem.*, 9, 481–490, 2007.
46. Pernak, J., Sobaszkiewicz, K., Mirska, I., Anti-microbial activities of ionic liquids. *Green Chem.*, 5, 52–56, 2003.
47. Kanematsu, H. *et al.*, Biofilm Formation of a Polymer Brush Coating with Ionic Liquids Compared to a Polymer Brush Coating with a Non-Ionic Liquid. *Coatings*, 8, 398, 2018.
48. Zeng, Y. *et al.*, Fabrication of zwitterionic polymer-functionalized MXene nanosheets for anti-bacterial and anti-biofouling applications. *Prog. Org. Coat.*, 183, 107727, 2023.
49. Elshaarawy, R.F.M., Mustafa, F.H.A., Herbst, A., Farag, A.E.M., Janiak, C., Surface functionalization of chitosan isolated from shrimp shells, using salicylaldehyde ionic liquids in exploration for novel economic and ecofriendly antibiofoulants. *RSC Adv.*, 6, 20901–20915, 2016.
50. Wang, P. *et al.*, *In situ* generation of zwitterionic-functionalized liquid metal-based polydimethylsiloxane antifouling coatings with self-healing ability. *Prog. Org. Coat.*, 181, 107604, 2023.
51. Zhao, W., Ye, Q., Hu, H., Wang, X., Zhou, F., Fabrication of binary components based on a poly(ionic liquid) through "grafting" and "clicking" and their synergistic antifouling activity. *RSC Adv.*, 5, 100347–100353, 2015.

52. He, B. *et al.*, Grafting embedded poly(ionic liquid) brushes on biomimetic sharklet resin surface for anti-biofouling applications. *Prog. Org. Coat.*, 157, 106298, 2021.
53. Reddy, G.K.K., Nancharaiah, Y.V., Venugopalan, V.P., Long alkyl-chain imidazolium ionic liquids: Antibiofilm activity against phototrophic biofilms. *Colloids Surf. B.*, 155, 487–496, 2017.
54. *Waterborne underwater hull cleaning of Naval ships*, Naval Sea Systems Command, 2006, S9086-CQ-STM-010 Revision 5 Published by direction of Commanderhttps://maritime.org/doc/nstm/ch081.pdf.
55. Liu, H., Ma, Z., Yang, W., Pei, X., Zhou, F., Facile preparation of structured zwitterionic polymer substrate via sub-surface initiated atom transfer radical polymerization and its synergistic marine antifouling investigation. *Eur. Polym. J.*, 112, 146–152, 2019.
56. Elshaarawy, R.F.M., Mustafa, F.H.A., Sofy, A.R., Hmed, A.A., Janiak, C., A new synthetic antifouling coatings integrated novel aminothiazole-functionalized ionic liquids motifs with enhanced antibacterial performance. *J. Environ. Chem. Eng.*, 7, 102800, 2019.
57. Qin, J. *et al.*, Synthesis of Pyrrolidinium-Type Poly(ionic liquid) Membranes for Antibacterial Applications. *ACS Appl. Mater. Interfaces*, 9, 10504–10511, 2017.
58. Liu, Q., Singh, A., Liu, L., Amino Acid-Based Zwitterionic Poly(serine methacrylate) as an Antifouling Material. *Biomacromolecules*, 14, 226–231, 2013.
59. Livi, S. *et al.*, Antibacterial surface based on new epoxy-amine networks from ionic liquid monomers. *Eur. Polym. J.*, 116, 56–64, 2019.
60. Lin, P.-A., Cheng, C.-H., Hsieh, K.-T., Lin, J.-C., Effect of alkyl chain length and fluorine content on the surface characteristics and antibacterial activity of surfaces grafted with brushes containing quaternized ammonium and fluoro-containing monomers. *Colloids Surf. B*, 202, 111674, 2021.
61. Carson, L. *et al.*, Antibiofilm activities of 1-alkyl-3-methylimidazolium chloride ionic liquids. *Green Chem.*, 11, 492–497, 2009.
62. Cieniecka-Rosłonkiewicz, A. *et al.*, Synthesis, anti-microbial activities and anti-electrostatic properties of phosphonium-based ionic liquids. *Green Chem.*, 7, 855–862, 2005.
63. Safety Data Sheet Hempaguard X8 89949 Base. *Hempel*, 2023, https://www.hempel.com/en-sea/products/hempaguard-x8-89940-89940.
64. Baiju, S.K., Martin, B.J., Fredericks, R., Raghavan, H., De Silva, K., Cowan, M.G., Anti Fouling Properties of Phosphonium Ionic Liquid Coatings in the Marine Environment. *Polymers*, 15, 3677–3687, 2023.

4

Inorganic Nanomaterial Coating to Prevent Biofouling

Anju T. R.[1*], **Mariat George**[2] **and Rose Mary Jose**[1]

[1]*Department of Biotechnology, Newman College, Thodupuzha, Kerala, India*
[2]*Department of Botany, Newman College, Thodupuzha, Kerala, India*

Abstract

Biofouling occurs when a submerged artificial surface gets colonized by micro- or macroorganisms and biofilm produced by them resulting in deterioration of the surface. This process is a matter of grave concern in many fields where immersed synthetic materials are used for human benefit like marine structures, medical implants or devices, and food processing. Strategies to overcome biofouling using nontoxic antifouling agents have attracted a lot of research and the advent of nanotechnology helped in addressing the issue in an eco-friendly and sustainable manner. Nanomaterials of desirable properties (like fouling resistance, fouling release, or fouling degrading) were considered as the most suitable candidates for antifouling coating. The antifouling nanomaterials may either form a hydrophilic surface that holds a barrier layer of water thereby preventing the adsorption of foulants or form antimicrobial characteristics that kill the biofoulants. A wide range of nanomaterials are identified with biocidal action which can inhibit microbes and biofilm formation and, hence, can be explored as excellent antifouling agents. Inorganic nanoparticles have great utility as an antifouling coating material due to their durability, antimicrobial activities, and hydrophobicity, which are essential features in industries like maritime, biomedical, and food processing. Among various inorganic nanoparticles, silver nanoparticles, carbon nanotubes, graphene, and oxides of titanium and zinc are highly recommended for the prevention of biofouling. Many inorganic metals and their oxides like silver, gold, selenium, titanium, copper, iron, and zinc are used as coating material for biomedical devices due to their favorable properties. A lot of research is also taking place to formulate green synthesis of these metallic nanoparticles to make them more acceptable for wide industrial applications without causing any possible threat due to the

**Corresponding author*: anju.tr@newmancollege.ac.in

Hari Murthy, Vinay Jha Pillai, Kukatlapalli Pradeep Kumar and Matthew Cowan (eds.) Novel Anti-Corrosion and Anti-Fouling Coatings and Thin Films, (77–108) © 2024 Scrivener Publishing LLC

leaching of metal to the water body and subsequent agglomeration. In addition to inorganic nanoparticles, nowadays, novel combinations of inorganic polymer composite coating materials are also formulated with better features. Thus, the use of inorganic nanomaterial and various polymer-inorganic composite–based antifouling coating can overcome biofouling in a nontoxic manner and can be considered an eco-friendly alternative to toxic coating materials.

Keywords: Biofilm, silver nanoparticles, graphenes, carbon nanotubes, oxides of Ti, oxides of Zn

4.1 Introduction

Biofouling is a natural process that occurs in aquatic environments where a variety of organisms including bacteria, algae, barnacles, and other marine invertebrates attach and grow on surfaces submerged in water. It is a complex phenomenon that can occur on natural and artificial surfaces in freshwater and marine environments and has been studied for centuries. Biofouling has significant economic and environmental consequences, particularly for industries that rely on equipment and infrastructure that are constantly exposed to the water, such as shipping, aquaculture, and offshore oil and gas industries.

The process of biofouling begins when a surface is submerged in water and exposed to the aquatic environment. The surface initially becomes coated with a layer of organic and inorganic materials, known as the conditioning film. This conditioning film provides a surface for the attachment of microbial and macrofouling organisms. Microbial organisms, such as bacteria and fungi, are the first to colonize the surface and they secrete substances that can facilitate the attachment of larger organisms, such as algae, barnacles, and other marine invertebrates [1]. Once attached, these organisms can grow and develop into complex communities, known as biofilms, which can have a profound impact on the surface they colonize. Biofilms are formed by different strains of microorganisms that adhere to a surface and produce a protective layer of extracellular matrix, which helps to protect the microorganisms from environmental stressors, such as ultraviolet (UV) radiation and predation. Biofilms can also impede the flow of water and increase drag, which can cause higher fuel consumption and less efficiency in marine vehicles and infrastructure. The different stages of the biofouling process are depicted in Figure 4.1 [2].

Biofouling can have significant economic consequences for industries that rely on equipment and infrastructure that are constantly exposed to the water. The shipping industry is particularly vulnerable to biofouling, as the hulls of ships get coated with a layer of marine organisms, which can increase drag and reduce fuel efficiency. This can lead to increased fuel costs, decreased speed, and increased greenhouse gas emissions. The off-shore oil and gas industry is also susceptible to biofouling, as structures such as pipelines and platforms can be colonized by marine organisms, which can lead to corrosion and structural damage. Aquaculture facilities are also vulnerable to biofouling, as nets, cages, and other equipment can be coated with marine organisms, which can reduce water flow and impede the growth of cultured organisms. Biofouling also affects the biomedical and food processing industries.

Biofouling can also have significant environmental consequences. The presence of macrofouling organisms, such as barnacles and mussels, can alter the physical structure of marine ecosystems and disrupt the balance of natural communities. Invasive species can also be introduced through biofouling, as they can attach to ships and other marine equipment and be transported to new environments, where they can compete with and displace native species.

To mitigate the impacts of biofouling, a variety of strategies have been developed. One of the most common strategies is the use of antifouling coatings, which are intended to prevent the attachment and growth of marine organisms on submerged surfaces [3]. Antifouling coatings can be made from a variety of materials, including copper, zinc, and silicone, and can be applied to a range of surfaces, including hulls, nets, and pipes. Other strategies include the use of physical barriers, such as screens and

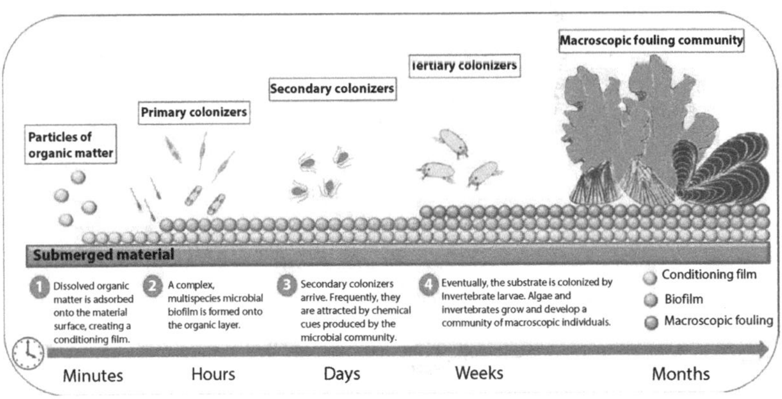

Figure 4.1 Schematics of biofouling process (reprinted with permission from Martín-Rodríguez *et al.*, 2015).

mesh, to prevent the attachment of macrofouling organisms and the use of biological controls, such as the introduction of non-native species that can compete with and displace biofouling organisms.

4.2 Major Industries Affected by Biofouling

Biofouling is a widespread problem that affects various industries, including maritime, food processing, and biomedical sectors (Figure 4.2) [4]. The accumulation of biofouling organisms on submerged surfaces poses significant challenges and impacts the efficiency, safety, and costs associated with operations in these industries.

Maritime Industry: In the maritime industry, biofouling affects ships, offshore platforms, and underwater structures. The growth of biofouling organisms on ship hulls increases drag, reducing vessel speed and fuel efficiency. This leads to higher fuel consumption, resulting in increased operational costs and environmental impact. Biofouling can also compromise

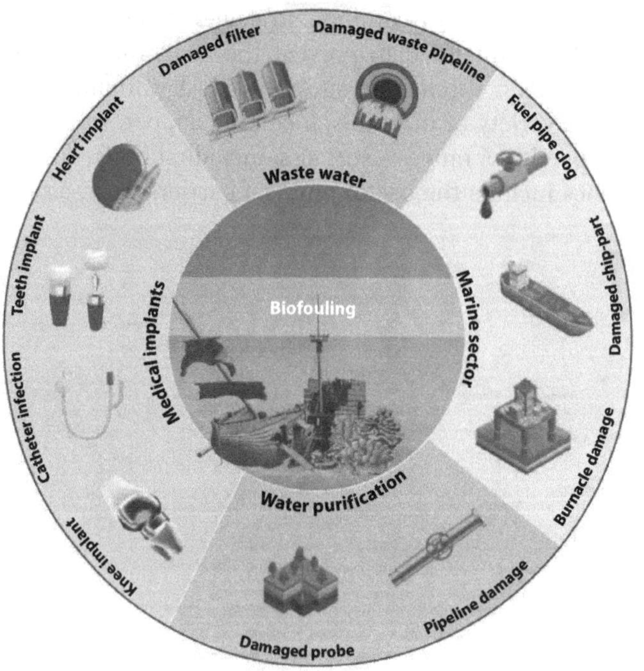

Figure 4.2 Schematic overview of the damaging effect of biofilm-mediated biofouling (reprinted with permission from Patra *et al.*, 2022).

maneuverability and stability, posing safety risks. Regular maintenance, cleaning, and the application of antifouling coatings are necessary to minimize these effects [3].

Food Processing Industry: Biofouling is a concern in the food processing industry, particularly in facilities involved in aquatic product processing, aquaculture, and food production. Equipment such as pipelines, heat exchangers, and storage tanks can become fouled by microorganisms and biofilms. This can lead to contamination, reduced heat transfer efficiency, increased energy consumption, and equipment corrosion. Stringent cleaning and sanitation protocols are necessary to prevent biofouling and to maintain food safety and quality standards.

Biomedical Industry: In the biomedical sector, biofouling impacts medical devices, implants, and equipment used in research and healthcare settings. Biofilms formed on medical implants, such as catheters or prosthetic devices, can cause infections and impair their functionality. Biofouling can also affect laboratory equipment, reducing the accuracy and reliability of experiments. Regular cleaning, sterilization, and the development of biofouling-resistant materials are essential to ensure patient safety and enhance the effectiveness of biomedical devices.

To reduce the effects of biofouling, various strategies are employed. Antifouling coatings are commonly used to inhibit the settlement and growth of fouling organisms. These coatings may release biocides or have physical properties that make it difficult for organisms to attach. Nontoxic and environmentally friendly alternatives are being actively researched and developed to minimize the ecological impact.

4.3 Conventional Antifouling Coatings and their Demerits

Antifouling coatings are used to protect submerged surfaces, such as ship hulls, offshore structures, and underwater equipment, from the accumulation of marine organisms, known as biofouling. Conventional antifouling coatings have been widely used for decades to prevent the attachment and growth of organisms, but they come with certain demerits that need to be considered. The major drawbacks of conventional antifouling coatings are discussed below.

Environmental Impact: Conventional antifouling coatings often contain toxic compounds, such as organotin compounds like tributyltin (TBT), copper-based compounds, or biocides, which are released into the water during the coating's lifetime. These substances can harm marine ecosystems, affecting both marine organisms and nontarget organisms, leading to bioaccumulation and biomagnification of toxins in the food chain.

Toxicity Concerns: The release of toxic compounds from conventional antifouling coatings raises concerns for human health and safety [5]. The toxicity of some antifouling compounds can pose risks to workers during application, maintenance, and disposal. Moreover, recreational activities such as swimming or diving in waters coated with toxic antifouling paints can expose individuals to these harmful substances.

Short Life Span: Conventional antifouling coatings typically have a limited life span, ranging from several months to a few years, depending on factors such as coating quality, water conditions, and vessel speed [6]. The coatings gradually wear off or lose their effectiveness, necessitating frequent reapplication, which increases maintenance costs and generates more waste.

Resistance Development: Over time, marine organisms can develop resistance to the active ingredients present in conventional antifouling coatings. This resistance may be due to genetic adaptation or the presence of tolerant strains, rendering the coatings less effective in preventing biofouling [7]. As a result, the frequency of coating application and the use of higher concentrations of toxic substances may be required to achieve the desired antifouling effect, exacerbating environmental and health concerns.

Fouling on Dry-Docked Vessels: When ships are dry-docked or stored out of the water for extended periods, conventional antifouling coatings can be less effective in preventing fouling. These coatings are primarily designed to work in the presence of water, and their efficacy decreases when exposed to air. As a result, ships may experience increased biofouling during dry-docking that necessitates additional cleaning and maintenance before returning to service [8].

Nonbiodegradability: Many conventional antifouling coatings contain nonbiodegradable materials, such as microplastics or persistent organic pollutants (POPs), which can persist in the environment for long periods. These materials can accumulate in sediments and marine organisms, leading to long-term ecological and environmental impacts [9].

Regulatory Restrictions: Due to the environmental and health concerns associated with conventional antifouling coatings, various international and national regulations have been implemented to restrict the use of certain toxic compounds. For example, the International Maritime Organization (IMO) has banned the use of TBT-based coatings on ships [10]. Compliance with these regulations may require alternative antifouling strategies, posing additional challenges for the maritime industry.

These drawbacks in the conventional antifouling compounds compelled industries to adopt alternatives. The promising qualities of nanomaterials ensure them as a suitable replacement for conventional coating materials to prevent biofouling.

4.4 Nanomaterials as Antifoulants: Properties and Mechanisms

Biofouling is a widespread issue that impacts the efficiency, performance, and durability of submerged surfaces [11]. Conventional antifouling coatings have limitations, such as environmental concerns and resistance development. Nanomaterials offer promising solutions as antifoulants due to their unique properties and versatile applications.

Nanomaterials, defined as materials with dimensions on the nanometer scale, have attracted significant interest in the development of antifouling coatings. Their distinct properties, such as large surface area, tunable surface chemistry, controlled release capabilities, photocatalytic activity, and enhanced mechanical properties, make them promising candidates for combating biofouling. The main properties of nanomaterials that make them unique in its properties are the following.

Large Surface Area: Nanomaterials exhibit a high surface-to-volume ratio, providing an extensive contact area for interactions with fouling organisms [12]. This property enhances the effectiveness of antifouling coatings by increasing the chances of disrupting organism attachment.

Tunable Surface Chemistry: Nanomaterials can be engineered to have specific surface properties, such as charge, hydrophilicity, and surface energy [12]. By modifying these properties, nanomaterials can create surfaces that repel fouling organisms or hinder their attachment, thus preventing biofouling.

Controlled Release of Active Agents: Nanomaterials can encapsulate or carry antifouling agents thereby facilitating their controlled and sustained release over time [13]. This controlled release mechanism improves the longevity and effectiveness of antifouling coatings while reducing environmental concerns associated with conventional coatings.

Photocatalytic Activity: Certain nanomaterials, such as titanium dioxide (TiO_2) nanoparticles, possess photocatalytic properties. When exposed to light, they generate reactive oxygen species (ROS) with antimicrobial and antifouling properties [14]. Photocatalytic nanomaterials offer a self-cleaning mechanism by continuously degrading organic matter on the coating surface.

Enhanced Mechanical Properties: Incorporating nanoparticles into coating matrices enhances the mechanical strength and durability of coatings. Nanomaterials improve adhesion to substrates, resist abrasion and erosion, and enhance resistance to fouling organism removal.

Mechanisms of Nanomaterial-Based Antifouling Action: Nanomaterial-based antifouling agents exhibit various mechanisms of action against foulants and biofilm formed by bacteria, depending on the material, its size, and other factors. The common modes of action are described here (Figure 4.3) [15].

Physical Barrier and Surface Energy Alteration: Nanomaterials can create physical barriers on coating surfaces, preventing fouling organisms from attaching. Modifying the surface energy of coatings using nanomaterials can render them less favorable for organism attachment and settlement [16].

Disruption of Biofilm Formation: Nanomaterials with antimicrobial properties can inhibit the formation of biofilms, which serve as the initial attachment point for fouling organisms. By disrupting biofilm formation, nanomaterials reduce the availability of suitable substrates for attachment, thus preventing fouling.

Controlled Release of Antifouling Agents: Nanomaterials can encapsulate antifouling agents allowing controlled release over time and reducing the overall toxicity and environmental impact [17].

Generation of ROS: Nanomaterials with photocatalytic properties generate ROS which possess antimicrobial properties. This disrupts the attachment and growth of fouling organisms [18].

Electric Fields and Surface Charge: Certain nanomaterials can be electrically charged or responsive to electric fields. By applying an electric field to the coating surface or incorporating charged nanoparticles, fouling

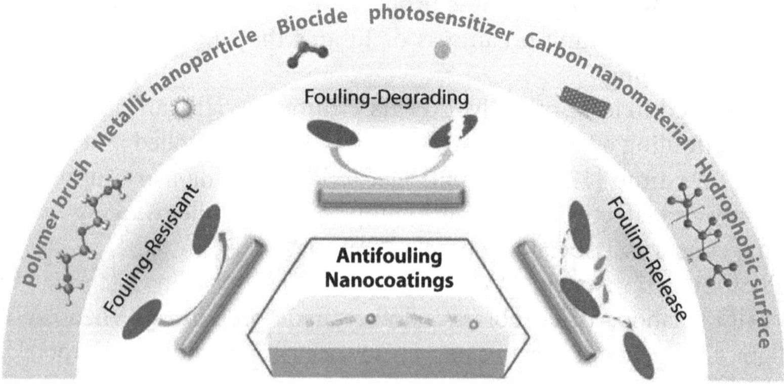

Figure 4.3 Common modes of actions of antifouling nanocoatings (reprinted with permission from Kumar *et al.*, 2021).

organisms can be repelled or disrupted, preventing their attachment and settlement [19].

4.5 Types of Nanomaterials Used as Antifoulants

Different nanomaterials are investigated for their antifouling properties and some of them like metal and metal oxide nanoparticles, carbon-based nanomaterials, polymer nanocomposites, and biological nanomaterials are found to be effective antifoulants.

Metal and Metal Oxide Nanoparticles: Silver (Ag), copper oxide (CuO), zinc oxide (ZnO), and titanium dioxide (TiO2) nanoparticles exhibit antimicrobial properties and can inhibit the growth of fouling organisms.

Carbon-Based Nanomaterials: Carbon nanotubes (CNTs), graphene, and graphene oxide (GO) have shown promise as antifouling agents [20]. They can inhibit biofilm formation, disrupt bacterial attachment, and enhance the mechanical properties of coatings.

Polymer Nanocomposites: Incorporating nanomaterials, such as clay nanoparticles, silica nanoparticles, or metallic nanoparticles, into polymer matrices creates nanocomposite coatings [21]. These coatings exhibit improved mechanical properties, controlled release of antifouling agents, and enhanced resistance to fouling.

Biological Nanomaterials: Natural nanomaterials derived from organisms, such as chitosan nanoparticles or liposomes, have demonstrated antifouling properties [22]. They can disrupt the attachment and growth of fouling organisms through physical and chemical interactions.

4.6 Inorganic Nanomaterials as Antifoulant

Inorganic nanomaterials have turned out to be potential candidates for antifouling applications due to their exclusive physicochemical properties and versatility. Traditional antifouling strategies often rely on toxic biocides, which raise environmental concerns. Inorganic nanomaterials offer a potential solution by providing environment-friendly and effective antifouling capabilities [23].

Metal oxide nanoparticles, such as TiO2 and ZnO, have received considerable attention for their photocatalytic properties. When exposed to light, these nanoparticles generate ROS that exhibit potent antimicrobial activity, effectively preventing the attachment and growth of fouling organisms [24]. This photocatalytic mechanism offers a sustainable and self-cleaning approach to antifouling. Silver nanoparticles have also demonstrated excellent antimicrobial properties. They can release silver ions, which exhibit strong biocidal activity against bacteria, algae, and other microorganisms [25]. However, their potential environmental impact and long-term efficacy need to be thoroughly evaluated to ensure their safe and sustainable use.

Carbon-based nanomaterials, such as CNTs and graphene-based materials, possess unique properties that make them attractive for antifouling applications. These materials can create a superhydrophobic or superoleophobic surface, repelling water- or oil-based fouling agents [26]. Moreover, their high surface area and mechanical strength enable the development of novel antifouling coatings and membranes with enhanced performance and durability.

Surface modification of nanoparticles with antifouling polymers or biomolecules is another approach to enhance their antifouling properties [27]. These surface modifications can create a repulsive surface, hinder the adhesion of fouling organisms, or disrupt biofilm formation. By tailoring the surface chemistry and topography of nanoparticles, researchers aim to optimize their antifouling efficacy and long-term stability. Despite the significant progress made in utilizing inorganic nanomaterials as antifoulants, there are still challenges to overcome [28]. Ensuring the long-term stability and effectiveness of these materials, understanding their potential environmental impact, and addressing scalability and cost considerations are critical for their successful commercialization. Inorganic nanomaterials hold great promise as antifoulants, providing environmentally friendly and effective solutions for combating biofouling. Figure 4.4 shows some common antifouling inorganic nanomaterials. Continued research and development efforts are necessary to further optimize their properties, understand their safety profiles, and translate them into practical applications across various industries.

4.6.1 Silver Nanoparticles

Quick advancement in science and technology innovations has motivated researchers to search for novel aspects of nanotechnology. The management of matter to make any one of its dimensions completely inside the range of the nanoscale (1–100) is known as nanotechnology. The interest in the study of silver nanoparticles to explore their different properties has increased recently because of their uniqueness and attractive physical, chemical, and biological properties. Silver nanoparticles (AgNPs) were distinct from other nanomaterials in their optical, electrical, and, most importantly, antimicrobial characteristics. The antibacterial property of

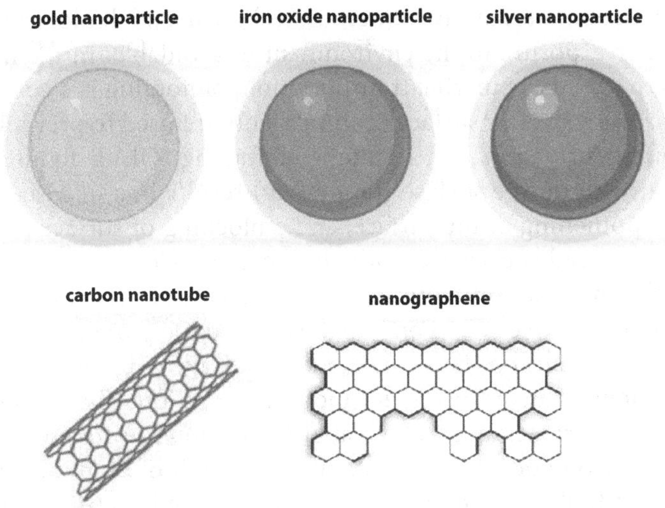

Figure 4.4 Few common antifouling inorganic nanomaterials (reprinted with modification and permission from Greene *et al.*, Nanomedicine in Pancreatic Cancer: Current Status and Future Opportunities for Overcoming Therapy Resistance. Cancers 2021, 13, 6175. https://doi.org/10.3390/cancers13246175).

AgNPs plays a crucial role in the development of many AgNP-based products. Its biological capabilities are mostly determined by variables such as size, shape, and surface coatings. As a result, the synthesis of AgNP with well-controlled morphological and physiochemical properties is critical, particularly for biomedical applications. Its antibacterial action is one of its many diverse applications and is of tremendous interest. Wine and water have been kept clean since ancient times thanks to the usage of silver containers. The therapeutic use of silver also has a long history. Around the time of World War II, interest in silver as a medicine started to diminish as a result of this and the creation of more easily accessible antibiotics like penicillin and cephalosporin. Despite this, the huge increase in the number of antibiotic-resistant bacterial strains and the discovery that silver nanoparticles (AgNPs) have high antibacterial action caused interest in silver to quickly re-emerge. According to a publication, AgNPs exhibit biocidal effects through the gradual release of Ag+ and a variety of mechanisms, including interactions with thiol groups in proteins and enzymes and DNA inhibition.

Silver Nanoparticles as Antifoulant

Environment-friendly nanoparticles are being developed for biological applications due to international ecological regulations. The growing usage of nanomaterials in numerous technical and biological sectors makes their exposure to the environment unavoidable. In the maritime sciences, particularly marine transportation, biofouling is a complicated problem. Many dyes or surface coating agents are used to prevent microbial biofilm formation and resultant biofouling. Of all fouling kinds, biofilm growth on the surface (biofouling) results in a decrease in pore size, a smoothening of the surface, and a blocking of surface functional groups. Increased energy use, operating costs, accelerated corrosion, and the development of antibiotic-resistant bacteria are the results, which have serious negative effects on both health and the economy. Preventing bacterial adherence on the surface, a crucial stage of colonization, is the most efficient strategy to restrict biofilm formation. Thus, stable antifouling nanoparticles must be created to continue to function over an extended period and possibly prevent the growth of biofilm. Designing effective antifouling materials requires controlling bacterial adhesion to nanoparticles by coating them with an antimicrobial ingredient. With minimal toxicity to mammalian cells, nanoscale silver materials displayed broad-spectrum antibacterial activity. The U.S. Environmental Protection Agency has certified colloidal silver as a disinfectant for use in hospitals and other medical facilities.

Fabricating AgNPs on a supporting material might provide a potential remedy for preventing the production of biofilms. AgNPs, a strong antibacterial agent, may prevent biofouling on nanomaterials and hinder the development of biofilms. Various techniques have been developed in recent years to produce AgNPs on a wide range of porous supporting materials, including microporous polymers, mesoporous silica, cellulose, CNTs, and nanofibers. Most often, impregnation and precipitation are used to create metal nanoparticles on solid supports. For the synthesis of metal nanoparticles, hydrothermal processes and sol-gel techniques have also been used. These chemical processes frequently produce combustible and/or poisonous by-products, which inescapably raise environmental and safety concerns. AgNPs' aggregation behavior in an aqueous medium, however, frequently compromises their antibacterial capabilities, and the dissolution of silver ions takes place over time, which causes environmental contamination in addition to a steady loss in efficiency. Moreover, mass transfer reactions cause these porous composite materials' activities to decline. To avoid pore diffusion and biofilm formation, it is therefore highly desirable to identify a simple and quick method for fabricating novel stable nanomaterials using AgNPs on nonporous materials.

4.6.2 Oxides of Ti and Zn

Titanium oxide (TiO_2) can be found as nanocrystals or nanodots and is recognized for its bactericidal activity and inhibition of new cell structures. Since titanium oxide has excellent photocatalytic properties, it is used to make antiseptic and antibacterial compositions, in addition to self-cleaning ceramics and glass, coating, printing ink, and other products. It is also used to make cosmetic products like sunscreen creams, whitening creams, morning and night creams, and skim milk, as well as in the paper industry to increase the opacity of paper. Due to their great stability, anticorrosive, and photocatalytic qualities, TiO_2 NPs are being produced in large quantities and employed widely [29].

Due to their long-term stability, compatibility with conventional materials, and high photocatalytic activity, which makes them suitable for self-cleaning surfaces, titanium dioxide nanoparticles (TiO_2NPs) are among the most adaptable nanoparticles used in construction and building materials. This final characteristic, when combined with hydrophilic characteristics in the presence of UV radiation, enables the fabrication of a protective coating against environmental pollution deposits. Both organic and inorganic pollutants that stain and deteriorate stone walls are broken down by TiO_2NPs. Along with the mechanism underlying their biocidal effects, researchers have also looked at their adaptability and endurance. TiO_2NPs act as a biocide by oxidizing the cell walls of microorganisms, which prevents or at least impedes their ability to breathe. This is especially true when exposed to UV light.

Toxic antifouling coatings made of metals like copper and TBT or biocides like silver or surfactants are used to counteract biofouling [29]. However, due to the risk of harming marine organisms and the chance of environmental contamination, it is not advised to apply these coatings. The International Maritime Organisation intends to fully get rid of paints based on Ti from all ships worldwide to reduce their usage, even though organotin-based coatings are the most effective for long life. One of the promising developments in new biofouling control techniques since 2003 is UV radiation treatment. In a recent study, Patil *et al.* suggested UV-C irradiation (254 nm) as a promising method for preventing biofouling in marine optical instruments [30]. The benefits of UV-C radiation include preventing abrasion-induced surface degradation. The drawback of this method is that it requires a lot of energy for UV irradiation and is therefore expensive. As a result, it was discovered that TiO_2 coatings are suitable for investigation due to their photocatalytic characteristics when exposed to UV light [31]. Li and Logan investigated how UV light affected bacterial adhesion on metal oxide–coated glass substrates and came to the conclusion that UV irradiating TiO_2-coated glass surfaces can be an efficient way to lessen bacterial adhesion. They studied the role of surface heterogeneity of manufactured metal oxide (TiO_2)–coated glass surfaces in determining bacterial and colloid adherence [32].

Zinc oxide plays a similar purpose in combating biofouling as titanium oxide does. When compared to their bulk condition, zinc oxide–based nanomaterials exhibit various physicochemical characteristics, such as decreased size, increased surface area, and improved reactivity and stability in the aquatic environment. Together, these characteristics demonstrate the potential influence of ZnO nanoparticles on marine ecosystems. Zinc oxide metallic nanomaterials' solubility is an essential factor in becoming hazardous. It mostly depends on interactions with organic molecules (proteins, amino acids, organic matter, and humic substances), which cover and spread metallic ions on nanomaterials [33]. Oxidative stress is the toxicity mechanism of ZnO nanoparticles, which causes oxidative damage, inflammation, and cell death. In addition, different morphologies of ZnO-based nanomaterials correspond to diverse hazardous effects. As an illustration, different shapes of ZnO nanoparticles with fixed size and surface area exhibit varying cytotoxicity and cell viability, with rod-shaped ZnO nanomaterials being relatively more hazardous than spherical ones. Physical and chemical characteristics are altered through surface modification with various functional groups or by direct morphological alteration. It may lead to greater nanomaterial dispersion in aqueous solutions. ZnO nanoparticles have drawn attention in comparison to other metal oxide–based nanoparticles because of their lower cost, lack of color, and UV-blocking properties and it is designated as a safe substance by the FDA (Food and Drug Administration, USA).

The following reasons make zinc oxide nanoparticles and nanorod coatings desirable nanostructures:

a) The fabrication process (sol-gel and hydrothermal methods) is simple and controlled.
b) Support-based ZnO nanorod coatings have good stability, minimal nanoparticle loss, and low toxicity.
c) A wide range of support materials may be used.
d) Low temperature (less than 100°Celsius) syntheses.

Bacillus species, Salmonella species, *Escherichia coli*, *Listeria monocytogenes*, and *Staphylococcus aureus* have been shown to be inhibited by ZnO nanorod coatings in freshwater [34]. The production of ROS (such as hydroxyl radicals, hydrogen peroxide, and superoxide ions) as a result of the generation of electron-hole pairs in the aquatic environment under visible light irradiation occurs with zinc oxide nanocoating which can degrade various contaminants. Damage to DNA, particularly base oxidation and strand breakage, damage to cell membranes, and increased oxidative stress are caused by ROS, which results in cell death. On the other hand, the zinc ions that are leached out bind to the pili of bacteria, which alters the bacterial growth cycle. In the lack of light, this method may be beneficial. Zinc ions have a lower rate of ion release than antifouling coatings, which has a positive impact on both target and nontarget organisms in the environment. Due to zinc oxide's exceptional durability in seawater (serving as a natural buffer with pH 8), zinc ions from nanocoating are released at a very slow pace (20 ppm/year under steady illumination). Moreover, the produced surface radical species have a short half-life and are confined to a small area, thus they only neutralize the biofouling microorganisms that are near the nanocoated surface [34].

4.6.3 Metals Like Gold, Copper, and Selenium

Gold nanoparticles (AuNPs) are an excellent alternative for antibacterial therapy and drug delivery because of their high surface area to volume ratio, susceptibility to surface modification, tiny size (less than 10 nm), inert nature, biocompatibility, and biosafety with simple clearance from tissue [35]. Similar benefits are offered by other metallic nanoparticles made of copper, zinc, titanium, cerium, and their corresponding oxides; nevertheless, these nanoparticles' detrimental effects on people and the environment exceed their benefits. AuNPs have been extensively studied to identify their beneficial roles, especially antimicrobial activity. The recent developments in nanotechnology, chemistry, and biotechnology enable the production of AuNPs using straightforward, affordable, and environment-friendly processes, as well as the fine-tuning and surface modification of those AuNPs for better and synergistic activity. Particularly in the case of addressing biofouling, there is enormous potential for creating a strong and extremely targeted antimicrobial approach [36].

When AuNPs are near the biofilm or co-incubated with the biofilm, they go through three crucial processes while acting as an antibacterial agent. These nanoparticles approach the biofilm surface from the bulk phase and first interact with it. The lipids, lipopolysaccharides (LPS), or proteins of the bacterial cell membrane interact with the nanoparticles based on their surface chemistry, charge, and hydrophobicity. The nanoparticles then enter the biofilm. The age of the biofilm, the surface chemistry and composition of the biofilm, and the size, surface charge, and concentration of the nanoparticles are a few variables that affect how well they penetrate [35]. Following penetration, the nanoparticles either move inside as a whole or as ions—gold ions ($Au+$) leached from nanoparticles—to interact with the cellular and biofilm constituents. Therefore, disrupting the bacterial and biofilm components is the basis of AuNPs' antibacterial effect.

A micronutrient called selenium (Se) is necessary for both plants and animals to function properly. Covalently linked organo-selenium compounds have also been shown to be an effective way to prevent biofouling in recent studies. Because selenium compounds can catalyze the production of superoxide radicals through non-enzymatic mechanisms, they effectively limit the formation of biofilms. Due to the catalytic nature of the process, the coatings act as a permanent surface treatment [37]. Selenium is employed in solar cells, rectifiers, photographic exposure meters, and xerography because of its well-known photoelectrical and semiconducting characteristics. Due to its distinctive qualities, such as high biological activity, bioavailability, low toxicity, good particle dispersion, and large surface area, selenium nanoparticles have been recognized as a potential material for numerous applications [38]. By serving as a catalyst for redox reactions involving ROS, selenium prevents the growth of bacterial biofilm.

For eukaryotic cells, selenium is a vitamin required for the formation of glutathione peroxidase. By converting hydrogen peroxide to water, glutathione peroxidase shields cellular components from hydrogen peroxide. Glutathione serves as an electron donor and is oxidized during the production of hydrogen peroxide. Selenium compounds have the ability to catalytically oxidize glutathione using oxygen as the electron acceptor without the help of glutathione peroxidase. The dissolved oxygen is changed into a free radical called superoxide, which has the ability to react with nearby molecules and produce other ROS like hydrogen peroxide and hydroxyl radicals. These ROS can all target different substances found in bacterial outer membranes. Cell lysis is frequently the result of the bacterial outer membrane being damaged by selenium. As a result, selenium that is adhered to a surface continuously produces superoxide to stop bacterial attachment and biofilm growth [39].

Several studies have evaluated copper's antibacterial capabilities and its ability to prevent the formation of biofilm in drinking water distribution systems. Copper has been employed in drinking water systems all over the world. The antibacterial action of copper surfaces includes membrane damage, copper intake (produced by copper ions dissolving off surfaces), oxidative stress damage, cell death, and DNA degradation. The antibacterial effect of copper is due to the surface's ion release, which damages the bacterial membrane and results in the loss of cytoplasmic content. The copper ions' production of ROS leads to increased damage to cell structures, including DNA destruction [40]. Temperature, alloy properties, humidity, bacterial species, direct contact between the bacteria and surface, and its oxidation status are a few of the variables that can affect how copper behaves [41]. According to scientific research, there has been a nearly 100% decrease in the growth of pathogenic bacteria, and there has been a 20% and 34% decrease in surface pollution on floors and drains as a result of direct contact with copper. In areas where the bacterial load is continuously renewed, such as slaughterhouses, Gomes *et al.* proposed that copper sheeting could be used to reduce persistent bacterial load after daily cleaning, prevent biofilm formation, lessen the viability of pathogenic bacteria, and decrease contamination [42]. The relationship between the buildup of copper ions in bacterial cells after they are released onto the surface and their mortality in 3-hour time has been investigated. Since initial bacterial attachment can start immediately once a bacterium makes contact with a surface, there is no proof that copper surfaces can prevent it [43]. But, it was observed that nonviable bacterial cells separate from surfaces after cell death; as cell attachment was not seen for hours following contact, implying the bactericidal activity of copper ions builds up in the cells [43, 44].

The methods of antimicrobial mode of action of copper surfaces involved in contact killing may vary depending on the microbes and their condition, including planktonic, adherent to a surface, or as multiple layered biofilms. Although direct contact between microbes and surfaces may not always be necessary, the mechanisms relating to the antimicrobial capabilities of copper surfaces are mostly attributed to the discharge of copper ions from metallic surfaces [42]. Bacterial harm and activity loss when exposed to copper have been attributed to a number of mechanisms, including damage to the cell envelope (both the outer and inner membrane), oxidative harm from ROS production, enzymatic inhibition, and nucleic acid degradation. However, environmental factors, mainly water, temperature, pH, the presence of other metals or ions, oxidation state, surface roughness, and the microbes present on the surface, may change the mode of action of copper and copper surfaces. The antibiofouling effects of copper are frequently linked to the copper ions produced, which harm cells by changing the structure and activity of proteins and enzymes. Moreover, the interaction of copper with biofilm is different from that with planktonic cells. Microorganisms in biofilms are shielded by an extracellular polymeric substance (EPS) matrix. Therefore, the impact of copper surfaces and copper ions will interfere with EPS in addition to focusing on microbial contact. Regardless of the industry or region of the economy where they are created, undesirable biofilm formation poses a significant economic burden.

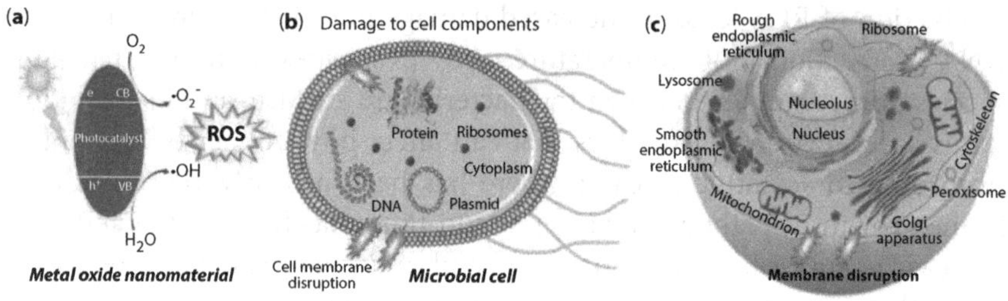

Figure 4.5 Probable antifouling mechanisms of metal oxide nanomaterials (a) generation of ROS; effect of ROS on (b) a prokaryotic cell, and (c) a eukaryotic cell (reprinted with permission from Kumar *et al.*, 2021).

Therefore, innovative and efficient biofilm prevention and control methods are required. Figure 4.5 describes the possible antifouling mechanisms of metal oxide nanomaterials [15].

4.6.4 Graphene

Graphene (Gr), the allotrope of carbon, has exclusive properties mainly due to its two-dimensional (2D) hexagonal lattice structure. It is a flat monolayer made up of single-atom-thick, 2D sheets of a hexagonally organized honeycomb lattice, and it is isolated from crystalline graphite [45]. In 1 mm of graphite, there are about 3 million layers of graphene, which, like graphite, is made solely of carbon atoms. Graphene is a 2D crystal that is only one atom thick compared to graphite, which has a three-dimensional crystalline structure. Only 0.30 nm thick, with just 0.1 nm separating each atom, the hexagonal honeycomb structure has the carbon atoms evenly dispersed throughout. The strongest, thinnest, and most heat- and electrical-conducting substance in the world is thought to be graphene. Some extraordinary qualities are bestowed by the simplicity of 100% pure carbon [46].

Since the material's discovery in 2004 and due to its distinct structural, surface area, and mechanical properties, graphene's uses and applications have attracted a lot of attention. Monolayer graphene, few-layer graphene (FLG), GO, reduced GO (RGO), graphene nanosheets (GNS), and graphene nanoribbons are some examples of graphene and its derivatives. GO is one of the most important chemical graphene derivatives among nanomaterials from the graphene family [47]. Graphene materials are widely employed in different industries because of their excellent physical and chemical capabilities. Examples include energy storage, nanoelectrical devices, batteries, and medicinal applications including antimicrobial, biosensors, cell imaging, drug delivery, and tissue engineering.

Graphene as Antifoulant: Cutting-edge surface engineering methods that alter surface characteristics at the nanoscale can provide tremendous antifouling potential. It has been shown that adding nanoparticles to marine antifouling paints has a significant impact on a surface's charge potential, hydrophobicity, and topography, as well as its antibacterial and anticorrosion capabilities. Among these newly developed nanoengineered antifouling paints, those containing carbon nanomaterials, such as CNTs and graphene, have shown promising results [48].

The most remarkable characteristics of graphene include its high heat conductivity, high electrical conductivity, high elasticity and flexibility, high hardness, high resistance, and cannot support the growth of bacteria. The exceptional physicochemical properties, antimicrobial activity, and biocompatibility of graphene and its derivatives make them promising candidates for a wide range of antimicrobial applications [49]. Although the mechanisms of action of this carbon substance are still not fully known, it is hypothesized that the rough edges of graphene sheets can cause membrane disruption and bacterial cell entrapment. In addition, it is thought that graphene causes oxidative stress by producing ROS, which damage the DNA and proteins of bacteria. Thus, adding graphene to marine paints can not only improve their mechanical strength and durability but also provide them with enhanced antifouling attributes [50].

Graphene (Gr), GO, and RGO are the three major forms of graphene materials that are used to create new antibacterial substances. The process of synthesis, which establishes the sheet number, thickness, and oxygen content, has a substantial impact on the antimicrobial activity of these materials [50]. Zou *et al.* comprehensively investigated the antibacterial potential of GO in both macrophages and animal models, demonstrating that GO is an effective antimicrobial nanomaterial both *in vitro* and *in vivo* against bacteria that have developed multidrug resistance like *Klebsiella pneumonia* (*K. pneumonia*), *Escherichia coli* (*E. coli*), and *Pseudomonas aeruginosa*, which are all capable of forming biofilms [51].

The effectiveness of GO as an antimicrobial agent is the subject of numerous studies, along with various surface modifications and functionalization with inorganic nanostructures, biomolecules, and polymers to decrease toxicity and increase effectiveness [52]. The development of novel, more potent antimicrobial coatings has more prospects now that graphene-based nanomaterials and composites with distinctive characteristics and broad-spectrum antibacterial action are being synthesized. Antimicrobial graphene nanocomposite coatings come in a wide variety and aim to modify some of their characteristics for conceivable niche applications. The emphasis is given to increasing the antimicrobial activity through synergistic effects.

Slate *et al.* discussed 2D carbon nanomaterials in surface coatings and the antibacterial efficacy of carbon-based nanomaterials. Through the use of a two-phase process, a ternary nanocomposite made of elastomeric silicone, GO sheets, and nanorods was made and Selim *et al.* fabricated an extremely hydrophobic antifouling coating (using solution casting) [53]. As a novel multifunctional antibacterial and antifungal material, Jaworski *et al.* created GO-based nanocomposites decorated with AgNPs [54].

Graphene also prevents *S. mutans* and *E. faecalis* from forming biofilms while leaving the bacteria unharmed. By using sonication exfoliation technology, Muthu *et al.* created hydrophobic, bacterial-repellent Gr coatings on hydrophilic glass surfaces reclaimed from pencil tubes [55]. Understanding the ways by which Gr nanomaterials prevent microbial growth is essential to improve their effectiveness. A variety of antimicrobial mechanisms and physicochemical properties that affect the antimicrobial activity of Gr nanomaterials, such as physical contact destruction, oxidative stress (ROS-dependent/independent), photoinduced antibacterial activity, controlled drug/metal ion release, and synergistic antibacterial activity, are identified [56]. The physical, structural (morphology: mono- and multilayer), electrical, and surface chemistry of graphene and its derivatives (Gr, GO, and RGO) is different. As a result, they interact with the microbial cells in diverse ways. When it comes to bacterial adherence, for instance, lateral size is crucial, but sharp edges can function as nanoknives. GO can increase the antibacterial ability by triggering oxidative stress with or without the generation of ROS. Staneva *et al.* categorized the main theories put out to account for Gr and its derivatives' antibacterial action as follows: the photothermal hypothesis, the electron transfer hypothesis, the entrapment theory, the oxidative stress hypothesis, and the membrane stress hypothesis [56].

It is accepted that the Graphene nanoparticles cause physical and chemical harm to the bacteria in order to exert their antibacterial activity. Their pointed edges make direct contact with bacterial membranes, which causes a damaging extraction of lipid molecules. Wrapping and photothermal ablation methods are further examples of such damage. Bacterial chemical damage is brought on by oxidative stress, which produces ROS and causes charge transfer. Gr nanoparticles are already recognized as potential materials for the production of antimicrobial coatings. Additional factors influencing the antimicrobial activity of Gr, Gr derivatives, and Gr composites used as the foundation for antimicrobial coatings are elastic modulus, graphene wettability, concentration, and orientation of the graphene nanoparticles on the surface, etc. It is remarkable that biofilm formation was boosted when low doses of GO were present but hindered when large doses of GO were present. The functions of dead cells inactivated by GO provide an explanation for these findings. A portion of the cells that are only inactivated at low GO concentrations act as a protective barrier and nutrition source for the remaining, biofilm-forming, live cells. Nearly all cells may be totally inactivated at high GO concentrations, and the production of biofilms will be significantly diminished [56].

Indicators of the biological activity of graphene materials and coatings include their chemical makeup, particle size, number of sheets, orientation, hydrophilic/hydrophobic balance, free energy, morphology, and roughness. The ability to modify these features may be a means of modifying the antibacterial effectiveness of coated materials. The physical and physicochemical interactions of physical and physicochemical graphene materials, such as contact destruction and the formation of ROS, have been proposed as antibacterial processes for graphene materials.

4.6.5 Carbon Nanotubes

CNT refers to cylindrical sheets of hexagonally arranged carbon atoms with lengths in the micrometer range and widths of a few nanometers. Single-walled CNTs (SWCNTs) and multi-walled CNTs (MWCNTs) are the two different forms of CNTs. While MWCNTs frequently consist of several to 10 concentric cylinders of graphitic shells, SWCNTs are essentially made of a single cylindrical graphene sheet. MWCNTs typically have an outer diameter of (2.5–100 nm), which is bigger than that of SWCNTs (0.6–2.4 nm), and MWCNTs are made up of many concentric SWCNT layers [57]. However, SWNTs have a more clearly defined diameter, whereas MWCNTs have a less stable nanostructure because of the possibility of structural flaws. Large aspect ratios, astonishingly high mechanical strength (such as tensile strengths 100 times greater than those of steel), equivalent thermal and electrical conductivities to copper, and incredibly high surface areas are the characteristics of CNTs [58]. When the tips of CNTs are opened, they can be considered nanocontainers that can hold a variety of molecules, including ions, fullerenes, porphyrins, and metals [59]. It was discovered that hydrophobic interactions were to blame for trapping these compounds inside CNTs. Such structures consist of a metal or a metal complex that may be used to build pharmaceuticals, particularly for diagnostic applications, and they may be introduced and developed as novel contrast agents for a number of imaging modalities [60].

CNTs have a concentric cylinder shape with a diameter in the nanoscale range (based on the number of walls) and a length that can range from several microns (100 m) to a few millimeters (approximately 4 mm). CNTs are potential nanomaterials for a variety of applications because of their distinctive qualities, including remarkable tensile strength, high thermal conductivity, and structural stability. These applications include those in the industrial, environmental, and medical domains. Carbon nanoparticles have found use as coatings for marine vessels and underwater structures and gadgets due to their antibacterial and antifouling capabilities as well as their exceptional mechanical qualities [61].

The functional aspects and performance of usual engineering materials used for environmental applications can be improved with the help of nanomaterials. It is generally known that one way to improve the performance of polymeric membranes is to incorporate carbon-based materials such as biochar, CNTs, graphene, mesoporous carbon nanoparticles, and carbon quantum dots into the membranes. CNTs stand out among them because of their potent antibacterial impact on a range of bacteria and viruses. ROS, physical piercing/trapping effect, hydrophobic-hydrophilic interactions, and electrostatic repulsions are the key causes of CNT's antibacterial capabilities. Previously, the antimicrobial was a super-hydrophobic material with enhanced biofouling reduction. CNT-polymer nanocomposites may operate as biocides to destroy both planktonic and surface-attached bacteria by stopping the adherence of microbes to prevent biofouling. Due to their potent antibacterial properties and ability to effectively remove biofouling, CNT nanocomposites have been proposed as maritime fouling-release coatings and surface coatings for medical equipment.

Recent research on various nanomaterials, in particular, CNTs, is exciting due to their versatile potential in environmental engineering applications. CNTs were first discovered by Iijima in 1991 [62], and they quickly gained popularity, making their distinct identity felt in nearly every important field of science and engineering. Many CNT-based research applications are currently on the verge of commercialization. CNT adsorbents, in particular, show outstanding sorption capacities for big biological pollutants (microorganisms) and may have the highest reported capabilities for modern media. CNTs can be used in desalination as membrane materials in addition to adsorption applications [63].

The size, functionalization, and dispersion of nanotubes, among other variables affect their antibacterial capabilities. Smaller particles can more easily bind to microbial cells and alter their membranes and metabolic activities because they have a larger surface-to-volume ratio. In this sense, in comparison to SWCNTs, MWCNTs appear to be less harmful to bacteria. The diameter, intrinsic stiffness, and potential van der Waal forces at the surface of MWCNTs are all bigger. All of this lessens MWCNTs' toxicity to bacterial cells. The antibacterial impact of thin MWCNTs with a lower diameter and length is also more potent [64].

High microbial immobilization abilities, inherent cytotoxicity, outstanding mechanical strength, and the capacity to create nanocomposites (NC) are the four characteristics of bulk CNTs. Bacteria and their spores have been successfully eradicated by CNTs' potent antibacterial properties. The mechanical properties of CNT nanomaterials, which operate as "nano-darts" to pierce bacterial membranes and cause cell death, are what primarily contribute to their antibacterial properties. The chemical modifications of CNTs, their concentration, diameter, surface area, and length also play an important role in their antimicrobial properties. It was discovered that when the concentration of MWCNTs-OH increased, the rate of oxygen uptake, the rate of nitrification and denitrification, and the rate of dehydrogenase activity all decreased in a microbial community of activated sludge [65].

Bacteria are mostly known to reside as biofilms or multispecies communities encased in an extracellular polymeric matrix (EPS), in nature. Compared to planktonic cells, cells in biofilms are substantially more resistant to a variety of harmful physicochemical impacts; however, the degree of adaptive response may vary. The EPS is crucial in minimizing the negative effects of CNTs, and the cells in the biofilm structure are less vulnerable to their effects. It is possible to include CNTs in EPS to stop nanoparticles from coming into touch with cell membranes directly. The creation of diverse composite materials is one of the possible approaches to biofouling control. Surface topography has a significant impact on microbial adherence and biofilm formation, and CNTs can be used to alter the surface of composite materials. High concentrations of CNTs treated with carboxyl and hydroxyl groups can reduce the efficiency of beneficial biofilms' capacity to digest contaminants in the marine environment. These groups have the ability to alter CNTs to produce composite materials with an antifouling function [66].

Thus, different inorganic nanomaterials have a high potential to inhibit microbial growth and biofilm formation and hence can be effectively utilized as antifouling agents. Considering the mechanism of action and its environmental impact, it is possible to select the right nanomaterial for antifouling applications. Inorganic nanomaterial coatings have shown great potential in preventing biofouling, offering an alternative to traditional antifouling strategies that often rely on toxic biocides. The unique physicochemical properties of inorganic nanomaterials, such as metal oxide nanoparticles, silver nanoparticles, carbon-based nanomaterials, and surface-modified nanoparticles, contribute to their antifouling capabilities. The mode of action of different inorganic nanomaterials to prevent biofouling is summarized in Table 4.1.

Table 4.1 Antifouling action of different inorganic nanomaterials.

Nanomaterial	Mode of action to prevent biofouling	References
Silver nanoparticles	Disruption of the cell wall; change the cell membrane properties; change membrane fluidity and the composition of the lipid bilayer and facilitate cell membrane dissolution; damage bacterial DNA; denaturation of ribosomes	[25]
Oxides of tin	Act as a biocide by oxidizing the cell walls of microorganisms, which prevents or at least impedes their ability to breathe	[29]
Oxides of zinc	Oxidative stress which causes oxidative damage, inflammation, and cell death	[33]
Gold nanoparticles	Nanoparticles interact with the cellular and biofilm constituents, lipids, LPS, or proteins of the bacterial cell membrane and change the membrane potential; decrease the ATP level	[35]
Selenium	Catalyze the production of superoxide radicals through non-enzymatic mechanisms; as a catalyst for redox reactions involving reactive oxygen species, selenium prevents the growth of bacterial biofilm	[37]
Copper nanoparticles	Membrane damage, cell death, and DNA degradation; the loss of cytoplasmic content. Production of reactive oxygen species leads to increased damage to cell structures, including DNA destruction.	[40]
Graphene	Oxidative stress by producing reactive oxygen species (ROS), which damage the DNA and proteins of bacteria; inactivate bacterial cell proliferation	[48]
Carbon nanotubes	ROS, physical piercing/trapping effect, hydrophobic-hydrophilic interactions, and electrostatic repulsions; cell disruption and the release of intracellular content.	[57]

In addition to conventional polymer and organic molecule coatings, researchers have investigated the use of organic/inorganic hybrid composites as antifouling agents. Considering the solubility of many metal oxide nanostructures, nowadays, researchers incorporate them to support substrates or stable coating matrices to improve stability, decrease environmental impact, and improve the antifouling property. Many metal oxides like ZnO and TiO_2 nanoparticles are incorporated into nanocomposite coatings. These types of inorganic polymer nanocomposites are also investigated extensively for their potential antifouling properties.

4.7 Impact and Challenges of Inorganic Nanomaterials as Antifoulants

Even though nanomaterials are an attractive option for mitigating fouling, there is a likelihood of toxicity as the production process for these uses toxic compounds. Multiple parameters need to be considered in order to ensure the environmental safety of antifouling compounds like the physiochemical and toxicological properties of the compounds used. Identification of the leaching and release rate to water along with the ecotoxicology of the degradation products needs to be evaluated. To address the growing concerns about the toxicity of nanoparticles, benign conditions must be used in their manufacturing. Excellent antimicrobials have been produced through the biological synthesis of nanoparticles *via* plants and microbes. A variety of microorganisms that are employed to make nanoparticles *via* biological synthesis routes are secure, biocompatible, and environmentally benign. The environmental impact of the use of nanomaterials can be addressed to a great extent by adopting green protocols for their synthesis and by selecting inert nanomaterials with the least biological interactions as antifouling compounds.

4.8 Conclusion

Biofouling poses a significant challenge for many industries including food, maritime, and biomedical industries. Conventional methods of combating biofouling rely on the use of biocidal agents, which deter or kill organisms in the vicinity. However, these methods are not environmentally friendly. In this chapter, alternative approaches to toxic antifouling paints are explored using inorganic nanomaterials that exhibit antifouling effects through mechanisms such as fouling resistance, release, or degradation. Photocatalytic nanostructure–based antifouling applications involve the use of photocatalytic nanostructures that generate ROS when exposed to light, effectively killing or inactivating micro- and macroorganisms. Photocatalytic nanocoatings are less damaging to aquatic life than conventional antifouling coatings because, unlike toxic biocides, these ROS are transient and primarily affect creatures near the coatings. The intrinsic qualities of inorganic nanoparticles have a role in the fouling degradation of inorganic polymer nanocomposites. The use of coatings with metal oxide nanostructures is thought to be a promising and environmentally benign method of preventing maritime fouling. However, only a few research studies have looked into the prevention of macrofouling in real-world settings, maybe as a result of the low efficacy of nanocomposite antifouling coatings or the difficulties in scaling up such coatings.

The optimal antifouling coating should be long-lasting, dependable, simple to use, affordable, environmentally friendly, and compatible with a range of substrates. The simultaneous fulfillment of all these needs is still quite difficult. Antifouling does not have a single chemistry or method that works for everyone. Instead, the most efficient method for creating a successful antifouling coating is probably to create a multifunctional coating that synergistically integrates numerous antifouling techniques. Given their efficiency and affordability, the use of inorganic polymers and their composite coatings represents a potentially workable solution for biofouling. To improve antifouling performance, particularly in terms of long-term durability in both static and dynamic situations, more study is required. To enable large-scale applications outside of the lab, coating processes must be improved in addition to efficiency. Technology-advanced antifouling coatings suitable for actual industrial needs and various materials like vessels, fishing nets, and pipes can be created by combining straightforward coating techniques with self-replenishing antifouling properties to increase durability and employing cross-linkers or catch-bonds to improve mechanical stability. In addition, it is essential to look into the environmental toxicity and potential effects of these materials.

References

1. Callow, J.A. and Callow, M.E., Trends in the development of environmentally friendly fouling-resistant marine coatings. *Nat. Commun.*, *2*, 1, 244, 2011.
2. Martín-Rodríguez, A.J., Babarro, M.F.J., Lahoz, F., Sansón, M., Martín, S.V., Norte, M., Fernández, J.J., From Broad-Spectrum Biocides to Quorum

Sensing Disruptors and Mussel Repellents: Antifouling Profile of Alkyl Triphenylphosphonium Salts. *PloS One*, 10, 4, e0123652, 2015.
3. Schultz, M.P., Bendick, J.A., Holm, E.R., Hertel, W.M., Economic impact of biofouling on a naval surface ship. *Biofouling, 27*, 1, 87–98, 2011.
4. Patra, A., Das, J., Agrawal, N.R., Kushwaha, G.S., Ghosh, M., Son, Y.O., Marine Antimicrobial Peptides-Based Strategies for Tackling Bacterial Biofilm and Biofouling Challenges. *Molecules*, 27, 7546, 2022.
5. Rosenberg, M., Ilić, K., Juganson, K., Ivask, A., Ahonen, M., Vrček, I.V., Kahru, A., Potential ecotoxicological effects of antimicrobial surface coatings: a literature survey backed up by analysis of market reports. *PeerJ, 7*, e6315, 2019.
6. Logan, K.P., Using a ship's propeller for hull condition monitoring. *Naval Eng. J.*, 124, 1, 71–87, 2012.
7. Wu, Z., Chan, B., Low, J., Chu, J.J.H., Hey, H.W.D., Tay, A., Microbial resistance to nanotechnologies: An important but understudied consideration using antimicrobial nanotechnologies in orthopaedic implants. *Bioact. Mater.*, 16, 249–270, 2022.
8. Cardarelli, N.F., *Controlled release pesticides formulations*, CRC press, 2018.
9. Lusher, A., Hollman, P., Mendoza-Hill, J., Microplastics in fisheries and aquaculture: status of knowledge on their occurrence and implications for aquatic organisms and food safety. FAO, 2017.
10. Champ, M.A., Economic and environmental impacts on ports and harbors from the convention to ban harmful marine anti-fouling systems. *Mar. Pollut. Bull., 46*, 8, 935–940, 2003.
11. Zhang, N., Jiang, B., Zhang, L., Huang, Z., Sun, Y., Zong, Y., Zhang, H., Low-pressure electroneutral loose nanofiltration membranes with polyphenol-inspired coatings for effective dye/divalent salt separation. *Chem. Eng. J., 359*, 1442–1452, 2019.
12. Wu, Q., Miao, W.S., Zhang, Y.D., Gao, H.J., Hui, D., Mechanical properties of nanomaterials: A review. *Nanotechnol. Rev., 9*, 1, 259–273, 2020.
13. Lvov, Y. and Abdullayev, E., Functional polymer–clay nanotube composites with sustained release of chemical agents. *Prog. Polym. Sci.*, 38, 10-11, 1690–1719, 2013.
14. Amiri, M.R., Alavi, M., Taran, M., Kahrizi, D., Antibacterial, antifungal, antiviral, and photocatalytic activities of TiO2 nanoparticles, nanocomposites, and bio-nanocomposites: Recent advances and challenges. *J. Public Health Res., 11*, 2, 22799036221104151, 2022.
15. Kumar, S., Ye, F., Dobretsov, S., Dutta, J., Nanocoating Is a New Way for Biofouling Prevention. *Front. Nanotechnol.*, 3, 771098, 2021.
16. Banerjee, I., Pangule, R.C., Kane, R.S., Antifouling coatings: recent developments in the design of surfaces that prevent fouling by proteins, bacteria, and marine organisms. *Adv. Mater., 23*, 6, 690–718, 2011.
17. Abdullayev, E. and Lvov, Y., Halloysite clay nanotubes for controlled release of protective agents. *J. Nanosci. Nanotechnol., 11*, 11, 10007–10026, 2011.

18. Jassby, D., Farner Budarz, J., Wiesner, M., Impact of aggregate size and structure on the photocatalytic properties of TiO2 and ZnO nanoparticles. *Environ. Sci. & Technol.*, 46, 13, 6934–6941, 2012.
19. Kumar, R. and Ismail, A.F., Fouling control on microfiltration/ultrafiltration membranes: Effects of morphology, hydrophilicity, and charge. *J. Appl. Polym. Sci.*, 132, 21, 2015.
20. Thamaraiselvan, C., Wang, J., James, D.K., Narkhede, P., Singh, S.P., Jassby, D., Arnusch, C.J., Laser-induced graphene and carbon nanotubes as conductive carbon-based materials in environmental technology. *Mater. Today*, *34*, 115–131, 2020.
21. Bogue, R., Nanocomposites: a review of technology and applications. *Assem. Autom.*, *31*, 2, 106–112, 2011.
22. Blecher, K., Nasir, A., Friedman, A., The growing role of nanotechnology in combating infectious disease. *Virulence*, *2*, 5, 395–401, 2011.
23. Selim, M.S., Shenashen, M.A., El-Safty, S.A., Higazy, S.A., Selim, M.M., Isago, H., Elmarakbi, A., Recent progress in marine foul-release polymeric nanocomposite coatings. *Prog. Mater. Sci.*, *87*, 1–32, 2017.
24. Lallo da Silva, B., Abuçafy, M.P., Berbel Manaia, E., Oshiro Junior, J.A., Chiari-Andréo, B.G., Pietro, R.C.R., Chiavacci, L.A., Relationship between structure and antimicrobial activity of zinc oxide nanoparticles: An overview. *Int. J. Nanomed.*, 9395–9410, 2019.
25. Siddiqi, K.S., Husen, A., Rao, R.A., A review on biosynthesis of silver nanoparticles and their biocidal properties. *J. Nanobiotechnol.*, *16*, 1, 1–28, 2018.
26. Al-Anzi, B.S. and Siang, O.C., Recent developments of carbon based nanomaterials and membranes for oily wastewater treatment. *RSC Adv.*, *7*, 34, 20981–20994, 2017.
27. Mahmoudpour, M., Jouyban, A., Soleymani, J., Rahimi, M., Rational design of smart nano-platforms based on antifouling-nanomaterials toward multifunctional bioanalysis. *Adv. Colloid Interface Sci.*, 102637, 2022.
28. Lau, W.J., Gray, S., Matsuura, T., Emadzadeh, D., Chen, J.P., Ismail, A.F., A review on polyamide thin film nanocomposite (TFN) membranes: History, applications, challenges and approaches. *Water Res.*, *80*, 306–324, 2015.
29. Shi, H., Magaye, R., Castranova, V., Zhao, J., Titanium dioxide nanoparticles: a review of current toxicological data. *Part Fibre Toxicol.*, 10, 15, 2013.
30. Patil, J.S., Kimoto, H., Kimoto, T., Saino, T., Ultraviolet radiation (UV-C): a potential tool for the control of biofouling on marine optical instruments. *Biofouling*, 23, 3-4, 215–30, 2007.
31. Dineshram, R., Subasri, R., Somaraju, K.R.C., Jayaraj, K., Vedaprakash, L., Ratnam, K., Venkatesan, R., Biofouling studies on nanoparticle-based metal oxide coatings on glass coupons exposed to marine environment. *Colloids Surf. B.*, *74*, 1, 75–83, 2009.

32. Li, B., and Logan, B.E., The impact of ultraviolet light on bacterial adhesion to glass and metal oxide-coated surface. *Colloids Surf. B.*, 41, 2-3, 153–61, 2005.
33. Bondarenko, O., Juganson, K., Ivask, A., Kasemets, K., Mortimer, M., Kahru, A., Toxicity of Ag, CuO and ZnO nanoparticles to selected environmentally relevant test organisms and mammalian cells *in vitro*: a critical review. *Arch. Toxicol.*, *87*, 1181–1200, 2013.
34. Singh, D., Rehman, N., Pandey, A., Nanotechnology: the Alternative and Efficient Solution to Biofouling in the Aquaculture Industry. *Appl. Biochem. Biotechnol.*, 1–16, 2023.
35. Ikuma, K., Decho, A.W., Lau, B.L., When nanoparticles meet biofilms—interactions guiding the environmental fate and accumulation of nanoparticles. *Front. Microbiol.*, 6, 591, 2015.
36. Nikolova, M.P. and Chavali, M.S., Metal oxide nanoparticles as biomedical materials. *Biomimetics*, *5*, 2, 27, 2020.
37. Vercellino, T., Morse, A., Tran, P., Song, L., Hamood, A., Reid, T., Moseley, T., Attachment of organo-selenium to polyamide composite reverse osmosis membranes to inhibit biofilm formation of S. aureus and E. coli. *Desalination*, *309*, 291–295, 2013.
38. Bisht, N., Phalswal, P., Khanna, P.K., Selenium nanoparticles: a review on synthesis and biomedical applications. *Mater. Adv.*, 3, 1415–1431, 2022.
39. Low, D., Hamood, A., Reid, T., Mosley, T., Tran, P., Song, L., Morse, A., Attachment of selenium to a reverse osmosis membrane to inhibit biofilm formation of S. aureus. *J. Membrane Sci.*, 378, 1-2, 171–178, 2011.
40. Grass, G., Rensing, C., Solioz, M., Metallic copper as an antimicrobial surface. *Appl. Environ. Microbiol.*, 77, 5, 1541–1547, 2011.
41. Bleichert, P., Espírito Santo, C., Hanczaruk, M., Meyer, H., Grass, G., Inactivation of bacterial and viral biothreat agents on metallic copper surfaces. *Biometals*, 27, 1179–1189, 2014.
42. Gomes, I.B., Simões, M., Simões, L.C., Copper surfaces in biofilm control. *Nanomaterials*, 10, 12, 2491, 2020.
43. Noyce, J.O., Michels, H., Keevil, C.W., Potential use of copper surfaces to reduce survival of epidemic meticillin-resistant Staphylococcus aureus in the healthcare environment. *J. Hosp. Infect.*, *63*, 3, 289–297, 2006.
44. Santo, C.E., Quaranta, D., Grass, G., Antimicrobial metallic copper surfaces kill Staphylococcus haemolyticus via membrane damage. *Microbiologyopen*, *1*, 1, 46–52, 2012.
45. Novoselov, K.S., Geim, A.K., Morozov, S.V., Jiang, D.E., Zhang, Y., Dubonos, S.V., Firsov, A.A., Electric field effect in atomically thin carbon films. *Science*, *306*, 5696, 666–669, 2004.
46. Nine, M.J., Cole, M.A., Tran, D.N., Losic, D., Graphene: a multipurpose material for protective coatings. *J. Mater. Chem. A*, 3, 24, 12580–12602, 2015.

47. Ou, L., Song, B., Liang, H., Liu, J., Feng, X., Deng, B., Shao, L., Toxicity of graphene-family nanoparticles: a general review of the origins and mechanisms. *Part. Fibre Toxicol.*, *13*, 1, 1–24, 2016.
48. Jin, H., Bing, W., Tian, L., Wang, P., Zhao, J., Combined effects of color and elastic modulus on antifouling performance: A study of graphene oxide/silicone rubber composite membranes. *Materials*, 12, 16, 2608, 2019.
49. Sattar, T., Current review on synthesis, composites and multifunctional properties of graphene. *Topics Curr. Chem.*, *377*, 1–45, 2019.
50. Zhang, T. and Tremblay, P.L., Graphene: an antibacterial agent or a promoter of bacterial proliferation? *IScience*, *23*, 12, 101787, 2020.
51. Zou, X., Zhang, L., Wang, Z., Luo, Y., Mechanisms of the antimicrobial activities of graphene materials. *J. Am. Chem. Soc.*, *138*, 7, 2064–2077, 2016.
52. Yousefi, M., Dadashpour, M., Hejazi, M., Hasanzadeh, M., Behnam, B., de la Guardia, M., Mokhtarzadeh, A., Anti-bacterial activity of graphene oxide as a new weapon nanomaterial to combat multidrug-resistance bacteria. *Mater. Sci. Eng. C*, *74*, 568–581, 2017.
53. Selim, M.S., Fatthallah, N.A., Higazy, S.A., Hao, Z., Jing Mo, P., A comparative study between two novel silicone/graphene-based nanostructured surfaces for maritime antifouling. *J. Colloid Interface Sci.*, 606, Pt 1, 367–383, 2022.
54. Jaworski, S., Wierzbicki, M., Sawosz, E. *et al.*, Graphene Oxide-Based Nanocomposites Decorated with Silver Nanoparticles as an Antibacterial Agent. *Nanoscale Res. Lett.*, 13, 1, 116, 2018.
55. Muthu, M., Gopal, J., Chun, S. *et al.*, Hydrophobic Bacteria-Repellant Graphene Coatings from Recycled Pencil Stubs. *Arab J. Sci. Eng*, 43, 241–249, 2018.
56. Staneva, A.D., Dimitar, K.D., Dilyana, N.G., Todorka, G.V., Antibiofouling Activity of Graphene Materials and Graphene-Based Antimicrobial Coatings. *Microorganisms.*, 9, 1839, 2021.
57. Serpell, C.J., Kostarelos, K., Davis, B.G., Can carbon nanotubes deliver on their promise in biology? Harnessing unique properties for unparalleled applications. *ACS Cent. Sci.*, *2*, 4, 190–200, 2016.
58. Ebbesen, T.W., *Carbon nanotubes: preparation and properties*, CRC press, 1996.
59. Koshino, M., Moon, P., Son, Y.W., Incommensurate double-walled carbon nanotubes as one-dimensional moiré crystals. *Phys. Rev. B*, 91, 3, 035405, 2015.
60. Ménard-Moyon, C., Kostarelos, K., Prato, M., Bianco, A., Functionalized carbon nanotubes for probing and modulating molecular functions. *Chem. & Biol.*, 17, 2, 107–115, 2010.
61. Gu, B.E., Huang, C.Y., Shen, T.H., Lee, Y.L., Effects of multiwall carbon nanotube addition on the corrosion resistance and underwater acoustic absorption properties of polyurethane coatings. *Prog. Organic Coat.*, 121, 226–235, 2018.

62. Iijima, S., Helical microtubules of graphitic carbon. *Nature*, *354*, 6348, 56–58, 1991.
63. Upadhyayula, V.K. and Gadhamshetty, V., Appreciating the role of carbon nanotube composites in preventing biofouling and promoting biofilms on material surfaces in environmental engineering: a review. *Biotechnol. Adv.*, *28*, 6, 802–816, 2010.
64. Jackson, P., Jacobsen, N.R., Baun, A., Birkedal, R., Kühnel, D., Jensen, K.A., Wallin, H., Bioaccumulation and ecotoxicity of carbon nanotubes. *Chem. Cent. J.*, 7, 1–21, 2013.
65. Sousa-Cardoso, F., Teixeira-Santos, R., Mergulhão, F.J., M., Antifouling Performance of Carbon-Based Coatings for Marine Applications: A Systematic Review. *Antibiotics (Basel)*, 11, 8, 1102, 2022.
66. Maksimova, Y., Zorina, A., Nesterova, L., Oxidative Stress Response and *E. coli* Biofilm Formation under the Effect of Pristine and Modified Carbon Nanotubes. *Microorganisms*, 11, 5, 1221, 2023.

5

Thin Film Transparent Conducting Oxides and its Anticorrosion and Surface Protection Applications: A Review

Balaprakash V.[1]*, Thangavel K.[1], Mahitha Mohan[1], Gowrisankar P.[2] and Sakthivel R.[3]

[1]Hindusthan College of Arts & Science, Coimbatore, Tamil Nadu, India
[2]Muthayammal College of Arts and Science, Rasipuram, Tamil Nadu, India
[3]PSG College of Arts & Science, Coimbatore, Tamil Nadu, India

Abstract
Wide energy bandgap (2.7–4.6 eV) nanostructured transparent conducting oxide (TCO) thin films typically exhibit high electrical conductivity (10^2–10^4 S/m), optical transmittance (in visible region), and reflectance (in IR region). Semiconducting nanostructured transparent films are important functional oxides that have been extensively used in a range of applications such as solar cells, light-emitting diodes (LED), liquid crystal displays (LCD), gas sensors, heat reflectors, protective coating, light transparent electrodes, and photocathode in photoelectron-chemical cells. For nearly all optoelectronic devices including solar cells, it is necessary to use transparent electrodes containing thin film of a TCO. Contemporary evolution in the photovoltaic and optoelectronic industries has shown a rapid increase in the demand for thin film transparent electrodes. This chapter describes the properties, necessities, and relative study on principal TCO materials and finally summarizes the anticorrosion and surface protection applications of doped zinc oxide (ZnO)–based TCO materials against stainless steel.

Keywords: Electrical conductivity, optical transmittance, transparent electrodes, optoelectronic devices, anticorrosion, surface protection

**Corresponding author*: km.balaprakash@gmail.com

Hari Murthy, Vinay Jha Pillai, Kukatlapalli Pradeep Kumar and Matthew Cowan (eds.) Novel Anti-Corrosion and Anti-Fouling Coatings and Thin Films, (109–128) © 2024 Scrivener Publishing LLC

5.1 About Transparent Conducting Oxides

TCOs belong to a variety of materials that yield extensive estimations of electrical conductivity while keeping up an extraordinary transmission in the perceptible extent of the electromagnetic range [1]. TCOs are commonly fabricated with a thin film fabrication process. The glasses are about lossless channels of light energy, yet electrical protectors. Silicon (Si) and compound semiconductors are wavelength subordinate optical resistors (making compact electrons) as well as dopant subordinate electrical conveyors. TCOs have extremely versatile direct states with both these qualities and they demonstrate both transparent and conductive properties. Their conductivity can be modified from insulating by strategies for semiconducting to driving while adjusting their transparency. Where they can be conveyed as n-type and p-type conductors similarly open a wide variety of energy-efficient opto-electrical circuitries and various innovative applications [2]. Since the affirmation of this kind of material, a tremendous measure of creative work has gone into commercializing these nanostructured TCOs. These TCO films have a far-reaching extent of employment from heat mirror window coatings which control the transmission of infrared energy in and out of buildings [3], to their usages as the transparent electrode materials in solar cells, touch screen technology, and flat panel displays. This includes a liquid crystal display (LCD), organic light-emitting diodes (OLED), plasma screen display [4], and anticorrosion and surface protection applicant against stainless steel.

5.2 Electrical Properties of TCOs

TCOs possess a wide bandgap (Eg), similar to semiconducting oxides, and exhibit electric resistivity (ρ) ranging from 10^{-4} to 10^{11} Ω/cm. Doping has a direct impact on the conductivity of TCOs, attained *via* oxygen voids or external dopants. In the absence of doping agents, these oxide materials perform as perfect actual insulators with an electrical resistivity of more than 10^{8} Ω/cm. A significant portion of TCOs are usually n-type semiconductor materials, and their electrical conductivity is purely connected with the flux of the electron per cm^{-3} (n) in the conduction band as well as its carrier mobility (μ), described as $\sigma = \mu\, n\, q$, where q denotes the elementary charge. The carrier mobility is mathematically represented by

$$\mu = \frac{q\tau}{m^{*}} \tag{5.1}$$

The mobility (μ) is determined by mean time among Moller scattering (τ) as well as the effective electron mass (m^*), but it is constrained due to the inverse relationship between n and τ. The wide energy bandgap ($Eg > 3$ eV) between the valence and conduction bands prevents thermal occupancy of the conduction band at room temperature, rendering stoichiometric crystalline TCOs excellent insulators [5]. Researchers have proposed various models to elucidate the attributes of TCOs and their electron mobility, which are interconnected with the energy bandgap [6].

Intrinsic TCO materials owe their density of conducting electrons to accidentally imparted donor centers, such as oxygen vacancies or metallic interstitials, located near the conduction band. At ambient temperature, these excess electrons are thermally ionized and inhabit the conduction band. Extrinsic doping agents play a vital part in occupying the filled conduction band, and few dopants have shown particularly noteworthy effects. For example, interstitial hydrogen (H^+) has been suggested to be responsible for carrier electrons in ZnO [7].

Regarding tin oxide (SnO_2), interstitial Sn and oxygen vacancies have been shown to influence the defect structure and revamp the precise material into an inherent n-type semiconductor [7, 8]. Due to this, released electrons, which are not recompensed by acceptor-like intrinsic deformities, prevent to create direct optical shift in the visible electromagnetic spectrum as a result of the significant energy gap between the electrochemical potential level and the energy level of the essential abandoned circumstances.

The conductivity (σ) of TCOs is inherently restricted for two primary reasons. First, the carrier density (n), as well as carrier mobility (μ), is unable to autonomously extend for real TCOs through comparatively extraordinary carrier either electrons or holes concentration. When the density of electrons is high, carrier transportation is predominantly restricted through charge carrier scattering by ionization in the lattice, which involves Coulomb contacts among electrons as well as doping elements. Increasing the doping agent concentration reduces carrier movement to a point where the conductivity does not improve further. Besides, optical transmission at the nearby infrared edge reduces due to excess doping. As the dopant agent agglomeration increases further, the resistivity of the sample touches a lower boundary beyond which it does not fall off, as the window of the optical transparency shrinks. Ionized impurity scattering has an impact on the mobility and resistivity of TCO materials (ZnO, ITO, and SnO_2) for carrier concentrations greater than 10^{20} cm^{-3} according to Bellingham *et al.* [9]. Additionally, Ellmer [10, 11] showed that the resistivity and mobility of ZnO films made in a variety of ways remained essentially constant at values of around 2×10^{-4} Ω/cm and 50 cm^2/Vs, respectively. The greatest carrier concentration for ITO films was approximately 1.5×10^{21} cm^{-3}, and comparable restrictions on conductivity and mobility were noted [12]. Numerous semiconductors exhibit this phenomenon often [5].

The presence of ionized dopant particles homogeneously dispersed in the semiconductor causes scattering, leading to a sole detrimental effect on transportability. All modern TCO elements, containing doped and un-doped binary, ternary, and quaternary combinations, face these same constraints. Ultimately, unique samples exhibited a resistivity of less than or equal to 1×10^{-4} Ω/cm.

Regarding the limitations on conductivity mentioned earlier, a high dopant concentration can cause clustering of dopant molecules [13], resulting in an increased scattering frequency. Particularly in severely doped semiconductors with filled conduction bands, this clustering effect may also result in non-parabolicity in the conduction band [14].

5.3 Optical Properties of TCO

As was previously indicated, efficient thin films of TCOs would have low resistivity and a low absorption coefficient in the UV-vis-NIR spectrum. Transmission in the near ultraviolet area is constrained by the bandgap energy (Eg), where photons with energies greater than Eg are absorbed. The reflection at the Langmuir wave frequency causes another transmission edge to appear in the NIR area. Theoretically, TCOs with extensive bandgaps should not absorb light in the UV-vis-NIR region, but in practice, there are not any perfect TCO thin films, and even if they were, reflections and interferences would reduce the transmission, preventing complete transparency over a wide range.

The refractive index (n), attenuation coefficient (k), energy bandgap (Eg), and film shape and form of nanostructured TCOs all affect their optical properties, including transmission (T), reflection (R), and absorption (A). Shape and form factors include things like film thickness, coating area thickness homogeneity, and film surface roughness. Geometry is an external element, whereas T, R, and A are intrinsic attributes determined by the chemical composition of the material and the solid's structure. Due to the Moss-Burstein effect, where the energy bandgap (Eg) value increases at higher carrier density, there is a negative correlation between the carrier thickness and the location of the infrared (IR) absorption edge. However, there is a progressive correlation between the carrier thickness and the UV absorption edge. As a result, the conductivities and transmission bounds of the TCO are connected.

The varied optical boundaries of the TCO film itself as well as the optical properties of the substrate influence the thickness of the UV-vis transmission window of a TCO film made on a very transparent substrate. Common substrate indices of refraction (n_{sub}) for fused silica or quartz glass are around 1.46, and they are 1.61 for other types of glass. Since the extinction constant of the substrate (k_{sub}) is approximately 10^{-7}, any visible light absorption would predominantly take place in the film, where $k_{film} > k_{sub}$. If the wavelength or thickness is altered, multiple interference fringes may occur for films that have a density of more than 100 nm, leading to a range of T values. Maximum transmission (T_{max}) equals substrate transmission as k_{film} approaches zero, giving T_{max} values of around 95% and 92% for films made on silica and glass, respectively [15–19]. The minimum sample transparency (T_{min}) of an air circumstance is stated as follows:

$$T_{min} = \frac{4n^2 n_{sub}}{(1+n^2).(n^2+n_{sub}^2)} \tag{5.2}$$

Utmost every TCO has estimations of 'n' in the visible region that vary from 1.8–2.9, T_{min} will be in the range of 0.8–0.52. T_{min} remains estimated through the equation:

$$T_{min} = 0.051n^2 - 0.545n + 1.654 \tag{5.3}$$

T_{min} increases concerning wavelength as n in the visible area decreases proportionally to wavelength, but it will not reach over 0.8. Exactly once the film attenuation coefficient is not insignificant and it impacts the transparency, $T_{max} < T_{sub}$ also T_{min} furthermore reduces. As a result of minimizing the film thickness of TCO, T is better improved, and also the sheet resistance drops. The amount of flux consumed in the nanostructured film (A), which combines its optical and electrical properties, is determined by the relation shown below.

$$A \approx 1 - e^{-\frac{\alpha}{\sigma R}} \tag{5.4}$$

Figure 5.1 illustrates the absorbed percentage of power for TCOs at a wavelength of around 400 nm, with k ~ 0.02, as a component of conductivity for three illustrative R_S values. When a low R_S value is desired for a specified σ, it necessitates the use of thick films. Similarly, lower conductivity requires significantly thicker films, leading to a greater waste of irradiative power. The relationship between the thickness of the film and conductivity for the three illustrative R_S values is depicted in Figure 5.2.

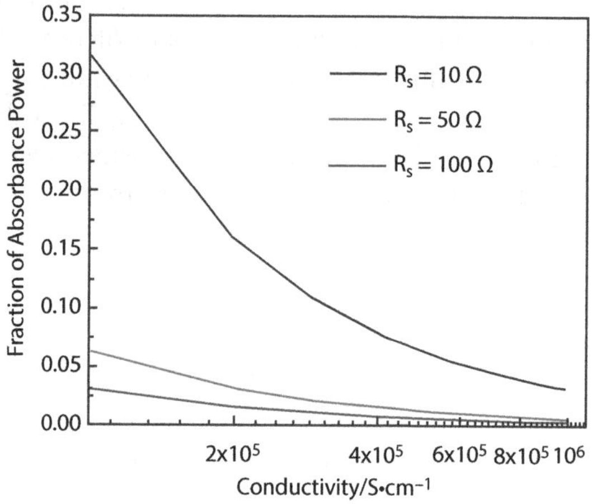

Figure 5.1 Fraction of absorbed power as a function of TCO conductivity.

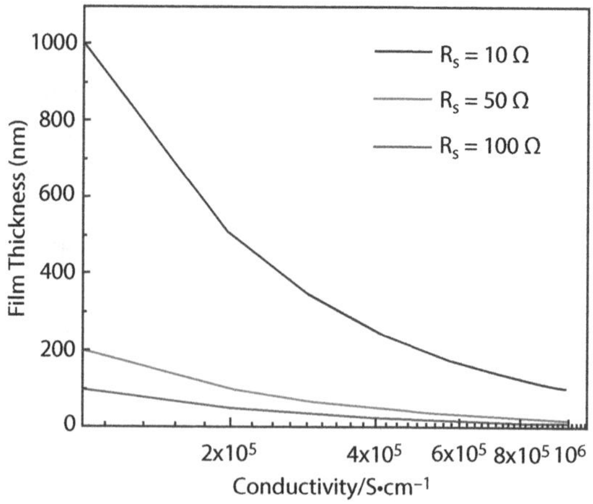

Figure 5.2 TCO Thin film thickness as a function of film conductivity.

By utilizing films with higher conductivity, implementations demanding the most reduced sheet resistance (R_S) will bring about denser films and higher absorbed fractions. Currently, only high-quality indium tin oxide (ITO) meets the requirement of having the absorbed fraction of power below 10% also $R_S = 10\ \Omega/cm$. However, for films with lower extinction coefficients (k), worse electrical conduction can be used, such as when $k = 0.002$ as a replacement for 0.02. As a result, the absorbed power (A) is reduced by approximately eight times, making denser films possible. The radiation transition retention is determined by the combination of film thickness, conductivity, and extinction coefficient.

Reflection and interference should also be taken into account in the overall Transmission (T). This component depends on the film density, the refractive indices of the thin film, and the substrate. For T and R, Cisneros [15] offered a broad phrase.

5.4 Need of TCO for Solar Cells and Optoelectronic Devices

ITO, aluminum-doped zinc oxide (AZO), zinc-doped indium oxide (IZO), and nickel-doped zinc oxide (NZO) are just a few examples of the several varieties of TCOs that are available. Due to their exceptional conductivity and transparency in the visible range of the electromagnetic spectrum, these doped metal oxides have found significant use in optoelectronic devices including solar cells (SC) and LCDs. The optical and electrical conductivity and nanostructural and topographical properties of TCO films are crucial to control and achieve the desired characteristics of these devices. In nanostructured solar cells, TCO layers act as a significant part, impacting more than 40% of SCs performance through their electrical and optical properties, and their manufacturing costs constitute approximately 30–35% of the overall device cost [20].

For photovoltaic (PV) applications, incident light must penetrate through the TCO layer with no optical misfortunes. Additionally, the photo-generated transporters in the active layers of the PV cell, such as hydrogenated amorphous silicon (a-Si:H), must be accumulated efficiently from the TCO layer. Furthermore, due to the low thickness of thin film solar cells (TFSC), the TCO layer must scatter incident light [21] to increase the optical path length within the active layers, facilitating "light trapping" [21]. A typical example of TCO coated on a glass substrate is depicted in Figure 5.3.

Figure 5.3 TCO on a glass substrate.

5.5 Requirements of TCO

The complete list of all TCO requirements for solar cells and optoelectronic devices is presented below [22]:

(i) High overall transmittance in the visible (Vis) region of the solar spectrum (>80–95%).
(ii) Huge impart transparency in the visible range (>30–35%).
(iii) Large surface roughness (root mean square [rms] > 35 nm) to scatter the occurrence light and enable light trapping.
(iv) Strong adhesion to the glass medium (>100 kg/cm^2).
(v) Noble film consistency as well as little porosity.
(vi) Huge crystalline grain dimension (>20 nm).
(vii) Least volume or specific electrical resistivity (<10^{-3} Ω/cm).

5.6 Commonly Used TCO Materials

TCOs are thin films of binary or ternary mixtures that are optically transparent and include a few metallic elements. While the development of novel TCO materials is mostly driven by the demands of specific applications, low resistivity and low optical retention are invariably important criteria. In handling the work of creating sophisticated TCO that may fulfill the requirements, there are essentially two methods. Doping known binary compounds is the key strategy. TCO with different components can surge the density of conducting electrons. Binary mixtures, such as CdO, ZnO, In_2O_3, and SnO_2, can be used to generate a thin layer with the appropriate resistivity, transparency, and atomic concentration for transparent

Table 5.1 List of TCO binary compounds and impurity dopants.

Si. no.	TCO binary compounds	Dopants
1.	ZnO	Al, Ga, B, H, In, F, Si, Ge, Ti, Zr, Mg, Ni, As, Y, F, V, and Sc
2.	In_2O_3	Sn, Ge, Zr, Hf, Nb, Mo, F, Ti, Ta, W, Te, Mg, W, and Zr
3.	CdO	In and Sn
4.	SnO_2	Sb, F, As, Nb, and Ta

electrode applications [23]. Because of their sensitivity to high temperatures, impurity doping is required to produce thin films with improved properties [24]. These complementary mixtures can be easily doped with an alternative impurity agent influence, and the synthetic organization in the nanostructured thin film can be successfully managed for greater optical transparency. Table 5.1 lists the matched compounds and doping agents that can be used to create nanostructured TCOs.

5.7 Application of TCOs

TCOs have been broadly utilized as a part of the optoelectronic industry as a fundamental segment in solar cells, smart windows, touch panels, flat panel displays (FPD), LCDs, gas sensors, and OLED [25–33]. Usually for high-accuracy applications, TCO has both low electrical resistivity ($\rho<10^{-4}$ Ω/cm) and high optical transmittance (>80%) for the visible region [34–36]. Figure 5.4 shows some typical applications of TCO.

5.8 Anti-Corrosion and Surface Protection Application of Doped Zinc Oxide–Based TCO Materials

The origin of corrosion is due to moisture and climate around the subject, continuous accumulation of dust particles in various parts of automobiles, chloride compounds in coastal areas, and salts such as sodium chloride (NaCl), calcium chloride ($CaCl_2$), and magnesium chloride (MgCl) [37]. Metals and alloy surfaces fail when they are exposed to a corrosive environment. Providing thin film coating over the metals and alloy surface detains and disallows rapid corrosion as well as protecting the surface from

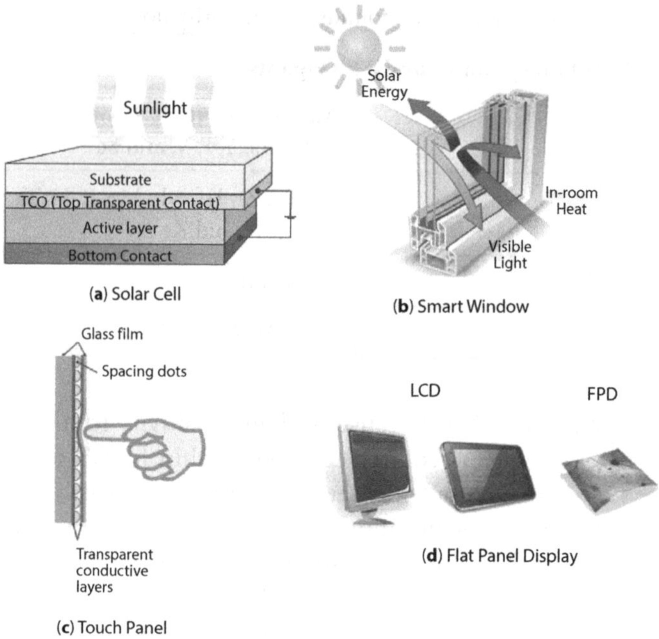

Figure 5.4 Some typical applications of TCO.

corrosion [38–40]. Corrosion in metals and alloys creates significant disputes as corrosion exhibits damage to various metal components in industries [41, 42]. Thin film coating is an art that renders complete protection to material surface from stress and corrosive circumstances in numerous fields including marine, aerospace, and automotive industries [43]. Conducting polymers and transition metal oxides is the optimistic applicant because their heavy bonding nanostructure withstands different varieties of applications [44–48]. There are various metals doped with ZnO for attaining desired properties; among the all nickel-doped ZnO (NZO) thin film coating exhibits superior anti-corrosion and surface protection properties.

5.9 NZO Coating Over Stainless Steel

5.9.1 Materials Used for Synthesis

The materials used in this study are the following:

(i) Zinc acetate dihydrate [$C_4H_6O_4Zn.2H_2O$] of Mw = 219.5 (analytical reagent).

(ii) Nickel acetate tetrahydrate [$Ni(OCOCH_3)_2.4H_2O$] of M_W = 248.84 (guaranteed reagent).
(iii) Monoethanolamine (MEA) [C_2H_7NO] of M_W = 105.14 (analytical reagent).
(iv) Ethanol [C_2H_6O] of M_W = 46.07 (analytical reagent).

5.9.2 Synthesis of Nickel-Doped ZnO Nanostructured Thin Films

In the present investigation, the sol was produced by utilizing analytical regent zinc acetate dihydrate ($C_4H_6O_4Zn.2H_2O$), ethanol (C_2H_6O), and MEA (C_2H_7NO). Initially, zinc acetate dihydrate (ZnAc) was liquefied in ethanol. Then, MEA was supplemented dropwise while stirring and the mixture was heated under reflux at 60°C for half an hour. The concentration of ZnAc in the solution was retained at 0.5 moles for each liter, and the dissolution of zinc acetate dihydrate in MEA was achieved at a mole-to-mole ratio of 1 (MEA to Zn = 1). To introduce a doping agent, 0.5 mole of nickel acetate tetrahydrate [$Ni(OCOCH_3)_2.4H_2O$] was added to the base solution. The mole-to-mole ratio of the dopant in the synthesized solution (Ni/Zn) was varied from 0 to 1 at.%.

After stirring the resulting solution at 60°C for 3 hours, a clear colorless, and homogeneous solution was obtained, which acts as the coating solution after cooling to ambient room temperature. On a stainless steel substrate, a dip coating method was carried out. After each deposition, the coated films were heated to 200°C for 10 min in an air environment to dry them. Once the wanted density was achieved by reiterating the dip coating procedure, the anticipated films were subjected to post-annealing in an air atmosphere at 450°C for 4 hours to fine-tune their characteristics. The preparation of nanostructured NZO thin films is illustrated in Figure 5.5.

5.9.3 Structural Properties of NZO Thin Films Coated Over Stainless Steel

The structural behavior of the anticipated NZO nanostructured thin films was inspected using an x-ray diffractometer (XRD). The prepared films displayed a prominent c-axis crystalline orientation, and sharp peaks were observed in the (101) plane. All the prepared films exhibited a polycrystalline nature, with a consistent grain size observed throughout the sample. The 1 at.% NZO nanostructured transparent thin films annealed at 450°C exhibited the highest peak intensities, which were detected between 30° to

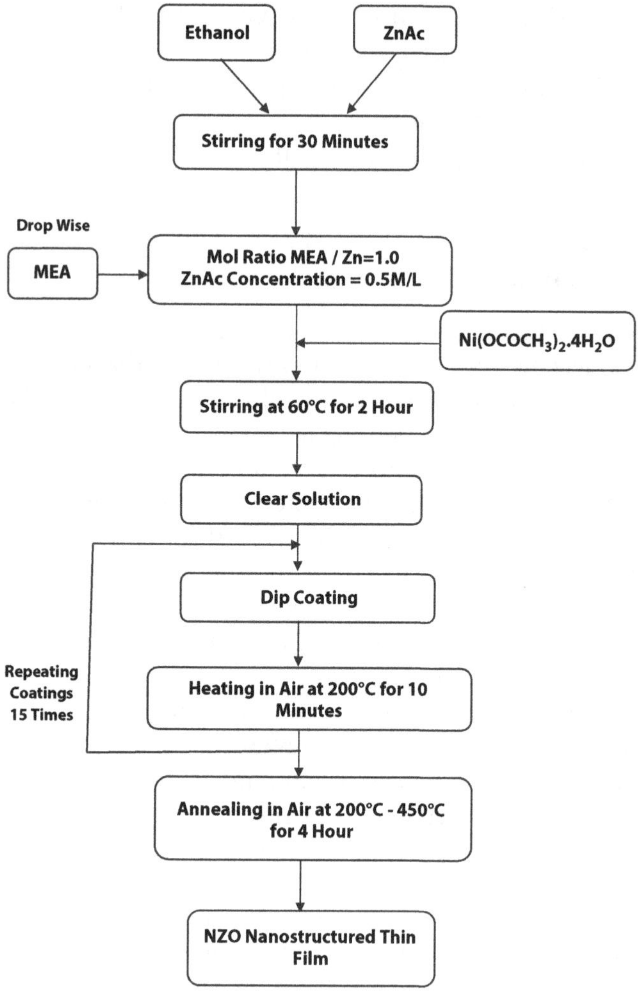

Figure 5.5 Flow chart represents the preparation process of NZO thin films.

36° in the 2θ position. The resulting typical crystalline size was found to be 16 nm, with a dominant peak at 36.07°.

The extremely superior intensity and sharp peaks showed that the anticipated nanostructured Ni-doped ZnO thin transparent films were well-crystalline in nature. The incorporation of Ni into ZnO resulted in a structural transformation of the NZO thin film, potentially generating various stresses arising from the differences in ion size between Ni and Zn. Additionally, higher annealing temperatures further improved the crystallinity of the sample, as evidenced by the increased intensity of the

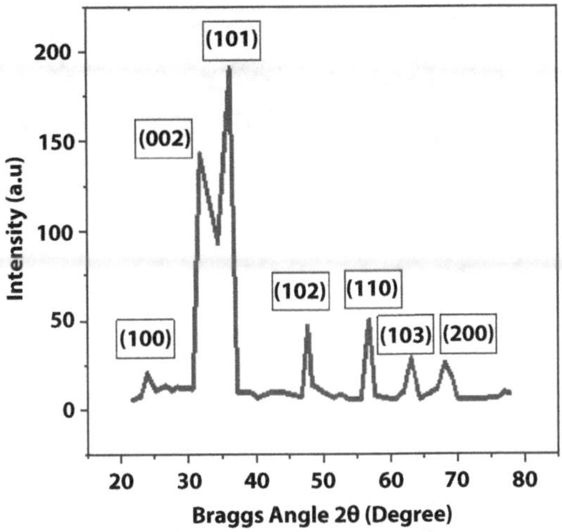

Figure 5.6 XRD image of 1 at.% NZO thin films annealed at 450°C.

(101) peak and reduced intensity of other peaks. This may be ascribed to the discrimination of Ni in the grain limits for predominant doping concentrations.

Notably, the 1 at.% Ni concentration exhibits the high-intensity XRD diffraction peak in the (101) direction, while the peak intensity starts to decrease with concentrations of dopant exceeding 1 at.%, indicating that increased doping concentrations adversely affected the crystallinity of the films due to stress induced in order to the variation in ionic molecule size among Zn and Ni and the isolation of Ni in grain limits. The XRD representation of thin films of 1 at.% NZO that were annealed at 450°C is shown in Figure 5.6.

5.9.4 Morphological Properties of NZO Thin Films Coated Over Stainless Steel

The surface topography of the synthesized NZO nanostructured thin films is predicted by employing scanning electron microscopy (SEM). Particle size further decreases if the annealing temperature is further increased. Figure 5.7 displays a SEM interpretation of 1 at.% NZO nanostructured thin films that have undergone 450°C heat treatment. Uniform-shaped spherical grains are observed in the entire sample region. It is furthermore perceived that the amount of the dopant increases the intermolecular space decreases and the molecules become closer thus the conductivity

Figure 5.7 SEM image of 1 at.% NZO nanostructured transparent thin films heat-treated at 450°C.

increases and it also favors the ferromagnetic property of ZnO. It exhibits a hexagonal structure. The grain size increases as the doping concentration increases but all are at the nanoscale. The 1 at.% NZO nanostructured thin films exhibit uniform grain dimension over the entire surface and particles are bound closely with each other. For the thin films with 1 at.% NZO nanostructures that have been heat-treated at 450°C, the minimum grain size of 39 nm is noted.

5.9.5 Optical Properties of NZO Thin Films Coated Over Stainless Steel

The absorbance spectrum of 1 at.% NZO samples were taken in the span of 200–800 nm wavelength for analyses of the application of the NZO thin film. All the measurements are measured at room temperature. The graph is obtained for plotting nanometers on the x-axis and absorbance on the y-axis. As transparent conductors in thin film PV cells, nickel-doped nanostructured ZnO thin films are widely employed. The prepared films provided have lessthan 10% absorbance. Comparatively, 1% nickel-doped ZnO thin film shows low absorbance. Based on the observation of UV results, prepared thin films are suitable for optoelectronic devices. Figure 5.8 shows the UV absorbance spectrum of 1 at.% nanostructured NZO thin films that have undergone a 450°C heat treatment. The optical studies

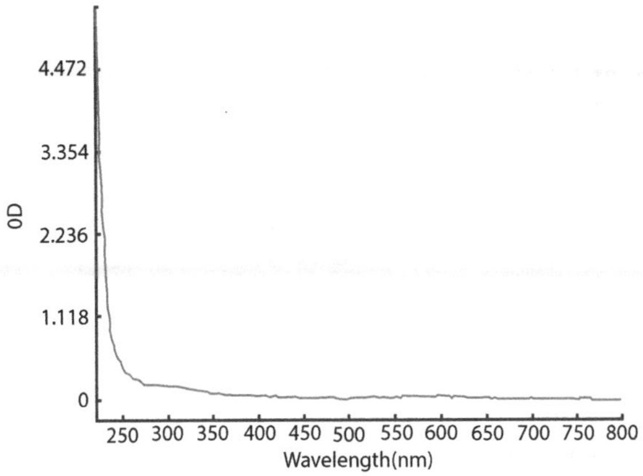

Figure 5.8 UV absorbance spectrum of 1 at.% NZO nanostructured thin films annealed at 450°C.

show that the maximum range of absorbance occurs at the visible region. The absorbance increases as the dopant concentration increases.

5.9.6 Anticorrosion and Surface Protection Applications of NZO Thin Films

The anticipated nanostructural and topographical characteristics of the 1 at.% NZO nanostructured thin films guarantee that all the preferred samples are least affected by anticorrosion agents and it yields an extreme surface protection nature for the stainless steel. The resistance to corrosion of the anticipated stainless steel samples can be estimated *via* acquainted polarization plots. Figure 5.9 shows anodic and cathodic polarization curves for the sample exposed to 4 at.% sodium chloride(NaCl) solution after a day exposure time period. Prominently, the coating solution can deliver parting of the stainless steel base with a corrosive environment. As illustrated in Figure 5.9, the anodic polarization graphs are examined through passive regions instead of sample, presenting that the passive layers of the coating had noticeably formulated over the surface region of steel once they were exposed to the seawater or sodium chloride environment. Besides, a momentous swing remains identified in corrosion potential to a desired positive path which divulges that the dissolution of anodic metallic iron was commendably impeded by the coating materials. The measured corrosive potential of the sample is 1.23 V and the hardness value is 234.04.

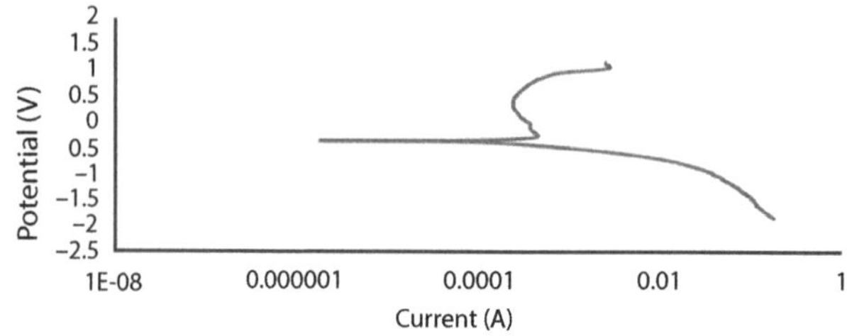

Figure 5.9 Anodic and cathodic polarization curves of 1 at.% NZO nanostructured thin films post heat treated at 450°C.

5.10 Conclusion

In the present study, various properties of TCOs are summarized, and the preparation of NZO thin films over stainless steel is discussed. The structural behavior of the prepared samples exhibits a polycrystalline nature and identical grain size is observed throughout the sample. Morphological analyses show that the sample grains are all of the same size, with the 1 at.% NZO nanostructured thin films that were heat treated at 450°C having the smallest grain size of 39 nm. Optical studies of the sample confirm the synthesized film has the least optical absorbance. The corrosive potential of the prepared sample is 1.23 V and the hardness value is 234.04. These properties make the prepared films suitable for optoelectronic, anticorrosion, and surface protection applications.

References

1. Edwards, P.P., Porch, A., Jones, M.O., Morgan, D.V., Perks, R.M., Basic materials physics of transparent conducting oxides. *Dalton Trans*, https://doi.org/10.1039/B408864F.
2. Stadler, A., Transparent Conducting Oxides – An Up - To - Date Overview. *Materials*, 2012, https://doi.org/10.3390/ma5040661.
3. Gelin, K., Roos, A., Geotti-Bianchini, F., Nijnatten, P.V., Thermal emissivity of coated glazing – Simulation versus measurements. *Opt. Mater.*, 2005, https://doi.org/10.1016/j.optmat.2004.09.007.
4. Ginley, D.S. and Bright, C., Transparent Conducting Oxides. *MRS Bull.*, 2000, https://doi.org/10.1557/mrs2000.256.
5. Ellmer, K., Klein, A., Rech, B. (Eds.), *Transparent Conductive Zinc Oxide. Springer Series in Materials Science*, Springer, Heidelberg, Berlin, 2008, http://dx.doi.org/10.1007/978-3-540-73612-7.

6. Mizoguchi, H. and Woodward, P.M., Electronic Structure Studies of Main Group Oxides Possessing Edge-Sharing Octahedra: Implication for the Design of Transparent Conducting Oxides. *Chem. Mater.*, 2004, https://doi.org/10.1021/cm049249w.
7. Van de Walle, C.G., Hydrogen as a Cause of Doping in Zinc Oxide. *Phys. Rev. Lett.*, 2000, https://doi.org/10.1103/PhysRevLett.85.1012.
8. Kilic, C. and Zunger, A., Origins of Coexistence of Conductivity and Transparency in SnO_2. *Phys. Rev. Lett.*, 2002, https://doi.org/10.1103/PhysRevLett.88.095501.
9. Bellingham, J.R., Phillips, W.A., Adkins, C.J., Electrical and optical properties of amorphous indium oxide. *J. Phys. Condens. Matter*, 1990, https://doi.org/10.1088/0953-8984/2/28/011.
10. Ellmer, K., Magnetron sputtering of transparent conductive zinc oxide: relation between the sputtering parameters and the electronic properties. *J. Phys. D: Appl. Phys.*, 2000, https://doi.org/10.1088/0022-3727/33/4/201.
11. Ellmer, K., Resistivity of polycrystalline zinc oxide films: current status and physical limit. *J. Phys. D: Appl. Phys.*, 2001, https://doi.org/10.1088/0022-3727/34/21/301.
12. Frank, G. and Kostlin, H., Electrical properties and defect model of tin- doped indium oxide layers. *Appl. Phys.*, 1982, https://doi.org/10.1007/BF00619080.
13. Ebert, P., Zhang, T., Kluge, F., Simon, M., Zhang, Z., Urban, K., Importance of Many-Body Effects in the Clustering of Charged Zn Dopant Atoms in GaAs. *Phys. Rev. Lett.*, 1999, https://doi.org/10.1103/PhysRevLett.83.757.
14. Pisarkiewicz, T., Zakrzewska, K., Leja, E., Scattering of Charge Carriers in Transparent and Conducting Thin Oxide Films with a Non-Parabolic Conduction Band. *Thin Solid Films*, 1989, https://doi.org/10.1016/0040-6090(89)90892-4.
15. Cisneros, J., II, Optical characterization of dielectric and semiconductor thin films by use of transmission data. *Appl. Opt.*, 1998, https://doi.org/10.1364/AO.37.005262.
16. Casteleiro, C., Gomes, H.L., Stallinga, P., Bentes, L., Ayouchi, R., Schwarz, R., Study of trap states in zinc oxide (ZnO) thin films for electronic applications. *J. Non-Cryst. Solids*, 2008, https://doi.org/10.1016/j.jnoncrysol.2007.10.059.
17. Romero, R., Lopez, M.C., Leinen, D., Martin, F., Ramos-Barrado, J.R., Electrical Properties of the n-ZnO/c-Si heterojunction prepared by chemical spray pyrolysis. *Mater. Sci. Eng. B.*, 2004, https://doi.org/10.1016/j.mseb.2004.03.010.
18. Cannella, G., Principato, F., Foti, M., Garozzo, C., Lombardo, S., Capacitance study of thin film SnO_2:F/p-type a-Si:H heterojunctions. *Energy Proc.*, 2011, https://doi.org/10.1016/j.egypro.2011.01.009.
19. Minami, T., Transparent conducting oxide semiconductors for transparent electrodes. *Semicond. Sci. Technol.*, 2005, https://doi.org/10.1088/0268-1242/20/4/004.
20. Luque, A. and Hegedus, S., *Handbook of Photovoltaic Science and Engineering*, John Wiley & Sons Ltd, England, 2003.

21. Dumcenco, D.O., Chen, Y.M., Huang, Y.S., Firszt, F., Legowski, S., Meczynska, H., Marasek, A., Tiong, K.K., Contactless electroreflectance and photoluminescence characterization of $Zn_{0.68}Be_{0.06}Mg_{0.26}$ Se crystalline alloys. *J. Alloys Compd.*, 2010, https://doi.org/10.1016/j.jallcom.2009.10.234.
22. Rowell, M.W. and McGehee, M.D., Transparent electrode requirements for thin film solar cell modules. *Energy Environ. Sci.*, 2011, https://doi.org/10.1039/c0ee00373e.
23. Jayaraj, M.K., Antony, A., Ramachandran., M., Transparent conducting zinc oxide tin films prepared by off-axis rf magnetron sputtering. *Bull. Mater. Sci.*, 2002, https://doi.org/10.1007/BF02711158.
24. Minami, T., Nanto, H., Takata, S., Optical Properties of Aluminum Doped Zinc Oxide Thin Films Prepared by RF Magnetron Sputtering. *Jpn. J. Appl. Phys.*, 1985, https://doi.org/10.1143/JJAP.24.L605.
25. Znaidi, L., Touam, T., Vrel, D., Souded, N., Yahia, S.B., Brinza, O., Fischer, A., Boudrioua, A., AZO Thin Films by Sol-Gel Process for Integrated Optics. *Coatings*, 2013, https://doi.org/10.3390/coatings3030126.
26. Rakesh, A., Afre., Sharma, N., Sharon, M., Sharon, M., Transparent conducting oxide films for various applications: A review. *Rev. Adv. Mater. Sci.*, 2018, https://doi.org/10.1515/rams-2018-0006.
27. Khan, F., Singh, V.S.N., Husain, M., Singh, P.K., Sol-gel derived hydrogen annealed ZnO: Al films for silicon solar cell application. *Sol. Energy Mater. Sol. Cells*, 2012, https://doi.org/10.1016/j.solmat.2011.04.024.
28. Suryawanshi, M.P., Agawane, G.L., Bhosale, S.M., Shin, S.W., Patil, P.S., Kim, J.H., Moholkar, A.V., CZTS based thin film solar cells: a status review. *Mater. Technol.*, 2013, https://doi.org/10.1179/1753555712Y.0000000038.
29. Caglar, M. and Yakuphanoglu, F., Structural and optical properties of copper doped ZnO films derived by sol-gel. *Appl. Surf. Sci.*, 2012, https://doi.org/10.1016/j.apsusc.2011.11.033.
30. Shelke, V., Sonawane, B.K., Bhole, M.P., Patil, D.S., Electrical and optical properties of transparent conducting tin doped ZnO thin films. *J. Mater. Sci. Mater. Electron.*, 2012, https://doi.org/10.1007/s10854-011-0462-2.
31. Wie, S.M., Hong, C.H., Oh, S.K., Cheong, W.S., Yoon, Y.J., Kwak, J.S., Fully crystallized ultrathin ITO films deposited by sputtering with *in-situ* electron beam irradiation for touch-sensitive screens. *Ceram. Int.*, 2014, https://doi.org/10.1016/j.ceramint.2014.03.145.
32. Slocombe, D., Porch, A., Pepper, M., Edwards, P.P., The Mott transition and optimal performance of transparent conducting oxides in thin-film solar cells. *Energy Environ. Sci.*, 2012, https://doi.org/10.1039/c1ee02585f.
33. Chen, M., Pei, Z.L., Sun, C., Gong, J., Huang, R.F., Wen, L.S., ZAO: an attractive potential substitute for ITO in flat display panels. *J. Mater. Sci. eng. B.*, 2001, https://doi.org/10.1016/S0921-5107(01)00584-0.
34. Calnan, S. and Tiwari, A.N., High mobility transparent conducting oxides for thin film solar cells. *Thin Solid Films*, 2010, https://doi.org/10.1016/j.tsf.2009.09.044.

35. Guillen, C. and Herrero, J., TCO/metal/TCO structures for energy and flexible electronics. *Thin Solid Films*, 2011, https://doi.org/10.1016/j.tsf.2011.06.091.
36. Cebulla, R., Wendt, R., Ellmer, K., Al-doped zinc oxide films deposited by simultaneous rf and dc excitation of a magnetron plasma: Relationships between plasma parameters and structural and electrical film properties. *J. Appl. Phys.*, 1998, https://doi.org/10.1063/1.366798.
37. Mahajan, V.N., Sable, M.J., Suhas, R.P., Metal oxides and its blended derivative's coating for anti-corrosion application. *Indian J. Eng. Mater. Sci.*, 2022, http://nopr.niscpr.res.in/handle/123456789/59475.
38. Shang, Y.L., Huo, L., Ling, J.Y., Liaoa, F.H., Ran, J.L., Xiu, L.M., Electrorheological property of M-doped (M = Na, Zr) nano-sized TiO_2 particle materials, influence of surface composition and microstructure. *Colloids. Surf. A: Physicochem Eng. Asp.*, 2008, https://doi.org/10.1016/j.colsurfa.2008.04.048.
39. Krauss, G., Advanced surface modification of steels. *J. Heat Treat.*, 1992, https://doi.org/10.1007/BF02833144.
40. An, Y., II, Du, H.Y., Wei, Y.H., Yang, N., Hou, L.F., Lin, W.M., Interfacial structure and mechanical properties of surface iron–nickel alloying layer in pure iron fabricated by surface mechanical attrition alloy treatment. *Mater. Des.*, 2013, https://doi.org/10.1016/j.matdes.2012.11.005.
41. Ajao, J.A., Phase transitions in some nickel-rich nickel–boron–titanium hard alloys. *J. Alloys Compod.*, 2009, https://doi.org/10.1016/j.jallcom.2009.12.091.
42. Shakoor, R.A., Kahraman, R., Waware, U.S., Wang, Y., Gao, W., Synthesis and properties of electrodeposited Ni-B-Zn ternary alloy coatings. *Int. J. Electrochem. Sci.*, 2014, https://doi.org/10.1016/S1452-3981(23)08186-5.
43. Fotovvati, B., Namdari, N., Dehghanghadikolae, A., On coating techniques for surface protection. *J.manuf. Mater. Process.*, 2019, https://doi.org/10.3390/jmmp3010028.
44. Anandhavelu, S., Dhanasekaran, V., Sethuraman, V., Park, H.J., Chitin and chitosan based hybrid nanocomposites for super capacitor applications. *J. Nanosci. Nanotechnol.*, 2017, https://doi.org/10.1166/jnn.2017.12721.
45. Lin, S., Shih, H., Mansfeld, F., Corrosion protection of aluminum alloys and metal matrix composites by polymer coatings. *Corros. Sci.*, 1992, https://doi.org/10.1016/0010-938X(92)90176-4.
46. Singh, B.P., Jena, B.K., Bhattacharjee, S., Besra, L., Development of oxidation and corrosion resistance hydrophobic graphene oxide-polymer composite coating on copper. *Surf. Coat. Technol.*, 2013, https://doi.org/10.1016/j.surfcoat.2013.06.004.
47. Miao, J., Zhang, L.-C., Lin, H., A novel kind of thin film composite nanofiltration membrane with sulfated chitosan as the active layer material. *Chem. Eng. Sci.*, 2013, https://doi.org/10.1016/j.ces.2012.10.015.
48. Miao, J., Lin, H., Wang, W., Zhang, L.-C., Amphoteric composite membranes for nanofiltration prepared from sulfated chitosan crosslinked with hexamethylene diisocyanate. *J. Chem. Eng.*, 2013, https://doi.org/10.1016/j.cej.2013.08.085.

6

Integrated Anticorrosion and Antifouling Coatings

Ankita Kumari[1], Nirmala Kumari Jangid[1*], Sudesh Kumar[1], Rekha Sharma[1] and Navjeet Kaur[2]

[1]Department of Chemistry, Banasthali Vidyapith, Banasthali, Rajasthan, India
[2]Department of Chemistry & Division of Research and Development, Lovely Professional University, Phagwara, Punjab, India

Abstract

The issue of surface corrosion and biofouling of underwater apparatus has grown significantly with the marine industry's quick development. Current demands cannot be met by conventional anticorrosion or antifouling coatings by a single function. Integrated antifouling or anticorrosion coatings (IAACs) now face new opportunities and challenges as a result of recent advancements. To better inform readers on how their detrimental effects can be reduced, the mechanisms of corrosion and biofouling are explained in this chapter. The bulk of IAACs are made by combining anticorrosion or antifouling units into a single coating, despite the fact that many IAACs contravene the abovementioned design concept because they already have these properties. This chapter highlights recently created IAACs, such as coatings made of polyanilines, coatings made of polymer brushes and polydopamine, amphiphilic polymers, copper and amorphous carbon, coatings made of nanocomposites, endospore-loaded sol-gel, and coatings that are inspired by biological systems. The benefits and drawbacks of these recently developed IAACs will be discussed because they may not yet be fully developed in terms of commercialization and practical applications. The discussion of IAACs' potential future development concludes. We anticipate that reading this chapter will aid readers in developing an understanding of the design principles of IAACs.

Keywords: Anticorrosion coatings, marine anticorrosion, antifouling marine, antifouling coatings

**Corresponding author*: nimalajangid.111@gmail.com

Hari Murthy, Vinay Jha Pillai, Kukatlapalli Pradeep Kumar and Matthew Cowan (eds.) Novel Anti-Corrosion and Anti-Fouling Coatings and Thin Films, (129–154) © 2024 Scrivener Publishing LLC

6.1 Introduction

Inappropriate organisms, such as bacteria, algae, and barnacles, attaching to wet surfaces is known as marine biofouling [1–4]. A micro-thick biofilm that shaped on surfaces was said to increase surface drag by up to 70% [5], and heavy calcareous fouling was said to decrease it by up to 86% at cruising speed [6]. More fuel needs to be consumed in order to maintain the set speed. However, excessive fuel use results in greenhouse gas emissions (like CO_2 and SO_2) that could increase the risk of global warming [7]. Ships can carry fouling organisms with them as they travel the world, which could lead to bio invasion when they reach a sea where there are no natural enemies [8]. For instance, in the 1980s, the comb jelly *Mnemiopsis leidyi* made its way to the Black Sea. The local anchovies were in danger of extinction because the comb jelly was such a formidable rival [9–14]. Additionally, biofouling can speed up surface corrosion. Fouling organisms induce localized changes in ions, oxygen, and pH, which damage coatings, enhance liquid conductivity, and promote chemical and electrochemical processes [15]. The term "biocorrosion" or "microbially influenced corrosion" (MIC) refers to this type of corrosion [16, 17]. MIC is estimated to cause 20% or less of aqueous system corrosion [18–21]. Thus, reducing marine biofouling and corrosion is essential for global maritime sector growth. Recently developed surface coating technologies can protect subsea surfaces from corrosion and biofouling at low cost [22, 23]. The antifouling performance of antifouling coatings is the main focus; the corrosion performance is disregarded [24, 25]. Similar to how anticorrosion coatings are typically not designed with antifouling performance in mind [26, 27]. The complexity of this situation makes it difficult to optimize the bonding between various coatings, which reduces coating effectiveness and raises maintenance costs. Therefore, the future of coatings that combine antifouling and anticorrosion properties is bright. Recent reviews on antifouling and anticorrosion coatings have been widely published. There have recently been reports of various IAACs using this design principle. Due to their built-in antifouling and anticorrosion features, many IAACs also violate the aforementioned design principle [28, 29]. The development of IAACs highlights the synthetic techniques, operating mechanisms, and current problems. Finally, it is discussed where IAACs stand right now and what the future holds [30, 31].

The anticorrosive coating and the antifouling coating both underwent modest formulation changes in the current work. Aluminum powder, which is renowned for enhancing the impermeability to moisture and its capacity to exclude UV radiation, was added to the pigment composition of the anticorrosive coating. The lamellar pigment particles might organize themselves horizontally and concentrate in the paint film's topmost layer. This "leafing" activity, which may be partially explained by the aluminum stearate layer on the particles, significantly lengthens the path that moisture must take through the film and might prevent skin blistering. Tritolyl phosphate, a hydrolyzable plasticizer, and polyvinyl methyl ether, a nonhydrolyzable but swellable plasticizer, respectively, were used to replace the nonhydrolyzable chlorinated paraffin plasticizer in the antifouling coating. The adjustments produced outstanding protection upgrades that, for the first time, met the standards of the Egyptian Standard Specifications.

New opportunities for protective polymer nanocomposites (PNC) coatings with antifouling anticorrosion, and self-healing properties have emerged as a result of recent advances in nanotechnology. The term "nanotechnology" refers to the study, production, and use of nanoparticle architectures, tubular structures, sheets, or plates with diameters of less than 100 nanometers (nm) in at least one dimension. Additionally covered are fresh applications and emerging market structures.

6.2 Mechanisms of Corrosion and Biofouling

6.2.1 Biofouling Mechanisms

Biofouling may affect any marine subsea surface; however, the severity depends on its properties. The general consensus is that biofouling is a three-stage process that is carried out in steps [32–34]. Since there are many oceans across the world, fouling organism species and adhesion behavior vary temporally and spatially [35], hence the biofouling process may not follow the steps shown [36]. Biofilm production is a key phase in the biofouling process, hence antifouling coatings prevent it [37–39]. TBT's superior antifouling properties were first noted by Der kerk's team in the 1950s, and its early applications took on a "free-associated form" in coatings. TBT-based coatings rose to the top of the antifouling coatings hierarchy over the ensuing decades (1950s–1980s). These coatings use the TBT moiety's gradual release to effectively control fouling. TBT's teratogenicity and lethality, however, pose a risk to the survival of organisms that are not its intended targets [40]. In light of this, the next iteration of antifouling coatings must be both efficient and environmentally friendly. In light of this, the next iteration of antifouling coatings must be both efficient and environmentally friendly.

6.2.2 Corrosion Mechanisms

Complex variables cause marine corrosion. Steel is a substance that is frequently found in marine environments, and the following reaction model can explain how steel corrodes in seawater [41].

Microorganisms, media, and metals cause MIC (Figure 6.1c). The two types of biocorrosion—MIC with oxygen and MIC without oxygen—can be categorized in accordance with the various oxygen conditions of MIC [42].

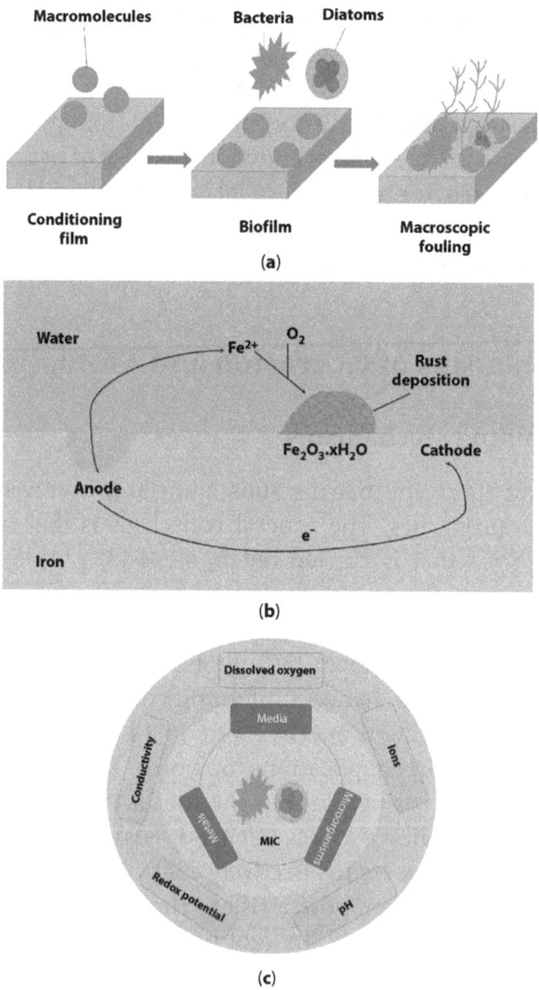

Figure 6.1 (a) Diagram of the various stages of biofouling. (b) Mechanism of iron corrosion in water. (c) Microbially influenced corrosion (MIC) [143] (figures redrawn).

In the former, there are MICs for extracellular electron transfer [43], metabolite MICs [44, 45], and biodegradation [46], while in the latter, there are MICs for metal oxidation [47], oxygen concentration cells [48], and corrosive metabolites [49].

6.3 Recent Developments in Integrated Anticorrosion Coatings and Antifouling Coatings

This idea allows existing antifouling and anticorrosion methods to create new IAACs. Nature-inspired slippery liquid-infused porous surfaces (SLIPS) and superhydrophobic surfaces are inherently antifouling and anticorrosion. As a result, rather than using additional components to achieve these properties, emphasis is placed on promoting them in their design. This section introduces and discusses recent developments in IAACs.

6.3.1 Polyaniline-Based Coatings

PANI has a widespread application in marine anticorrosion because of its distinct electrochemical characteristics and superior stability [50, 51]. Numerous anticorrosion mechanisms for PANI have been put forth by researchers, including cathodic separation, anodic protection, release of corrosion inhibitors, electric field effect, and isolation [52, 53]. IAACs are created by combining PANI with antifouling units, such as nanomaterials, organic biocides, and antibacterial epoxies, to overcome this limitation [54–56]. Another method for enhancing PANI's antifouling performance is the synthesis of PANI derivatives [57, 58].

6.3.1.1 PANI Coatings/Organic Biocides

Organic bromine, a powerful biocide prevalent in daily life, may be employed in marine antifouling [59]. According to the research mentioned above, a significant factor affecting the properties of coatings is their bromine content. The potential toxicity of organic bromine to unintended marine organisms is unknown, despite the fact that it is a biocide that is relatively eco-friendly. Additionally, a thorough investigation into Br-PANI degradation in aquatic environments is necessary. This method's low yield is another drawback, which might prevent some applications.

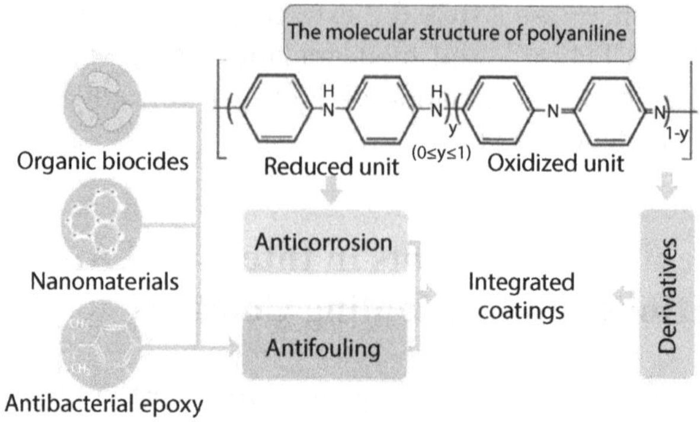

Figure 6.2 Design of PANI-based IAACs is depicted schematically [143] (figure redrawn).

6.3.1.2 PANI/Nanoparticle Composites

A hexagonal lattice of sp_2 hybridized carbon atoms makes up the 2D nanomaterial known as graphene. Graphene's optical, electrical, mechanical, and antibacterial capabilities improve coatings' mechanical and antifouling performance [60, 61]. Thus, graphene and PANI in IAACs have garnered interest recently [62–64]. The investigation of other nanoparticles, such as manganese dioxide, titanium dioxide, and silver, has not received as much attention in recent studies on PANI-based IAACs as graphene.

6.3.1.3 Antibacterial Epoxy/PANI Composites

TBEP's hydrophobicity reduces fouling organism adherence, and its functional groups kill bacteria. TBEP and PANI were thus combined to create IAACs. However, this additional bonding step might result in higher costs and less effective coating. Therefore, promoting the TBEP's natural bonding strength is a promising strategy. Due to inorganic coverage and antifoulant leaching, these coatings may not provide long-term antifouling [65]. The effectiveness of their antifouling properties must therefore be assessed in a long-term marine field.

6.3.1.4 Epoxy Coatings /PANI Derivatives

PmAP serves as a barrier and prevents corrosion in epoxy coatings, but too much of it can cause serious PmAP aggregation in the epoxy. The 5 wt% PmAP coatings' anticorrosion abilities consequently declined. Additionally, it was discovered as epoxy's ECPANOA content increased,

Table 6.1 Comparing properties of anticorrosion and antifouling PANI-based coatings [143].

Coatings	Antibacterial rate (%)		Charge transfer resistance Rct (Ω/cm^2)	Anticorrosion test period	Reference
	E. coli	*B. Subtilis*			
Br-PANI	87.2	98.5	2.03×10^9	100 days	[53]
PGO-PANI	None	None	2.60×10^6	192 hours	[55]
CAPNOA	94.43	95.83	1.81×10^{11}	55 days	[58]
GO-PANI	None	None	8.00×10^5	192 hours	[54]
TBEP-PANI	93.17	94.22	2.34×10^{11}	70 days	[56]
PmAP	97.20	97.20	2.26×10^{11}	50 days	[58]

so did the sterilization ratio. However, as ECPANOA increased, the anticorrosion capacity decreased, which may be due to poor dispersion [66]. However, CPANOA's anticorrosion ability was more comparable to PANI's, and the slight decline in anticorrosion ability is tolerable.

6.3.2 Copper/Amorphous Carbon Coatings

Due to its antibacterial properties [67–69], Cu can be added to amorphous carbon to improve its antifouling properties [70–74]. The anticorrosion ability, however, did not stick with this pattern. Coatings need Cu for antifouling and anticorrosion. Icorr may fluctuate because of hardness and roughness synergy. Additionally, there are a number of drawbacks that might limit how widely this technique is used [75]. Further research is needed because there are currently no assessments of the environmental hazard of copper carbon coatings.

6.3.3 Polydopamine/Polymer Brushes

Due to mussel sticky proteins, a biopolymer polydopamine (PDA) was created. PDA has been used to change many surfaces [76] and can cling to practically any substrate [77–82]. Polymer brushes and PDAs were combined to make IAACs [83]. The addition of PDA and organosilicon did, however, increase the corrosion resistance. PDA/Si/PSB coatings may adhere to SS, cotton cloth, and wood due to PDA's strong adhesive capabilities. For creating such coatings, Zhang *et al.* created an additional method [84]. IAACs' performance and functionality are enhanced by grafting polymer brushes, which give them good expandability. Despite the potential of these coatings, there is currently no research on how they might be used in marine environments. These coatings typically have a hydrophilic surface, but studies have shown that this surface type cannot sustain antifouling activity over an extended period of time. So, in the future, a thorough evaluation of the coatings' durability is necessary. In addition, PDA use in coatings is both complicated and expensive.

6.3.4 Amphiphilic Polymers

Hydrophobic, hydrophilic, and amphiphilic antifouling coatings are classified by surface wettability [85]. Therefore, the initial biofouling can be stopped by hydrophilic coatings. In addition, coatings like hydrogels typically have weak mechanical properties [86–88]. FSPA-4 (15.4 wt %) had the greatest BSA and HFg removal ratio. The best antifouling and anticorrosion characteristics were therefore displayed by FSPA-4. Such coatings hold great promise for applications in the future due to their straightforward fabrication process and excellent performance. Because natural lacquers are friendly to the environment, they have a bright future [89]. To achieve desired properties, controlling the F content in coatings is a significant challenge. Before applying such coatings, it is also necessary to evaluate them in marine field tests.

6.3.5 Endospore-Loaded Sol-Gel Coatings

They function well as barriers and exhibit good compatibility [90–92]. A similar pattern was seen in the EIS measurements, where the biotic coatings outperformed the abiotic coatings in terms of corrosion resistance. Additionally, the reduced biofouling was a factor in the improved corrosion resistance. Suleiman *et al.* later looked into the potential of various bacterial endospore types [93]. The corrosion inhibitors reduced antifouling, contrary to previous investigations. The fact that these coatings' antifouling abilities are positively correlated with bacterial endospore activity presents another challenge.

6.3.6 Corrosion-Inhibiting Hydrophobic Coatings

The results showed that pH had stronger antifouling capabilities than PL and had a higher MBT than PL due to its superior corrosion resistance [94].

According to EIS studies, the coating could prevent 97% of chlorella during settling and had good anticorrosion capabilities. Curcumin was used by Deng *et al.* to make a polybenzoxazine coating [95]. They respect the environment as well. These hydrophobic coatings, however, often have low surface energies, indicating that their connections to their substrates are frail.

6.3.7 Nanocomposites

Numerous desirable properties can be improved by adding nanomaterials to conventional coatings [96, 97]. Reactive oxygen species (ROS) produced by ZnO can cause oxidative stress, cell aging, and the expulsion of intracellular components, ultimately causing the death of bacteria [98–103]. Thus, epoxy coating fouling resistance may depend on the curing agent choice [104]. EPZ-1 had fewer fouling organisms than EPC following a 9-month marine field test. It should be noted, though, that too many ZnO nanoparticles in coatings could cause them to aggregate, which would reduce their antifouling and anticorrosion properties. Therefore, it is crucial to enhance ZnO nanoparticle dispersion in coatings [105–109]. They have a tendency to group together in coatings, which reduces the antifouling activity. A technique to get around this restriction was just developed by Liu *et al.* [107]. To prevent aggregation, they specifically loaded silver nanoparticles onto graphene nanosheets, greatly improving the coatings' antifouling properties. For loading PANI and silver nanoparticles, Liu *et al.* reported the use of attapulgite (ATP). The anticorossion and antialgal properties of the composite coatings were distinct. Chlorella and *B. subtilis* were successfully repelled by the ZnO-g-C_3N_4/epoxy. The complexity and expense of some preparation techniques, however, may limit their usefulness. Therefore, simple and affordable synthetic methods are required in the upcoming.

6.3.8 Bioinspired Coatings

6.3.8.1 Superhydrophobic Surfaces

Artificial substitutes are employed for anti-sticking, anticorrosion, oil-water separation, and drag reduction [110–112]. They were developed in response to the lotus phenomenon [113]. In the event of an external disturbance, these states can change to another [114–120] with different substrates, and broad compatibility is needed for further development [121]. The hydrothermal process produced a layered double hydroxide (LDH) with a rough surface to trap air. Triethoxy(octyl)silane (TTOS) was then added to the LDH to give it superhydrophilicity. Therefore, further research is needed. Hence, the study of simple and affordable production methods is helpful for practical applications and commercial production [122–129].

6.3.8.2 Slippery Liquid-Infused Porous Surfaces (SLIPS)

In tropical regions, *nepenthes* plants, which are carnivorous plants. Insects can easily fall into a *nepenthes* jar because it is so slippery. *Nepenthes* then consumes insects. This lubricity is due to the porous structures and lubricants of the peristome [130]. The fat layer blocks the chemical signals of the polluting organisms [131] and reduces their accumulation [132–133]. That is why it is believed to withstand the adhesion of almost any substance. According to SLIPS, they have better anti-sludge properties than typical hydrophobic surfaces. [134]. These findings suggest that SLIPS have excellent antifouling properties. Due to flow shear stress, which is a persistently important problem in moving water lubrication loss, this decrease is actually expected [135–139]. The coating was created by coating the steel surface with a mixture of PDMS, triethoxyvinylsilane, and SiO_2 and drying in an oven. This reduces lubricant leaching. According to 14-day EIS studies, both organogels outperformed Q235 steel in preventing corrosion [140]. Electrodeposition produced superhydrophobic surfaces with properties similar to Bayberry and Hyacinth. From the above samples, *Bacillus* sp. occurred in 9.2%, 2.0%, 1.4%, 0.7%, and 0.3%. Determining the appropriate surface morphology is essential to improve performance.

6.4 PNCs Nanocoatings (Polymer Nanocomposites Coatings)

Recent developments in nanotechnology have developed inventive improvements in tried-and-true methods of preventing corrosion at the metallic and/or electrolyte interface. Nanomaterials expressing various shapes and dimensions have been used to create PNC coatings and other PNC reinforced with natural fibers for flexible applications where flame retardancy and thermal conductivity are essential. These techniques include *in situ* polymerization, solution exfoliation, and melt intercalation. Phase separation, hybridization phases, intercalation, and exfoliative PNC nanocoatings are some of the ways that PNC micro-architectures are categorized.

Based on how well they are distributed in the host polymeric matrix, nanomaterials can reduce the cost of increasing the anticorrosive properties of PNC coatings. The incorporation of nanoparticles into polymeric matrix offers environmentally safe ways to increase the robustness and integrity of PNC coating systems. These substances have proven to be extremely effective in repairing invisible micro-deterioration that occurs in protective coatings. However, in the near future, efficient, environmentally friendly, and straightforward methods for creating PNC coatings with much-improved wettability, antifouling, hydrophilicity, and anticorrosion capabilities are anticipated. Along with improved corrosion barriers, extended endurance in harsh conditions is crucial for the application of these PNC coatings.

6.4.1 Biocides Used as Antifouling Agents

Based on the available data, it was discovered that some antifouling paint formulas that are now approved for commercial usage worldwide use up to six biocides simultaneously. The biocides most often discovered in these formulations were copper pyrithione, cuprous oxide, zinc pyrithione, DCOIT, zineb, and cuprous thiocyanate. Copper pyrithione, zineb, zinc pyrithione, DCOIT, and tralopyril, however, should be further examined in upcoming environmental and ecotoxicological investigations due to their importance and the paucity of knowledge about them. Medetomidine should also be evaluated since, along with tralopyril, antifouling paint makers have promoted it as a fast-degrading, non-metallic, and environmentally benign alternative. Unexpectedly, antifouling paints containing TBT were still listed for sale and are still usable today, proving that this extremely hazardous material is still an issue for the environment. These results can be used as a proxy for biocides that may be utilized by antifouling systems and subsequently discharged into the aquatic environment, albeit only partially representing the existing general condition.

6.4.2 Poly(2-Hydroxyethyl Methacrylate) (PHEMA) Hydrogel Film as Antifouling Alternative

Microfouling on a hydrophilic hydrogel was lower in laboratory bioassays utilizing marine creatures compared to certain more hydrophobic surfaces. The creation of a fouling-resistant material may benefit from the hydrophilic surface present in PHEMA. The efficacy of the coating in the maritime environment was enhanced by the addition of benzalkonium chloride (BCl) to PHEMA hydrogel, and visibly clean surfaces were consistently observed for a duration of 12 weeks—longer than what is necessary for sensor applications. The surfaces created were mostly microfouling-free, and one sample remained so for 5 months. It has been suggested that the performance of PHEMArBCl as a fouling-resistant material is due to BCl's dual activity as an antibacterial agent and a surfactant. The surface qualities of the hydrogel and the alteration of these properties by the addition of BCl are what cause the PHEMArBCl combination's overall effectiveness. PHEMArBCl, a potential antifouling agent, is less expensive and has a substantially lower environmental effect than copper or tributyltin oxide (TBT).

6.5 Marine Environment Durability Test of IAACs

Liu *et al.* [141] created a zwitterion polymer covering that performed well against the red algae *Porphyridium* sp. In some instances, the samples were also examined through optical or SEM microscopy to measure the degree of surface corrosion or microfouling. IAACs are a new tactic, so there is not much research on them yet. IAACs were taken into account to list the test specifics. In addition, more publications are anticipated to appear in the future. The length of the studies' test periods is another drawback of salinities [142]. Second, a longer test period will more clearly show the coatings' shortcomings (such as their weak mechanical and bonding strengths). In general, there is still much work to be done before IAACs are used outside.

6.6 Evaluation of Various IAACs and Difficulties

IAACs that combine antifouling and anticorrosion units and those that have built-in antifouling and anticorrosion functions are the two categories into which the IAACs are mentioned. The following is a description of the current issues with using these IAACs:

(i) Several coatings' constrained application range. For instance, plasma-enhanced chemical vapor deposition, spraying, magnetron sputtering, etc., are frequently used to create GLD/DLC coatings. For some large-scale and unique pipeline walls, these deposition techniques might not be appropriate.

(ii) Expensive, time-consuming, and complicated fabrication techniques. In recent years, manufacturing methods for surfaces with bioinspired design have advanced quickly. However, some of them (such as soft lithography and 3D printing) are ineffective for creating replicas with large areas.
(iii) The vast majority of assays in use today can only be performed in laboratories.
(iv) Low bonding force. A crucial property to improve is the bonding power between substrates and coatings.
(v) Environmental risk assessments are lacking. Marine organisms that are not the intended target may be at risk from nanomaterials and organic biocides.
(vi) In real-world applications, bioinspired superhydrophobic surfaces and SLIPS typically have unacceptable durability due to their insufficient mechanical strength and stability. Future research should focus on stable SLIPS and robust superhydrophobic surfaces.

Despite the fact that there are no perfect IAACs, production and practical applications will benefit from simple, affordable, and efficient synthesis techniques. Such IAACs are also inexpensive and simple to apply to a variety of substrates. Combination techniques demand additional fabrication processes and cost than IAACs with integrated antifouling and anticorrosion. Therefore, additional research is needed. Hopefully, more creative methods will be created to get around the aforementioned restrictions.

6.7 Conclusion and Outlook

Introduced in this chapter are the performance, synthesis methods, and current IAAC development issues. IAACs have many benefits over conventional antifouling or anticorrosion coatings that only serve one purpose and have a lot of potential in real-world applications. Therefore, more diverse combinations would start to appear in the upcoming. Despite their limitations, these bioinspired IAACs can serve as a source of inspiration for new research or IAAC development. For instance, in order to survive in the marine environment, marine life must possess antifouling and anticorrosion properties. We can design new IAACs by having an understanding of how marine life functions. IAACs are still in their early stages overall.

Consider that as materials science and synthesis techniques advance, they will become more sophisticated.

Recent advancements in nanotechnology have led to new breakthroughs in the techniques used to prevent corrosion at the metallic and/or electrolyte interfacial region. By reducing the porosity and irregularities that arise at the diffusion routes of corrosive substances, nanomaterials' incorporation in PNC coatings has significantly improved their barricading and protective performances. The incorporation of nanoparticles into polymeric matrix offers environmentally safe ways to increase the robustness and integrity of PNC coating systems. The incorporation of modest amounts of nanoparticles in PNC coatings generally results in improved barrier, protective behavior toward corrosion resistance, and a decrease in coating inclination toward delamination and blistering. These substances have proven to be extremely effective in repairing invisible micro-deterioration that occurs in protective coatings. Along with improved corrosion barriers, extended endurance in harsh conditions is crucial for the application of these PNC coatings. In order to create highly effective PNC protective coatings that can be used in a variety of corrosive, foul-smelling, and abrasive environments and achieve prolonged durability during service relative to improved shelf-life, further advancements in architectural properties and synthesis parameters are essential.

Conflict of Interest

No conflict of Interest

Research Funding

No funding Available

Acknowledgements

It is my proud privilege and special appreciation toward the Department of Chemistry, Vigyan Mandir, Banasthali Vidyapith, Rajasthan. We are also thankful to the reviewers for their most valuable suggestions, guidance, and constructive criticism that made it possible to get published.

References

1. Jin, H., Tian, L., Bing, W., Zhao, J., Ren, L., Toward the application of graphene for combating marine biofouling. *Adv. Sustain. Syst.*, 5, 1, 2000076, 2021.
2. Tian, L., Yin, Y., Jin, H., Bing, W., Jin, E., Zhao, J., Ren, L., Novel marine antifouling coatings inspired by corals. *Mater. Today Chem.*, 17, 100294, 2020.
3. Hu, P., Xie, Q., Ma, C., Zhang, G., Silicone-based fouling-release coatings for marine antifouling. *Langmuir.*, 36, 9, 2170–2183, 2020.
4. Yan, H., Wu, Q., Yu, C., Zhao, T., Liu, M., Recent progress of biomimetic antifouling surfaces in marine. *Adv. Mater. Interfaces*, 7, 20, 2000966, 2020.
5. Schultz, M.P., Walker, J.M., Steppe, C.N., Flack, K.A., Impact of diatomaceous biofilms on the frictional drag of fouling-release coatings. *Biofouling*, 31, 9-10, 759–773, 2015.
6. Schultz, M.P., Effects of coating roughness and biofouling on ship resistance and powering. *Biofouling*, 23, 5, 331–341, 2007.
7. Selim, M.S., Shenashen, M.A., El-Safty, S.A., Higazy, S.A., Selim, M.M., Isago, H., Elmarakbi, A., Recent progress in marine foul-release polymeric nanocomposite coatings. *Prog. Mater. Sci.*, 87, 1–32, 2017.
8. Davidson, I.C., Smith, G., Ashton, G.V., Ruiz, G.M., Scianni, C., An experimental test of stationary lay-up periods and simulated transit on biofouling accumulation and transfer on ships. *Biofouling*, 36, 4, 455–466, 2020.
9. Galil, B.S., Marchini, A., Occhipinti-Ambrogi, A., Mare Nostrum, mare quod invaditur—the history of bioinvasions in the Mediterranean Sea, in: *Histories of Bioinvasions in the Mediterranean*, pp. 21–49, 2018.
10. Venkatesan, R., Kadiyam, J., SenthilKumar, P., Lavanya, R., Vedaprakash, L., Marine biofouling on moored buoys and sensors in the Northern Indian Ocean. *Mar. Technol. Soc. J.*, 51, 2, 22–30, 2017.
11. Reyes, R., del Norte-Campos, A., Añasco, N.C., Santander-de Leon, S.M.S., Biofouling development in marine fish farm influenced by net colour, immersion period and environmental conditions. *Aquac. Res.*, 51, 8, 3129–3138, 20202020.
12. Nacorda, H.M., Austero, N.M., Pagdilao, C.R., Tan, K.S., Azanza, R.V., Marine biofouling communities of Manila South Harbor, Philippines. *ASEAN J. Sci. Technol. Dev.*, 35, 1-2, 115–123, 2018.
13. Xie, Q., Pan, J., Ma, C., Zhang, G., Dynamic surface antifouling: mechanism and systems. *Soft Matter*, 15, 6, 1087–1107, 2019.
14. Chen, L., Duan, Y., Cui, M., Huang, R., Su, R., Qi, W., He, Z., Biomimetic surface coatings for marine antifouling: Natural antifoulants, synthetic polymers and surface microtopography. *Sci. Total Environ.*, 766, 144469, 2021.
15. Videla, H.A., Prevention and control of biocorrosion. *Int. Biodeterior. Biodegrad.*, 49, 4, 259–270, 2002.
16. Little, B.J., Blackwood, D.J., Hinks, J., Lauro, F.M., Marsili, E., Okamoto, A., Rice, S.A., Wade, S.A., Flemming, H.C., Microbially influenced corrosion—Any progress? *Corros. Sci.*, 170, 108641, 2020.

17. Usher, K.M., Kaksonen, A.H., Cole, I., Marney, D., Critical review: microbially influenced corrosion of buried carbon steel pipes. *Int. biodeterior. biodegrad*, 93, 84–106, 2014.
18. Pica, D., Bloecher, N., Dell'Anno, A., Bellucci, A., Pinto, T., Pola, L., Puce, S., Dynamics of a biofouling community in finfish aquaculture: A case study from the South Adriatic Sea. *Biofouling*, 35, 6, 696–709, 2019.
19. Gangopadhyay, S. and Mahanwar, P.A., Recent developments in the volatile corrosion inhibitor (VCI) coatings for metal: a review. *J. Coat. Technol. Res.*, 15, 789–807, 2018.
20. Hou, B., Li, X., Ma, X., Du, C., Zhang, D., Zheng, M., Xu, W., Lu, D., Ma, F., The cost of corrosion in China. *NPJ Mater. Degrad.*, 1, 1, 4, 2017.
21. Saranya, J., Lavanya, K., Haritha Kiranmai, M., Subbiah, R., Zarrouk, A., Chitra, S., Quinoxaline derivatives as anticorrosion additives for metals. *Corros. Rev.*, 39, 2, 79–92, 2021.
22. Ding, R., Chen, S., Lv, J., Zhang, W., Zhao, X.D., Liu, J., Wang, X., Gui, T.J., Li, B.J., Tang, Y.Z., Li, W.H., Study on graphene modified organic anti-corrosion coatings: A comprehensive review. *J. Alloys Compd.*, 806, 611–635, 2019.
23. Vazirinasab, E., Jafari, R., Momen, G., Application of superhydrophobic coatings as a corrosion barrier: A review. *Surf. Coat. Technol.*, 341, 40–56, 2018.
24. Nazeer, A.A. and Madkour, M., Potential use of smart coatings for corrosion protection of metals and alloys: A review. *J. Mol. Liq.*, 253, 11–22, 2018.
25. Wang, R., Zhou, T., Liu, J., Zhang, X., Yang, J., Hu, W., Liu, L., Designing novel anti-biofouling coatings on titanium based on the ferroelectric-induced strategy. *Mater. Des.*, 203, 109584, 2021.
26. Pourhashem, S., Saba, F., Duan, J., Rashidi, A., Guan, F., Nezhad, E.G., Hou, B., Polymer/Inorganic nanocomposite coatings with superior corrosion protection performance: A review. *J. Ind. Eng. Chem.*, *88*, 29–57, 2020.
27. Ren, B., Chen, Y., Li, Y., Li, W., Gao, S., Li, H., Cao, R., Rational design of metallic anti-corrosion coatings based on zinc gluconate@ ZIF-8. *J. Chem. Eng.*, 384, 123389, 2020.
28. Jin, H., Tian, L., Bing, W., Zhao, J., Ren, L., Bioinspired marine antifouling coatings: Status, prospects, and future. *Prog. Mater. Sci.*, 124, 100889, 2022.
29. Selim, M.S., El-Safty, S.A., Shenashen, M.A., Higazy, S.A., Elmarakbi, A., Progress in biomimetic leverages for marine antifouling using nanocomposite coatings. *J. Mater. Chem. B.*, 8, 17, 3701–3732, 2020.
30. Cui, G., Bi, Z., Zhang, R., Liu, J., Yu, X., Li, Z., A comprehensive review on graphene-based anti-corrosive coatings. *J. Chem. Eng.*, 373, 104–121, 2019.
31. Deyab, M.A., Anticorrosion properties of nanocomposites coatings: A critical review. *J. Mol. Liq.*, 313, 113533, 2020.
32. Halder, P., Hossain, N., Pramanik, B.K., Bhuiyan, M.A., Engineered topographies and hydrodynamics in relation to biofouling control—a review. *Environ. Sci. Pollut. Res.*, 28, 40678–40692, 2021.

33. Ferrari, M., Benedetti, A., Santini, E., Ravera, F., Liggieri, L., Guzman, E., Cirisano, F., Biofouling control by superhydrophobic surfaces in shallow euphotic seawater. *Colloids Surf. A Physicochem. Eng.*, 480, 369–375, 2015.
34. More, T.T., Yadav, J.S.S., Yan, S., Tyagi, R.D., Surampalli, R.Y., Extracellular polymeric substances of bacteria and their potential environmental applications. *J. Environ. Manage.*, 144, 1–25, 2014.
35. Salama, A.J., Satheesh, S., Balqadi, A.A., Development of biofouling communities on nylon net panels submerged in the central Red Sea: effects of season and depth. *Thalassas: An Int. J. Marine Sci.*, 34, 199–208, 2018.
36. Wassick, A., Hunsucker, K.Z., Swain, G., Measuring the recruitment and growth of biofouling communities using clear recruitment panels. *Biofouling*, 1–18, 2023.
37. Jin, H., Zhang, T., Bing, W., Dong, S., Tian, L., Antifouling performance and mechanism of elastic graphene–silicone rubber composite membranes. *J. Mater. Chem. B.*, 7, 3, 488–497, 2019.
38. Batista-Andrade, J.A., Caldas, S.S., Batista, R.M., Castro, I.B., Fillmann, G., Primel, E.G., From TBT to booster biocides: levels and impacts of antifouling along coastal areas of Panama. *Environ. Pollut.*, 234, 243–252, 2018.
39. Qi, Y., Liang, W., Miao, Q., Lin, H., An, H., Liu, Y., Zuo, S., Ma, H., Corrosion behavior and antifouling ability of Cu-Zn-Al/Zn-Al composite coating on Q235 steel. *Surf. Coat. Technol.*, 405, 126614, 2021.
40. Bryan, G.W. and Gibbs, P.E., Impact of low concentrations of tributyltin (TBT) on marine organisms: a review. *Met. Ecotoxicol. Concepts Appl.*, 323–361, 2020.
41. Ahmad, Z., *Principles of corrosion engineering and corrosion control*, Elsevier, 2006.
42. Jia, R., Unsal, T., Xu, D., Lekbach, Y., Gu, T., Microbiologically influenced corrosion and current mitigation strategies: A state of the art review. *Int. Biodeterior. Biodegrad.*, 137, 42–58, 2019.
43. Gu, T., New understandings of biocorrosion mechanisms and their classifications. *J. Microb. Biochem. Technol.*, 2012.
44. Jia, R., Yang, D., Xu, D., Gu, T., Anaerobic corrosion of 304 stainless steel caused by the Pseudomonas aeruginosa biofilm. *Front. Microbiol.*, 8, 2335, 2017.
45. Jia, R., Tan, J.L., Jin, P., Blackwood, D.J., Xu, D., Gu, T., Effects of biogenic H_2S on the microbiologically influenced corrosion of C1018 carbon steel by sulfate reducing Desulfovibrio vulgaris biofilm. *Corros. Sci.*, 130, 1–11, 2018.
46. Jia, R., Yang, D., Abd Rahman, H.B., Gu, T., An enhanced oil recovery polymer promoted microbial growth and accelerated microbiologically influenced corrosion against carbon steel. *Corros. Sci.*, 139, 301–308, 2018.
47. Wang, H., Ju, L.K., Castaneda, H., Cheng, G., Newby, B.M.Z., Corrosion of carbon steel C1010 in the presence of iron oxidizing bacteria Acidithiobacillus ferrooxidans. *Corros. Sci.*, 89, 250–257, 2014.

48. Skovhus, T.L., Enning, D. and Lee, J.S. eds., *Microbiologically influenced corrosion in the upstream oil and gas industry*. CRC press, 2017.
49. Gu, T., Rastegar, S.O., Mousavi, S.M., Li, M., Zhou, M., Advances in bioleaching for recovery of metals and bioremediation of fuel ash and sewage sludge. *Bioresour. Technol.*, 261, 428–440, 2018.
50. Jangid, N.K., Chauhan, N.P.S., Punjabi, P.B., Novel dye-substituted polyanilines: conducting and antimicrobial properties. *Polym. Bull.*, 71, 2611–2630, 2014.
51. Jangid, N.K., Chauhan, N.P., Ameta, C., Meghwal, K., Ameta, R., Punjabi, P.B., Synthesis and characterization of functionalized polyaniline having methyl violet as pendant groups. *J. Macromol. Sci. A.*, 51, 8, 625–632, 2014.
52. Gao, F., Mu, J., Bi, Z., Wang, S., Li, Z., Recent advances of polyaniline composites in anticorrosive coatings: A review. *Prog. Org. Coat.*, 151, 106071, 2021.
53. Cai, W., Wang, J., Quan, X., Zhao, S., Wang, Z., Antifouling and anticorrosion properties of one-pot synthesized dedoped bromo-substituted polyaniline and its composite coatings. *Surf. Coat. Technol.*, 334, 7–18, 2018.
54. Fazli-Shokouhi, S., Nasirpouri, F., Khatamian, M., Polyaniline-modified graphene oxide nanocomposites in epoxy coatings for enhancing the anticorrosion and antifouling properties. *J. Coat. Technol. Res.*, 16, 983–997, 2019.
55. Fazli-Shokouhi, S., Nasirpouri, F., Khatamian, M., Epoxy-matrix polyaniline/p-phenylenediamine-functionalised graphene oxide coatings with dual anti-corrosion and anti-fouling performance. *RSC Adv.*, 11, 19, 11627–11641, 2021.
56. Wu, C., Wang, J., Song, S., Wang, Z., Zhao, S., Antifouling and anticorrosion performance of the composite coating made of tetrabromobisphenol- A epoxy and polyaniline nanowires. *Prog. Org. Coat.*, 148, 105888, 2020.
57. Quan, X., Wang, J., Zhao, S., Cai, W., Wang, Z., Wang, S., Cui, X., Improved antibacterial, antifouling and corrosion protective performance of epoxy coatings with poly (m-aminophenol). *Prog. Org. Coat.*, 115, 9–17, 2018.
58. Gao, M., Quan, X., Wang, J., Wang, Z., Preparation and characterization of coatings incorporated with poly (aniline-co-nitroaniline) nanoparticles having antifouling and anticorrosion behavior. *Ind. Eng. Chem. Res.*, 59, 51, 22173–22186, 2020.
59. Ashraf, M.A., Ullah, S., Ahmad, I., Qureshi, A.K., Balkhair, K.S., Abdur Rehman, M., Green biocides, a promising technology: current and future applications to industry and industrial processes. *J. Sci. Food Agric.*, 94, 3, 388–403, 2014.
60. Bano, S., Mahmood, A., Kim, S.J., Lee, K.H., Graphene oxide modified polyamide nanofiltration membrane with improved flux and antifouling properties. *J. Mater. Chem. A.*, 3, 5, 2065–2071, 2015.
61. Anand, A., Unnikrishnan, B., Mao, J.Y., Lin, H.J., Huang, C.C., Graphene-based nanofiltration membranes for improving salt rejection, water flux and antifouling–A review. *Desalination.*, 429, 119–133, 2018.

62. Kakaei, K., Esrafili, M.D., Ehsani, A., Graphene and anticorrosive properties. *Interface Sci. Technol.*, 27, 303–337, 2019.
63. George, J.S., Paduvilan, J.K., Salim, N., Sunarso, J., Kalarikkal, N., Hameed, N., Thomas, S., Advances and future outlook in epoxy/graphene composites for anticorrosive applications. *Prog. Org. Coat.*, 162, 106571, 2022.
64. Zhou, S., Wu, Y., Zhao, W., Yu, J., Jiang, F., Wu, Y., Ma, L., Designing reduced graphene oxide/zinc rich epoxy composite coatings for improving the anticorrosion performance of carbon steel substrate. *Mater. Des.*, 169, 107694, 2019.
65. Zhang, L., Xu, J., Tang, Y., Hou, J., Yu, L., Gao, C., A novel long-lasting antifouling membrane modified with bifunctional capsaicin-mimic moieties via in situ polymerization for efficient water purification. *J. Mater. Chem. A.*, 4, 26, 10352–10362, 2016.
66. Kar, P., Pradhan, N.C., Adhikari, B., Isomeric effects on structures and properties of polyaminophenols synthesized in basic medium. *J. Macromol. Sci. A*, 47, 3, 282–290, 2010.
67. Toh, C.T., Zhang, H., Lin, J., Mayorov, A.S., Wang, Y.P., Orofeo, C.M., Ferry, D.B., Andersen, H., Kakenov, N., Guo, Z., Abidi, I.H., Synthesis and properties of free-standing monolayer amorphous carbon. *Nature*, 577, 7789, 199–203, 2020.
68. Liu, Y., Suo, X., Wang, Z., Gong, Y., Wang, X., Li, H., Developing polyimide-copper antifouling coatings with capsule structures for sustainable release of copper. *Mater. Des.*, 130, 285–293, 2017.
69. Cerchier, P., Pezzato, L., Gennari, C., Moschin, E., Moro, I., Dabalà, M., PEO coating containing copper: A promising anticorrosive and antifouling coating for seawater application of AA 7075. *Surf. Coat. Technol.*, 393, 125774, 2020.
70. Kamil, M.P., Suhartono, T., Ko, Y.G., Corrosion behavior of plasma electrolysis layer cross-linked with a conductive polymer coating. *J. Mater. Res. Technol.*, 15, 4672–4682, 2021.
71. Zhang, J., Wang, Y., Zhou, S., Wang, Y., Wang, C., Guo, W., Lu, X., Wang, L., Tailoring self-lubricating, wear-resistance, anticorrosion and antifouling properties of Ti/(Cu, MoS2)-DLC coating in marine environment by controlling the content of Cu dopant. *Tribol. Int.*, 143, 106029, 2020.
72. Huang, M., Zhang, X., Ke, P., Wang, A., Graphite-like carbon films by high power impulse magnetron sputtering. *Appl. Surf. Sci.*, 283, 321–326, 2013.
73. Dwivedi, N., Kumar, S., Malik, H.K., Strange hardness characteristic of hydrogenated diamond-like carbon thin film by plasma enhanced chemical vapor deposition process. *Appl. Phys. Lett.*, 102, 1, 2013.
74. Tyagi, A., Walia, R.S., Murtaza, Q., Pandey, S.M., Tyagi, P.K., Bajaj, B., A critical review of diamond like carbon coating for wear resistance applications. *Int. J. Refract. Met. Hard Mater.*, 78, 107–122, 2019.
75. Yang, K., Chen, J., Zheng, L., Zheng, B., Chen, Y., Chen, X., Bai, W., Jian, R., Wei, F., Xu, Y., Urushiol titanium polymer-based composites coatings for anti-corrosion and antifouling in marine spray splash zones. *J. Appl. Polym. Sci.*, 138, 34, 50861, 2021.

76. Wang, Z., Jiang, X., Cheng, X., Lau, C.H., Shao, L., Mussel-inspired hybrid coatings that transform membrane hydrophobicity into high hydrophilicity and underwater superoleophobicity for oil-in-water emulsion separation. *ACS Appl. Mater. Interfaces*, 7, 18, 9534–9545, 2015.
77. Prieto-López, L.O., Herbeck-Engel, P., Yang, L., Wu, Q., Li, J., Cui, J., When Ultimate Adhesive Mechanism Meets Ultimate Anti-Fouling Surfaces—Polydopamine Versus SLIPS: Which One Prevails? *Adv. Mater. Interfaces*, 7, 18, 2000876, 2020.
78. Zhang, C., He, Y., Li, F., Di, H., Zhang, L., Zhan, Y., h-BN decorated with Fe_3O_4 nanoparticles through mussel-inspired chemistry of dopamine for reinforcing anticorrosion performance of epoxy coatings. *J. Alloys Compd.*, 685, 743–751, 2016.
79. Xia, Z., Liu, G., Dong, Y., Zhang, Y., Anticorrosive epoxy coatings based on polydopamine modified molybdenum disulfide. *Prog. Org. Coat.*, 133, 154–160, 2019.
80. Ding, Z., Fatollahi-Fard, F., Kwon, I.S., Pistorius, P.C., Bettinger, C.J., Polydopamine nanomembranes as adhesion layers for improved corrosion resistance in low carbon steel. *Adv. Eng. Mater.*, 20, 11, 1800621, 2018.
81. Feng, C. and Huang, X., Polymer brushes: efficient synthesis and applications. *Acc. Chem. Res.*, 51, 9, 2314–2323, 2018.
82. Ma, S., Zhang, X., Yu, B., Zhou, F., Brushing up functional materials. *NPG Asia Mater*, 11, 1, 24, 2019.
84. Zhang, B., Yan, Q., Yuan, S., Zhuang, X., Zhang, F., Enhanced antifouling and anticorrosion properties of stainless steel by biomimetic anchoring PEGDMA-cross-linking polycationic brushes. *Ind. Eng. Chem. Res.*, 58, 17, 7107–7119, 2019.
85. Asri, L.A., Crismaru, M., Roest, S., Chen, Y., Ivashenko, O., Rudolf, P., Tiller, J.C., van der Mei, H.C., Loontjens, T.J., Busscher, H.J., A shape-adaptive, antibacterial-coating of immobilized quaternary-ammonium compounds tethered on hyperbranched polyurea and its mechanism of action. *Adv. Funct. Mater.*, 24, 3, 346–355, 2014.
86. Guo, H., Chen, P., Tian, S., Ma, Y., Li, Q., Wen, C., Yang, J., Zhang, L., Amphiphilic marine antifouling coatings based on a hydrophilic polyvinylpyrrolidone and hydrophobic fluorine–silicon-containing block copolymer. *Langmuir.*, 36, 48, 14573–14581, 20202020.
87. Zhang, R., Li, Y., Su, Y., Zhao, X., Liu, Y., Fan, X., Ma, T., Jiang, Z., Engineering amphiphilic nanofiltration membrane surfaces with a multi-defense mechanism for improved antifouling performances. *J. Mater. Chem. A*, 4, 20, 7892–7902, 2016.
88. Zhu, B., Liu, Z., Liu, J., Yang, Y., Meng, Y., Yu, F., Jiang, L., Wei, G., Zhang, Z., Preparation of fluorinated/silanized polyacrylates amphiphilic polymers and their anticorrosion and antifouling performance. *Prog. Org. Coat.*, 140, 105510, 2020.

89. Chen, Y., Zhang, G., Zhang, G., Ma, C., Rapid curing and self-stratifying lacquer coating with antifouling and anticorrosive properties. *J. Chem. Eng.*, 421, 129755, 2021.
90. Alcantara-Garcia, A., Garcia-Casas, A., Jimenez-Morales, A., The effect of the organosilane content on the barrier features of sol-gel anticorrosive coatings applied on carbon steel. *Prog. Org. Coat.*, 139, 105418, 2020.
91. Alibakhshi, E., Akbarian, M., Ramezanzadeh, M., Ramezanzadeh, B., Mahdavian, M., Evaluation of the corrosion protection performance of mild steel coated with hybrid sol-gel silane coating in 3.5 wt.% NaCl solution. *Prog. Org. Coat.*, 123, 190–200, 2018.
92. Eduok, U., Suleiman, R., Gittens, J., Khaled, M., Smith, T.J., Akid, R., El Ali, B., Khalil, A., Anticorrosion/antifouling properties of bacterial spore-loaded sol–gel type coating for mild steel in saline marine condition: a case of thermophilic strain of Bacillus licheniformis. *RSC Adv.*, 5, 114, 93818–93830, 2015.
93. Suleiman, R., Gittens, J., Khaled, M., Smith, T.J., A kid, R., El Ali, B., Khalil, A., Assessing the anticorrosion and antifouling performances of a sol–gel coating mixed with corrosion inhibitors and immobilised bacterial endospores. *Arab J. Sci. Eng.*, 42, 4327–4338, 2017.
94. Zhang, H., Liang, T., Liu, Y., Misra, R.D.K., Zhao, Y., Low-surface-free-energy GO/FSiAC coating with self-healing function for anticorrosion and antifouling applications. *Surf. Coat. Technol.*, 425, 127690, 2021.
95. Deng, Y., Xia, L., Song, G.L., Zhao, Y., Zhang, Y., Xu, Y., Zheng, D., Development of a curcumin-based antifouling and anticorrosion sustainable polybenzoxazine resin composite coating. *Compos. B: Eng.*, 225, 109263, 2021.
96. Prasai, D., Tuberquia, J.C., Harl, R.R., Jennings, G.K., Rogers, B.R., Bolotin, K.I., Correction to graphene: corrosion-inhibiting coating. *ACS nano*, 6, 5, 4540–4540, 2012.
97. Böhm, S., Graphene against corrosion. *Nat. Nanotechnol.*, 9, 10, 741–742, 2014.
98. Sirelkhatim, A., Mahmud, S., Seeni, A., Kaus, N.H.M., Ann, L.C., Bakhori, S.K.M., Hasan, H., Mohamad, D., Review on zinc oxide nanoparticles: antibacterial activity and toxicity mechanism. *Nanomicro Lett.*, 7, 219–242, 2015.
99. Yong, H.E., Krishnamoorthy, K., Hyun, K.T., Kim, S.J., Preparation of ZnO nanopaint for marine antifouling applications. *J. Ind. Eng. Chem.*, 29, 39–42, 2015.
100. Al-Naamani, L., Dobretsov, S., Dutta, J., Burgess, J.G., Chitosan-zinc oxide nanocomposite coatings for the prevention of marine biofouling. *Chemosphere*, 168, 408–417, 2017.
101. Selim, M.S., Yang, H., Wang, F.Q., Fatthallah, N.A., Huang, Y., Kuga, S., Silicone/ZnO nanorod composite coating as a marine antifouling surface. *Appl. Surf. Sci.*, 466, 40 50, 2019.

102. Arukalam, I.O., Oguzie, E.E., Li, Y., Fabrication of FDTS-modified PDMS-ZnO nanocomposite hydrophobic coating with anti-fouling capability for corrosion protection of Q235 steel. *J. Colloid Interface Sci.*, 484, 220–228, 2016.
103. Saravanan, P., Duraibabu, D., Jayamoorthy, K., Suresh, S., Kumar, S.A., Twin applications of tetra-functional epoxy monomers for anticorrosion and antifouling studies. *Silicon*, 10, 555–565, 2018.
104. Verma, S., Das, S., Mohanty, S., Nayak, S.K., Development of multifunctional polydimethylsiloxane (PDMS) -epoxy-zinc oxide nanocomposite coatings for marine applications. *Polym. Adv. Technol.*, 30, 9, 2275–2300, 2019.
105. Uzoma, P.C., Liu, F., Xu, L., Zhang, Z., Han, E.H., Ke, W., Arukalam, I.O., Superhydrophobicity, conductivity and anticorrosion of robust siloxane-acrylic coatings modified with graphene nanosheets. *Prog. Org. Coat.*, 127, 239–251, 2019.
106. Ma, Y., Zhang, Y., Liu, J., Ge, Y., Yan, X., Sun, Y., Wu, J., Zhang, P., GO-modified double-walled polyurea microcapsules/epoxy composites for marine anticorrosive self-healing coating. *Mater. Des.*, 189, 108547, 2020.
107. Liu, Z., Tian, S., Li, Q., Wang, J., Pu, J., Wang, G., Zhao, W., Feng, F., Qin, J., Ren, L., Integrated dual-functional ORMOSIL coatings with AgNPs@ rGO nanocomposite for corrosion resistance and antifouling applications. *ACS Sustain. Chem. Eng.*, 8, 17, 6786–6797, 2020.
108. Liu, X., Wang, N., Hou, B., Multifunctional Ag-based ternary nanocomposite incorporated epoxy coating as antibacterial, anticorrosion and antifouling agent. *Corros. Eng. Sci. Technol.*, 56, 2, 144–153, 2021.
109. Lv, Y., Zheng, Y., Zhu, H., Wu, Y., Designing a dual-functional material with barrier anti-corrosion and photocatalytic antifouling properties using g-C_3N_4 nanosheet with ZnO nanoring. *J. Mater Sci. Technol.*, 10656-69, 2022.
110. Bing, W., Wang, H., Tian, L., Zhao, J., Jin, H., Du, W., Ren, L., Small structure, large effect: Functional surfaces inspired by salvinia leaves. *Small Struct.*, 2, 9, 2100079, 2021.
111. Mohamed, A.M., Abdullah, A.M., Younan, N.A., Corrosion behavior of superhydrophobic surfaces: A review. *Arab. J. Chem.*, 8, 6, 749–765, 2015.
112. Jeevahan, J., Chandrasekaran, M., Britto Joseph, G., Durairaj, R.B., Mageshwaran, G.J.J.O.C.T., Superhydrophobic surfaces: a review on fundamentals, applications, and challenges. *J. Coat. Technol. Res.*, 15, 231–250, 2018.
113. Fang, W., Guo, H.Y., Li, B., Li, Q., Feng, X.Q., Revisiting the critical condition for the Cassie–Wenzel transition on micropillar-structured surfaces. *Langmuir*, 34, 13, 3838–3844, 2018.
114. Celia, E., Darmanin, T., de Givenchy, E.T., Amigoni, S., Guittard, F., Recent advances in designing superhydrophobic surfaces. *J. Colloid Interface Sci.*, 402, 1–18, 2013.

115. Guo, C., Ding, H., Xie, M., Zhang, H., Hong, X., Sun, L., Ding, F., Multifunctional superamphiphobic fluorinated silica with a core-shell structure for anti-fouling and anti-corrosion applications. *Colloids Surf. A: Physicochem.*, 615, 126155, 2021.
116. Yang, Z., Wang, L., Sun, W., Li, S., Zhu, T., Liu, W., Liu, G., Superhydrophobic epoxy coating modified by fluorographene used for anti-corrosion and self-cleaning. *Appl. Surf. Sci.*, 401, 146–155, 2017.
117. Xu, S., Wang, Q., Wang, N., Zheng, X., Fabrication of superhydrophobic green surfaces with good self-cleaning, chemical stability and anti-corrosion properties. *J. Mater. Sci.*, 54, 13006–13016, 2019.
118. Selim, M.S., Fatthallah, N.A., Higazy, S.A., Hao, Z., Mo, P.J., A comparative study between two novel silicone/graphene-based nanostructured surfaces for maritime antifouling. *J. Colloid Interface Sci.*, 606, 367–383, 2022.
119. Xia, X.C., Cao, X.K., Cai, G.Y., Jiang, D., Zhang, F., Dong, Z.H., Underwater superoleophobic composite coating characteristic of durable antifouling and anticorrosion properties in marine environment. *Colloids Surf. A: Physicochem. Eng.*, *628*, 127323, 2021.
120. Wang, L., Xiao, X., Liu, E., Yu, S., Yin, X., Wang, J., Zhu, G., Li, Q., Li, J., Fabrication of superhydrophobic needle-like Ca-P coating with anti-fouling and anti-corrosion properties on AZ31 magnesium alloy. *Colloids Surf. A: Physicochem. Eng.*, 620, 126568, 2021.
121. Xiang, Y., He, Y., Zhang, W., Li, B., Li, H., Wang, Y., Yin, X., Tang, W., Li, Z., He, Z., Superhydrophobic LDH/TTOS composite surface based on microstructure for the anti-corrosion, anti-fouling and oil-water separation application. *Colloids Surf. A: Physicochem. Eng.*, 622, 126558, 2021.
122. Zhou, S., Zhu, X., Yan, Q., One-step electrochemical deposition to achieve superhydrophobic cobalt incorporated amorphous carbon-based film with self-cleaning and anti-corrosion. *Surf. Interface Anal.*, 50, 3, B290–296, 2018.
123. Lv, Y. and Liu, M., Corrosion and fouling behaviours of copper-based superhydrophobic coating. *Surf. Eng.*, 35, 6, 542–549, 2019.
124. Li, M., Liu, N., Chen, J., Li, Q., Simple construction based on epoxy-bonded super-hydrophobic anti-corrosion coating. *J. Taiwan Inst. Chem. Eng.*, 95, 682–691, 2019.
125. Wang, D., Sun, Q., Hokkanen, M.J., Zhang, C., Lin, F.Y., Liu, Q., Zhu, S.P., Zhou, T., Chang, Q., He, B., Zhou, Q., Design of robust superhydrophobic surfaces. *Nature*, 582, 7810, 55–59, 2020.
126. Song, S., Yan, H., Cai, M., Huang, Y., Fan, X., Zhu, M., Constructing mechanochemical durable superhydrophobic composite coating towards superior anticorrosion. *Adv. Mater. Technol.*, 7, 6, 2101223, 2022.
127. Guo, R., Guo, X., Pei, H., Liu, W., Guo, W., Fang, M., Liu, N., Mo, Z., Polyphenylene sulfide hydrophobic composite coating with high stability, corrosion resistance and antifouling performance. *Surf. Interfaces*, 27101577, 2021.

128. Rasitha, T.P., Krishna, D.N.G., Thinaharan, C., Vanithakumari, S.C., Philip, J., Optimization of coating parameters for fabrication of robust superhydrophobic (SHP) aluminum and evaluation of corrosion performance in aggressive medium. *Prog. Org. Coat.*, 172, 107076, 2022.
129. Liu, F., Du, H., Han, Y., Wang, C., Liu, Z., Wang, H., Recent Progress in the Fabrication and Characteristics of Self-Repairing Superhydrophobic Surfaces. *Adv. Mater. Interfaces*, 8, 17, 2100228, 2021.
130. Chen, H., Zhang, P., Zhang, L., Liu, H., Jiang, Y., Zhang, D., Han, Z., Jiang, L., Continuous directional water transport on the peristome surface of Nepenthes alata. *Nature*, 532, 7597, 85–89, 2016.
131. Epstein, A.K., Wong, T.S., Belisle, R.A., Boggs, E.M., Aizenberg, J., Liquid-infused structured surfaces with exceptional anti-biofouling performance. *Proc. Natl. Acad. Sci. U S A.*, 109, 33, 13182–13187, 2012.
132. MacCallum, N., Howell, C., Kim, P., Sun, D., Friedlander, R., Ranisau, J., Ahanotu, O., Lin, J.J., Vena, A., Hatton, B., Wong, T.S., Liquid-infused silicone as a biofouling-free medical material. *ACS Biomater. Sci. Eng.*, 1, 1, 43–51, 2015.
133. Amini, S., Kolle, S., Petrone, L., Ahanotu, O., Sunny, S., Sutanto, C.N., Hoon, S., Cohen, L., Weaver, J.C., Aizenberg, J., Vogel, N., Preventing mussel adhesion using lubricant-infused materials. *Science*, 357, 6352, 668–673, 2017.
134. Li, D., Lin, Z., Zhu, J., Yu, J., Liu, J., Liu, Z., Chen, R., Liu, Q., Liu, P., Wang, J., An engineering-oriented approach to construct rough micro/nano-structures for anticorrosion and antifouling application. *Colloids Surf. A: Physicochem. Eng.*, 621, 126590, 2021.
135. Preston, D.J., Song, Y., Lu, Z., Antao, D.S., Wang, E.N., Design of lubricant infused surfaces. *ACS Appl. Mater. Interfaces*, 9, 48, 42383–42392, 2017.
136. Li, Y., Chen, R., Zhang, L., Liu, Q., Yu, J., Liu, J., Liu, P., Wang, J., Fabrication and antifouling behavior research of self-healing lubricant impregnated films with dynamic surfaces. *Colloids Surf. A: Physicochem. Eng.*, 582, 123865, 2019.
137. Ouyang, Y., Zhao, J., Qiu, R., Hu, S., Niu, H., Zhang, Y., Chen, M., Nanowall enclosed architecture infused by lubricant: A bio-inspired strategy for inhibiting bio-adhesion and bio-corrosion on stainless steel. *Surf. Coat. Technol.*, 381, 125143, 2020.
138. Liang, Y., Li, C., Wang, P., Zhang, D., Fabrication of a robust slippery liquid infused porous surface on Q235 carbon steel for inhibiting microbiologically influenced corrosion. *Colloids Surf. A: Physicochem. Eng.*, 631, 127696, 2021.
139. Liang, Y., Wang, P., Zhang, D., Designing a highly stable slippery organogel on Q235 carbon steel for inhibiting microbiologically influenced corrosion. *ACS Appl. Bio Mater.*, 4, 8, 6056–6064, 2021.
140. Yang, Z., He, X., Chang, J., Yuan, C., Bai, X., Facile fabrication of fluorine-free slippery lubricant-infused cerium stearate surfaces for marine antifouling and anticorrosion application. *Surf. Coat. Technol.*, 415, 127136, 2021.

141. Liu, H., Ma, Z., Yang, W., Pei, X., Zhou, F., Facile preparation of structured zwitterionic polymer substrate via sub-surface initiated atom transfer radical polymerization and its synergistic marine antifouling investigation. *Eur. Polym. J.*, 112, 146–152, 2019.
142. Lim, C.S., Neo, M.L., Zainul Rahim, S.Z., Andre, R.D.C.T., Teo, S.L.M., Investigating the effect of temporal variation on fouling settlement using a rapid assay method. *J. Coat. Technol.*, 16, 761–769, 2019.
143. Jin, H., Wang, J., Tian, L., Gao, M., Zhao, J., Ren, L., Recent advances in emerging integrated antifouling and anticorrosion coatings. *Mater. Des.*, 213, 110307, 2022.

7

EIS Study of Anticorrosive Nanocomposite Films

Vandana Shinde

Department of Physics, School of Physical Sciences, Kavayitri Bahinabai Chaudhari North Maharashtra University, Jalgaon, India

Abstract

Nanocomposite coatings were formed by combining the inorganic nanoform dispersed phase in the organic matrix phase. The main goal of such nanocomposite is to enhance required properties. As per types of dispersed and matrix phases, the specific properties that are under study are improved. The corrosion of metallic parts is a big issue now a days. Hence, the requirement is to discover the new coatings which will inhibit the corrosion effectively. Accordingly, several groups reported many research publications that explore a variety of coatings for corrosion prevention. Recently, new organic-inorganic nanocomposite–based corrosion-protective coatings have the main attention due to their excellent corrosion resistance and mechanical behavior. Two-dimensional nanostructures synthesized from dispersed phases in nanodimension form in microscopic matrix phase or nanostructured matrix phase are utilized to enhance corrosion protection ability. Among the variety of coating materials, inorganic and organic nanocomposites including at least one phase or both phases in nanodimensions are the interest of corrosion scientists. Such coatings behave as a smart coating due to their ability to repair themselves caused by ion exchange between nanofillers and matrix.

The anticorrosive properties of organic-inorganic nanocomposite coatings dominantly are affected by several factors such as a variety of nanoparticles, their morphology, dimensions, surface area to volume ratio, % of nano form dispersed phase in the matrix phase, the chemical structure of matrix and dispersed phase, and the interaction of dispersed phase with matrix phase. Furthermore, corrosion protection performances are also influenced by the dispersion and synthesis methods of both phases. This corrosion protection mechanism is based on physical/chemical/mechanical surface interactions between both phases and it is

Email: vdashinde@gmail.com; ORCID ID: 0000-0001-6117-191X

Hari Murthy, Vinay Jha Pillai, Kukatlapalli Pradeep Kumar and Matthew Cowan (eds.) Novel Anti-Corrosion and Anti-Fouling Coatings and Thin Films, (155–210) © 2024 Scrivener Publishing LLC

ion exchange interactions with corrosive media. To evaluate these mechanisms and achieve, this goal corrosion testing is required. The characterization of these coatings informs their physical and chemical properties before their actual applications. The actual application performances of the nanocomposite thin films or coatings on substrates are estimated by electrochemical impedance spectroscopy (EIS) at the laboratory level. EIS is a uniquely powerful tool that gives the answers to the following questions.

1. How do these organic-inorganic nanocomposite coatings provide actual performance?
2. Why do these organic-inorganic nanocomposite coatings provide better performance than ordinary coatings?
3. What are the realistic working mechanisms of these coatings?

This chapter covers the importance of organic-inorganic nanocomposite coatings, a literature survey of the fabrication of these coatings its corrosion testing by EIS. Further along with some basics of the EIS technique, its importance, evaluation of corrosion performance, and investigation of the working mechanism of nanocomposite coatings during the corrosion process.

Keywords: Anticorrosive coating, nanocomposites, electrochemical impedance spectroscopy (EIS), principles and mechanism

7.1 Importance of Organic-Inorganic Nanocomposite Coatings

In industry, a variety of metallic and alloy parts are required to make machinery and or tools since they are durable and possess excellent mechanical properties. But as per the tendency of metals and alloys, they were stabilized in oxides. These are its original states which are thermodynamically stable forms formed via oxidation due to interactions with surroundings. These oxidation reactions result in corrosive products on surfaces of metals and alloys which alter the properties of surfaces of metals and alloys through the corrosion process [1, 2].

Due to the chemical reactions with the environment, the deterioration of metals and alloys observed in this process is termed as 'corrosion' of materials. This corrosion process converts the metallic surfaces that are immersed in a corrosive medium to form insoluble compounds such as oxides, nitrides, sulfides, or chlorides of respective metals/alloys. This is the undesirable deterioration of a metal or an alloy since an interaction of the metallic materials with its environment changes the main properties of the metals/alloys. Therefore, its protection or prevention is required for a lifelong and economical application of metallic equipment in industrial processes [3].

Mainly, metallic parts from iron, copper, aluminum, and its alloys have been utilized in various petrochemical, nuclear power and pharmaceutical industries for storing various types of acidic, alkaline, and salt solutions due to their good strength. However, due to its active nature with corrosive medium, it experiences corrosion. Every year, losses due to corrosion are a number of millions [3, 4].

Initially, to avoid this corrosion, protective coatings or paintings are used. These coatings and paints synthesized organic coatings, primers, varnishes, etc. are polymeric types. But, difficulties arise as coatings fail due to frictions observed during machinability and workability processes. Furthermore, if coated surfaces are exposed to seawater, acidic, alkaline medium, pollute gasses, etc., then may be the porous nature of the coating allows it to enter the corrosive ions inside the coating and it fails or deteriorates [3].

Traditional Coatings Versus Organic-Inorganic Nanocomposite Coatings

Anciently, 2000 years ago, fire gilding and silvering mercury-based processes were employed for thin coating on jewelry, statues, antiques, optical parts, metallic gears, and decorative items with gold, silver, and mercury glue. But, industrial growth and development explore a variety of alloys that were utilized in technology as engineering materials. The alloying process reduces crevice corrosion but pitting corrosion may be observed in some alloys. Therefore, researchers are turning toward a variety of organic and galvanizing coatings which are different than traditional coatings. Polymeric epoxy, polyester, lubrications, nylon, urethane, etc., were used for corrosion protection. Initially, galvanizing, electrochemical coatings, vapor deposition, conversion coating, and thermal spraying surface coating methods were employed. These methods have some advantages as well as disadvantages. Nowadays, various coating methods are employed including sol-gel, drop-casting, dip coating, electrospraying, electrospinning, and layer-by-layer deposition with some advantages and disadvantages. Simultaneously, surface conversion methods such as phosphating and chromating via anodizing are also utilized in some industrial applications. But, all these coatings and coating methods are not able to create any perfect anticorrosive coatings. Hence, composite materials can be a substitute for corrosion protection coatings explored by many research groups.

Furthermore, its limited life motivated some researchers to use it as one phase among dispersed or matrix phases in nanoform. Different types of organic-organic, inorganic-inorganic, or organic-inorganic nanocomposites are utilized as an anticorrosive coating on metallic and alloying parts for corrosion prevention. Various factors such as the selection of phases, their types, their nanodimensions, chemical and physical properties of both phases, their morphology, percentage compositions, and synthesis process parameters affect the anticorrosive performance of nanocomposite coatings [2, 5–11].

Need for Organic-Inorganic Nanocomposites

Among the variety of coating combinations of nanocomposites, long durability and smart healing nature are generally observed in the case of organic-inorganic nanocomposite coatings. Smart coatings are the coating that will repair itself after pinhole to the coating. Organic conductive polymers as a matrix phase while metal oxides as a dispersed phase in nanoform-made nanocomposites are observed as smart coatings against anticorrosive applications. Obliviously, its lifetime depends on various factors related to coating and concentrations of corrosive medium [2, 5, 10, 11].

Therefore, recent research groups have explored various coatings, and their corrosion protection performances were evaluated in various corrosive media [2, 5–10, 12–17]; a few of them are surveyed in Table 7.1.

Need of Corrosion Testing

Complete corrosion prevention is not possible, but to reduce its losses, many alternatives are utilized such as anticorrosive smart coatings/paints, anticorrosive pigments in coatings, anodic/cathodic protection, and smart inhibitors.

For corrosion testing, many methods are available in the literature. For evaluation of corrosion protection performances, some corrosion testing methods are laboratory tests, salt spray, and acid fog tests which are industry standard American Society for Testing and Materials (ASTM) tests [18] accepted worldwide for reliable metallic and alloying parts [19]. These tests are accredited by international standards on (ISO) corrosion protection for different alloys. Quality control and quality assurance are the most important. On the other hand, researchers and scientists use various corrosion testing methods such as the use of inhibitors, immersion, and visualization tests. But, the use of inhibitors is not suitable every time metallic parts are not in contact with corrosive solutions. Furthermore, the use of inhibitors in the sea creates some problems such as concentrations of inhibitors producing toxicity and other living organisms in seawater affecting the overall natural cycle. Hence, coatings or paints are better compared to inhibitors.

Table 7.1 Corrosion protection performances of a few nanocomposite coatings published.

Sr. no.	Nanocomposite coatings	Corrosive solution	Corrosion tests	Reference
1.	Electroplated pure Zn and Zn-graphene oxide	3.5 wt% NaCl aqueous solution	Potentiometric, potentiodynamic polarization, EIS, and salt spray tests	[9]
2.	Nanocomposites of cloisite 30B nanoplates and polyaniline emeraldine base particles reinforced with Epoxy-siloxane	Atmospheric agents	Adhesion test (pull-off), and impact tests. Q-UV test, AFM, and color spectroscopy	[6]
3.	Nitrogen and fluorine-doped graphene oxide layers in between polyaniline (PANI) layer–type composites	H_2SO_4 and NaOH	EIS and salt spray tests	[7]
4.	PANI, ZnO, Zn, and graphene composite coating on aluminum alloy	3.5 % NaCl	Potentiodynamic polarization and EIS	[8]
5.	CeO_2-GO/epoxy nanocomposite	Hybrid solutions	EIS, Tafel plots	[17]

Laboratory-based immersion corrosion tests along with electrochemical techniques such as potentiodynamic polarization, linear polarization, measurement of open circuit potential, cyclic polarization, cyclic voltammetry (CV), and electrochemical impedance spectroscopy (EIS) methods are widely used [20–23]. After these tests, by varying immersion times, inspection of coatings by microscopic and spectroscopic levels is necessary to understand the morphological and chemical changes of coated surfaces, respectively. In all such ways, one can predict coating life and durability in various corrosive environments. Among all the electrochemical techniques, the EIS is one very interesting and precise technique to predict the corrosion performance of coatings [21, 24].

The life prediction of nanocomposite coatings including insulating and conducting can be approximated with the help of the small-duration EIS study to the long-duration industrial corrosion test study [5, 22, 25–29]. Really, EIS is a remarkable tool to evaluate the corrosion prevention process of various coatings in different corrosive environments or media and provides corrosion rate, inhibition efficiency, etc. Therefore, for the inspection of coatings, the EIS technique is utilized, which indirectly helps with its selection, optimization, and suitability for particular applications. Throughout, immersion/exposure testing in various aqueous and nonaqueous atmospheres due to the redox behavior of nanocomposite coatings of physical and chemical properties is changing. Therefore, investigating the corrosion mechanism offered by these coatings has been the most interesting in recent years. Corrosive anions change the surface characteristics of corrosion-protective coatings because of their electroactive nature. During the immersion test due to the absorption of corrosive ions inside the pores of coatings pinholes or cracks like defects/imperfections are produced on the surface of the coatings. Many research groups have explored the EIS technique for the determination of corrosion-protective performances of nanocomposite coatings especially for understanding corrosion mechanisms [30–34]. Hence, the next section of this chapter describes the detailed EIS technique.

7.2 EIS: Brief Basic Principle and Introductions

7.2.1 Brief History

Oliver Heviside introduced the concept of electrical impedance in 1880. He created the concepts of inductance, capacitance, and impedance in the case of electrical circuits for the first time. Actually, during the work related to the calculation of the dielectric constants for aqueous electrolytes, Nernst used a Wheatstone electrical bridge. Later on, a few people such as J. Hopkinson and E. Wilson (1897), J. Dewar - J. A. Fleming (1898), C. H. Ayres (1902), A. Finkelstein (1902), E. Warburg (1901) applied this Nernst approach. One interesting observation of the literature the EIS concept was first time applied in the field of cytology and physiology in the duration of 1930–1940. Furthermore, Cole and Cole's brother show a frequency-dependent dielectric constant in 1941, representing a depressed semicircle admittance plane plot, now called a complex impedance plane plot. Simultaneously, Frumkin and Randles use this method for the study of kinetic reactions that occur near the vicinity of the electrode. Before the 1950s, impedance commenced to be utilized to understand the complex process of the systems. Up to the year ~1970, this technique has been applied for the study of metallic dissolution and corrosion process. Furthermore, in the duration of 1971–1990, this is used for analysis of mixed conductors and redox materials. Later on from 1991 to various fields, this technique is used for the investigation of the interaction of heterogeneous surfaces and their behavior with surroundings such as sensing hazardous gasses, biological enzymes, and corrosion protection. Simultaneously, it is also used for the characterization of supercapacitor batteries and fuel cells like energy devices [24].

7.2.2 Significance of EIS

Various electrochemical techniques are utilized for the characterization of modern coatings especially the measurement of currents, electrical charges passing through the electrolytic system, or electrode potentials with respective time and frequency of input signal which represents the coating's behavior or variation in properties of coatings. EIS is one extraordinary technique like optical spectroscopy since it presents electric impedance which varies according to the frequency of the signal at a constant potential. In optical spectroscopy, the source of electromagnetic radiation represents sinusoidal oscillations at various frequencies, phases, and amplitudes. Therefore, representing optical parameters as a function of frequency domain means wave number, frequency, or wavelength and uses Fourier transform (FT) of the optical signal [35].

Similarly, to understand the EIS, some knowledge of mathematics such as the FT, Laplace transform, and complex numbers is necessary. But, EIS has several advantages. Basically, it gives useful qualitative and quantitative information about the system that can be further analyzed [29]. It contains analyzable information at each frequency. It gives qualitative and quantitative information of modern coatings utilized in various applications. EIS is not enough to characterize complicated systems but other studies such as microscopic level surface morphology, structure analysis, compositions of coating, and other DC electrochemical characterization are required, to understand carefully EIS analysis. EIS study includes the modeling of electrical circuits to EIS spectra which helps to understand systems mechanism.

EIS provides a lot of information, but it does not provide information on all physical and chemical processes. Generally, EIS is used to explain the slow reaction process related to the system well. It reveals the rate of chemical reactions via resistances and capacitances. It also tells *in situ* real surface areas during the reactions that occur near the vicinity of electrodes. Hence it is a very sensitive technique but careful expertise will answer all electrochemical reactions [24, 29, 36, 37].

From the literature, it is observed that the deconvolution of impedance data is described. In the 1920s, the EIS technique was performed on biological samples such as blood and plasma/cell membranes to find the dielectric properties and impedimetric performances [22, 24, 27]. Later in the 1950s, impedance studies were employed for anodic dissolution studies. Therefore, impedance measurements developed significantly for studying corrosion, deposition of films and their growth, and some other electrochemical processes. In reality, thin film/coating on electrode substrates whose surfaces are not homogeneous causes complications during the analysis of the impedance responses in terms of consequential physical properties or parameters [24, 33, 36]. Many researchers have used local impedance spectroscopy for studying heterogeneous complex electrode surfaces.

Therefore, the EIS technique became popular in understanding the electrochemical processes and as a consequence in electrochemical engineering. Simultaneously, developments in instrumentation due to innovative electronics/control systems increased its performance and sensitivity, and its use in various fields opened an era with wide scope in practical/industrial applications [29, 36, 37]. Furthermore, Frumkin and Grahame discovered the double-layer theory; Randles and Warburg exposed an equivalent circuit (EC) modeling approach for the presentation of impedance data and plots.

Therefore, EIS became important in various fields for the investigation of various physical mechanisms. Electrochemical reactions such as diffusion and adsorption processes via the porosity of coatings were explained by various electrochemists Gerisher, Eppelboin, de Levie, and Newman. This electrochemical kinetics shows the non-uniform current and potential distribution which creates dispersions in an impedance study of systems. Kramers-Kronig relations and nonlinear complex regression like various mathematical and statistical methods were employed for analyzing the existing systems with EIS theories [36]. These methods reveal relations between dependent variables and single or multiple independent variables.

During the past 40 years, due to developments in electrochemical equipment and computer sciences, good selective/sensitive potentiostat and galvanostat devices were built up that are capable of providing advanced quality, improved controlled, noise-reduced automated impedance measurement data as compared to the earlier instrument.

Thus, the achievements completely changed the branch of electrochemistry as a consequence of impedance spectroscopy. Hence, it can be employed in various fields. In solid-state or suspension forms, samples are analyzed by a similar technique explored as dielectric spectroscopy which has been generally used for evaluations of conduction mechanisms since from year 1950. Furthermore, this study has begun to understand the corrosion behavior of biomedical implants inside a human body [36, 38].

If electrolytic cells are in solid-state form, then interphase between electrode and electrolyte is significantly more complicated than liquid-state electrolytic cells or wet cells since polycrystalline solid electrolytic cells differentiate their aspects. Due to the applied voltage, polarization of each interface is observed. Moreover, the polarization rate mainly depends on the signal frequency and sample nature. At a certain polarization rate near the polarized region, a chemical reaction in between the interface of atmosphere, electrode surface, and solid electrolyte slows down but is faster near the grain boundaries of solid electrolyte.

In this way, the importance of electrochemistry changed from concentration/time domain to frequency domain studies at small AC signals. During this study, electric double layers and associated intrinsic reactance are measured by relaxation time or dispersion. For heterogeneous systems, these electrical responses vary according to ionic species, charges on it, type of electrolyte, nature, microstructure, and texture of electrode or substrate. Hence, impedance spectroscopy is a comparatively interesting and dominant technique for indemnification of electrical properties of bulk or film form samples as well as the interface between electrode and electrolyte. Furthermore, it also tells the kinetics and dynamics of diffusion reactions through a dry or wet system of any type of sample such as ionic or dielectric systems [24, 29, 36], although we shall mainly focus, in this chapter, on inorganic-organic nanocomposite–based sample coatings which are made by at least one phase in nanodimension form.

EIS has been used in several applications on that basis; it is employed in various types of analyses given as follows:

1. Interfacial Reactions: These are oxidation-reduction reactions occurring near the vicinity of electrodes, via ion or electron adsorption processes, mainly due to forced mass transfer. Such reactions involve the transfer kinetics of homogeneous reactions carried in the solution or electrolytic bath associated with redox processes.
2. Effect of Geometry of Electrodes: Electrodes or substrates are of various shapes; hence, as a consequence, linear, spherical, and cylindrical mass transfer is observed in localized volumes in between electrodes and electrolyte layer, which provides the solution resistance, generally in the case of porous electrodes.
3. Claims in Power Sources: To understand the various phenomena related to energy storage and/or energy conversion in various devices such as batteries, fuel cells,

supercapacitors, and membranes, the EIS technique is a popular test [39–41]. In the case of corrosion testing of coatings and paints, it is also required. In some chemical applications, water electrolysis, Cl_2 evolution such as electrocatalytic processes, nature and behaviors of electroactive polymers, self-assembled layers, biomembranes, sensing mechanisms, and semiconductors behaviors are also investigated by the EIS technique [29].

We all are aware of recent developments in various fields such as rechargeable or high-power-density energy storage devices, and a modernization in electrochemical sensing elements in ecological systems. Furthermore, energy efficiency control and the introduction of fuel cells to avoid the Carnot inefficiency in noncatalytic energy conversion are also the main interests of researchers. All these phenomena are dominantly analyzed by EIS [41]. Therefore, in all these fields, a significant number of publications including the EIS study is observed from 1996. From surveying the number of publications and a number of citations from the year 1993 in this field, a histogram is plotted and shown in Figure 7.1.

In electrochemical energy engineering, to avoid corrosive aqueous solutions, researchers now prefer solid-state technology which is handiness

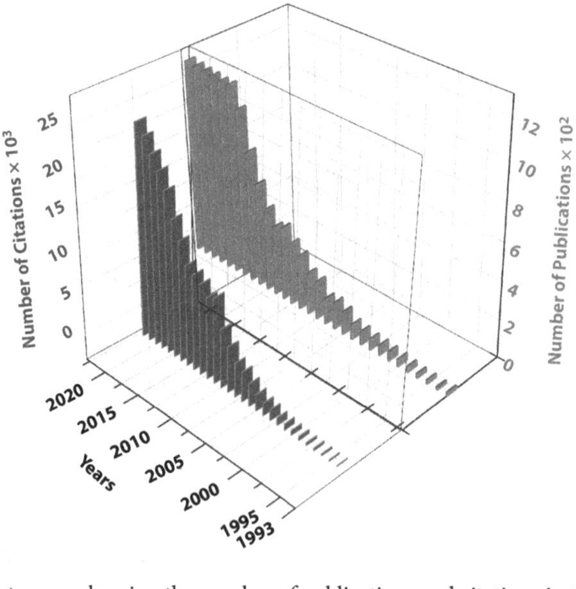

Figure 7.1 Histogram showing the number of publications and citations in the years 1993–2022.

and safety. Hence, for the analysis of the various systems, interfaces with solid-solid or solid-liquid always contain ionic conductors in solid and generally perform at room temperature, which makes this technique significant for materials scientists. Contact between interfaces with surrounding environments modifies the physical properties and chemical nature of coatings due to crystallographic, and compositional level varies, and, as a result, electrical properties also change abruptly. Therefore, charges are distributed heterogeneously in the system, and the electrical conductivity of a system decreases [29, 31, 42, 43].

7.2.3 Basic Principle

In situ electrical methods are employed for the study of an electrochemical system, such as polarography and CV since the electrical responses can be related to the surface reaction kinetics. EIS is a frequency response technique, in which by application of a potential to an electrochemical electrode system, the corresponding sinusoidal current is measured [44].

Usually, an EIS spectrum of the system that is under study is established by sweeping the frequency of the signal and by measuring the impedance offered by the system for the transportation of charges at every time interval, because impedance measurement shows a nonlinear nature during the polarization of electrochemical electrodes due to a small amplitude signal perturbing the system. The linear region of polarization curve is confined, this is referred to as a pseudo-linear system. Furthermore, the output response will oscillate at the input frequency and show a shift in phase without harmonics.

From the input and output signals, determination of the cell impedance is possible and later its analysis is performed. Mathematically, the input potential signal using Cartesian and polar variables is given as

$$\vec{V}(t) = V_0 cos(\omega t) = V_0 exp(j\omega t) \tag{7.1}$$

where V_0 is the signal amplitude of the input potential with varying frequency at a certain time t, and ω is the angular frequency. In addition, corresponding current I_0 and phase shift ϕ are as follows:

$$\vec{I}(t) = I_0 cos(\omega t - \phi) = I_0 exp\{i(\omega t - \phi)\} \tag{7.2}$$

where "*i*" is the imaginary number (see Figure 7.2).

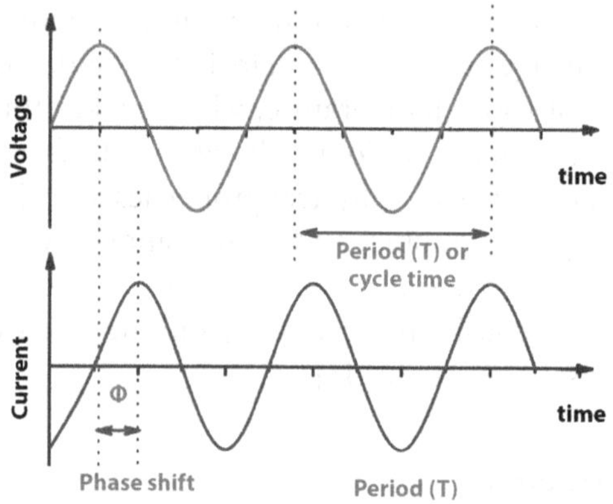

Figure 7.2 Potential and current variation with time showing phase shift.

The complex impedance according to Ohm's law is given as

$$Z = \left(\frac{\vec{V}}{\vec{I}}\right)_t = \frac{V_0 cos(\omega t)}{I_0 cos(\omega t - \phi)} = |Z_0| \frac{cos(\omega t)}{cos(\omega t - \phi)} = |Z_0| exp(i\phi)$$
$$= |Z_0|(cos\phi + isin\phi) = Z_r + iZi \tag{7.3}$$

Here, Z_r and Z_i are the real and imaginary parts of the impedance, respectively.

The Nyquist plot displays graphically complex impedance by plotting the negative of the the imaginary part of the impedance on the ordinate axis, while the real part of the impedance is on the abscissa axis [36].

Always, the EIS curve is interpreted modeling an equivalent circuit model created by resistors, capacitors, and other elements either series or parallel or its various combinations. A sample "Nyquist plot" is shown in Figure 7.3. This shows that each data point of the Nyquist graph indicates the impedance value of the system at a certain frequency value of frequency range [45]. The Nyquist graph shows that the measured data readings from low-frequency data points are on the right side to the higher-frequency data points on the left side. In the Nyquist plot, the impedance can be symbolized as a vector (arrow) of length |Z| which indicates its magnitude. The angle made by this vector and the x-axis is ϕ (=argZ). The shortcoming of Nyquist plots is any data point on the plot that cannot tell the frequency value that was employed to note down that data point [4, 24, 29, 37, 46, 47].

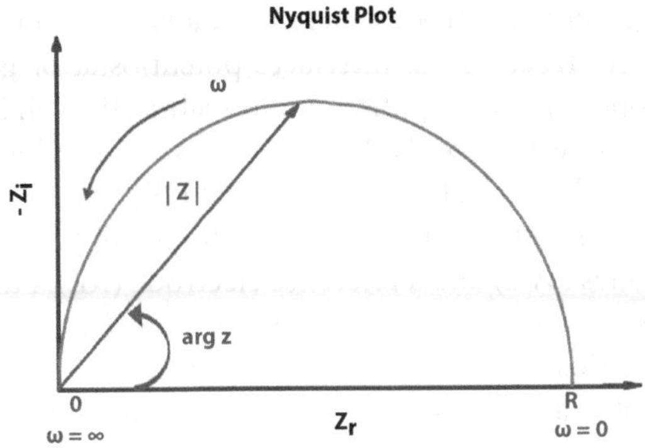

Figure 7.3 Nyquist plot.

7.2.4 Impedance Measurement Techniques

Before the invention of the digital computer, during all electrochemical experiments, the measured analog signals conditioning and its analysis were represented either in the time or frequency domain. Furthermore, for determination of the impedance of such signal, ac-coupled bridges and Lissajous figures were used which provide information in analog format. The frequency domain impedance measurement experiment involves the low-amplitude ac excitation signal with variable frequency as the independent variable. On the other hand, in the time domain impedance measurement experiment, time is an independent variable. In addition, impedance versus frequency behavior can be obtained using Laplace or FT methods. In this way, digital processing is needed in the time domain, which is common in analysis nowadays, while frequency domain methods are traditional analog methods. Sine wave excitation to electrochemical cells is used to understand the mechanistic and kinetic reactions that are observed during different applications. Macdonald, Gabrielli, and McKubre attempted such a response of electrochemical reactions for the first time. Basically, the input of the system is the current as an excitation function and the output signal is the voltage, and then the transfer function is the impedance of the system. However, during ac excitation amplitude and phase of current changes, the response of the system is impedance which is a complex quantity [48].

7.2.4.1 Instrumentation

The basic impedance spectroscopy experiment consists of an electrochemical cell connected to an electrochemical interface potentiostat or galvanostat together with an AC generator/analyzer [see Figure 7.4 (a) and (b)]. Basically, electrochemical interfaces potentiostat or galvanostat can be built up by operational amplifiers and resistors. The cell is connected to a frequency response analyzer (FRA) or phase-sensitive (lock-in) amplifier directly or indirectly through the potentiostat [29].

A potentiostat maintains the constant potential difference between the working electrodes (WE) and a reference electrode (RE). Furthermore, the RE is joined to the inverting input of the operational amplifier [24].

A basic potentiostat can enclose two operational amplifiers: one as a potential controller and the other as a current follower. A control amplifier maintains the essential voltage and current across the electrolytic system when it is connected in potentiostatic and galvanostatic mode, respectively [24, 37].

A basic potentiostat for impedance measurements needs a voltage adder to superimpose the sinusoidal signal onto an imposed potential, as shown in Figure 7.5. A potentiostat is used in most setups but is not always necessary [24]. It is used for potential control and to provide high input impedance if needed. It also eliminates would-be bandwidth problems. The AC generator/analyzer is either an FRA or a lock-in amplifier, which is used to generate the perturbation waveform and analyze the response [49]. The type

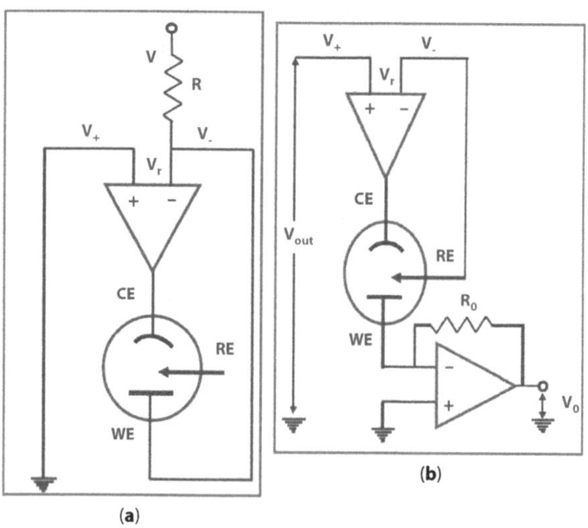

Figure 7.4 (a) A simple scheme for controlling the current through a working electrode (reproduced with permission from [24]). (b) Potentiostat with the current measurement by use of a current follower (reproduced with permission from [24]).

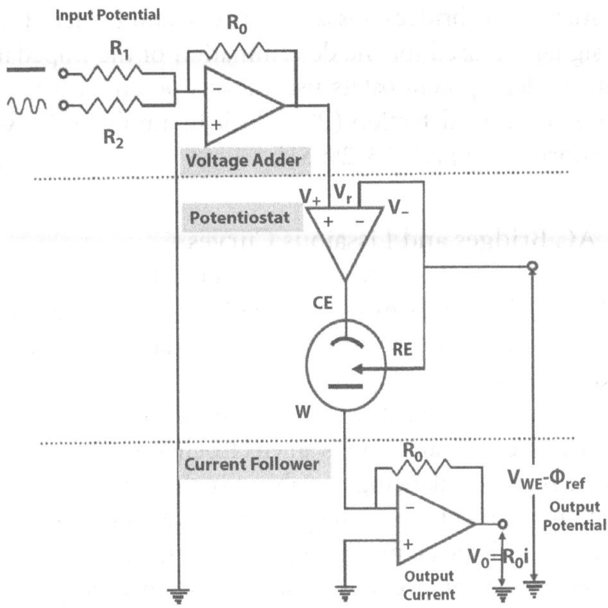

Figure 7.5 Potentiostat for EIS measurement (reproduced with permission from [24]).

of AC generator/analyzer to use depends on the intended experiment. Most instrumentation allows for measuring frequencies between 100 kHz and a few megahertz down to a few hertz since, for most kinetic studies, the information of interest can be obtained using a more moderate frequency range. High-frequency data can also contain artifacts caused by instrumentation, such as RE, potentiostat, and connecting cables. Low-frequency measurements require significant time to perform and the system measured must display a high stability.

Low-frequency limit is selected on the basis of acquisition time and res olution of their improvement. Generally, 12 to 15 points per decade and a lower-frequency limit of 40 Hz are acceptable to achieve the good accuracy and time duration required for data measurement [24, 50].

7.2.4.2 Comparisons Between Various Techniques for Impedance Measurement

During EIS measurements, an insignificant sinusoidal voltage signal having restricted amplitude and varying frequency perturbed the system along with the DC electrical program. Generally, various other methods are also used for the measurement of system impedance. In earlier days, techniques such as AC bridges, Lissajous curves, and Laplace transform of an applied signal are used for the determination of the impedance of the system. But, modern potentiostats use various potential plans for a WE such as phase-sensitive detection (PSD)/lock-in amplifier, FRAs, and fast Fourier transform (FFT) [10, 23, 29, 51].

7.2.4.2.1 AC Bridges and Lissajous Curves

AC bridge- and Lissajous curve-related methods were used in earlier days. These methods provide exact information but are very laborious and time-spending hence these are rarely used nowadays for impedance measurements of the system.

Basically, in an AC bridge, the main circuit is an ordinary Wheatstone bridge that possesses one arm and is an electrochemical system along with the additional regular balancing variables, such as resistance or capacitance. In such instrumentations, the potentiometer supplies the dc potential at the center and the ac potential given by an external source signal. The double-layer capacitance was measured using a dropping mercury electrode (DME). Bridges were compensated at the DME surface areas at the same time from the beginning of the voltage drop. AC potential perturbation to the electrochemical system is responsible for showing phase shift in current as compared to voltage and hence voltage lags behind the current. Therefore, the dual channel oscilloscope shows its I-V characteristics mean, Lissajous curves, which vary according to phase shift.

If the phase angle difference, φ, in between current and voltage is zero, then the real and imaginary impedance plot shows a straight line at 45°. Furthermore, if the phase difference is 90°, a semicircle is observed in impedance; for intermediate phase angle shifts, in between 0° and 360°, ellipses are shown in Figure 7.6. The amplitude ratio of voltage to current signals provides the modulus of the impedance, |Z|. The phase angle φ is calculated from the tilt of the ellipses at different phase angles. These |Z| and φ were determined from the real and imaginary components of the impedance measured at all frequency ranges and are useful for the investigation of particular applications. Sometimes, the nonlinearity in such signals is responsible for the formation of asymmetric impedance curves. Therefore, due to these limitations, these methods use frequency analyzers or lock-in amplifiers [29, 32].

7.2.4.2.2 Lock-In Amplifiers for Phase-Sensitive Detection

Automatic determination of the impedance is possible using two additional methods which are included in potentiostat. The PSD is employed

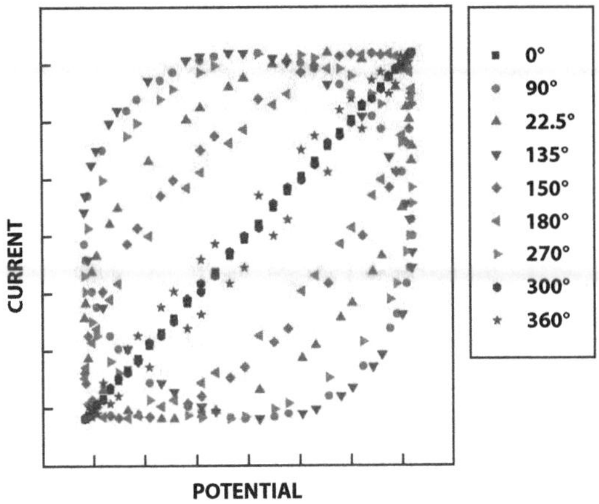

Figure 7.6 Lissajous curves for various phase differences; the values of φ in degrees are shown in the legend.

in lock-in amplifiers. EG&G Princeton Applied Research explored acceptable commercial instruments using lock-in amplifiers. In this method, the square wave signal that measured a signal of similar frequency ω is multiplied. This square wave signal is nothing but an infinite Fourier series made by the sum of sine functions with odd harmonics of fundamental frequency and has unit amplitude. Thus, both signals possess different phases. The initial phase shift may be assumed zero since both signals are multiplied by a similar square signal.

In practice, as both the potential and current signals are multiplied by a similar square wave, its initial phase shift is considered a zero. Normally, a square wave generator is situated indoors of the lock-in amplifier which generates an AC signal and is employed for impedance analysis. The resulting signal of this process includes a stable signal and harmonics of the fundamental signal. These constant signals provide the phase of the output signal having an amplitude that is the same as that of the measured signal.

The average periodic signal is input to the low-pass filter, which gives a constant integrating low-frequency signal.

Actually, the measured signal is a mixture of two signals, whose phases are 0° and 90°, separating the real and imaginary components of the measured impedance. The block diagram of the two-phase lock-in amplifier is given in Figure 7.7.

If the measured signal includes harmonic signals, then outputs are decreased. Lock-in amplifiers work in a frequency range of 0.5 to ~10^5 Hz, with precision of ~0.1 to 0.2%. But, recent sophisticated lock-in amplifiers

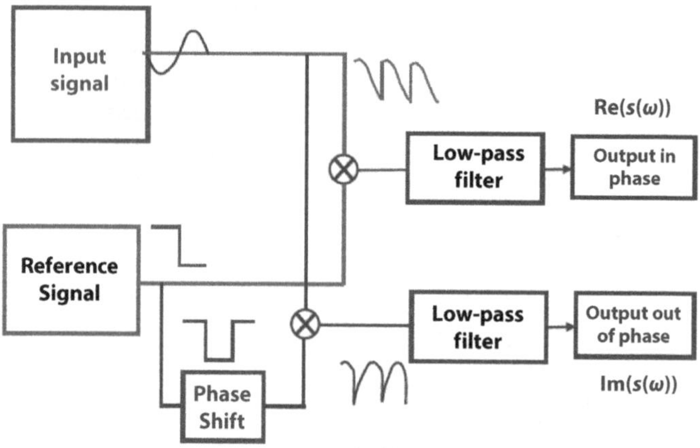

Figure 7.7 Diagram of lock-in-amplifier (reproduced with permission from [24]).

permit the computerized measurements with the help of a microprocessor [29, 36].

7.2.4.2.3 Frequency Response Analyzers

Alternatively, for measurement of system impedance at varying frequencies, FRAs are also used, as shown in Figure 7.8. FRA works according to the association of the measured and the reference signal. A cosine and sine signal from a frequency generator of the same frequency multiplied by the measured signal. This process continued resulting in all these products being integrated during many wave periods. A two-phase lock-in amplifier gives integration of n wave cycles. FT is the mathematical operation that creates

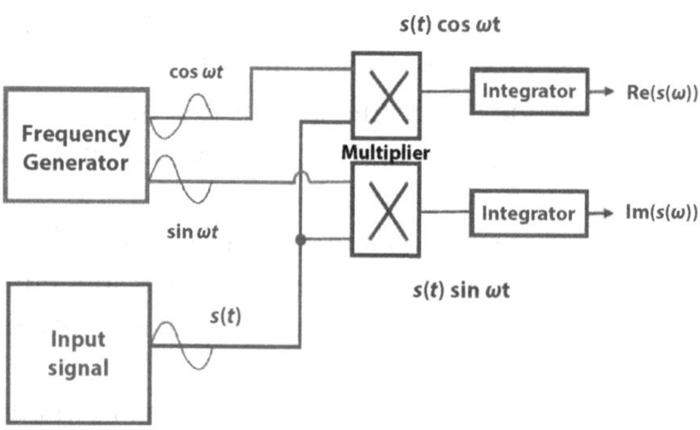

Figure 7.8 Diagram of frequency response analyzer (reproduced with permission from [24]).

real and imaginary components of the measured signal with less noise [29, 36].

For both the potential and current signals, FT operation is performed, and its ratio gives the impedance of the system. The working of an FRA is shown in Figure 7.8.

FRAs were widespread in electrochemistry, hence Solartron, and many companies, such as PAR–AMETEK, Oak Ridge, Tennessee, Metrohm Autolab, Utrecht, Netherlands, ZAHNER, Kronach, Germany, Scribner Associates, Southern Pines, North Carolina, and ZiveZcon, Seoul, Korea, frequently use FRA in instrumentation. It firmly removes all the harmonics, existing in the input signal. FRAs correlate the signals $\sin(k\omega t + \varphi)$ with $\sin(\omega t)$ or with $\cos(\omega t)$ resulting in zero for $k > 1$. The benefit of this correlation is mainly noise reduction or unwanted frequency since it affects the integration time [1].

Hence, the output signal reduces with the varying frequency, and the number of integration cycles changes. All calculations are computerized and performed by recent FRAs. FRAs have a broad ~12-decade frequency range with eminent precision [29]. Evans was given a comparison between the lock-in amplifier and FRA [24, 31, 34, 52–55] which is shown in Table 7.2.

Table 7.2 Comparison of lock-in amplifier and frequency response analyzer [29, 56] (reproduced with permission from [36]).

Lock-in amplifier	Frequency response analyzer
Advantages	**Advantages**
High sensitive	Rapid analysis
Capable of noise removal well	Capable of working at a wide frequency range
Reduces harmonic distortion	Removes harmonic distortion effectively
Suppresses dc noise easily	Direct output to an external device
Relatively low cost	Easy standalone measurement
Disadvantages	**Disadvantages**
Slower analysis	Limited sensitive
Small frequency range for working	Limited noise removal
Standalone readings difficulty	Higher cost

7.2.4.2.4 FT Methods

The FT method is a mathematical operation performed on signals of a series of varying frequencies resulting in a system output response. The transfer function of impedance, of the measured system, is determined by FT operation on perturbation signal. Fast FT (FFT) is a rapid and competent algorithm, which is capable of computing all FT signals. But, for this, the number of points required is 2k (here, k is an integer). The invention of the FFT algorithm gives discrete FT (DFT), which does not require such criteria and the true power of performance of computers increases significantly. Generally, an excitation signal can be used as a pulse or step, noise, or sum of sine waves.

But, the important point is that the total low signal should be applied to the system to maintain the working of the system in a linear manner.

Although FFT method provide reliable output with improving the signal to noise ratio very fast and hence it is employed for measurement of impedance as a function of time. But the drawback of this method is applying single frequency its response at certain frequencies is inefficient. In addition, other Laplace transform–type perturbation methods are also employed for the measurement of impedance responses of the system [24, 29, 37].

7.2.4.2.5 Dynamic EIS

EIS is used at a constant potential or current, referred to as a stable or stationary method. However, CV is a technique in which the potential cycle repeats with time and it is also capable of measuring the impedance of the system, called dynamic EIS.

Some researchers called this potentiodynamic EIS (PDEIS) or time-resolved impedance spectroscopy. During slow cycling of electrodes, normally, the sum of odd harmonic signals is applied in this method.

With the help of a lock-in amplifier or FRA, it is probable to get system impedance during the AC or DC potential cycle.

At a low scan rate of voltammetric sweep, an AC signal can be superimposed. In addition, the AC response of the system is measured with frequency in a particular time duration which is a function of potential. Repeating this experiment a number of times by varying frequencies of signals, a complete impedance spectrum can be achieved. For individual cycles, complex admittance is usually recorded at each experiment. But, for this type of experiment, a low scan rate and the lowest frequency were seen. If FFT is used in time windows then it is called pseudostationary and during each window at varying frequencies of signal impedance is measured. Therefore, for performing such an experiment, the steady time interval is acceptable. Several research groups developed its DEIS instrument along with its software for controlling as well as the acquisition of impedance data [29, 36]. Keithley KUSB3116 Data Acquisition Module–based DEIS system's block diagram is explained in Figure 7.9.

The digital-to-analog converter produces the addition of sine waves with total amplitude. A linear sweep generator provides a dc signal, which is combined with an ac signal and given as an input to the potentiostat. The analog amplifier amplifies the current from the potentiostat which is a potential drop against the resistor. The output of potentiostat, is AC form and after removal of DC component which is amplified. Furthermore, these three signals, current, ac potential, and DC potential signals, were measured by analog-to-digital converters (ADC). This ADC samples the current and potential one by one and then adjusts its phase. This is required for measuring the impedance of the cell. In the case of nonstationary systems which are carried slowly with time due to potential sweep, then its impedance measurement is possible by FT-EIS. Furthermore, it is also capable to identify and compute time change as well as nonlinear noise in impedance measurement. Generally, this signal includes frequencies ranging from 0.1 to 10^4 Hz having a root mean square (RMS) amplitude of 10 mV. This experiment determines the noise level at frequency change and indirectly calculates deviations in the impedance of the system.

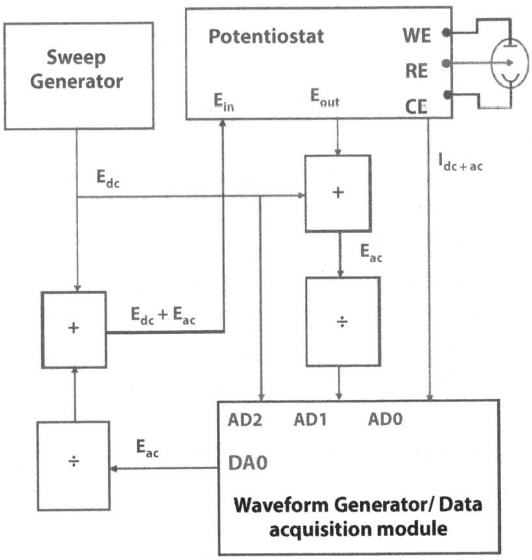

Figure 7.9 Block diagram of dynamic electrochemical impedance spectroscopy system (reproduced with permission from [24]).

This deviation is useful for the identification of linear and nonlinear noise/distortion and amplitude elevation at sudden changes in frequencies which informs the system performance [24, 29, 36, 37].

This understanding is most important in the case of immersion studies of coating since so many chemical changes occur inside the cell or system. In this way, the purpose of FFT analysis is to prevent leakage, data collection, and synchronization with wave periods. The frequencies are applied at the same time hence FT methods are faster and efficient for measuring impedance when potential cycling. Hence, the interest in the FT method is continuously growing as a consequence of achieving popularity in this measurement and even in spectroscopy.

Overall, due to this method, nowadays, EIS is promising to carry out impedance studies of complex systems and all various fields using modern instrumentation based on FRAs, lock-in amplifiers, and FT operations [29, 31, 37, 57].

7.3 EIS Data Analysis: Fitting the Equivalent Electronic Circuit, Introduction of the Various Circuit Parameters Its Physics

7.3.1 EIS Data Introduction

Instead of ac input voltage and radial frequency ω, dc voltage modulation in terms of "offset voltage" or "offset electrochemical potential" is also used in impedance spectroscopy for studying electrochemical processes/reactions. But, "dielectric spectroscopy" or "modulus spectroscopy" is always carried out only with ac modulation without dc offset voltage [24].

The impedance data are described by three different graphs. The Nyquist plot is also called the Argand diagram. It is the graph of the real versus imaginary component (Z' vs. Z") of impedance explained in the earlier section along with its drawback. Hence, the same data in terms of frequency as an independent variable are displayed using two Bode plots shown in Figure 7.10(b) whose dependent variable on the y-axis is [47]

(i) the logarithm of the modulus of the impedance (log |Z|) and
(ii) the phase angle (θ)

versus the logarithm of the frequency (log f), as shown in Figure 7.10. The Nyquist and Bode plots provide several chemical parameters that are

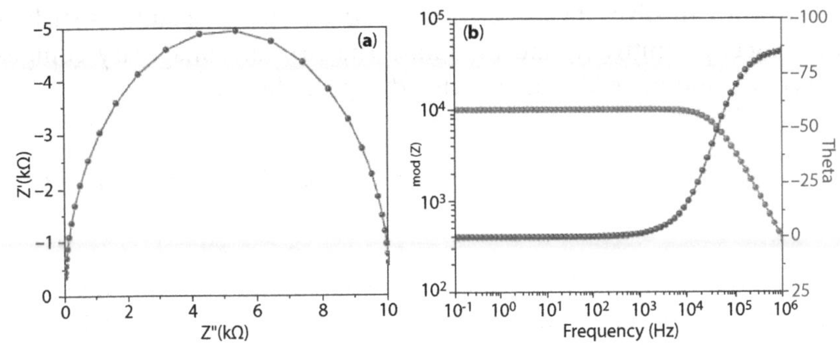

Figure 7.10 Representations of complex impedance data as a function of AC frequency: (a) Z' vs. Z" (b) modulus of impedance and phase angle vs. frequency.

related to the electrochemical cell or system and can be obtained with no or only minor calculations.

Appearance of frequency responses from the Bode plots one can find out information about "dielectric" effects of bulk sample or thin film. These impedance representations convey the interfacial high-to-low impedance values of the system at low to high frequency. The dielectric analysis of the system provides the bulk sample information at higher frequencies and its impact on the low-frequency impedance has to be analyzed. The resulting plot is imaginary (loss factor ε") vs. real(ε') portions of permittivity/dielectric constant (ε) of a material medium, called the "Cole-Cole" plot [36].

This Cole-Cole plot can be identified from the measured data over a range of frequencies. Lossy dielectric also shows a semicircle nature in the Cole-Cole plot, intercepting the real (ε') axis at two points which gives the information on the polarizability of the sample. The dielectric displacement, $\vec{D}$, and the field vector, $\vec{E}$, show phase difference which is responsible for the displacement of the electric polarization, $\vec{P}$. Hence, dispersion of the dielectric constant with frequency was observed. The phase difference originates from the following factors:

(a) electric resistivity, directing to a response component in a phase accordingly E;
(b) relaxation effects, in dielectrics, permanent dipole orientation linked with the polarization component of P; and
(c) high-frequency resonance effects observed due to atomic and electronic polarization [36, 58, 59].

This study involves the associations between various quantities such as impedance, permittivity, and modulus. The real and imaginary components of the modulus M* can be defined as given below:

$$M^* = \frac{1}{\varepsilon^*} = M' - iM'' = i\omega\varepsilon_0 Z^* = -\omega\varepsilon_0 Z_{Im} + i\omega\varepsilon_0 Z_{Real} \quad (7.4)$$

This study employs DC voltage modulation, hence sometimes, it is referred to as "offset voltage" or "offset electrochemical potential", alternatively named "dielectric spectroscopy" or "modulus spectroscopy". EIS is always employed to explain an impedance analysis of the system and is successfully performed only with AC modulation in the absence of a DC offset voltage as shown in Figure 7.10 [24, 36].

7.3.1.1 *Dielectric Spectroscopy*

Basically, the dielectric study focuses on two fundamental quantities of dielectric materials, which involve the variation of permittivity ε and conductivity σ (indirectly resistivity ρ) with the change in time, temperature, and frequency ω. As we are all aware, permittivity and conductivity are the two main parameters that provide its electric energy and charge storage-transport capabilities. These quantities are related to the molecular activity of solids. An "ideal dielectric" is generally an insulator without free charges but it is able to store electrical charges and indirect energy. The relative permittivity ε of this concept is related to material polarization density P [C/m^2]. Basically, it is an electrical dipole moment [C/m] per unit volume [m^3]. By application of an electric field V, polarization varies and is given by the Debye equation as [60]

$$P = (\varepsilon - 1)\varepsilon_0 V \quad (7.5)$$

Depending on the type of sample and the frequency of the applied signal, polarization can be observed in various types. An atomic and electronic polarization is the insignificant translational motion of the electronic cloud at an approximately THz high-frequency range. An orientational or dipolar polarization is observed in permanently polar molecules due to rotational motion observed at an approximately kHz to MHz frequency range. Meanwhile, an ionic-type polarization is due to the movement of ions with each other at an approximately Hz to kHz region of frequency.

The dielectric studies involve the measurement of "complex permittivity" ε^*, which is like a complex impedance Z^* [Figure 7.10(a)]. Similar to a complex impedance, complex permittivity also includes "real" permittivity, ε', which is often called "permittivity" or "dielectric constant" and imaginary permittivity, ε'', is "loss factor" expressed as

$$\varepsilon^* = \varepsilon' - i\varepsilon'' \tag{7.6}$$

The permittivity, ε', symbolizes the alignment of dipoles in dielectric, related to the charge storage element, and loss factor ε'' indicates the conduction of ions which is a reciprocal of real and imaginary parts of Z. These real permittivity and loss factor can be estimated from sample resistance R, conductivity σ, resistivity ρ, and capacitance C, which can be measured from a fundamental experiment (Figure 7.11) given below [36]

$$\varepsilon' = \frac{Cd}{\varepsilon_0 A} \text{ and } \varepsilon'' = \frac{d}{RA\omega\varepsilon_0} = \frac{\sigma}{\omega\varepsilon_0} = \frac{1}{\rho\omega\varepsilon_0} \tag{7.7}$$

The permittivity and conductivity of dielectric sample depend on various experimental parameters such as temperature, externally applied AC frequency signal, and magnitude of the applied voltage. The real permittivity is always not appropriate as the "dielectric constant". This quantity should always be a specific value at a certain frequency and ambient temperature. But, generally, at room temperature, it is agreed. Thus, this concept for a material which is under study is also unclear. Therefore, careful measurement at standard temperature with the assumption of impedance or dielectric constant dependency on the applied electric field is required. For

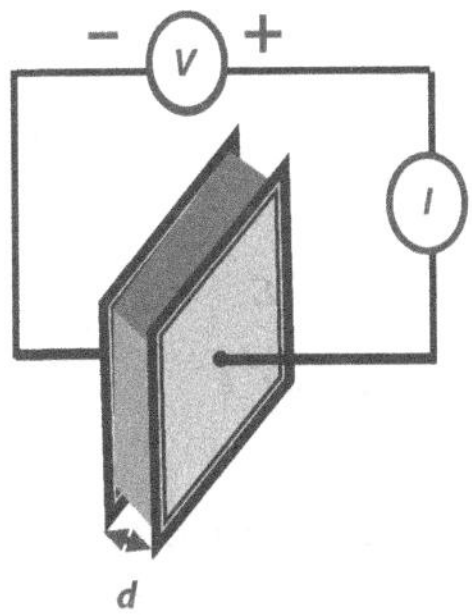

Figure 7.11 Fundamental experimental setup for measurement of dielectric constant.

example, in the case of a wet system, the conductivity of the electrolyte also affects the impedance measurement. In the case of a dry cell, the impedance of the cell is dominated at a higher frequency and it is a function of the dielectric constant. Hence, the same information is achieved from both dielectric and impedance spectroscopies but data analysis and approach to data analysis are totally logically different. Although both spectroscopies are different, mainly, energy storage and corrosion exchange capabilities of ions via redox reactions. The idea of dielectric relaxation was initially launched by Maxwell and later elaborated by Debye. Debye also described that the time is needed for the reorientation of dipoles from dielectric according to an external ac signal. During his experiment, he put the concept of relaxation of the system after removal of excitation signal and relaxation time, $\tau= 1/2\pi\omega_c$ which is inversely related to certain critical relaxation frequency value ω_c. Thus, dielectric spectroscopy aimed to find out the relaxation time and frequency-dependent complex permittivity ε^* of the sample from the measured data. This measurement is useful to find out the capacitance *C(w)*, permittivity $\varepsilon'(\omega)$, and conductivity $\sigma(\omega)$ of bulk samples or thick films. Debye model is a traditional symbolization of a dielectric system, where from high-frequency permittivity $\varepsilon\infty$ also called as high frequency capacitance denoted as C_{hf}. The transition observed to low-frequency permittivity ε_{LF} called low frequency capacitance denoted as C_{lf} and obviously, here $C_{lf}> C_{hf}$. Thus, at both high- and low-frequency terminals, this response is completely capacitive. Permittivity changes through this transition, and as a result, the permittivity increment is given by [36]

$$\Delta\varepsilon = \varepsilon_{LF} - \varepsilon_{\infty} \tag{7.8}$$

In addition, the complex permittivity of the system or cell which is a function of frequency is given as

$$\varepsilon^*(\omega) = \varepsilon' - i\varepsilon'' = \varepsilon_{\infty} + \frac{\Delta\varepsilon}{1+i\omega\tau} \tag{7.9}$$

$$\text{where}\quad \varepsilon'(\omega) = \varepsilon_{\infty} + \frac{\Delta\varepsilon}{1+\omega^2\tau^2} \quad \text{and} \quad \varepsilon'(\omega) = \frac{\Delta\varepsilon\omega\tau}{1+\omega^2\tau^2}$$

In the early 1970s, Macdonald and Boukamp used nonlinear complex curve fitting techniques for impedance data. This curve fitting or regression is based on the least square principle. This modeling uses the equivalent electrical circuits which turn into the fundamental approach for elucidation of impedance data. Therefore, this approach related to experiments such as electrodeposition and corrosion studies with impedance measurements became popular. The idea of the generalized transfer function for impedance spectroscopy was brought by Gabrielli. Later in the 1920s, the confirmation of electrochemical impedance data is explained using the Kramers-Kronig relations. For determinations of electrochemical complex impedance (Z^*), modulus (M^*), and permittivity (e^*), such various quantities of the sample/coatings or films are basically estimated by supplying an AC potential with varying the frequency of input signal at AC or DC certain excitation. According to interest of study this excitation signal is optimizes or required. Overall, dielectric, modulus, and impedance studies and their analysis possess nearly about same working principles, and hence, all these studies are called a part of universal broadband EIS [36].

Various conductive mechanisms through the analyzed system or cell are studied by these techniques. These conductive mechanisms may be a single process or a combination or sequences of various ionic, electronic, and particle conduction processes and their relative concentrations related to chemical or redox reactions. The conduction of ions/electrons or charges is observed in the bulk part of the sample, in the vicinity of the interfaces of the sample and the electrodes, through a series of reactions via electron exchange. Relaxation frequencies can be found from significant changes in complex impedance, modulus, and permittivity graphs.

However, the basic difference between "dielectric" and" impedance" spectroscopies is explained in the literature on the basis of their instrumentation and application. The dielectric method is characterized by using higher AC voltage amplitudes, lack of DC voltage perturbation, independent temperature modulation, and always working the frequencies above 1 kHz. That is, these measurements are carried out at several optimized discrete frequencies for investigation of "dielectric" properties of high resistive materials [36].

Meanwhile, impedance spectroscopy studies the charge and electrochemical material–exchange kinetic reactions that take place near the electrode-sample interfaces. This technique actively employs wide frequency ranges from MHz to mHz DC modulation and or uses low ac voltage amplitudes to keep signal response linear.

The majority of EIS studies focused on the examinations of double-layer interface and Faradaic reactions and their kinetics. On the opposite side, dielectric spectroscopy aimed to investigate impacts at varying frequencies on the bulk sample and its impedance analysis including dielectric relaxations due to changes in the capacitance of energy-storage samples or devices [24, 29, 36, 37, 57].

7.3.2 Electrochemical Experiment and Transportation of Charges

The basic electrochemical experiment required an electrochemical cell with at least two electrodes, a cathode and an anode. Among these, one is a sample while another is an inert electrode that completes the circuit. This cell contains the electrolytic solution and applied voltage difference, *V*, supplied across the two electrodes, as shown in Figure 7.12.

Current flows through an electrolytic cell by applying the electric field and completing the circuit. As a result, electromotive force eE transports electrons through the conducting electrodes, and ionic migration is observed in the electrolyte. Metals are rich in free electrons which are transferred to the conduction bands. A current that is observed to flow and measured by potentiostat is connected to an electrolytic cell. This current flows through the electrolyte and dissociates the ionic solution which may be aqueous or nonaqueous. In general, the average velocity of electrophoretic ions migration in the aqueous electrolyte is ~10 mm/sec observed [36].

The resistance to ionic migration in the aqueous electrolytic medium at a certain frequency range during the measurement of impedance is one

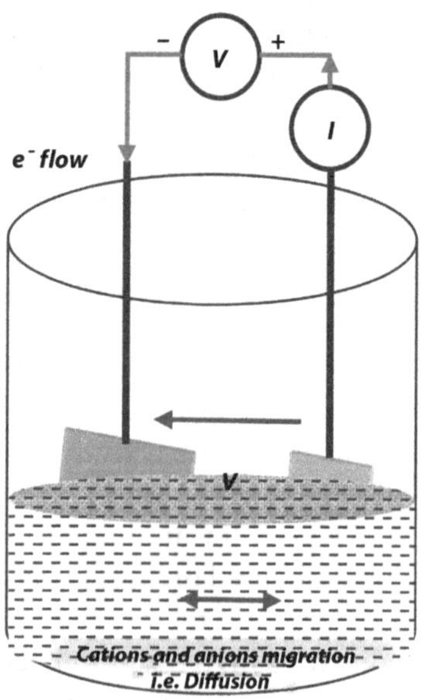

Figure 7.12 The basic electrochemical experiment with material transport directions.

type of "solution resistance" component R_{SOL} (Figure 7.13) which is generally very small. Current flow in the electrolytic solution is observed mainly transported by cation and anion migration. This mainly depends on the applied electrochemical potential difference $\Delta\phi$.

These cation and anion migration processes are nothing but the electroactive species of type *i (say)*. According to the Kohlrausch equation, the flux of the charged ions J_i [mol/sec] due to electrochemical potential gradient $\Delta\phi/\Delta x$ is given as [29, 36]

$$J_i = \frac{\Delta V(x)}{\Delta x} AF \sum z_i u_i C_i^* = \frac{\Delta \phi(x)}{\Delta x} A \sum \sigma_i \tag{7.10}$$

where A is the surface area of WE, z_i is the total charge of ionic species of certain one type i, F is the Faraday's constant (96500 C/mol), C^*_i is the concentration (mol/cm^3) of moving ionic species in electrolyte, and σ_i is the mobility constant of ionic species and given as

$$\sigma_i ;= F \sum z_i u_i C^* = \text{conductivity of solution ionic specis [Sm/cm]}.$$

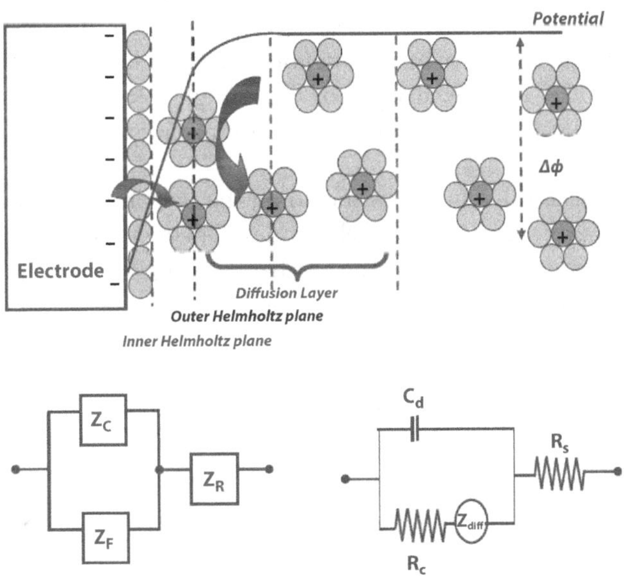

Figure 7.13 (a) Interfacial electrochemical reaction with diffusion and double-layer components; (b) Representative electrical circuits.

Electrochemical analysis can be performed on a variety of samples. In experiments, aqueous and nonaqueous solutions including even micron-sized colloidal particles can be studied. In electrolytic cells due to charges, a double layer of "counter ions" gathers and surrounds each particle and then it is named "macro-ions". These colloidal particles also migrate by driving force from the applied electric field and are responsible for the solution conductivity. Hence, composite materials and/or solid ionic electrolytes are also selected as electrolytes in electrochemical studies. Various polymers also possess ionic and electronic conductance; hence, they act as mixed conductors. Furthermore, semiconductors possess low conductivity due to small forbidden energy gaps that resist electrons from entering conduction bands as compared to metals. Therefore, by adding impurities/doping in semiconductors, localized energy centers can be created, and thus, finally, conduction can be increased [36, 53–55].

The interface between electrodes and electrolytes isolates the electrodes from the electrolyte; at this site, electronic and ionic charges are accumulated via charge carrier shift due to the application of an electric field. These electric charges transfer across the interface and an electrochemical reaction is carried near each electrode. The ionic species discharge at the interface is estimated. Accordingly, the electrochemical behavior of the ionic species releases or accepts an electron at a voltage, and the electrode is electrochemically polarized. The electrode potential quantifies these extra-charged negative electrons at the surface of the electrode. The discharge processes create an approximately nanometer-thick interfacial layer called the "Helmholtz" interfacial layer of the sample [36]. Thus, the further neighboring electrode can be represented as "charge transfer" resistance R_{CT} [see Figure 7.13(b)].

Thus, the electrolysis process including the depletion or accumulation of charges may occur next to the electrodes. During these processes, additional and some neural ionic species from electrolytes are utilized and discharged. Furthermore, a concentration gradient is observed in electrolytic cells and is responsible for the diffusion process. Hence, number of the ionic species in the electrolyte solution and the vicinity of the electrodes are different. However, all ionic species are trying to keep a uniform

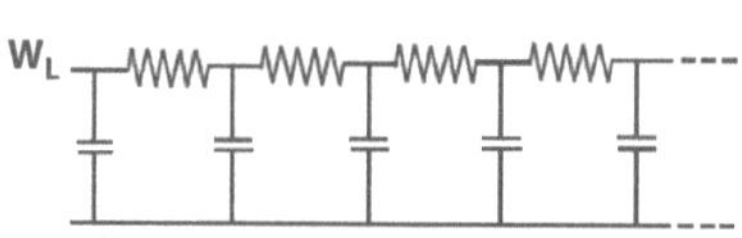

Figure 7.14 The equivalent circuit for Warburg impedance.

distribution of ions which means concentration in the electrolytic system. As a result, the species consumption occurs near the electrode continually. Thus always at the interfaces, solution ionic species endeavor to continuously deplete ionic species via diffusions [31, 36].

Hence, the concentration of bulk electrolytes reduces to the concentration near the electrode-solution interfaces. This concentration gradient enhances the diffusion of mass transport and a thick diffusion layer is originated. This diffusion layer is measured by the complex diffusion impedance element Z_{DIFF}. In this way, at the electrode potential, a variety of ionic species are adsorbed at the interfaces but not involved in electrochemical reactions. Such accumulation of the opposite polarity charges and ionic species forms a "double-layer capacitance" denoted as C_{DL}. In a conductive electrolytic solution, immediately after inserting the electrode, an electric double layer is produced. After wetting electrodes in various types of electrolytes, whether fluid or gel, in the vicinity of electrode, distribution of interface charges is non-uniform. Hence an electrochemical potential gradient is observed [36].

In electrolytic cell ionic species after dissociations migrating, some ionic species discharging, diffusing toward the electrode and creating parallel electrical path like plate layers inside cell it is the origin of capacitive element. Hence, this double-layer capacitance is in parallel to the charge transfer resistance during the diffusion process, as shown in Figure 7.13(b).

Explanation of interfacial and double-layer capacitance of electrochemical cells is evaluated using theories described by various electrochemists such as Maxwell-Wagner, Helmholtz, Gouy-Chapman, Stern, and Grahame. They put forth their own theory in electrochemistry.

In some electrolytic cells rather than the electrochemical double-layer process via diffusion chemisorptions and specific adsorption according to chemical affinity, Coulomb forces were also observed. These processes are observed near or on the electrode surface due to applied potential. All these variables are calculated from the particular electrochemical nature of participating ionic species. The voltage difference between the anode and cathode along with the current passing through the cell is measured and analyzed later.

In the case of a heterogeneous electrochemical process, the mass transfer between sample, electrolyte, and electrode has been controlled properly by selecting optimized parameters. In the case of homogenous and isotropic samples, interfacial effects are absent, and current matches with the field, so Ohm's law is absolutely valid. In an anion-free bulk solution, the electroneutrality condition, $\sum zC_i^* = 0$, is applicable, and the summation over net charges is equal to zero [29, 36, 37].

At the boundaries of interfaces, near to space charge areas electroneutrality is not observed.

Generally, it is assumed that in dilute solution after the application of an electric field, the flux of ionic species due to diffusion and migration of species and current is set, earlier is compelled by the concentration of gradient.

In some cases, ionic species transport from one site to another involving diffusion, migration, and convection which are gradients of concentration, potential, and pressure respectively, which give the resultant current. These can be given by Nernst-Planck in mathematical equation form [36, 61]

$$I = \begin{cases} -R_G TF \sum (z_i \Delta C_i^* u_i) - F^2 \sum (z_i^2 C_i^* u_i \Delta\phi) + Fv \sum (z_i C_i^*) \\ I = -F \sum (z_i \Delta C_i^* D_i) - F \sum (z_i \sigma_i \Delta\phi) + Fv \sum (z_i C_i^*) \end{cases} \tag{7.11}$$

where by the Nernst-Einstein equation, the mobility constant of ions is u_i and it is directly proportional to the diffusion coefficient D_i and ideal gas constant R_G given as

$$u_i = D_i R_G T \text{ and conductivity of electrolyte is, } \sigma_i = F\ \Sigma z_i C_i^* u_i \tag{7.12}$$

By theory, mobility is a feature of drag force on particles in viscous samples which is the influence of an applied electric field [24, 29, 31–34, 36].

7.3.3 The Prospect of Electrochemistry

Thus, electrochemistry has undergone a remarkable revolution in the past century. Recently, it has not only been an interest of academics in understanding the thermodynamic properties of electrolytes or systems but it also has dominant industrial applications such as electrolysis in the case of corrosion prevention and manufacturing/testing of commercial batteries. Although these two applications are obviously different, they have the same base root and they have developed along with each other in the research field via an electrochemical process. Several factors are responsible for the growth of this technique including reproducibility, dynamic experiment, reliability, and precise instrumentation. Thus, this technique is capable of studying many properties and parameters related to the electrochemical system.

Thus, now it arises a separate branch of electrochemistry which plays a dominant role in many fundamental research as well as industrial areas such as the investigation of new bulk and nanoscale compounds, biological fields, corrosion testing, energy, and sensor applications. Practically, all these applications related to chemical reactions are mainly associated with charge separation and transportation whether the system is homogeneous or heterogeneous, and whether electrode surfaces are rough or smooth. These reactions are combinations of half-reactions for keeping electroneutrality. Obviously, there are redox reactions including oxidations and reductions observed in electrolytic cells or in the vicinity of electrodes. These electrodes are connected via conducting cables and paths inside the solution of the cell. Furthermore, it promotes the charge transfer from one terminal to the opposite electrode. According to the cell configuration, the multiplication of two electrode reactions is divided. During the energy release in the case of batteries, the resultant free energy change at both electrodes is negative. While during electrolysis, energy is given to electrodes, and at that situation, the resultant free energy at electrodes is positive. Therefore, electrochemists use this technique for investigations of each elementary phenomenon related to redox reactions, extracting its data, and analyzing it for industrial applications [43, 44, 62].

7.3.4 Uncertainty in Impedance Data Analysis

Although EIS has several controversies, it occupies a significant role in the research field. The main intention of impedance data analysis is to identify the physical and chemical processes that take place in an electrolytic cell using the interpretation equivalent circuit components, which are modeled to the impedance response of the system received due to the application of external voltage perturbation at a certain AC frequency range. This impedance response due to external perturbation indirectly indicates the physical/mechanical structure and chemical nature of coatings or analyzed cells or samples [36].

Historically, EIS employed two systems—(i) electrically conductive and (ii) highly resistive systems. The purpose of these studies in both systems is to understand the electrochemical interfacial mass transport reactions and bulk relaxation facts.

The traditional focus of impedance spectroscopy is to know the Faradaic processes and mass transport carried due to diffusion and convection results. Meanwhile, the dielectric spectroscopic study focuses on material relaxations carried out by charge holding or releasing ability, which depends on system temperature. Temperature driven migration and conduction of carriers observed and as a result low frequency polarization processes are less significant and high frequency polarization means relaxation became dominant [62].

Hence, interfacial polarization is not the main priority of dielectric scientists while it is the interest of impedance scientists.

Thus, performing the experiment, data collection, and interpretation in both studies is lifted significantly nowadays, but it is one challenge for them. Often, it leads to delusion in the case of frequency relaxation and merging of data/spectrum due to some hidden process and indirect data analysis.

In a single universal broadband, EIS amalgamation of both techniques is a big challenge since there are interfacial effects. For example, an aqueous cell double-layer charging frequency can be allocated as a critical frequency of 100 Hz which, in the case of a highly resistive bulk sample, would be 1 megohm [46].

Hence, initially, we understand the fundamental physical and chemical properties at the surface as well as the bulk level of analyzed samples. After comprising electrochemical data interpretation one can analyze the probable interfacial kinetic reactions. Huge problems in impedance data analysis are only due to experimental parameters, which restrict diffusion mass transport, and interface effects during the impedance data measurement. As a consequence, laboratory results and real-life applications show discrepancies.

Furthermore, sometimes, unknown systems have been studied raising the difficulty in data analysis and may involve wrong interpretations. Hence, preliminary information of system is necessary before EIS studies.

Nevertheless, EIS is the leading multipurpose technique popular in the science and technology field. It is totally unsuitable to use without its fundamental knowledge. Hence, interest in sample types such as conductive or highly resistive and their proper physical/chemical properties such as their geometric sample size, structure, reactivity, and probable applications, all this information is necessary to perform the EIS experiment, data responses to applied field and interpretation [36, 37, 57].

7.3.5 Data Fitting of EIS Spectra

In the earlier section, we have seen that EIS interpretation is not straightforward. Here, EIS response curves fit with the help of equivalent circuit models. These fittings are performed on the basis of the least square principle and deconvolution is obtained. Even though graphical fitting by some mathematical functions which are combinations of various electrical components is simpler and sufficient, its conformity is a must. Hence, precision is required for the overall experiment and data fitting, and then finally, accuracy will be achieved. This all fitting is carried with the software's now a day's which are supplied by various potentiostat developing suppliers. Due to a lack of sufficient knowledge, few research groups utilize these techniques [40].

7.3.5.1 Nonlinear Least Squares Method for EIS Fitting

EIS response curves are nonlinear types; hence, fitting these curves by modeling appropriate electrical models on the basis of the least squares method is possible due to EIS analysis software via the nonlinear least squares (NLLS) method and it is properly deconvoluted. Obtaining reliable system parameters means that magnitudes of circuit components are necessary; hence, the NLLS Levenberg-Marquardt algorithm is the base of various impedance data analysis software. This algorithm also uses the chi-squared test, which is a certain value indicative of fitting accuracy to experiment responses and mathematically expressed as [34]

$$\chi^2 = \sum_{i=1}^{n} [(y_i - f(x_i))/\sigma_i]^2 \tag{7.13}$$

where σ_i represents the standard deviation value of measured data points, y_i is the measured data point, and $f(x_i)$ is the known function that is optimized or matches the fitting to the experimental data points, respectively. The value of χ^2 should be as low as possible as "goodness of fit" means accurate fitting.

Impedance is a complex quantity at varying frequencies; hence, for modeling the impedance responses, the complex nonlinear least squares (CNLS) method is employed. Thus, the object function, S, which is the sum of squares errors, is mathematically described as [41]

$$S = \sum_{i=1}^{n} w_i \left\{ \left[Z_{re,i} - Z_{re}(\omega_i, \alpha_k)\right]^2 + \left[Z_{im,i} - Z_{im}(\omega_i, \alpha_k)\right]^2 \right\} \tag{7.14}$$

Here, $Z_{re,I} + iZ_{im,i}$ is the recorded complex impedance value at a certain frequency ω_i (here, $i = -1$ is used for differentiation); its corresponding model function which is suited to fit is $Z(\omega_i,\alpha_k) = Z_{re}(\omega_i,\alpha_k) + iZ_{im}(\omega_i,\alpha_k)$. This model function which is a mathematical equation of the resultant of the equivalent circuit can be changed by adjusting the circuit components, such as resistance, capacitance, and Warburg either in series and/or parallel connections, $\alpha_k(k = 1,2,\ldots\ldots M)$; and its corresponding weight factors, ω_i, are the reciprocal of the square of the vector length of the impedance of the system.

The difference between measured and fitted impedances also involves two parts due to their complex quantity as a change in the real and imaginary part of impedance, called residuals in CNLS-fi, also Δ_{re} and Δ_{im}, these values should be least. Very small residuals indicate that the fit is good [34]. These residuals in real and imaginary components of impedance are given as

$$\Delta_{re} = \frac{Z_{re,i} - Z_{re}(\omega_i,\alpha_k)}{|Z(\omega_i,\alpha_k)|} \text{ and } \Delta_{im} = \frac{Z_{im,i} - Z_{im}(\omega_i,\alpha_k)}{|Z(\omega_i,\alpha_k)|} \tag{7.15}$$

Hence, for optimized and acceptable fitting, these residuals should be spread over the full wide frequency range. The first step of impedance curve fitting is the selection of the proper equivalent circuit, and the second step is the approximation of initial values of all the circuit elements of the equivalent circuit model [32–34, 36].

This initial value of circuit parameters is a tricky job in the analysis of equivalent circuit model fitting to measured data. Well knowledge about components or elements behavior in that circuit helps the estimation of these initial values.

If starting values are not far away from real values, then the optimized proper fit is not possible and it diverts in analysis. Starting values within the 10 of the true values and then fitting starts is better for finding out model components [36].

If the exact semicircle is recorded in the impedance spectrum, then charge transfer resistance, R_{ct}, is nothing but the intercepts of the semicircle on the real Z axis, and it is the corresponding capacitance of double layer, C_{dl}, is calculated using the relation of the maximum frequency, ω_{max} given as, $\omega_{max} = (R_{ct}C_{dl})^{-1}$. For every situations, exact semicircles are not observed leads to complications in component value measurements. But, it estimates the basic rules involved in data analysis. This is the main part of the spectrum where one element dominates at a certain frequency region which is related to particular reactions observed at that frequency range. Hence, the value of the element is properly calculated. Generally, the impedance of resistors dominates the low-frequency spectrum. While a capacitor tends to zero at high frequency, and it tends to infinity, means it is very large, at low frequency [40]. The specific resistors can be identified from the horizontal region of the Bode curve. On the basis of estimated values of parameters/components, the software program adjusts the other values of the remaining components of the circuit model. This process repeats until chi-squared or residues of impedance values least. Moreover, improper selection of the circuit model, bad starting values, and noise in recorded impedance responses caused difficulties in the fitting. Hence, the choice of model and optimizations of parameters depends on practice and expertise [24, 36, 37, 50, 53, 57].

7.3.5.2 Deconvolution of Spectra

If the Nyquist curve is not looking an exact semicircle and it displays a depressed semicircle or a broad arc, then that response could be explained using various RC combinations such as the two or more separate parallel RCs in series or a continuous distribution of parallel CPEs with R_s in series. Thus, for a single response, there could be many circuit models that may fit the response; however, analysis is based on deconvolution of the imaginary impedance [34].

The main purpose of deconvolution is to reduce the complexity of response and resolve that merged possible processes. As a consequence, the explanation of the Nyquist plot of the impedance becomes easy using the discrete various parallel or series time constants (RC), according to the discrete relaxation. Furthermore, peak relaxation time, τ_p, is calculated from the approximate frequency region by the relation, $\omega_p=\tau_p^{-1}$. Thus, results can be employed to construct a suitable circuit model and finally approximation of the starting assessment of the circuit components for ensuing impedance curve fittings [21, 26, 34, 35, 37, 63].

7.3.6 Equivalent Circuits and Its Components

7.3.6.1 Equivalent Circuit

Instead of manual extraction of parameters one by one, the whole spectra can be analyzed simultaneously using computer calculations. An equivalent circuit EC is an electrical circuit model constructed by combining passive electrical components either in parallel or series combinations, which

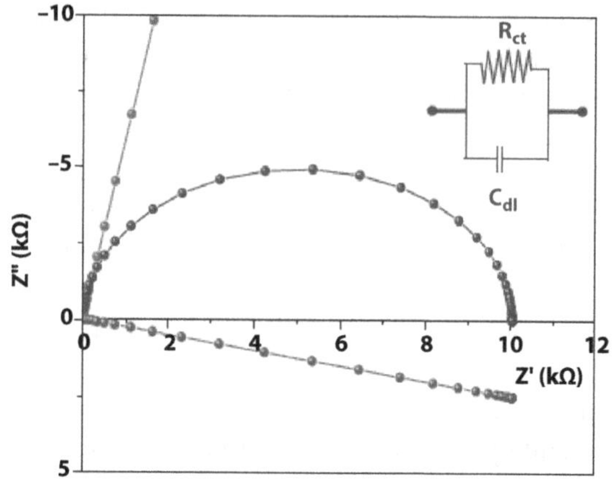

Figure 7.15 Nyquist plot showing the impedance spectra of an R/CPE electric circuit model.

are employed to build up the impedance response curves of circuits, which display the same impedance as the system's overall frequency ranges.

After the selection of EC "fitting" of that EC model to the recorded curve was carried out using any of the available modeling software. The values of the electrical components of the selected model are optimized by the software program such that the modeling curve and the recorded curve show the minimum deviation. The fitting accuracy can be checked graphically first and from the "weight sum of squares" (calc modulus weighting) (or chi-squared parameter (χ^2). Simultaneously, fitting can be checked from the % relative error of each circuit component; it should be less than 5%. Furthermore, modeling of the equivalent circuit to the measured impedance spectrum based on the perception of the type of impedance is projected in the present system. In addition, judgment of recorded response is reliable with modeling circuit, and assessment of correlated circuit components is precise [20, 29, 36].

Many electrochemical processes occurring in the system can be described using circuit elements such as those available in any electronics store. The frequently utilized circuit components and their impedance expression are listed in Table 7.3. A standard resistor (R) is used for indications of solution resistance and charge transfer resistance. Charge and discharge behavior is revealed by a capacitor (C). An inductor (L) also depicts for showing the activities that resist changes in the current. Other frequency dependent elements are also employed such as Warburg impedance (W) which is number of combinations of R and C, used to fit impedance related to

Table 7.3 Impedance components and their impedance expression [26, 47].

Element	Impedance expression Z=
Resistance (R)	R
Capacitance (C)	$(jwC)^{-1}$
Inductance (L)	jwL
Warburg (W)	$Y_0(jw)^{-0.5}$
Constant Phase Element (CPE)	$Y_0(jw)^{-n}$

diffusion processes and the Constant Phase Element (CPE) is an popular component possess the exponent n ($-1 \leq n \leq 1$) used to demonstrate distortion in surface roughness and accumulation or adsorption ionic species [24, 26, 47, 50, 64].

7.3.6.2 *Electrolyte Resistance*

The impedance of an electrochemical system is the resistance offered by an electrolyte called solution resistance. Generally, three-electrode geometry, namely, a working electrode, a counter electrode, and a reference electrode connected with a potentiostat, compensates for this resistance by a solution between the counter and REs. However, for the modeling system, the solution resistance between the RE and the WE must be considered and modeled. On the ion concentration, ion types, ambient temperature, and the electrode area through which current flows, the resistance of a solution depends [37, 46, 47, 55]. If area A and length *l* carry a uniform current, the resistance is generally defined as [46]

$$R = \rho \frac{l}{A} \tag{7.16}$$

Here, ρ is the resistivity offered by a solution which is the reciprocal of conductivity, *k*, of the solution with unit Siemens per meter (S/m), and then solution resistance is given as [46]

$$R = \frac{1}{k}\frac{l}{A} \Rightarrow k = \frac{1}{RA} \tag{7.17}$$

The majority of electrochemical systems unluckily do not show uniform current distribution throughout an electrolyte. The major problem for the calculation of solution resistance is measuring the current flow through the area of the electrolyte. Thus, the calculation of this resistance through ionic conductance is out of scope. Hence, researchers do not calculate solution resistance from an electrochemical system but it computes by fitting a circuit of impedance data [24, 29, 36, 50, 56].

7.3.6.3 Double-Layer Capacitance

Generally, in between the interface of an electrode and its adjacent electrolyte, an electrical double layer of ions and charges is formed. This double layer adsorbs on the electrode surface which isolates the charged electrode and charged ions by forming a separation of approximately a few angstroms called double-layer capacitance. Electrode potential, ions concentrations in electrolyte, types of ions, oxide layers on electrodes, roughness of electrode surface, impurities in electrodes and/or solutions, impurity adsorption temperature, such several factors decide the double-layer capacitance [24, 29, 36, 37, 47, 51, 57, 65].

7.3.6.4 Charge Transfer Resistance

By applying the electric field electrons, transfer across the electrode/electrolyte interface and opposition to this electron flow is called charge transfer resistance, R_{ct}. Mainly, it is influenced by reaction type, the surface of the electrode, and the applied electrode potential. If overpotential rises, then charge transfer resistance reduces [34–36, 47, 57].

An analogous type of opposition to charge transfer is observed in the case of a kinetically controlled single electrochemical process. During this process, the single kinetic reaction takes place at equilibrium in the absence of mixed potential reactions [31]. Electrons from the cathode are reacting with the metal and diffusion of metal ions is observed into the electrolyte. Due to metal oxidization charges are transfer in electrolytic cell via a particular velocity. Obviously, this velocity of charges is affected by reaction type, concentration of dissociative ionic species, applied potential, and ambient temperature [24, 35, 36].

The potential and the current of the electrolytic cell are related to charge flow, and the number of electrons supplied by the cathode is based on Faraday's law. If the concentration in the bulk and the electrode surface is similar, then the Butler-Volmer equation describing the situation became easy. Due to the oxidation-reduction reaction, charge transfer occurs and polarization is observed in the system as a consequence charge flows are resisted [36, 66]. For the reduction of diffusion layer thickness and the concentration gradient via polarization, the solution should be kept continuously stirring [65]. At the small overpotential, η, the electrochemical system achieves the equilibrium. This expression given below and using this expression exchange current density can be compute using the known value of R_{ct} [31, 35, 46, 55, 67].

$$R_{ct} = \frac{RT}{nFi_0} \tag{7.18}$$

7.3.6.5 *Frequency-Dependent Elements*

7.3.6.5.1 Warburg Resistance

It is one type of mass transfer resistance of the electrochemical process, it is frequency-dependent, and it includes both resistive and capacitive parts, as shown in Figure 7.14. The impedance of the Warburg resistance [$Z_w(\omega)$] is mathematically given below [24, 29, 31, 35, 51, 57, 58]

$$Z_w(\omega) = (1-i)\sigma\omega^{-1/2} \tag{7.19}$$

where $\sigma = \dfrac{1}{nFA\sqrt{2}}\left(\dfrac{\beta_O}{D_O^{\frac{1}{2}}} - \dfrac{\beta_R}{D_R^{\frac{1}{2}}}\right) =$ Warburg Coefficient, $i = \sqrt{-1}$

D_O, D_R are the diffusion coefficients during oxidation and reduction.

β_O, β_R are the derivatives of the voltage drop with respect to concentration during oxidation and reduction. Basically, it indicates that if the impedance of the diffusion process and the length of diffusion increases, then the impedance also increases [34]. At high frequencies, this diffusion path is observed to be short, and hence, low Warburg impedance is found and vice versa. The Warburg impedance can be shown as resistance connected with a capacitance in series, as shown in Figure 7.14.

In Nyquist spectra when the impedance of both real and imaginary components became equal, then the 45° line is a Warburg impedance which is followed by the semicircle which is modeled by keeping charge transfer resistance parallel to the double-layer capacitance. On a Bode Plot, this Warburg impedance displays a phase shift of 45° indicating that reactant species unable to go distant means diffusion process is slow. The reactants are capable of diffusing far away at low frequencies. Since a low-frequency system perturbed less, as a result, the diffusion process is enhanced, and

Warburg impedance increases. This Warburg impedance is significantly observed when the thickness of the diffusion layer approaches infinity, meaning a thick diffusion layer. At lower frequencies, moderate diffusion is observed, but at this case, the diffusion layer is bounded, then impedance not stick to Equation 7.18. This bounded Warburg impedance is given as

$$Z_{BW} = \sigma\omega^{-1/2}(1-i)\tanh\left(\delta\left(\frac{i\omega}{D}\right)^{1/2}\right) \tag{7.20}$$

The bounded Warburg indicates that the constraint means it is "finite" Warburg. At high frequencies means when, w→∞, thickness of the diffusion layer δ→∞, then $\tanh\left(\delta\left(\frac{i\omega}{D}\right)^{1/2}\right) \rightarrow 1$ and Equation 7.19 turns to the infinite Warburg impedance which is given in Equation 7.20. Occasionally, these equations are expressed in terms of an admittance quantity [34, 47, 54, 58, 65]

$$Y_0 = \frac{1}{\sigma\sqrt{2}} \tag{7.21}$$

7.3.6.5.2 Constant Phase Element

Impedance responses of real systems involve instead of pure capacitor the CPE which is a sensitive and non-instinctive circuit component. In general, its nature is not an exact semicircle. However, for some practical systems, this plot observed the compressed circle looks line arc of a circle whose center lies anywhere below the abscissa. In Figure 7.15, the Nyquist spectrum of a circuit including parallel combinations of R and CPE showing the center of this semicircle is situated at (1–n)×90° below the real Z-axis. And its impedance is given below [34]

$$Z_{CPE}(\omega) = q^{-1}(i\omega)^{-n} \tag{7.22}$$

Here, q has numerical values, and the phase shift is described by exponent n which ranges between –1 and 1. Hence, extreme values of n = 1, 0, -1, which represent that the CPE plays a role of various circuit elements such as C, R, and L, correspondingly. If n = 0.5, then it shows nature like a Warburg impedance [34, 47]. The physical meanings of the coefficient q for various n values are tabulated in Table 7.4.

Table 7.4 CPE functions according to quotient values of coefficient q variation [70, 72, 73].

n	Role of CPE	q (meaning)	Units
1	Capacitance	C	$F = \Omega^{-1}s$
0	Resistance	R^{-1}	Ω^{-1}
-1	Inductance	L^{-1}	$H^{-1} = \Omega^{-1}s^{-1}$
0.5	Warburg element	s^{-1}	$\Omega\, s^{-1}$

A bounded CPE explains the impedance of a bounded homogeneous diffusion layer. Its impedance is mathematically expressed as [34]

$$Z_{BCPE}(\omega) = q^{-1}(i\omega)^{-n} tanh R_0 q(i\omega)^n \tag{7.23}$$

where q, n, and R_0 are the factors of the bounded CPE component [34, 47].

If the surface of the electrode is rough or shows charge carriers delivering or collecting, then distortion of the capacitance is observed, and that situation $n = 1-\varepsilon$, here ε lies in between 0 and 0.2. When $n = 0.5\pm\varepsilon$, with $0 < \varepsilon \leq 0.1$, then CPE informs the diffusion process and in the overall process, Fick's second law deviates. And if $n = 0\pm\varepsilon$, here $0 < \varepsilon \leq 0.2$, the CPE indicates the indistinct resistance. For every $n < 0$, the CPE explains an inductive means of charge accumulation. Therefore, due to such variation in the n factor, the function of the CPE varied and it is called the generalized component [68–71]. Its role changes according to electrode surface roughness, varying diffusion thickness or ionic species concentration, non-uniform current distribution in the system, and a variation in reaction rates and non-homogeneous reaction rates near the electrode surface or interface [24, 32–34, 47, 48].

Structural Circuits

Various types of equivalent circuits are utilized for describing the EIS spectra suggested by various electrochemists/physicists. Among them, Randles, Maxwell, Voigt, and ladder structures/circuits are used.

Randles Structure/Model

This circuit includes the solution resistance, R_s (Ohm), which indicates the uncompensated resistance, R_p (Ohm), employed to suggest the polarization resistance or charge transfer resistance, CPE plays various elements which is explained in the earlier section. Then, the impedance of the Randles circuit is given as [24, 34, 47, 54, 55, 57]

$$Z = R_s + \frac{1}{\frac{1}{R_p} + i\omega C_{dl}} = R_s + \frac{R_p}{1 + i\tau\omega} \tag{7.24}$$

Maxwell's Structure/Model

For modeling parallel processes, Maxwell's structure is always employed and it is typically RC circuits in parallel.

Voigt's Structure/Model

Voigt's model includes several RC components in series and is commonly useful for the explanation of solid electrochemical systems/cells [34]. In this system, the interface between electrode and electrolyte is not perfect in contact since during electrode reaction along with ionic species, some gas species also may be observed near the vicinity of the interface. Therefore, at a high-frequency range, solution resistance creates two arcs suggesting charges moving inside the bulk and across the grain boundaries [24, 34, 35, 47, 54].

Ladder Structure/Model

The ladder model also includes series/parallel connections of capacitor and resistor components; thus, it encloses consecutively several kernels [20, 24, 26]. The impedance of this type of model is given as [34, 47, 67]

$$Z(\omega) = Z_1(\omega) + \{Z_2^{-1}(\omega) + [Z_3(\omega) + (Z_4^{-1}(\omega) + \ldots\ldots)^{-1}]^{-1}\}^{-1} \tag{7.25}$$

$$Z(\omega) = R_1 + \cfrac{1}{i\omega C_1 + \cfrac{1}{R_2 + \cfrac{1}{i\omega C_2 + \ldots\ldots\ldots}}} \tag{7.26}$$

7.4 EIS in Anticorrosive Modern Nanocomposite Coatings and Assessment of Corrosion Protection Performance

7.4.1 Introduction

Generally, polymer-incorporated nanoparticles of metals and/or polymer-dispersed nanostructured metal oxides are studied [11, 47]. Numerous research groups made a try to examine the scientific prevention of corrosion by nanocomposite coatings on various types of metal substrates [49, 74–82]. The incorporation of a nanostructured dispersed phase enhances anticorrosive characteristics. V. Shinde *et al.* investigated electrodeposition and corrosion studies of poly(o-toluidine) and poly(2, 5-dimethylaniline) coatings on low-carbon steel and copper in an aqueous electrolytic bath of sodium tartrate and sodium oxalate. Generally, the purpose of incorporation of electrolytic ion constituents in the polymer framework is to improve its required properties which are an interest of study. Furthermore, we also studied during the polymerization process from monomer o-toluidine on a stainless steel electrode, the dissolution of the substrate, and the formation of the passive oxide layer. But, the dissolution rate of the substrate decreases due to reason that the monomer is involved in the passivation reaction by forming an inhibitor layer via an adsorption mechanism [83–93].

Shinde and Patil [86] have investigated types of monomer(s) involved in the polymerization process of aniline and derivatives of anilines on low-carbon steel and it affects the dissolution of substrate and polymerization with the help of XPS of the samples. Actually, very interesting results were observed when we analyzed the interphases formed on the LCS substrate during the early stages of galvanostatic growth of conducting polyaniline and its derivatives. In this article, XPS analysis of the low-carbon steel surfaces at the different stages of electrochemical polymerization of o-anisidine and aniline reveals that in the passivation process, monomer(s) participate. Type of monomer associated with the process of passivation and simultaneously it also may prefer the phase of conducting polymers. Groups attached to monomers and their inductive effects are responsible for its adsorption on substrate. This adsorption forming inhibition thin layers on substrates which prevent dissolution. Since groups of monomer(s) and iron substrate shows dominant adsorption mechanism, as per chemistry this mechanism obeying Pearson's hard and soft acid-base principle [86]. As a consequence, passivation duration decreases in the case of o-anisidine due to the methoxy group with the compassion of aniline [86].

We also studied the EIS spectrum of poly(ethylaniline) (POEA) coating on a copper substrate as a function of immersion time [25, 94]. Results surely prove that as an increase in immersion time for POEA coating in 3% NaCl then due to aggressive corrosion reactions, the unique circuit model is deficient for fitting the impedance responses of all immersion samples. An increase in immersion duration and additional time constant (RC) is required for the evaluation of the corrosion prevention functioning of coating. The EIS response and its analysis from the impedance parameters with the various immersions were studied. Further water uptake inside the coating, area of delamination, diffusion constant, and dielectric constant of the coating were affected, which were calculated. The corrosion performance of POEA coating observed through various process such as water penetration via pores of coating, staring of corrosion process underneath of coating. Finally, creation and accumulation of corrosion product which diffuses coating from substrate. And degradation observed in coatings [25].

Incorporations of nanodimensional metal ions or metal oxide particles enhance corrosion protection performances since they will act as smart coating or self-healing coatings. Many reported such self-healing performances of nanocomposite coating against corrosion [14–16].

7.4.2 Corrosion Mechanism

The corrosion protection mechanism of conducting polymeric coatings on metals and alloys depends on its type of phase and metal/alloy substrates. Furthermore, the phase, structure, and quality of coating on certain substrates are mainly affected by several process parameters such as ionic species of electrolytes, their concentrations, pH of electrolytes, experimental conditions, synthesis techniques, etc. [20, 63, 95, 96]. Various corrosion mechanisms of CPs on oxidizable substrates explained by many researchers. These are brainy self-curative ability through anions releasing mechanism, the moving the oxidation potential via electrochemical interface in which oxidizable metal alloy covers by noble metal plating, oxidizable metallic substrates ennobles by oxidized state of conducting polymers which keeps metallic surfaces in passive form by forming oxide layer and stopping further dissolutions. Thus, various corrosion-protective mechanisms are dominantly barrier effects and its main origin is classic electrochemical kinetics theory [11, 12, 14–16].

Michalik and Rohwerder [97] showed that the CPs coating causes a shift in the electrochemical interface towards a more positive direction on the oxidation potential scale. With the help of mass spectroscopy, they proved that the heavy oxygen reduction process at the metal-polymer interface is responsible for the protection of metal surfaces through shifting corrosion potential to a noble direction. Ultimately, O_2 reduction on the polymer is essential since it holds the polymer in an oxidized state. Furthermore, this oxidized state of polymer kept it in conducting form and able to smart healing capability for corrosion protection via redox conversion of polymeric phases [11, 14–16, 63, 98–100].

The fact that, the O_2 reduction can occur in the vicinity of the polymer surface due to production of the anionic OH^- species. As a consequence coating are dealing with disbondment and delaminate from the substrate, this is the indication of failure of the coating. The ennobling of the metal surfaces is due to the conversion of the phase of conducting polymers [97]. But, up to a certain duration, it is sustaining since this coating is not perfectly impermeable [94].

Similar results were also highlighted in the pioneering study by DeBerry [101]. Furthermore, analogous results were proved in the case of polyaniline-coated steel in the work of Elsenbaumer *et al.* [103] using XPS and Auger analysis. Wessling proposed the passivation mechanism of PANI-coated Fe via iron oxide interphase and several rehearsers also reported similar passivation mechanism [63, 95, 99, 100, 102]. Figure 7.16 shows the redox conversion of protonated emeraldine salt (ES) to reduced emeraldine base (EB) and vice versa due to the availability of types of ions and corrosion mechanism. The numerous ferrous and nonferrous metals such as Fe, Zn, Al, and Mg coated by PANI/ES (ES synthesized using p-toluenesulfonic acid) were mixed in a polyvinyl butyral binder and allowed in the humid air. Its detailed explanation through the ennobling effect is reported by Williams and McMurray [38].

The corrosion mechanism offered by conducting or insulating type modern coatings is evaluated by the EIS method. A number of articles, representing similar studies, are observed in the literature. Li and co-workers reviewed the study of PANI primer–coated steel by EIS. Tallman *et al.* reviewed the redox behavior of PANI coatings and explained the protection mechanism [95, 98]. In the nanocomposite coatings, polymeric matrix–embedded inorganic metal or metal oxide in nanodimension takes sites as a dispersed phase; hence, its suitability for corrosion protection is important [27].

Significant challenges in the improvement of modern coatings are observed nowadays, which ensure excellent adhesion and show perfect barrier properties; simultaneously, they should be environmentally friendly. In addition, irrespective of the nature of the substrate, it is expected that coatings covers completely the metallic body parts and the buried parts and prevents them from scratch-resistant edges.

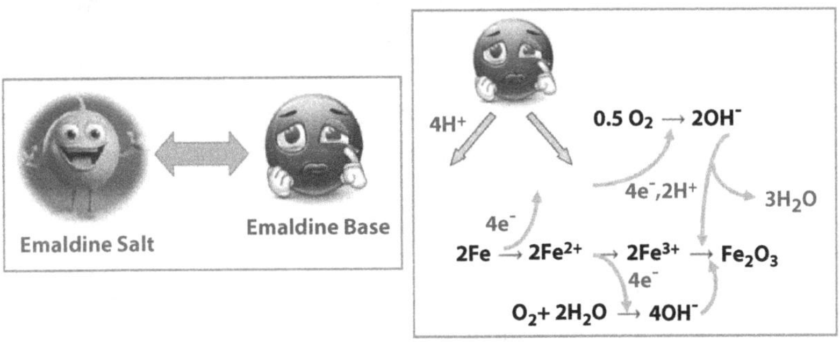

Figure 7.16 Proposed passivation mechanism of PANI-coated Fe (reproduced with permission from [100]).

Very recently, Muresan enlightened challenges and reviewed these nanocomposite coatings for corrosion protection [28]. As shown by some examples, EIS is an effective technique for assessing the behavior of undamaged and scratched coating samples within a short time. It provides quantitative facts as compared to the classical aging assessments [27]. However, this EIS technique requires precise controlling of the experimental parameters. Furthermore, the interpretation of the results for a description of the system, which according to interested researchers, needs an appropriate choice of the electrical equivalent circuits and its truthful fitting to experimental responses [27]. Overall, equivalent circuit modeling in the EIS technique really provides a valuable probable corrosion protection mechanism. However, the variability of equivalent circuit models is confusing, and the corrosion mechanism is observed as doubtful. But, in reality, EIS is one strong technique explaining the possible corrosion mechanism kinetically [25, 94].

7.5 Conclusions

This chapter provides discussions on inorganic-organic nanocomposite coating on metal alloys that acts as a corrosion-protective coating along with its need, its status, and its superior behavior. For evaluation of its corrosion protection performance, EIS is one effective technique. It assesses so many interesting electrochemical reactions and their related parameters. Therefore, it was observed that the EIS study using the electrochemical equivalent circuit modeling informs various circuit components which helps to find out the lifetime of coating and physical/chemical process, parameters, and anticorrosive mechanism. Corrosion field tests of various novel inorganic-organic nanocomposite coatings on corrosive surfaces by EIS are essential in future. Hence, this chapter focuses EIS basis and its data analysis which are helpful to beginner researchers. In addition, various fields such as sensors and their real-time performances, energy generation/energy storage processes, and various electronic properties along with the help of various spectroscopic techniques have also been tested and will evaluated correctly in the future using the EIS technique.

References

1. Abu-Thabit, N.Y. and Makhlouf, A.S.H., Recent Advances in Nanocomposite Coatings for Corrosion Protection Applications, in: *Handbook of Nanoceramic and Nanocomposite Coatings and Materials*, pp. 515–549, 2015.

2. De Riccardis, M.F. and Re, M., Conducting nanocomposite coatings, in: *Nanomaterials-Based Coatings*, pp. 73–117, 2019.
3. Fontana, M.G., Corrosion Engineering And Corrosion Science. *Corrosion*, 19, 6, 199t–204t, 1963.
4. Baboian, R., Corrosion—A Problem of International Importance, in: *Corrosion Testing and Evaluation: Silver Anniversary Volume*, p. 7.
5. Honarvar Nazari, M., Zhang, Y., Mahmoodi, A., Xu, G., Yu, J., Wu, J., Shi, X., Nanocomposite organic coatings for corrosion protection of metals: A review of recent advances. *Prog. Org. Coat.*, 162, 106573, 2022.
6. Mohammadkhah, S., Sarabi, A.A., Eivaz Mohammadloo, H., Ghamsarizade, R., Improvement of active/passive anti-corrosion/weathering properties of epoxy-siloxane structure via cloisite 30B/polyaniline inclusion as new hybrid nanocomposite coatings. *Colloids Surf. A: Physicochem. Eng. Asp.*, 666, 131297, 2023.
7. An, H., Liu, K., Wang, S., Jiang, C., Liang, S., Xiao, J., Zhao, X., Sun, Z., Cao, C., Gao, Y., Enhanced Corrosion Resistance of Waterborne Epoxy Coatings by Polyaniline Nanorods and Nitrogen and Fluorine Dual-Doped Graphene Oxide Composites. *ACS Appl. Nano Mater.*, 2023.
8. Meng, X., Hou, L., Jin, H., Li, W., Wang, S., Wang, Z., An, J., Wen, C., Ji, G., Xu, X., Zhou, D., Study on corrosion protection properties of PANI/ZnO/Zn/Graphene coating on aluminum alloy. *Diamond Relat. Mater.*, 136, 110067, 2023.
9. Karimi Azar, M.M., Shooshtari Gugtapeh, H., Rezaei, M., Evaluation of corrosion protection performance of electroplated zinc and zinc-graphene oxide nanocomposite coatings in air saturated 3.5 wt. % NaCl solution. *Colloids Surf. A: Physicochem. Eng. Asp.*, 601, 125051, 2020.
10. Zvonkina, I. and Soucek, M., Inorganic–organic hybrid coatings: common and new approaches. *Curr. Opin. Chem. Eng.*, 11, 123–127, 2016.
11. Eftekhari, A., *Nanostructured Conductive Polymers*, John Wiley and Sons, 2010.
12. Mirabedini, S.M., Farnood, R.R., Esfandeh, M., Zareanshahraki, F., Rajabi, P., Nanocomposite coatings comprising APS-treated linseed oil-embedded polyurea-formaldehyde microcapsules and nanoclay, part 2: Self-healing and corrosion resistance properties. *Prog. Org. Coat.*, 142, 105592, 2020.
13. Markovic, G. and Visakh, P.M., Polymer blends: State of art, in: *Recent Developments in Polymer Macro, Micro and Nano Blends: Preparation and Characterisation*, pp. 1–15, 2017.
14. Habib, S., Fayyed, E., Shakoor, R.A., Kahraman, R., Abdullah, A., Improved self-healing performance of polymeric nanocomposites reinforced with talc nanoparticles (TNPs) and urea-formaldehyde microcapsules (UFMCs). *Arabian J. Chem.*, 14, 2, 102926, 2021.
15. Ubaid, F., Radwan, A.B., Naeem, N., Shakoor, R.A., Ahmad, Z., Montemor, M.F., Kahraman, R., Abdullah, A.M., Soliman, A., Multifunctional self-healing

polymeric nanocomposite coatings for corrosion inhibition of steel. *Surf. Coat. Technol.*, 372, 121–133, 2019.
16. Idumah, C.I., Obele, C.M., Emmanuel, E.O., Hassan, A., Recently Emerging Nanotechnological Advancements in Polymer Nanocomposite Coatings for Anti-corrosion, Anti-fouling and Self-healing. *Surf. Interfaces*, 21, 100734, 2020.
17. Liu, X., Wu, Z., Lyu, Y., Li, T., Yang, H., Liu, Y., Liu, R., Xie, X., Lyu, K., Shah, S.P., Corrosion Resistance of CeO(2)-GO/Epoxy Nanocomposite Coating in Simulated Seawater and Concrete Pore Solutions. *Polymers*, 15, 12, 2602, 2023.
18. Corrosion Standards and Wear Standards - Standards Products - Standards & Publications - Products & Services.
19. Pereira, E., Br, E., Resende, A., De Medeiros, M.H.F., Meneghetti, L.C., Chloride accelerated test: influence of silica fume, water/binder ratio and concrete cover thickness. *Rev. IBRACON Estrut. Mater.*, 6, 4, 561–581, 2013.
20. Shinde, V.P. and Patil, P.P., A study on the electrochemical polymerization, characterization, and corrosion protection of o-toluidine on steel. *J. Solid State Electrochem.*, 17, 1, 29–41, 2013.
21. The Importance of Corrosion Testing, Sharretts Plating Company.
22. Chaudhari, S. and Patil, P.P., Corrosion-protection aspects of electrochemically synthesized poly(o-anisidine) coatings on mild steel: An electrochemical impedance spectroscopy study. *J. Appl. Polym. Sci.*, 106, 1, 400–410, 2007.
23. Indig, M.E. and Groot, C., Electrochemical Measurement of Corrosion. *Corrosion*, 26, 5, 171–176, 1970.
24. Orazem, M.E. and Tribollet, B., *Electrochemical Impedance Spectroscopy*, Wiley, 2008.
25. Shinde, V. and Patil, P.P., Evaluation of corrosion protection performance of poly(o-ethyl aniline) coated copper by electrochemical impedance spectroscopy. *Mater. Sci. Eng.: B*, 168, 1–3, 142–150, 2010.
26. Rammelt, U. and Reinhard, G., Application of electrochemical impedance spectroscopy (EIS) for characterizing the corrosion-protective performance of organic coatings on metals. *Prog. Org. Coat.*, 21, 2, 205–226, 1992.
27. Olivier, M.-G., Poelman, M., Olivier, M.-G., Poelman, M., Use of Electrochemical Impedance Spectroscopy (EIS) for the Evaluation of Electrocoatings Performances, in: *Recent Researches in Corrosion Evaluation and Protection*, 2012.
28. Muresan, L.M., Nanocomposite Coatings for Anti-Corrosion Properties of Metallic Substrates. *Mater. (Basel, Switzerland)*, 16, 14, 5092, 2023.
29. Lasia, A., Electrochemical Impedance Spectroscopy and its Applications. 32, 143–248, 1999.
30. Scully, J.R., Silverman, D.C., M.W.K. (Eds.), Electrochemical Impedance Analysis and Interpretation. *ASTM Special Technical Publication*, p. 1188, 1993.

31. de Rooij, D.M.R., *Electrochemical Methods: Fundamentals and Applications*, 2003.
32. Retter, U. and Lohse, H., Electrochemical Impedance Spectroscopy, in: *Electroanalytical Methods*, pp. 149–166, 2005.
33. Wang, S., Zhang, J., Gharbi, O., Vivier, V., M.G., M.E.O., Electrochemical impedance spectroscopy. *Nat. Rev. Methods Primers*, 1, 1, 1–1, 2021.
34. Yuan, X.Z., Song, C., Wang, H., Zhang, J., Electrochemical impedance spectroscopy in PEM fuel cells: Fundamentals and applications, in: *Electrochemical Impedance Spectroscopy in PEM Fuel Cells: Fundamentals and Applications*, pp. 1–420, 2010.
35. Lasia, A., *Electrochemical Impedance Spectroscopy and its Applications*, 2014.
36. Lvovich, V.F., *Impedance spectroscopy: applications to electrochemical and dielectric phenomena*, p. 353, 2012.
37. Macdonald, J.R. and Johnson, W.B., Fundamentals of Impedance Spectroscopy, in: *Impedance Spectroscopy*, pp. 1–26, 2005.
38. Williams, G., McMurray, H.N., Bennett, A., Inhibition of corrosion-driven organic coating delamination from a zinc surface using polyaniline pigments. *Mater. Corros.*, 65, 4, 401–409, 2014.
39. Orazem, M.E. and Tribollet, B., *Electrochemical Impedance Spectroscopy*, 2017.
40. Lazanas, A.C. and Prodromidis, M.I., Electrochemical Impedance Spectroscopy—A Tutorial. *ACS Meas. Sci. Au*, 3, 3, 162–193, 2023.
41. Barsoukov, E. and Macdonald, J.R., *Impedance Spectroscopy: Theory, Experiment, and Applications*, pp. 1–595, 2005.
42. Bonanos, N., Steele, B.C.H., Butler, E.P., Applications of Impedance Spectroscopy, in: *Impedance Spectroscopy*, pp. 205–537, 2005.
43. Barsoukov, E. and Macdonald, J.R. (Eds.), *Impedance Spectroscopy*, Wiley, 2018.
44. Kreysa, G. and Ken-ichiro Ota, R.F.S., *Electrochemical Dictionary*, Springer Berlin Heidelberg, 2012.
45. Ghali, E., *Corrosion Resistance of Aluminum and Magnesium Alloys: Understanding, Performance, and Testing*, 2010.
46. GAMRY instruments, Basics of EIS: Electrochemical Research-Impedance Gamry Instruments, 2021.
47. Walter, G.W., A review of impedance plot methods used for corrosion performance analysis of painted metals. *Corros. Sci.*, 26, 9, 681–703, 1986.
48. Raistrick, I.D., Franceschetti, D.R., Macdonald, J.R., Theory, in: *Impedance Spectroscopy*, pp. 27–128, 2005.
49. Shkirskiy, V., Keil, P., Hintze-Bruening, H., Leroux, F., Vialat, P., Lefèvre, G., Ogle, K., Volovitch, P., Factors Affecting MoO42- Inhibitor Release from Zn2Al Based Layered Double Hydroxide and Their Implication in Protecting Hot Dip Galvanized Steel by Means of Organic Coatings. *ACS Appl. Mater. Interfaces*, 7, 45, 25180–25192, 2015.

50. McKubre, M.C.H. and Macdonald, D.D., Measuring Techniques and Data Analysis, in: *Impedance Spectroscopy*, pp. 129–204, 2005.
51. Rosvall, M. and Sharp, M., A complete system for electrochemical impedance spectroscopy which combines FFT methods and staircase voltammetry. *Electrochem. Commun.*, 2, 5, 338–343, 2000, ISSN 1388-2481, https://doi.org/10.1016/S1388-2481(00)00032-1.(https://www.sciencedirect.com/science/article/pii/S1388248100000321)
52. Lvovich, V.F., *Impedance Spectroscopy: Applications to Electrochemical and Dielectric Phenomena*, 2012, Vadim F. Lvovich - Google Boeken.
53. Yadav, N. and Dhar, R., Impedance Spectroscopy, in: *Progress in Optical Science and Photonics*, vol. 13, pp. 515–540, 2021.
54. Diard, J.-P., Le Gorrec, B., Montella, C., Circuits made of Resistors and Capacitors, in: *Handbook of Electrochemical Impedance Spectroscopy*, 2011.
55. Application, A. and Eis, N., AUTOLAB Application Note- Electrochemical Impedance Spectroscopy (EIS): 5. Parameter Estimation, pp. 1–2.
56. Nishi, N., Electrochemical Impedance. *Rev. Polarogr.*, 65, 2, 87–88, 2019.
57. Heinze, J., Allen J. Bard and Larry F. Faulkner: Electrochemical Methods - Fundamentals and Applications. Wiley, New York 1980, 718 + XVIII S., Preis: £ 14.70. *Ber. Bunsenges. Physik. Chem.*, 85, 11, 1085–1086, 1981.
58. Lim, C., IV . Transport Phenomena Lecture 20: Warburg Impedance. *MIT OpenCourseWare*, pp. 1–7, 1899.
59. Barsoukov, E. and Macdonald, J.R. (Eds.), *Impedance Spectroscopy*, Wiley, 2005.
60. Goyal, S.C., Dielectric behaviour of solids. *Solid State Commun.*, 20, 3, 269–270, 1976.
61. Zarras, P., Stenger-Smith, J.D., Wei, Y. (Eds.), *Electroactive Polymers for Corrosion Control*, American Chemical Society, Washington, DC, 2003.
62. Kreysa, G., Ota, K., R.F.S. (Eds.), *Encyclopedia of Applied Electrochemistry*, Springer New York, 2014.
63. Ahmad, N. and MacDiarmid, A.G., Inhibition of corrosion of steels with the exploitation of conducting polymers. *Synth. Met.*, 78, 2, 103–110, 1996.
64. An, H., Liu, K., Wang, S., Jiang, C., Liang, S., Xiao, J., Zhao, X., Sun, Z., Cao, C., Gao, Y., Enhanced Corrosion Resistance of Waterborne Epoxy Coatings by Polyaniline Nanorods and Nitrogen and Fluorine Dual-Doped Graphene Oxide Composites. *ACS Appl. Nano Mater.*, 6, 14, 13250–13259, 2023.
65. Ahmed, R. and Reifsnider, K., Study of Influence of Electrode Geometry on Impedance Spectroscopy. *Int. J. Electrochem. Sci.*, 6, 1159–1174, 2011.
66. Douvartzidis, S., Dimoulas, G., Tsiakaras, P., Methane catalytic and electrocatalytic combustion over perovskite type oxides deposited on YSZ. *Stud. Surf. Sci. Catal.*, 119, 93–98, 1998.
67. Gamry, Redefining Electrochemical Measurement, 2022.
68. *Activation Overpotential*, pp. 1–26, Indian Institute of Technology, Kanpur, 2015.

69. Gamry, Physical Electrochemistry & Equivalent Circuit Elements (Part 2), vol. 2, Gamry Instruments, pp. 1–13, 2017.
70. Gateman, S.M., Gharbi, O., Gomes De Melo, H., Ngo, K., Turmine, M., Vivier, V., On the use of a constant phase element (CPE) in electrochemistry, 2022.
71. Helwig, N.E., Hong, S., Hsiao-wecksler, E.T., Spiro, M., *Modern Aspects of Electrochemistry №34*, J. O'M. Bockris, B.E. Conway, E. Ralph (Eds.), Modern Aspects of Electrochemistry (2002, Springer US) [10.1007_b114168] - libgen.li.pdf.
72. Vieira, D.S., Fernandes, P.R.G., Mukai, H., Zola, R.S., Lenzi, G.G., Lenzi, E.K., Surface Roughness Influence on CPE Parameters in Electrolytic Cells. *Int. J. Electrochem. Sci.*, 11, 9, 7775–7784, 2016.
73. Orazem, M.E., Electrochemical impedance spectroscopy: the journey to physical understanding. *J. Solid State Electrochem.*, 24, 9, 2151–2153, 2020.
74. Chen, Q., Tang, Y., Advincula, R.C., Superhydrophobic and anti-corrosion properties in thiol-ene graphene oxide nanocomposite coatings. *MRS Commun.*, 1–9, 2023.
75. Parihar, S. and Gaur, B., High performance self-healing polymeric nanocomposite coatings. *Prog. Organic Coatings*, 182, 107626, 2023.
76. Zhang, R., Guo, L., Obot, I.B., Anti-corrosive nanomaterials: Design, characterization, mechanisms and applications, in: *Anti-Corrosive Nanomaterials: Design, Characterization, Mechanisms and Applications*, pp. 1–336, 2023.
77. Xing, C., Song, X., Zhang, Z., Jiang, X., Yu, L., Anticorrosion Coatings from Poly (Aniline-co-2-Ethylaniline) Micro/Nanostructures. *J. Ocean Univ. China*, 18, 6, 1371–1381, 2019.
78. Deyab, M.A., De Riccardis, A., Mele, G., Novel epoxy/metal phthalocyanines nanocomposite coatings for corrosion protection of carbon steel. *J. Mol. Liq.*, 220, 513 517, 2016.
79. Pourhashem, S., Saba, F., Duan, J., Rashidi, A., Guan, F., Nezhad, E.G., Hou, B., Polymer/Inorganic nanocomposite coatings with superior corrosion protection performance: A review. *J. Ind. Eng. Chem.*, 88, 29–57, 2020.
80. Zandi-Zand, R., Ershad-Langroudi, A., Rahimi, A., Silica based organic–inorganic hybrid nanocomposite coatings for corrosion protection. *Prog. Org. Coat.*, 53, 4, 286–291, 2005.
81. Zhou, C., Lu, X., Xin, Z., Liu, J., Zhang, Y., Polybenzoxazine/SiO2 nanocomposite coatings for corrosion protection of mild steel. *Corros. Sci.*, 80, 269–275, 2014.
82. Chaudhari, S., Patil, P.P., Mandale, A.B., Patil, K.R., Sainkar, S.R., Use of poly(o-toluidine)/ZrO2 nanocomposite coatings for the corrosion protection of mild steel Use of Poly(o-toluidine)/ZrO 2 Nanocomposite Coatings for the Corrosion Protection of Mild Steel. *J. Appl. Polym. Sci.*, 106, 220–229, 2007.

83. Ding, C., Tai, Y., Wang, D., Tan, L., Fu, J., Superhydrophobic composite coating with active corrosion resistance for AZ31B magnesium alloy protection. *Chem. Eng. J.*, 357, 518–532, 2019.
84. Shinde, V., Gaikwad, A.B., Patil, P.P., Synthesis and characterization of corrosion protective poly(2,5-dimethylaniline) coatings on copper. *Appl. Surf. Sci.*, 253, 3, 2006.
85. Shinde, V., Sainkar, S.R., Gangal, S.A., Patil, P.P., Synthesis of corrosion inhibitive poly(2,5-dimethylaniline) coatings on low carbon steel. *J. Mater. Sci.*, 41, 10, 2006.
86. Shinde, V.P. and Patil, P.P., Investigation on role of monomer(s) during electrochemical polymerization of aniline and its derivatives on low carbon steel by XPS. *Electrochim. Acta*, 78, 2012.
87. Shinde, V. and Patil, P.P., Poly(o-toluidine) coatings on low carbon steel: Synthesis and characterisation. *Mater. Sci. Technol.*, 19, 11, 2003.
88. Shinde, V., Sainkar, S.R., Patil, P.P., Electrochemical synthesis and corrosion protection properties of poly(o-toluidine) coatings on low carbon steel. *J. Appl. Polym. Sci.*, 96, 3, 2005.
89. Shinde, V., Chaudhari, S., Patil, P.P., Sainkar, S.R., Electrochemical polymerization of 2,5-dimethylaniline on low carbon steel. *Mater. Chem. Phys.*, 82, 3, 2003.
90. Shinde, V., Gaikwad, A.B., Patil, P.P., Synthesis and corrosion protection study of poly(o-ethylaniline) coatings on copper. *Surf. Coat. Technol.*, 202, 12, 2591–2602, 2008.
91. Shinde, V., Mandale, A.B., Patil, K.R., Gaikwad, A.B., Patil, P.P., Poly(o-toluidine) coatings on copper: Electrochemical synthesis from aqueous media. *Surf. Coat. Technol.*, 200, 16–17, 2006.
92. Shinde, V., Sainkar, S.R., Patil, P.P., Corrosion protective poly(o-toluidine) coatings on copper. *Corros. Sci.*, 47, 6, 2005.
93. Shinde, V., Sainkar, S.R., Patil, P.P., Corrosion protection of low carbon steel by poly (O-toluidine) electrosynthesized from tartrate solution. *Trans. SAEST (Soc. Adv. Electrochem. Sci. Technol.)*, 40, 3, 2005.
94. Shinde, V.P., Study of water transport characteristics of poly(o-ethylaniline) coatings: corrosion mechanism. *Ionics*, 24, 2, 605–615, 2018.
95. Spinks, G.M., Dominis, A.J., Wallace, G.G., Tallman, D.E., Electroactive conducting polymers for corrosion control. *J. Solid State Electrochem.*, 6, 2, 85–100, 2001.
96. Williams, G. and McMurray, H.N., Polyaniline inhibition of filiform corrosion on organic coated AA2024-T3. *Electrochim. Acta*, 54, 17, 4245–4252, 2009.
97. Michalik, A. and Rohwerder, M., Conducting polymers for corrosion protection: A critical view. *Z. Phys. Chem.*, 219, 11, 1547–1559, 2005.
98. Tallman, D.E., Spinks, G., Dominis, A., Wallace, G.G., Electroactive conducting polymers for corrosion control. *J. Solid State Electrochem.*, 6, 2, 73–84, 2001.

99. MacDiarmid, A.G., Conducting Electronic Polymers by Non-Redox Processes, 1987.
100. Wessling, B., Passivation of metals by coating with polyaniline: Corrosion potential shift and morphological changes. *Adv. Mater.*, 6, 3, 226–228, 1994.
101. DeBerry, D.W., Modification of the Electrochemical and Corrosion Behavior of Stainless Steels with an Electroactive Coating. *J. Electrochem. Soc.*, 132, 5, 1022–1026, 1985.
102. Fahlman, M., Jasty, S., Epstein, A.J., Corrosion protection of iron/steel by emeraldine base polyaniline: an X-ray photoelectron spectroscopy study. *Synth. Met.*, 85, 1–3, 1323–1326, 1997.
103. Lu, W.-K., Elsenbaumer, R.L., Wessling, B., Corrosion protection of mild steel by coatings containing polyaniline. *Synth. Met.*, 71, 1–3, 2163–2166, 1995, https://doi.org/10.1016/0379-6779(94)03204-J.

8

Graphene-Based Coating on Mild Steel for Improving Anticorrosion and Microhardness Behavior: A Review

Sunita Dhar[1], Tapan Dash[1,2*], Ashok Kumar Sahu[3], Sushree Subhadarshinee Mohapatra[4], Nibedita Mohanty[4], Shubhra Bajpai[3], Tapan Kumar Rout[5] and Surendra Kumar Biswal[2]

[1]*Centurion University of Technology and Management, Odisha, India*
[2]*International PranaGraf Mintech Research Centre (IGMRC), Bhubaneswar, Odisha, India*
[3]*CSIR-Institute of Minerals and Materials Technology, Bhubaneswar, Odisha, India*
[4]*GIET University, Gunpur, Odisha, India*
[5]*Research & Development Department, Tata Steel Ltd, Jamshedpur, India*

Abstract
Mild steel is widely used for various construction *industries* and pipelines due to its low cost compared to other forms of steel. However, its poor anticorrosion behavior and limited mechanical properties are two main disadvantages in its potential applications. Graphene is one of the gifted carbon nanomaterials developed in this century. It is a wonderful two-dimensional carbon nanomaterial structured with a single layer of atoms and has attracted great interest in recent years due to its unique, mechanical, anticorrosion, electrical, optical, and catalytic properties. The literature claimed that graphene derivatives and their nanocomposites have efficient potential for protecting mild steel from corrosive environments and can enhance surface mechanical properties, especially hardness. An attempt has been made to study the overall properties, synthesis, and characterization of graphene derivatives. Anticorrosion behavior and hardness properties of graphene derivative-based coating on mild steel substrate have been presented in this review.

Keywords: Graphene coating, mild steel, anticorrosion behavior, surface hardness

**Corresponding author*: tapanphy@gmail.com

Hari Murthy, Vinay Jha Pillai, Kukatlapalli Pradeep Kumar and Matthew Cowan (eds.) Novel Anti-Corrosion and Anti-Fouling Coatings and Thin Films, (211–250) © 2024 Scrivener Publishing LLC

8.1 Introduction of Graphene

Graphene, a one-dimensional allotrope of carbon, has emerged as a promising substance in the field of science and technology. In the year of 2004, this novel material was discovered by Geim and Novoselov [1, 2]. They successfully isolated graphene by employing a simple yet ingenious method where exfoliation of single layers was done from a block of graphite using simple adhesive tape. This breakthrough was made possible due to the realization that graphene, although incredibly thin, possessed incredible strength and mechanical stability. For their groundbreaking discovery, they were awarded the Nobel Prize in Physics in the year of 2010. This prestigious accolade elevated the global recognition of graphene and stimulated further research in the field. Over time, researchers developed various techniques for its mass-scale production which include mechanical exfoliation, chemical vapor deposition (CVD), and epitaxial growth, among others. These procedures enabled the production techniques of graphene-based derivatives by controlling their properties which paved the way for their integration into various applications.

Following the discovery of graphene, researchers around the world began exploring its potential applications. The subsequent years witnessed extensive research into graphene-based materials, leading to several significant areas of applications. Its remarkable properties, including exceptional strength, anticorrosion properties, high electron mobility, and high heat conduction, have captured the attention of researchers worldwide [2–4]. Graphene's remarkable properties are due to its unique structure having hexagonally organized one atomic layer of carbon atoms. Graphene exhibits a two-dimensional nanostructure.

It is the most powerful substance known to mankind, with a tensile strength over 100 times greater than steel. Its lattice structure allows it to distribute stress evenly, making it incredibly robust. Graphene exhibits exceptional electrical conductivity, surpassing that of copper. The carbon atoms' arrangement allows electrons to move freely through the lattice, making it an excellent material for electronic applications; with superior thermal conductivity, graphene possesses outstanding thermal conductivity, even surpassing that of diamond. Its ability to efficiently transfer heat makes it highly desirable for applications in thermal management and electronics; optical properties: graphene absorbs only 2.3% of light across the entire visible spectrum, rendering it nearly transparent. This property, combined with its exceptional electrical conductivity, makes it an excellent candidate for optoelectronic devices. Graphene-based coatings provide corrosion protection for various materials. Graphene-based materials have found applications across numerous fields. In electronics, graphene transistors, flexible displays, and wearable devices have been developed. In energy storage, graphene has been explored for high-performance batteries and supercapacitors. Additionally, it has been investigated for applications in sensors, water purification, composite materials, biomedical devices, anti-corrosion materials, etc.

8.1.1 General Uses of Graphene Derivatives

Electronics and Optoelectronics

Graphene derivatives and their composites with other materials have shown immense promise in the field of electronics [5]. With its exceptional electrical conductivity, graphene can be used to develop high-speed transistors, flexible and transparent conductive films, and efficient energy storage devices like supercapacitors. Additionally, Materials based on graphene have the potential to revolutionize optoelectronics, photodetectors, solar cells, and light-emitting diodes (LEDs).

Energy Storage

The huge specific surface area (SSA) and strong electron mobility of graphene are highly useful for developing highly efficient energy storage devices [6]. Graphene-based materials have been explored for the development of supercapacitors with high energy density, rapid charge-discharge rates, and long cycle life. These advancements could revolutionize the energy storage industry, leading to faster-charging electric vehicles and more efficient renewable energy storage devices.

Composite Materials

Incorporating graphene-based materials into composite constructions can improve their mechanical and thermal qualities dramatically [7–9]. Graphene's exceptional strength and stiffness make it an excellent reinforcement material for composites, resulting in lightweight, potential materials with enhanced mechanical and electrical conductivity, as well as thermal stability. These composite materials find applications in the aerospace, automotive, and construction industries.

Coating Applications of Graphene-Based Materials

Graphene-based coatings have shown remarkable potential in providing corrosion protection for various materials [10, 11]. By forming a barrier against corrosive substances, graphene coatings can prevent the degradation of metals and alloys. The high permeability and excellent barrier properties of graphene help reduce the corrosion rate and extend the life span of infrastructure, such as bridges, pipelines, and offshore structures. Graphene-based coatings have exceptional lubricating properties due to their low friction coefficient and high mechanical strength [12–14]. These coatings can significantly reduce wear and friction in mechanical systems, leading to improved efficiency and durability. Applications range from automotive engines and gears to industrial machinery, where graphene coatings can minimize energy losses and enhance the life span of components. Graphene-based coatings have hydrophobic and oleophobic properties, making them ideal for creating self-cleaning surfaces and preventing the adhesion of water and oil-based substances [15, 16]. These coatings find applications in various sectors, including textiles, electronics, and automotive, where they can repel water, oil, and other contaminants, leading to easy cleaning and improved performance.

8.1.2 Different Types of Typical Graphene Derivatives

Among notable qualities of graphene derivatives, the large SSA is very important which gives ample opportunities for functionalization and modification. There are different types of graphene derivatives used to enhance specific properties for targeted applications. Let us explore some of the notable varieties.

Graphene Oxide (GO): It contains oxidizing graphene sheets [17]. It contains various oxygenated functional groups. It is potentially used in energy storage, catalysis, composite materials, etc.

Reduced Graphene Oxide (rGO): It is obtained by reducing GO, eliminating the majority of oxygen functional groups [18, 19]. This process restores the electrical conductivity of graphene, which makes rGO appropriate for use in electronics, sensors, and supercapacitors.

Graphene Nanoplatelets (GNPs): Graphene nanoplatelets are thin, platelet-like structures composed of multiple layers of graphene [20]. They have outstanding electrical and thermal conductivity as well as mechanical strength. GNPs find applications in composites, energy storage devices, coatings, and lubricants.

Graphene Quantum Dots (GQDs): These are small graphene fragments with dimensions less than 10 nm [21, 22]. Their unique size-dependent properties make them promising candidates for applications in optoelectronics, bioimaging, and quantum computing.

Graphene Nanoribbons (GNRs): These are tiny strips of graphene with nanometer widths and aspect ratios in the thousands [23]. They possess semiconducting properties, making them suitable for nanoelectronics, photovoltaics, and molecular sensing. These are just a few examples of the diverse range of graphene-based materials that researchers have developed. The continuous exploration and discovery of novel graphene derivatives open up endless possibilities for future applications.

8.2 Synthesis Mechanism of Graphene Derivatives

Graphene derivatives are produced basically by two basic mechanisms: top-down [24–38] and bottom-up approaches [39–58]. Tables 8.1 and 8.2 represent typical methods of graphene derivatives belonging to top-down and bottom-up approaches, respectively. The typical production methods of graphene derivatives are shown in the schematic representation in Figure 8.1 [58].

Table 8.1 Typical top-down methods reported in the literature for producing graphene derivatives.

Sl. no.	Method adopted for synthesis of graphene	Graphene derivative produced	Reference
1	Micromechanical exfoliation	Few layer graphene	[24]
2	Micromechanical exfoliation	Single-layer or multilayer graphene	[25]
3	Mechanical exfoliation by ball milling	Few-layer graphene or bilayer graphene	[26–28]
4	Hummer's method	Reduced graphene oxide	[29]
5	Electrochemical exfoliation	Multilayer graphene	[30]
6	Chemical exfoliation	Reduced graphene oxide	[31]
9	Chemical exfoliation	Reduced graphene oxide	[32, 33]
10	Chemical reduction	Reduced graphene oxide	[34, 35]
11	Liquid phase-exfoliation	Graphene	[36, 37]
12	Photoexfoliation	Single-layer graphene and N-layer graphene	[38]

Table 8.2 Typical bottom-up methods reported in the literature for producing graphene derivatives.

Sl. no.	Method adopted for the synthesis of graphene	Graphene derivative produced	Reference
1	Chemical vapor deposition (CVD)	Monolayer graphene	[39–44]
2	Chemical vapor deposition (CVD)	Few layer graphene	[45–48]
4	Unzipping of carbon nanotubes	Multiple layers	[49]
6	Epitaxial growth	Monolayer graphene	[50]
7	Epitaxial growth	Few layer graphene	[51]
9	Chemical synthesis	Few layer graphene	[52]
10	Chemical synthesis	Reduced graphene oxide	[53]
10	Pyrolysis	Monolayer graphene	[54]
11	Pyrolysis	Few layer graphene	[55]
12	Hydrothermal cathodic	Graphene	[56]
13	Anodic bonding	Single layer graphene	[57]

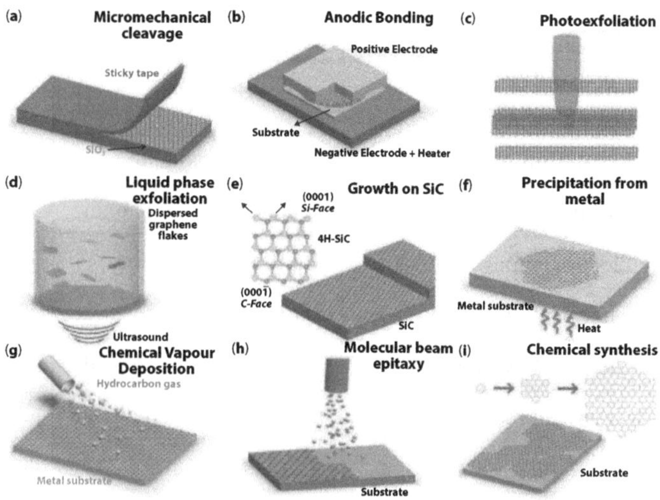

Figure 8.1 Schematic representation of typical production methods of graphene (reproduced with permission from [58], on permission of copyright © 2012 Elsevier Ltd.).

8.3 Typical Characterizations of Graphene Derivatives

The typical characterization techniques such as X-ray diffraction (XRD), Raman spectroscopy, scanning electron microscope (SEM), transmission electron microscope (TEM), selected area electron diffraction (SAED), Brunauer-Emmett-Teller (BET) SSA analysis, and electrical conductivity have been used by the researchers to evaluate and optimize the properties of graphene derivatives.

8.3.1 Evaluation of Properties of Graphene Derivatives by XRD

XRD characterization is used for evaluating crystallinity, constituent of phases, and spacing between two consecutive layers of graphene derivatives and other materials in general. XRD patterns of high pure graphite (HPG), chemically synthesized GO, and its reduction under the H_2 atmosphere to form rGO samples are presented in Figure 8.2. [59]. The characteristic peak C(002) for graphite is formed at $2\theta=26.8^0$. When graphite becomes oxidized, the diffracted peak C(001) is shifted to $2\theta=11.8^0$. Upon reduction of GO, the peak representing GO disappears, and C(002) as the major peak is formed at $\sim 2\theta=26.4^0$. Li *et al.* [60] studied the XRD results of GO and RGO samples as represented in Figure 8.3. They prepared GO by modified Hummers method that showed a C(001) peak at $2\theta=11.2^0$. The GO samples were reduced by immersing in ammonia water and scattered Zn powder for 1, 5, 10, 30, and 60 min to form rGO samples called rGO-1,

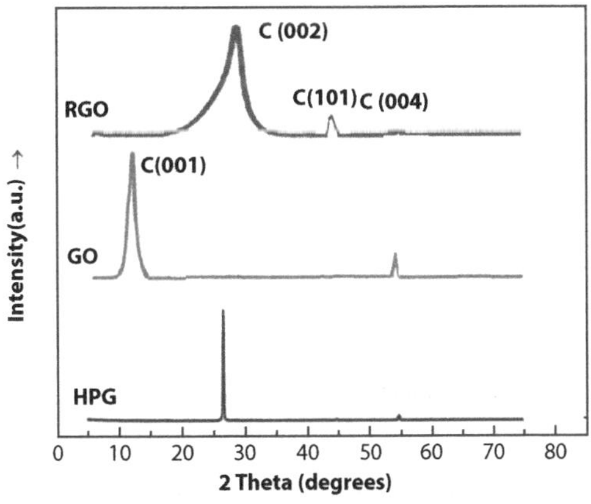

Figure 8.2 XRD analysis results for high pure graphite (HPG), graphene oxide (GO), and reduced graphene oxide (RGO) ([59], adapted on permission of copyright approval for free cultural works - under-creative commons).

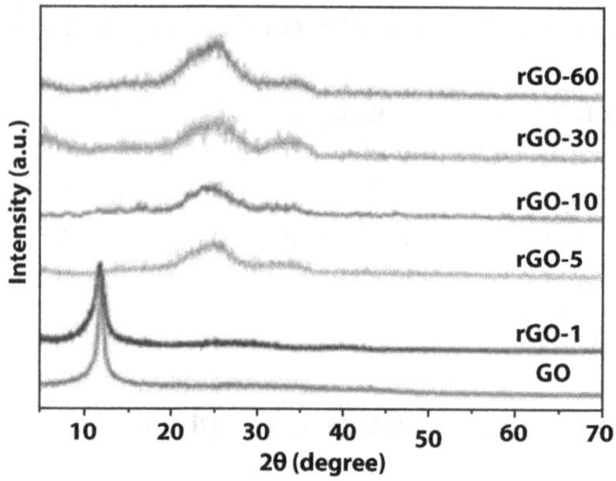

Figure 8.3 XRD patterns of rGO-1, rGO-5, rGO-10, rGO-30, and rGO-60 paper ([60], adapted on permission of copyright © 2015 Elsevier B.V.).

rGO-5, rGO-5, rGO-30, and rGO-60, respectively. They observed that the C(001) peak was completely removed and the C(002) peak appeared at around 2θ of 25^0. C(002) peak intensity increases as the duration for reduction of GO increases.

A similar kind of results was reported by Cui *et al.* [61]. Zhang *et al.* [62] reported similar XRD patterns (Figure 8.4) for graphite and GO. However, they obtained an XRD pattern of graphene without any peak.

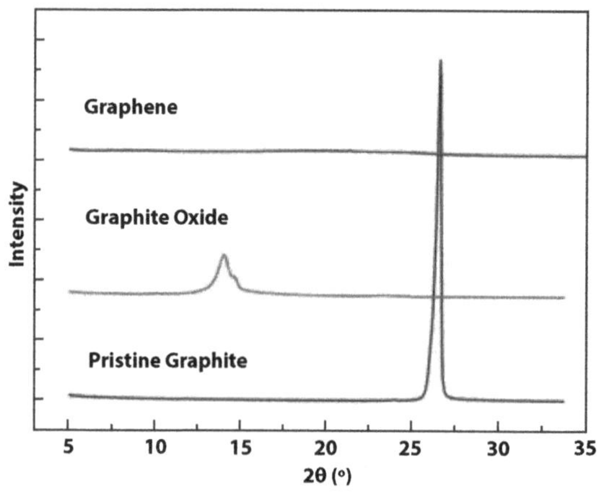

Figure 8.4 Comparison of XRD results of pristine graphite, graphite oxide, and graphene ([62], adapted on permission of copyright © 2010 Elsevier Ltd.).

8.3.2 Evaluation of Morphological Behavior of Graphene Derivatives by SEM

SEM characterization is used to analyze the morphological behavior of graphene derivatives. SEM results of GO and rGO prepared by Li *et al.* [60] showed the microstructure of multilayered graphene [Figure 8.5 (a–d)]. Zhang *et al.* [62] found graphene sheets of thickness up to 50 nm due to the interaction of polyethylene terephthalate (PET) and graphene. Graphene sheets with A lamellar type structure (lateral size < 1 μm) were found to be obtained by mechanical milling of graphite in 2-ethyl-hexanol and kerosene [63]. The graphene was also found to be aggregated. Mohan *et al.* reported flaky type morphology of graphite [Figure 8.6(a)] [64]. They chemically oxidized graphite to form GO and reduced using hydrobromic acid, hydrazine hydrate, and hydroiodic acid. After the reduction of GO to rGO, the surface was found to be corrugated because of the removal of oxygenated functional groups and rGO was found to be multilayered [Figure 8.6(b–d)]. Graphene developed on Cu substrates *via* CVD technique at different recipes [65]. SEM analysis of graphene developed on Cu substrates at different recipes such as recipe #3 (H_2 = 6 sccm, CH_4 = 30 sccm, chamber pressure-1 torr, growth time-16 min), recipe #7 (H_2 = 6 sccm, CH_4 = 10 sccm, chamber pressure-2 torr, growth time-16 min), and recipe #8 (H_2 = 6 sccm, CH_4 = 50 sccm, chamber pressure-10 torr, growth

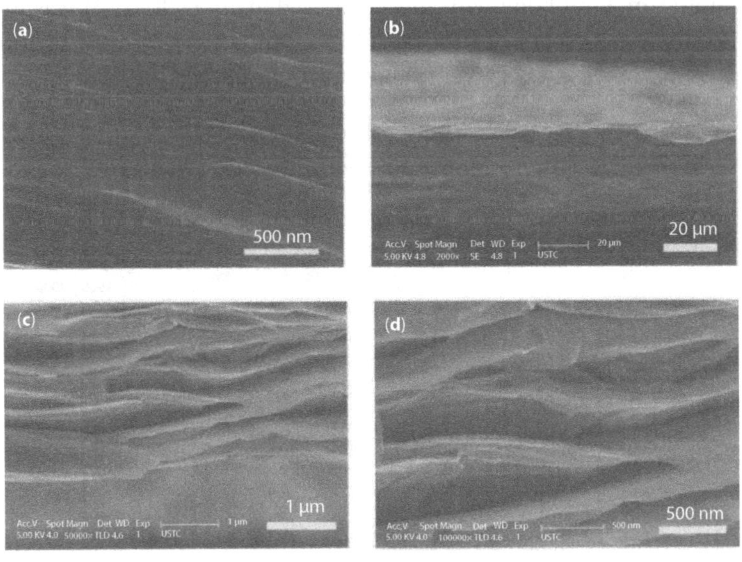

Figure 8.5 SEM images of (a) GO, and (b–d) rGO-30 ([60], adapted on permission of copyright © 2015 Elsevier B.V.).

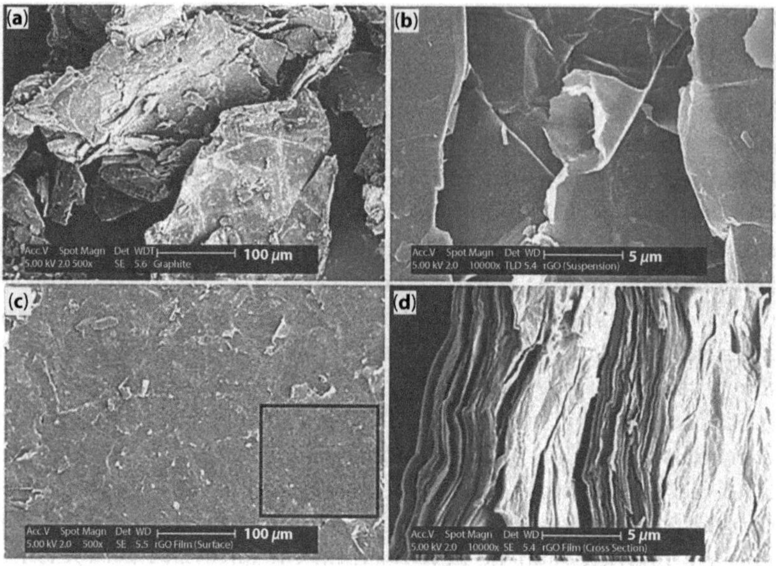

Figure 8.6 Morphological analysis by SEM: graphite (a), rGO (reduction by hydroiodic acid) suspension (b), surface (inside GO) (c), and cross-sectional view of rGO film (d) ([64], adapted on permission of copyright © 2014 Elsevier B.V.).

time-4 min) presented in Figure 8.7 (a), (c), and (e), respectively, clearly evident the development of graphene with bright small clusters of typically around 50 nm. Graphene with high cluster density (~30/μm^2) was grown by introducing 10 sccm of 95% N_2, 5% H_2, and 30 sccm of CH_4 into the synthesis chamber at 1000°C for 3 hours [Figure 8.7(g)].

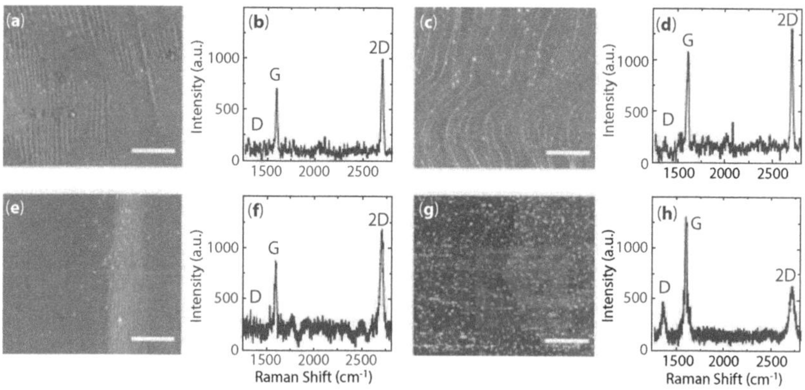

Figure 8.7 CVD grown graphene: SEM and Raman results of graphene (a and b) at recipe #3, (c and d) at recipe #7, (e and f) at recipe #8, and (g and h) at recipe used for developing high-density carbon clusters. Scale bars in SEM images are 3 μm. ([65], adapted on permission of copyright © 2014 Elsevier B.V.).

8.3.3 Evaluation of Morphological Behavior of Graphene Derivatives by TEM and SAED

TEM gives morphological analysis at the nanolevel. TEM analysis of HPG and reduced GO is represented in Figure 8.8 [59]. HPG shows dense and stacking of many flaky layers [Figure 8.8(a)]. TEM analysis results of reduced GO at different magnifications are presented in Figure 8.8 (b) and (c). TEM results show that reduced GO of a few layers confirms the successful exfoliation of graphene sheets by mechanical milling of HPG. Folding-type graphene sheets were found to be grown. SAED pattern [Figure 8.8 (d)] of rGO carried out on TEM image [Figure 8.8 (c)] shows a hexagonal structure. rGO prepared *via* a chemical reduction process showed the microstructure of a few layers ($n < 6$) which are stacked with each other with relatively less wrinkles and folding [66].

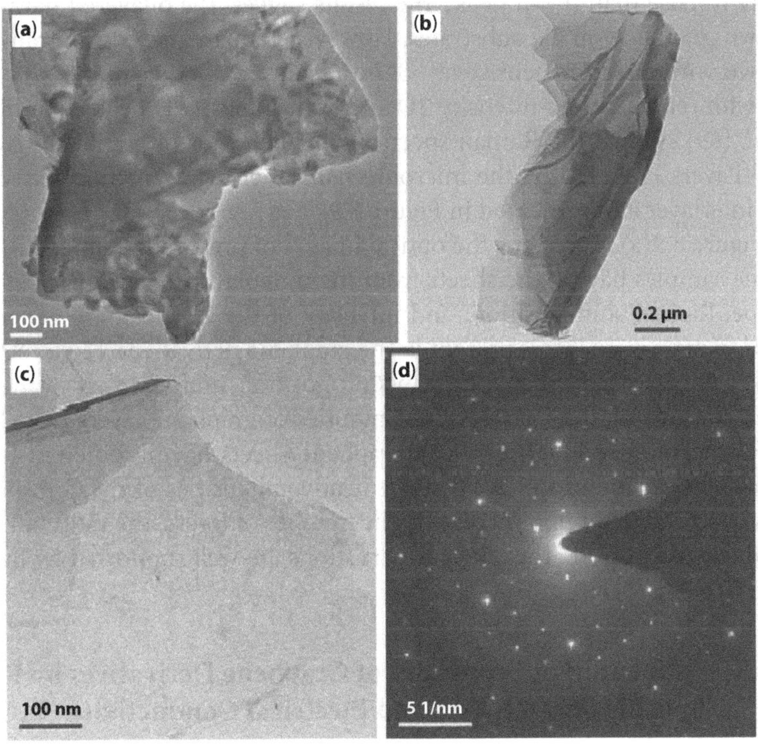

Figure 8.8 (a) TEM analysis results of high pure graphite. (b and c) TEM analysis results of reduced graphene oxide. (d) SAED analysis result of reduced graphene oxide carried out on TEM image (c) ([59], adapted on permission of copyright approval for free cultural works - under-creative commons).

Reduced GO prepared *via* polar solvents such as methanol, ethanol, and acetone shows a mostly highly disordered stacking porous structure with few layers of graphene sheets [67]. Luo *et al.* [68] confirmed the growth of single and bilayer graphene on Cu substrate *via* the CVD method.

8.3.4 Raman Spectroscopic Behavior of Graphene Derivatives

Raman spectroscopy is employed to evaluate the structure degree of disorders and the number of layers of graphene derivatives. Generally, micro-Raman spectroscopic graphene derivatives show three characteristic vibrational bands such as D (due to the first disorder peak of graphite), G (due to sp^2 carbon of graphite lattice), and 2D (due to the stacking layer of graphene). Raman analysis of graphene developed on substrates of Cu at different recipes such as recipe #3 [Figure 8.7(b)], #7 [Figure 8.7(d)], and #8 [Figure 8.7(f)] shows the low-intensity D peak [65]. The 2D peak was more intense than the G peak. The results suggest the bilayered nature of grown graphene on Cu substrates. However, Raman analysis of graphene grown with a cluster density of ~30/μm^2 showed a larger number of layers with relatively low-intensity 2D and high-intensity D peaks. Hwangbo *et al.* [69] studied the Raman spectroscopic behavior of graphene having multilayers prepared by the micromechanical cleaving method based on the interlayer and presented in Figure 8.9.

Figure 8.9(a) represents the optical images of prepared graphene membrane samples having the sheets from the monolayer to seven layers. The shape, Raman shift position, and intensity of the 2D peak are generally used to determine layers present in the graphene derivatives. Figure 8.9(b) shows the change of shape, Raman shift position, and intensity of the 2D peak concerning the change of the number of graphene layers from 1 to 7. Furthermore, in Figure 8.9(c), graphene sheets having different types of layers were found to have distinct relative intensities of the G peaks. It was observed that upon increasing the number of layers, the ratio of I_{2D}/I_G gradually decreased. The above observations are well supported by literature [70–72].

8.3.5 Evaluation of Properties of Graphene Derivatives by BET Specific Surface Area and Electrical Conductivity

Experimentally, it has been proved that SSA is related to different properties and structures of graphene derivatives such as interlayer spacing, packing density, electrical conductivity, etc. Generally, it has been observed that graphene derivatives with high SSA show high electrical

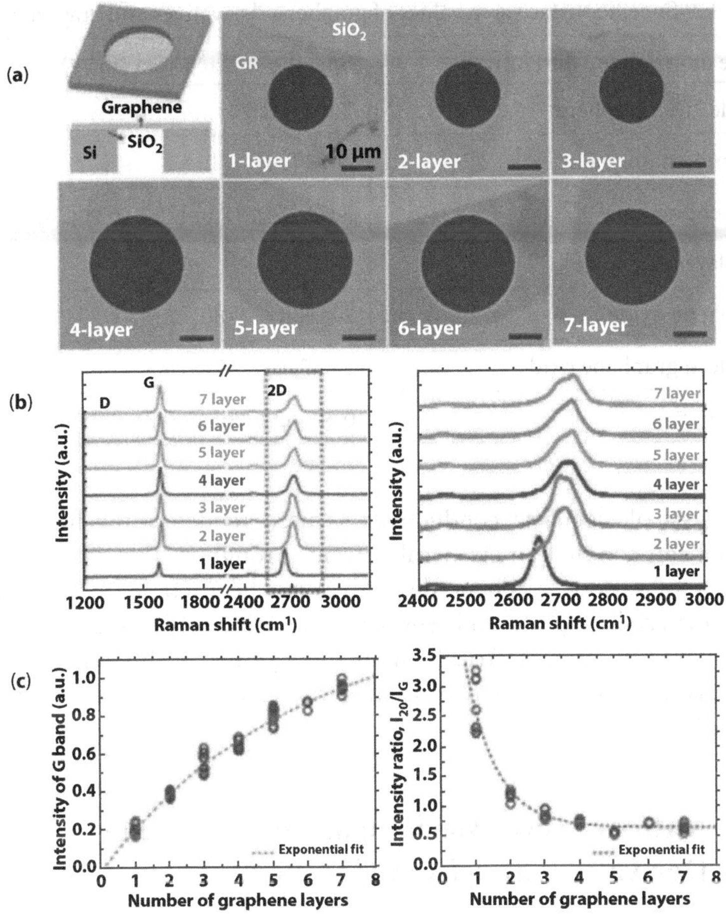

Figure 8.9 (a) Optical images of graphene membranes developed on $SiO_{2/}Si$ substrate. (b) A comparison of Raman spectroscopic results of different graphene membrane samples. (c) A comparison of the intensity of G peak and intensity ratios of I_{2D}/I_G of different graphene membrane samples ([69], adapted on permission of copyright © 2014 Elsevier Ltd.).

conductivity behavior. SSA is a direct measure of the degree of reduction of GO. Researchers report SSA of graphene derivatives. Still, there is a gap in comparative analysis for different graphene materials. The SSA of various graphene derivatives has been compared with their parent material, graphite, and represented in Table 8.3 [73–84]. The BET (Brunauer, Emmett, and Teller) analysis is commonly employed to study the SSA of materials. The decrease in the graphene layer can be corroborated by the increase in SSA. The single layer exhibits the highest SSA among graphene derivatives.

Table 8.3 Specific surface area values of graphene derivatives and graphite.

Type of graphene derivatives	Specific surface area (m^2/g)
Single layer graphene	2630 [73, 74]
Nanomesh graphene	2039 [74]
Bilayer graphene	2041 [74], 1230 [75], 1628 [76]
Few layer graphene	466 [74]
Graphene oxide	2–1000 [77–80]
Reduced graphene oxide	32.47–612.74 [81]
Graphite	0.50 [82], 7.6549 [83], 3.08 [84]

Analysis of electrical conductivity is used to determine how well a material can conduct an electrical current through it. The electrical conductivity of various graphene derivatives has been compared with their parent material, graphite, and represented in Table 8.4 [84–92]. According to Mohan *et al.* [93], electrical conductivity values of different graphene materials increase with increasing their respective specific surface values (Figure 8.10).

8.4 Anticorrosion Mechanisms of Graphene Derivatives and Their Composites

The corrosion protection mechanism of pure graphene coatings depends on structural integrity and irregularity and the typical mechanism is represented in Figure 8.11(a) [94]. It represents the prevention efficiency of a single-layer graphene coating with a highly ordered lattice arrangement. It does not allow penetration of oxidizing compounds and attack of corrosive media due to the high impermeability of single-layer graphene. In the case of graphene with defects, it was observed that graphene sheets lose their anticorrosion shielding toward corrosive media. The observation is well supported by the literature [95]. At the time of oxidation, electrons are liberated from the metal surface by releasing metal cations; as a result, metal oxides or hydroxides are formed over the substrate. The charges are transferred to the bulk metal through graphene layers due to their conducting nature, and as a result, the electrochemical reaction is unimpeded [96]. Figure 8.11(b) highlights the typical interaction stages of

Table 8.4 Electrical conductivity values of graphene derivatives and graphite.

Type of graphene derivatives	Electrical conductivity (S/m)	Synthesis method	Form	References
Single layer graphene	$1.46 \pm 0.82 \times 10^6$	CVD	Film	[85]
Bilayer graphene	2.54×10^5–6.401×10^5	Calculation	Sheet	[86]
Few layer graphene	3×10^5–9.8×10^5	Microwaves Assisted Exfoliation	Flakes	[87]
Multilayer graphene	2.76×10^5	The layer exchange technique	Film	[88]
Graphene oxide	4.57×10^{-6}	Chemical method	film	[89]
Reduced graphene oxide	45	Chemical reduction	Film	[90]
Reduced graphene oxide	29–800	Chemical reduction	Film	[91]
Reduced graphene oxide	200–3860	Chemical reduction	powder	[92]
Reduced graphene oxide	4.12×10^{-3}	Chemical method	film	[89]
Graphite	2.12×10^3	-	powder	[84]

pure graphene coatings such as initiation, intercalation, propagation, and, eventually, coating failure.

The protective mechanism for pure epoxy–graphene oxide (GO) (0.05%)-polypyrrole (PPy) nanocomposite coatings on mild steel reported by Zhu *et al.* [97] is represented by Figure 8.12. Epoxy provides a barrier

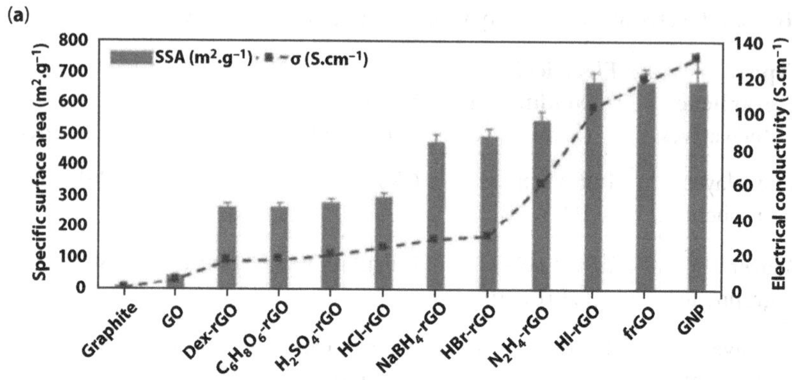

Figure 8.10 Comparison of electrical conductivity values vs. specific surface values of different graphene derivatives materials ([93], adapted on permission of copyright © 2020 Elsevier Ltd.).

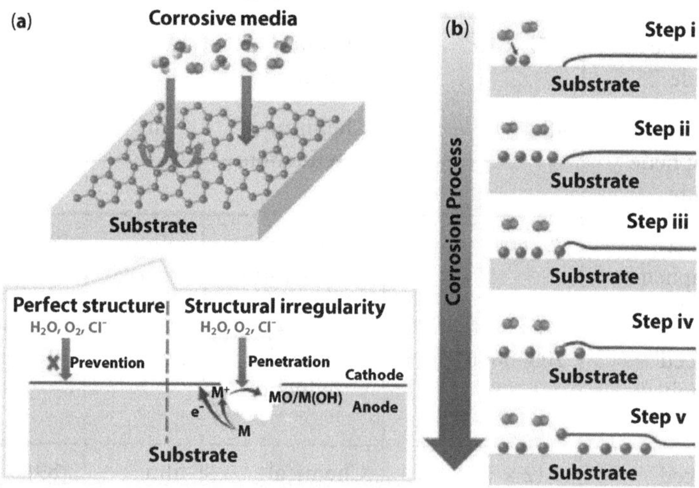

Figure 8.11 Illustrative view of the anticorrosion mechanisms for pure graphene coatings: (a) graphene with perfect structure and structural irregularity; (b) service stages of pure graphene coatings with O-intercalation ([94], adapted on permission of copyright © 2021 Elsevier Ltd.).

layer for corrosion. At the interface of coating and metal, the following corrosion reactions are observed

$$Fe - 2e^- \rightarrow Fe^{2+}$$

$$Fe^{2+} - e^- \rightarrow Fe^{3+}$$

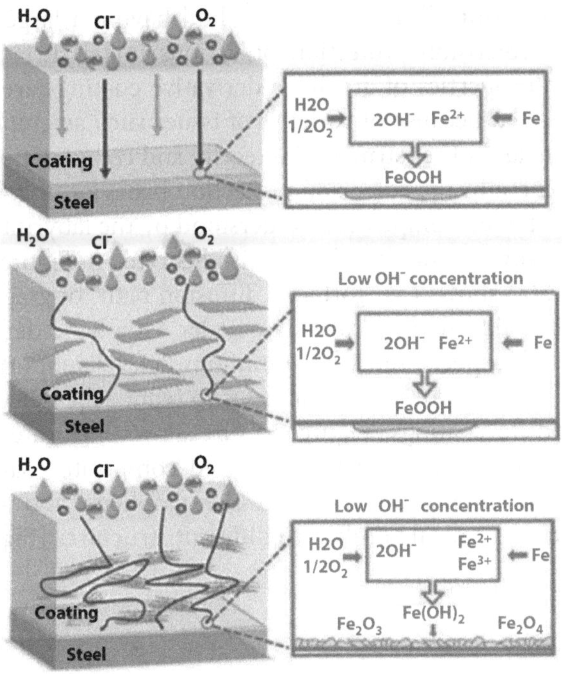

Figure 8.12 Protective mechanisms for (a) pure epoxy coating, (b) GO (0.05%)-PPy nanocomposite, and (c) GO (0.05%)-PPy nanocomposite ([97], adapted on permission of copyright © 2019 Elsevier B.V.).

$$O_2+H_2O \rightarrow 4OH^- - 4e^-$$

$$2H_2O+2Fe^{2+}+O_2 \rightarrow 2H^+ + 2FeOOH$$

Due to the above redox reactions, the coating over the substrate started degrading and lost its protective ability. However, in the case of coating by GO-PPy nanocomposite over mild steel, the impermeable property of coating was significantly enhanced to resist pitting corrosion. When the graphene nanosheets become perpendicular to the developed coating, a winding channel of diffusion for corrosive ions was developed, which resulted in delaying the formation of corrosion. Due to the presence of amino groups in PPy, additional cross-linking developed in the coating. This structure prevents in development of pores in the coating of epoxy, which helps to protect the surface from corrosion. The conducting PPy film by accepting the electrons forms passivate layers of Fe_2O_3 and Fe_3O_4 in the coating to enhance protection from corrosion of mild steel. Graphene prepared with a thin film *via* CVD method having a large surface area was used for corrosion protection of various metal surfaces [98]. The anticorrosion properties of graphene derivative coatings are found to be more complicated because of varieties of issues such as graphene quality, its orientation, and other structural integrity and complexities involved in graphene. Generally, the corrosion protection behavior of coatings by the graphene-reinforced composite is understood by the labyrinth effect [99], and the brief anticorrosion mechanism is illustrated in Figure 8.13 [94].

Figure 8.13(a) shows the "tortuous diffusion path" of graphene to provide impermeability toward corrosive media. Graphene derivatives with a relatively high aspect ratio with minimum defects support better anticorrosion behavior [100]. Graphene helps to reduce the defects and pores in the composite coatings. Figure 8.13(b) shows integrated nanostructures developed in the graphene-reinforced composite coating. Various chemo-physical interactions between the graphene and the polymer develop various chemical bonds with different structures [Figure 8.13(b)].

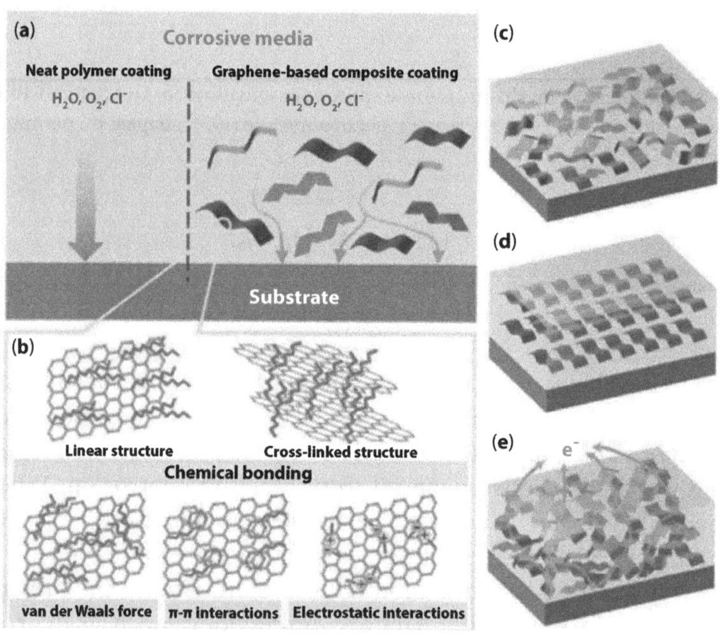

Figure 8.13 (a) The corrosion protection mechanisms of graphene-based composite coatings, (b) integrated nanostructures and the interfacial interaction developed in graphene/coating matrix, (c) dispersion of graphene sheets, (d) aligned graphene sheets, and (e) connected graphene sheets ([94], adapted on permission of copyright © 2021 Elsevier Ltd.).

Graphene significantly restricts the H_2O penetration and degradation of the polymer matrix [101]. The cross-linked graphene/polymer type of coating improved anticorrosion performance because of reducing inherent defects of graphene. Anticorrosion behavior of graphene-reinforced composites significantly depends on the nature of graphene in the composite such as its degree of dispersion [Figure 8.13(c)], alignment [Figure 8.13(d)], and randomness [Figure 8.13(e)].

Random distribution of graphene in the coating should be avoided. When GO-based composites are charged with corrosive media, GO becomes negatively charged because of the dissociation of carboxyl groups in water as a result ionic resistance develops, and Cl^- ions can be prevented by making coating positively charged or neutral. The basicity can be reduced by GO and prevents penetration of OH^- anions and corrosive media.

However, the corrosion prevention mechanism of graphene derivative-based coatings is specific and, for the typical purpose, it can be highlighted as follows:

(i) Hydrophobic graphene–based composite coatings: resist H_2O and the high interfacial energy of graphene protects metal substrates like mild steel from attacking corrosion.
(ii) Hydrophilic graphene–based composite coatings: For example, GO, H_2O is entrapped by the graphene sheets rather than the matrix of the coating.
(iii) Mild steel oxidation: it helps to reduce GO to form rGO which can resist anode sites and improve the corrosion protection efficiency of the metal substrate.

8.5 Anticorrosion Behavior of Graphene Derivative Coating on Mild Steel

When corrosive media such as water, oxygen, and chloride ions react with any metal, a time-consuming chemical or electrochemical process known as corrosion is developed, which results in the loss of the high hardness, high strength, and metallic luster properties of metal. Corrosion has a significant negative effect on progress of the society with unwanted financial burdens [102, 103]. Technologies used to preserve metals through cathodic protection, inhibitors used for corrosion, and coatings on the surface. Because of their ease of usage and effective protection, surface coatings are the most popular and cost-effective option [104–106]. Theoretically, a good anticorrosion coating on metal surfaces resists corrosive media from permeating onto the surface [107]. Conventional coatings on metals contain internal flaws and shrinkages caused by solvent volatilization, which reduces the barrier's resistance to corrosive media. As a result, maritime engineering equipment needs exceptionally thick coating materials with a thickness of around 300 to 500 μm [108]. Therefore, it is crucial to research and create new anticorrosive coatings that are durable and have a thin thickness.

A severe danger to the economy and industry is due to the degradation of metal components because of corrosion. Steel is the most popular and lucrative among the several metallic construction materials for businesses. Because of its mechanical qualities and inexpensive price, mild steel is utilized extensively across a variety of sectors. It is the most significant substance in the world and is recognized to have an impact on a nation's economic development. Even so, it has poor properties of hardness, wear resistance, and corrosion resistance compared to refractory materials. To expand the uses for steel, researchers utilize a variety of alloying techniques, compositions, coatings, and other techniques [109–111]. A coating is an excellent option among the many procedures since it is more efficient, uses less material, and is less expensive [111]. The mechanical and thermal load and corrosion resistance qualities of mild steel can all be enhanced by the coating. Due to its excellent formability and weldability, mild steel is frequently employed in engineering applications. Its employment is constrained by its weak saline medium corrosion resistance [112]. To perform better in corrosive circumstances, mild steel's corrosion resistance needs to be increased. Commercially available solutions for preventing mild steel from corroding include "corrosion inhibitors, cathodic protection, electrodeposited metallic coatings, organic coatings, and paints" [113–116]. An anticorrosive coating on mild steel is seen to be a practical and trustworthy way to preserve mild steel [111]. The substrate is protected from the corrosive environment by coatings that act as both a physical and chemical barrier. The usage of organic coatings is widespread in general, but their performance is constrained by the presence of hazardous compounds and porosity. The development of ecologically friendly coatings is thus continually desired. It can be required to safeguard the material from this damaging external activity if the substance's integrity is being harmed by the environment. Unfortunately, since corrosion cannot be completely stopped, the only option is to regulate it by slowing its kinetics or changing its mechanism.

Recently, the topic of shielding metals against corrosion has seen a lot of use for graphene due to its superior heat conduction, electron mobility, anticorrosion mechanical properties, etc. Graphene derivative–based coatings were found to significantly protect metals from corrosion as well as improve their surface properties [10, 117–119]. As the thinnest corrosion prevention coating, the defect-free single-layer graphene coating provides remarkable barrier performance for preventing oxygen, water, and other corrosive media from accessing the metal substrate's surface [120]. Due to their barrier qualities and low conductivity, newly developing 2D nanosheets that resemble graphene are now receiving research interest for the use of metal anticorrosion and oxidation [121, 122].

Chen *et al.* [123] investigated that due to its barrier effect, as protective layers, graphene/GO-based films are employed to protect metals from corrosion attack. The graphene-based coatings on the surfaces of metals including steel, nickel, and copper are found to be advantageous in several investigations. By applying techniques like CVD, electrophoretic deposition (EPD), and laser manufacturing processes, graphene-based coating on metal surfaces can be successfully deposited. It can also be transmitted to the surface utilizing mechanical transfer methods [124–127]. Due to its ease of usage, the film quality it produces, and its ability to cover vast surfaces, CVD is the preparation process that is most frequently utilized. The bonding strength between graphene and metal is greatly influenced by the distance between the two, and the architectures of graphene films grown on various metals vary. Since any metal substance may be coated with high-quality graphene sheets, additional methods, such as mechanical transfer, have reportedly been used [128].

However, some subsequent research has revealed that graphene derivatives coating on metals adversely affects protecting bare metals from corrosion, whereas, on the other hand, encouraging observations have been recorded about enhancing the anticorrosion behavior of metals upon coating by graphene derivatives [129]. These significant variances in outcomes are caused due to incomplete surface covering and various defects like wrinkles and cracks. The defects present in the coating limit the anticorrosion behavior of metals. As a result, numerous studies claimed that using graphene sheets for developing anticorrosion behavior for metals is impractical because of the challenges that arise in eliminating flaws during production methods [130, 131]. Other investigations have attempted to enhance the graphene sheets' capacity for protection by using an atomic layer precipitation technique to correct flaws [56]. According to certain studies, graphene's strong electrical conductivity encourages metal electrochemical corrosion [132]. Corrosion microcells are formed by graphene and metal in the areas that have been damaged, as evidenced. In these circumstances, anode served by metals such as

Cu, Ni, Al, Mg, and Fe encourages galvanic corrosion. Because of their poor connection with the adjacent metal, the as-grown graphene layers also exhibit minimal adhesion [129]. Modifying the functionalization of graphene and surface chemistry might enhance coating adherence and quality of coating [11, 133].

Due to the exceptional qualities of epoxy resins (EP) such as chemical resistance, toughness, minimal shrinkage on cure, mechanical, corrosive resistance, and great adhesion on a variety of substrates including mild steel [134–136], they are frequently employed for critical anticorrosion coatings. Due to rising governmental regulations limiting the atmospheric release of organic solvents, waterborne epoxy resins (WEP) have recently received considerable attention [137]. Due to WEP coating's low barrier qualities against corrosive media, it is difficult for it to attain optimal corrosion resistance [138]. To improve the barrier performance and corrosion protection capabilities of aqueous coatings, researchers took a variety of steps. Particularly, the inclusion of nanofillers such as graphene [10, 11], ZnO [139], TiO_2 [140], ZrO_2 [141], and PANi [142] has been thoroughly explored.

Yu *et al.* [143] published the results of another investigation on the effectiveness of epoxy coating in preventing corrosion on mild steel surfaces. In this work, the researchers produced a coating of epoxy matrix reinforced with nanosheets of GO-titanium oxide. It was observed that epoxy resin added with GO–titanium oxide increased its effectiveness in inhibiting corrosion on mild steel. Polyurethane (PU)/graphene composite coatings with varying weight percentages of graphene were created by Qiu *et al.* [144], and corrosion protection performance tests were conducted on them. They discovered that adding 4 and 8 wt% of graphene to polyurethane coatings improved their anticorrosion and adhesion capabilities. Pourhashem *et al.* [145] evaluated the influence of GO in the epoxy matrix for shielding from corrosive media on metal substrates by modifying the GO surface with three different silane coupling agents.

Sodani *et al.* [146] showed the effectiveness of calcium nitrite-based materials in protecting against the formation of defects and corrosion behavior in silica fume cement concrete beams by enhancing the corrosion resistance of reinforcing steel. Inhibitors like nitrite-based materials improved the anticorrosion property of steel against chloride and sulfate conditions [144]. A low corrosion rate with nearly 90% efficiency on mild steel was achieved by the benzoxazine compound based on benzene sulfonamide which was substantially lower than the permitted rate of corrosion especially against acidic media and high temperatures [147]. Against 15% HCl acidifying solution attack on the steel surface, the corrosion prevention performance of expired metformin was taken into consideration [148]. On mild steel, a coating functionalized with yttrium(III) oxide showed superior corrosion resistance [149]. It has been claimed that 304 stainless steel mesh covered with super hydrophobic polydimethylsiloxane (PDMS) may withstand corrosive environments [150]. Researchers have studied alumina (Al_2O_3), zinc-nickel (Zn-Ni), zinc-nickel-alumina (Zn-Ni-Al_2O_3), zinc–zinc oxide–yttrium oxide (Zn-ZnO-Y_2O_3), and graphene derivative–based coatings on mild steel can significantly enhance its anticorrosion behavior, hardness, and resistance toward wear [151–155]. The literature claimed that graphene derivatives may shield mild steel substrates from physical and chemical degradation and improve mechanical, thermal, physical, and anticorrosion qualities. Park and Park [156] observed that coating GO on mild steel did not influence its anticorrosion behavior. According to Kirkland *et al.* [157], various metals require distinct roles from graphene coatings in terms of corrosion resistance. It was reported that iron-phosphorus/reduced GO could enhance surface hardness and resistance to erosion and corrosion of mild steel [158]. The corrosion of steel in 15% HCl solution was reported to be inhibited by dopamine-functionalized GO (DA-GO) [159], octanoate-functionalized GO (SO-GO) [160], and *N*, *N*′-bis-(2-aminoethyl)piperazine–functionalized GO (NAEP-GO) [161]. It has been demonstrated that zinc chromate enhanced by a GO layer improves the protection of mild steel against corrosion [162].

Pu *et al.* [127] successfully coated graphene sheets using the CVD technique on the substrate of bare steel and stainless steel with plated nickel. While the SEM photographs of the steel sample demonstrated that graphene sheets could not completely cover the substrate, the SEM examination of the nickel-plated steel sample revealed a 100% graphitization rate. Nickel-plated steel coated with graphene showed a corrosion current of one-sixth value compared to bare substrate. The laser alloying procedure of Ye *et al.* [163] created a layer of nickel/iron on the substrate of carbon steel, which was then treated by laser irradiation to produce graphene sheets. The anti-corrosion performance of graphene-coated carbon steel in 3.5% NaCl (aq) solutions revealed a remarkable improvement in anticorrosive properties against stainless steel. Quezada-Renterıa *et al.* [164] developed an anticorrosive coating of reduced GO on carbon steel *via* the EPD technique. They demonstrated a significant decrease in corrosion rate as compared to bare substrate.

In a variety of corrosive situations, graphene derivatives have been utilized to protect metals from corrosion [165, 166]. Coating of graphene derivatives offers the advantages of being inexpensive and environmentally friendly compared to conventional techniques of coating. Graphene has excellent mechanical strength, a wide SSA, excellent resistance to chemical attack having high impermeability. Due to its lack of functions, in aqueous solution, graphene is found to be not soluble and is thus employed in anticorrosion coatings [165]. Functionalized GOs are effective corrosion inhibitors and are more soluble than GO [167–170]. According to Gupta *et al.* [171], mild steel coated with GO/amino- and diaminoazobenzene functionalization has remarkable abilities to significantly resist corrosion of the substrate (in 1 M HCl). It has been characterized that GO-reinforced diazo- and diaminopyridine can be used effectively for improving the anticorrosion property of mild steel substrate in 1 M HCl [172]. EPD is regarded as a workable method for producing graphene or GO coatings from colloidal suspensions [173]. GO coatings on NdFeB substrate showed greater corrosion resistance in corrosive media of NaCl solution [174]. Another researcher also reported on the anticorrosion effectiveness of GO coating made by EPD on mild steel [175]. In the other literature [157], the influence of graphene derivatives on improving the anticorrosion behavior of metal substrates was studied. Mohapatra *et al.* reported that mild steel substrate offers more resistance to corrosion after A_2O_3/GO (0.5 wt%) coating on its surface [176].

Zhu *et al.* [97] synthesized nanocomposites of GO to polypyrrole (PPy) at two mass ratios of 1:1 and 2:1 called GP and G2P, respectively, *via* an *in situ* process to grow. They added GP and G2P to the aqueous epoxy emulsion matrix in the presence of an aqueous curing agent to develop nanocomposite coating materials. GP of 2.52, 12.6, 25.2, 126, and 252 mg was added to the epoxy matrix called $GP_{0.01\%}$, $GP_{0.05\%}$, $GP_{0.1\%}$, $GP_{0.5\%}$, and $GP_{1\%}$ respectively and G2P of 12.6, 25.2, and 126 mg was added to epoxy matrix called as $G2P_{0.05\%}$, $G2P_{0.1\%}$, and $G2P_{0.5\%}$, respectively. By painting method, they coated these nanocomposite coating materials on mild steels and studied their anticorrosion performances (exposed to 3.5 wt% NaCl for 48 hours). The electrochemical parameters measured for mild steel with or without coatings are represented in Table 8.5 [98, 177, 158]. Generally, the material exhibits a lower corrosion current (i_{corr}) and higher corrosion potential (E_{corr}) and shows higher corrosion resistance. $GP_{0.05\%}$ coating showed the lowest i_{corr} (2.725×10^{-8} A/cm^2) and the lowest corrosion rate (core), which confirmed its excellent corrosion inhibition performance in comparison to other coatings. According to Ding *et al.* [177], sodium lignin sulfonate (LAI)–graphene(G)–epoxy–coated steel showed more positive E_{corr} and much lower score values compared to epoxy-coated steel and bare steel after 96-hour treatment in 3.5 wt% NaCl solution and results are presented in Table 8.5. They also determined that the corrosion protection

Table 8.5 The electrochemical parameter results obtained for steel with or without coatings (3.5 wt% NaCl solution).

Sample	Corrosion potential (E_{corr}) (mv)	Corrosion current (i_{corr}) (A/cm²)	Anodic Tafel slope (b_a) (mV/del)	Cathodictafel slope (b_c) (mV/de)	Inhibition/ protection efficiency (IE (%))	Corrosion rate (n_{corr}) (mm/ year)	References
Bare steel	−719.1	8.269×10^{-6}	52.02	−110.03	-	0.101	[98]
Pure epoxy-coated	−577.9	4.686×10^{-8}	75.52	−71.53	99.43	5.58×10^{-4}	[98]
$PPy_{0.05\%}$-coated	−638.7	3.523×10^{-8}	74.282	−74.048	99.57	4.14×10^{-4}	[98]
$GO_{0.05\%}$-coated	−602.7	3.883×10^{-8}	74.79	−60.54	99.53	4.93×10^{-4}	[98]
$G2P_{0.05\%}$-coated	−634.8	6.414×10^{-8}	74.24	−73.12	99.22	7.53×10^{-4}	[98]
$GP_{0.05\%}$-coated	−625.2	2.677×10^{-8}	74.80	−73.59	99.67	3.15×10^{-4}	[98]
Bare steel	−696.0	1.74×10^{-6}	78.0	-670.0	-	-	[177]
Pure epoxy-coated	−622.0	6.83×10^{-7}	128	−147	96.1	-	[177]
LAI-G/ epoxy-coated	−552.0	1.12×10^{-7}	231	−36	99.4	-	[177]
Bare mild steel	−149	3.894×10^{-6}	-	-	-	0.090 ± 0.007	[158]
Fe-P–coated	−136	1.447×10^{-6}	-	-	-	0.033 ± 0.004	[158]
Fe-P/rGO–coated	−29	1.145×10^{-6}	-	-	-	0.026 ± 0.003	[158]

efficiency of LAI-G/epoxy–coated steel was 99.4%, which was found relatively higher than pure epoxy-coated steel (96.1%). Debasish *et al.* [158] reported that iron-phosphorus/reduced GO–coated mild steel showed relatively higher positive E_{corr} and relatively lower i_{corr} and lower corrosion rate values compared to iron-phosphorus–coated mild steel and bare mild steel (immersed in 3.5 wt% NaCl solution).

8.6 Microhardness Behavior of Graphene Derivative Coating on Mild Steel

The wonderful material graphene is an outstanding material having super mechanical properties especially microhardness [28, 178]. Literature reported that graphene/rGO or GO can be used as a coating material for metals in general to improve physical, mechanical, heat conduction, and resistance to corrosion properties of metal substrates including mild steel. Iron-phosphorus/reduced GO–coated mild steel showed excellent microhardness, erosion, and anticorrosion properties compared to bare mild steel according to the work of Debasish *et al.* [158]. Microhardness is found to be improved as a consequence of the improvement of the anticorrosion behavior of mild steel. They developed more than 66% of the microhardness of mild steel (~214 VHN) by coating it with iron-phosphorus(Fe-P). The Vickers hardness (VHV) of mild steel enhanced more with the reinforcement of rGo in Fe-P. Two and 4 vol. % rGO in Fe-P increased the microhardness of mild steel up to more than 111% (452 VHN) and 115% (461 VHV), respectively. The increased microhardness was ascribed to the homogenous distribution of rGO nanosheets in Fe-P in the mild steel composite coating.

The anticorrosion and physical characteristics of mild steel substrate are improved by coatings made of GO-reinforced zinc chromate [162], DA-GO [159], SO-GO [179], NAEP-GO [161], and others. Alumina (Al_2O_3) shows potential industrial applications in developing excellent adhesion and advanced coatings with good hardness for mild steel substrate and other metals [180–182]. While alumina-graphene–coated (1.5 wt%) mild steel exhibited the hardness and Young's modulus values of 250 VHN and 270 GPa, respectively, mild steel showed hardness and Young's modulus values of 168 VHN and 190 GPa, respectively [183]. There has been very less number of research work done in the literature on graphene reinforcement in Al_2O_3 matrix for producing new coating materials for enhancing the characteristics of mild steel substrate. Preferably, graphene-reinforced alumina is coated on mild by a nontransferred plasma torch technique. When a mass ratio of 1:1 of GO and Al_2O_3 was coated on steel, it showed an improved microhardness of 500 VHN [184] in comparison to the microhardness of bare mild steel, i.e., 160 ± 06 VHN. The Al_2O_3-GO (0.5 wt%) composite–coated mild steel showed improved microhardness of 230 VHN compared to microhardness of mild steel of 175 ± 14 VHN due to developed microstructure and well dispersion of GO in the coating matrix [176].

When Al_2O_3/CNT (carbon nanotube) coated on steel, a decrease in Young's modulus values (from 41 GPa for Al_2O_3 to 9 GPa having 25% C) upon increasing carbon content was reported [185]. When C wt% increased from 0 to 14 wt%, it was observed that the microhardness was obtained to be the same which is nearly 1000 MPa. However, a decrease in hardness was observed for steel coated with Al_2O_3/CNT, having more than 14% C [185]. The hardness of Al_2O_3-graphene (4 wt%)–coated mild steel was 370 VHN, compared to 172 VHN for naked mild steel [151]. When graphene was well distributed in the coating matrix of Al_2O_3, it significantly enhanced the microhardness of mild steel.

8.7 Conclusions

Mild steel is a very popular material for various domestic and industrial applications because of its low cost and various excellent properties. However, its application becomes limited in comparison to refractory matcrials duc to its poor properties of hardness, wear resistance, and corrosion resistance. Different composites and alloys with mild steel and coating on mild steel have been addressed by researchers to improve the properties of mild steel. Among the above forms of modification of properties of mild steel, the coating process is found to be a cheap, fast, and efficient one for enhancing the properties of mild steel. Graphene is one of the gifted carbon nanomaterials that has unique, mechanical, anticorrosion, electrical, optical, and catalytic properties. From the literature survey, it has been observed that graphene or its derivatives have efficient potential for protecting mild steel from corrosive environments and can improve significantly its surface mechanical properties, especially hardness. Graphene-reinforced ceramic composites such as alumina are found a typical coating material that plays a vital role in enhancing the surface properties of mild steel.

References

1. Geim, A.K., Graphene prehistory. *Phys. Scr.*, 146, 014003, 2012.
2. Geim., A.K. and Novoselov, K.S., The rise of graphene. *Nat. Mater.*, 6, 183, 2017.
3. Balandin, A.A., Ghosh, S., Bao, W., Calizo, I., Teweldebrhan, D., Miao, F., Lau, C.N., Superior thermal conductivity of single-layer graphene. *Nano. Lett.*, 8, 902, 2008.
4. Zhu, Y., Murali, S., Cai, W., Li, X., Suk, J.W., Potts, J.R., Ruoff, R.S., Graphene and graphene oxide: synthesis, properties, and applications. *Adv. Mater.*, 22, 3906, 2010.
5. Kralj, M., Krivacic, S., Ivanisevi, I., Zubak, M., Supina, A., Marcius, M., Halasz, I., Kassal, P., Conductive inks based on melamine intercalated graphene nanosheets for inkjet printed flexible electronics. *Nanomaterials*, 12, 2936, 2012.
6. Samorì, P., Kinloch, I., Feng, X., Palermo, V., Graphene-based nanocomposites for structural and functional applications: using 2-dimensional materials in a 3-dimensional world. *2D Mater.*, 2, 030205, 2015.
7. Wang, R., Zhuo, D., Weng, Z., Wu, L., Cheng, X., Zhou, Y., Wang, J., Xuan, B., A novel nano silica/graphene oxide hybrid and its flame retarding epoxy resin with simultaneously improved mechanical, thermal conductivity, and dielectric properties. *J. Mater. Chem. A.*, 3, 9826, 2015.
8. Guo, W. and Chen, G., Fabrication of graphene/epoxy resin composites with much enhanced thermal conductivity via ball milling technique. *J. Appl. Polym. Sci.*, 131, 40565, 2014.
9. Kim, J., Yim, B.S., Kim, J.M., Kim, J., The effects of functionalized graphene nanosheets on the thermal and mechanical properties of epoxy composites for anisotropic conductive adhesives (ACAs). *Microelectron. Reliabil.*, 52, 595, 2012.
10. Cui, G., Bi, Z., Zhang, R., Liu, J., Yu, X., Li, Z., A comprehensive review on graphene-based anti-corrosive coatings. *Chem. Eng. J.*, 373, 104, 2019.
11. Nine, M.J., Cole, M.A., Tran, D.N.H., Losic, D., Graphene: A multipurpose material for protective coatings. *J. Mater. Chem. A.*, 3, 12580, 2015.
12. Li, Y., Wang, S., Wang, Q., Enhancement of tribological properties of polymer composites reinforced by functionalized graphene. *Compos. B: Eng.*, 120, 83, 2017.
13. Song, H.J. and Li, N., Frictional behavior of oxide graphene nanosheets as water-base lubricant additive. *Appl. Phys. A: Mater. Sci. Process.*, 105, 827, 2011.
14. Naghdi, S., Jaleh, B., Ehsani, A., Electrophoretic deposition of graphene oxide on aluminum: Characterization, low thermal annealing, surface and anticorrosive properties. *Bull. Chem. Soc. Jpn.*, 88, 722, 2015.

15. Bhushan, B., Natural and biomimetic artificial surfaces for superhydrophobicity, self-cleaning, low adhesion, and drag reduction. *Prog. Mater. Sci.*, 56, 1, 2011.
16. Zhang, L., Li, H., Lai, X., Su, X., Liang, T., Zeng, X., Thiolated graphene-based superhydrophobic sponges for oil-water separation. *Chem. Eng. J.*, 316, 736, 2017.
17. Singh, R.K., Kumar, R., Sing, D.P., Graphene oxide: strategies for synthesis, reduction and frontier applications. *RSC Adv.*, 6, 64993, 2016.
18. Hummers Jr., W.S. and Offeman, R.E., Preparation of graphitic oxide. *J. Am. Chem. Soc.*, 80, 1339, 1958.
19. Guerrero-Contreras, J. and Caballero-Briones, F., Graphene oxide powders with different oxidation degree, prepared by synthesis variations of the Hummers method. *Mater. Chem. Phys.*, 153, 209, 2015.
20. Chiang, A.L., Lin, S.H., Huang, W.G., Hsieh, W.Y., Shen, M.Y., Characterization and properties of graphene nanoplatelets/XNBR nanocomposites. *Polym. Polym. Compos.*, 26, 59, 2018.
21. Tian, P., Tang, L., Teng, K.S., Lau, S.P., Graphene quantum dots from chemistry to applications. *Mater. Today Chem.*, 10, 221–258, 2018.
22. Li, M., Chen, T., Gooding, J.J., Liu, J., Review of carbon and graphene quantum dots for sensing. *ACS Sens.*, 4, 1732, 2019.
23. Tian, C., Miao, W., Zhao, L., Wang, J., Graphene nanoribbons: Current status and challenges as quasi-one-dimensional nanomaterials. *Rev. Phys.*, 10, 1, 2023.
24. Novoselov, K.S., Geim, A.K., Morozov, S.V., Jiang, D., Zhang, Y., Dubonos, S.V., Grigorieva, I.V., Firsov, A.A., Electric field effect in atomically thin carbon films. *Science*, 306, 666, 2004.
25. Lu, X., Yu, M., Huang, H., Ruoff, R.S., Tailoring graphite with the goal of achieving single sheets. *Nanotechnol.*, 10, 269, 1999.
26. Gunasekaran, R.K., Jayasankar, K., Dash, T., Dash Shear-force-dominated dual-drive planetary ball milling for scalable production of graphene and its electrocatalytic application with Pd nanostructures. *RSC Advances*, 6, 20067, 2016.
27. Palei, B.B., Dash, T., Biswal, S.K., Reduced graphene oxide synthesis by dry planetary ball milling technique under hydrogen atmosphere. *IOP Conf. Ser.: Mater. Sci. Eng.*, 872, 1, 2020.
28. Palei, B.B., Dash, T., Biswal, S.K., Graphene reinforced aluminum nanocomposites: synthesis, characterization and properties. *J. Mater. Sci.*, 57, 8544, 2022.
29. Jo, J., Lee, S., Gim, J., Song, J., Kim, S., Mathew, V., Alfaruqi, M.H., Kim, S., Lim,J.,Kim, J.Facile synthesisof reduced graphene oxide by modified Hummer'smethod as anode material for Li-, Na- and K-ionsecondary batteries. *R. Soc. Open Sci.*, 6, 1, 2019.

30. Das, S., Ghosh, C.K., Sarkar, C.K., Roy, S., Facile synthesis of multi-layer graphene by electrochemical exfoliation using organic solvent. *Nanotechnol. Rev.*, 7, 6, 2018.
31. Gebreegziabher, G.G., Asemahegne, A.S., Ayele, D.W., Dhakshnamoorthy, M., Kumar, A., One-step synthesis and characterization of reduced graphene oxide using chemical exfoliation method. *Mater. Today Chem.*, 12, 233, 2019.
32. Sadek, R., Sharawi, M.S., Dubois, C., Tantawy, H., Chaouki, J., Superior quality chemically reduced graphene oxide for high performance EMI shielding materials. *RSC Adv.*, 12, 22608, 2022.
33. Guex, L.G., Sacchi, B., Peuvot, K.F., Andersson, R.L., Pourrahimi, A.M., Ström, V., Farris, S., Olsson, R.T., Experimental review: chemical reduction of graphene oxide (GO) to reduced graphene oxide (rGO) by aqueous chemistry. *Nanoscale*, 9, 9562, 2017.
34. Fan, X.B., Peng, W.C., Li, Y., Li, X., Deoxygenation of exfoliated graphite oxide under alkaline conditions: a green route to graphene preparation. *Adv. Mater.*, 20, 4490, 2008.
35. Liu, Y.Z., Li, Y.F., Yang, Y.G., Wen, Y.F., Wang, M.Z., Reduction of graphene oxide by thiourea. *J. Nanosci. Nanotechnol.*, 11, 10082, 2011.
36. Li, Z., Young, R.J., Backes, C., Zhao, W., Zhang, X., Zhukov, A.A., Tillotson, E., Mechanisms of liquid-phase exfoliation for the production of graphene. *ACS Nano*, 14, 9, 10976, 2020.
37. Li, L., Zhou, M., Jin, L., Liu, L., Mo, Y., Li, X., Mo, Z., Liu, Z., You, S., Zhu, H., Research progress of the liquid-phase exfoliation and stable dispersion mechanism and method of graphene. *Front. Mater.*, 6, 325, 2019.
38. Dhar, S., Barman, A.R., Ni, G.X., Wang, X., Xu, X.F., Zheng, Y., Tripathy, S., Ariando, Rusydi, A., Loh, K.P., Rubhausen, M., Neto, A.H.C., Őzyilmaz, B., Venkatesan, T., A new route to graphene layers by selective laser ablation. *AIP Adv.*, 1, 022109–1, 2011.
39. Wu, F., Tian, H., Shen, Y., Hou, Z., Ren, J., Gou, G., Sun, Y., Yang, Y., Ren, T., Vertical MoS2 transistors with sub-1-nm gate lengths. *Nature*, 603, 259, 2022.
40. Al-Hagri, A., Li, R., Rajput, N.S., Lu, C., Cong, S., Sloyan, K., Almahri, M.A., Tamalampudi, S.R., Chiesa, M., Ghaferi, A., Direct growth of single-layer terminated vertical graphene array on germanium by plasma enhanced chemical vapor deposition. *Carbon*, 155, 320, 2019.
41. Ying, H., Wang, W., Liu, W., Wang, L., Chen, S., Layer-by-layer synthesis of bilayer and multilayer graphene on Cu foil utilizing the catalytic activity of cobalt nano-powders. *Carbon*, 146, 549, 2019.
42. Huet, B. and Raskin, J.-P., Role of Cu foil *in-situ* annealing in controlling the size and thickness of CVD graphene domains. *Carbon*, 129, 270, 2018.
43. Yoshihara, N. and Noda, M., Chemical etching of copper foils for single-layer graphene growth by chemical vapor deposition. *Chem. Phys. Lett.*, 685, 40, 2017.

44. Mendoza, C., Caldas, P., Freire, F., Da Costa, M.M., Growth of single-layer graphene on Ge (1 0 0) by chemical vapor deposition. *Appl. Surf. Sci.*, 447, 816, 2018.
45. Prakash, G., Capano, M.A., Bolen, M.L., Zemlyanov, D., Reifenberger, R.G., AFM study of ridges in few-layer epitaxial graphene grown on the carbon-face of 4H–SiC. *Carbon*, 48, 2383, 2010.
46. Gong, P., Ye, Z., Yuan, L., Egberts, P., Evaluation of wetting transparency and surface energy of pristine and aged graphene through nanoscale friction. *Carbon*, 132, 749, 2018.
47. Li, N., Wang, Z., Zhao, K., Shi, Z., Gu, Z., Xu, S., Large scale synthesis of N-doped multi-layered graphene sheets by simple arc-discharge method. *Carbon*, 48, 255, 2009.
48. Karmakar, S., Kulkarni, N.V., Nawale, A.B., Lalla, N.P., Mishra, R., Sathe, V.G., Bhoraskar, S.V., Das, A.K., A novel approach towards selective bulk synthesis of few-layer graphenes in an electric arc. *J. Phys. D: Appl. Phys.*, 42, 115201, 2009.
49. Bhuyan, M.S.A., Uddin, M.N., Islam, M.M., Bipasha, F.A., Hossain, S.S., Synthesis of graphene. *Int. Nano. Lett.*, 6, 65, 2016.
50. Yu, K., Zhao, W., Wu, X., Zhuang, J., Hu, X., Zhang, Q., Sun, J., Xu, T., Chai, Y., Ding, F., Sun, L., *In situ* atomic-scale observation of monolayer graphene growth from SiC. *Nano Res.*, 11, 2809, 2018.
51. Hibino, H., Kageshima, H., Nagaseet, M., Epitaxial few-layer graphene: towards single crystal growth. *J. Phys. D: Appl. Phys.*, 43, 374005, 2010.
52. Hu, T., Chen, J., Lu, X., Chen, J., Chen, Z., Fu, J., Chen, Y., Synthesis of few-layer graphene sheets from waste expanded polystyrene by dense Fe cluster catalysis. *ACS Omega*, 5, 4075, 2020.
53. Abdolhosseinzadeh, S., Asgharzadeh, H., Kim, H.S., Fast and fully-scalable synthesis ofreduced graphene oxide. *Sci. Rep.*, 5, 1, 2015.
54. Zhang, H., Ding, F., Li, H., Fei Qu, F., Meng, H., Gu, H., Controlled synthesis of monolayer graphene with a high quality by pyrolysis of silicon carbide. *Mater. Lett.*, 244, 171, 2019.
55. Okoroanyanwu, U., Bhardwaj, A., Watkins, J.J., Large area millisecond preparation of high-quality, few-layer graphene films on arbitrary substrates via xenon flash lamp photothermal pyrolysis and their application for high-performance micro-supercapacitors. *ACS Appl. Mater.*, 15, 13495, 2023.
56. Tomai, T., Nakayasu, Y., Okamura, Y., Ishiguro, S., Tamura, N., Katahira, S., Honma, I., Bottom-up synthesis of graphene via hydrothermal cathodic reduction. *Carbon*, 158, 131, 2020.
57. Moldt, T., Eckmann, A., Klar, P., Morozov, S.V., Zhukov, A.A., Novoselov, K.S., Casiraghi, C., High-yield production and transfer of graphene flakes obtained by anodic bonding. *ACS Nano*, 5, 77002011.
58. Bonaccorso, F., Lombardo, A., Hasan, T., Sun, Z., Colombo, L., Ferrari, A.C. Production and processing of graphene and 2d crystals. *Mater. Today*, 15, 564, 2012.

59. Palei, B.B., Dash, T., Biswal, S.K., Reduced graphene oxide synthesis by dry planetary ball milling technique under hydrogen atmosphere. *IOP Conf. Ser.: Mater. Sci. Eng.*, 872, 012158, 2020, doi: 10.1088/1757-899X/872/1/012158.
60. Li, Y.F., Liu, Y.Z., Zhang, W.K., Guo, C.Y., Chen, C.M., Green synthesis of reduced graphene oxide paper using Zn powder for supercapacitors. *Mater. Lett.*, 157, 273, 2015.
61. Cui, P., Lee, J., Hwang, E., Lee, H., One-pot reduction of graphene oxide at subzero temperatures. *Chem. Commun.*, 45, 12370, 2011.
62. Zhang, H.B., Zheng, W.G., Yan, Q., Yang, Y., Wang, J.W., Lu, Z.H., Ji, G.Y., Yu, Z.Z., Electrically conductive polyethylene terephthalate/graphene nanocomposites prepared by melt compounding. *Polym.*, 51, 1191, 2010.
63. Al-Sherbini, A.S., Bakr, M., Ghoneim, I., Saad, M., Exfoliation of graphene sheets via high energy wet milling of graphite in 2-ethylhexanol and kerosene. *J. Adv. Res.*, 8, 209, 2017.
64. Mohan, V.B., Brown, R., Jayaraman, K., Bhattacharyya, D., Characterisation of reduced graphene oxide: Effects of reduction variables on electrical conductivity. *Mater. Sci. Eng. B.*, 193, 49, 2015.
65. Lee, J., Zheng, X., Roberts, R.C., Feng, P.X.L., Scanning electron microscopy characterization of structural features in suspended and non-suspended graphene by customized CVD growth. *Diam. Relat. Mater.*, 54, 64, 2015.
66. Shalaby, A., Nihtianova, D., Markov, P., Staneva, A.D., Iordanova, R.S., Dimitriev, Y.B., Structural analysis of reduced graphene oxide by transmission electron microscopy. *Bulg. Chem. Commun.*, 47, 291, 2015.
67. Liu, Y.Z., Chen, C.M., Li, Y.F., Li, X.M., Konga, Q.Q., Wanga, M.Z., Crumpled reduced graphene oxide by flame-induced reduction of graphite oxide for supercapacitive energy storage. *J. Mater. Chem. A.*, 2, 5730, 2014.
68. Luo, B., Chen, B., Wang, A., Geng, D., Xu, J., Wang, H., Zhang, Z., Peng, L., Xu, Z., Yu, G., Chemical vapor deposition of bilayer graphene with layer-resolved growth through dynamic pressure control. *J. Mater. Chem. C.*, 4, 7464, 2016.
69. Hwangbo, Y., Lee, C.K., Mag-Isa, A.E., Jang, J.W., Lee, H.J., Lee, S.B., Kim, S.S., Kim, J.H., Interlayer non-coupled optical properties for determining the number of layers in arbitrarily stacked multilayer graphenes. *Carbon.*, 77, 454, 2014.
70. Ferrari, A.C., Meyer, J.C., Scardaci, V., Casiraghi, C., Lazzeri, M., Mauri, F., Piscanec, S., Jiang, D., Novoselov, K.S., Roth, S., Geim,A.K.,Raman spectrum of graphene and graphene layers. *Phys. Rev. Lett.*, 97, 187401, 2006.
71. Graf, D., Molitor, F., Ensslin, K., Stampfer, C., Jungen, A., Hierold, C., Wirtz, L., Spatially resolved raman spectroscopy of single- and few-layer graphene. *Nano Lett.*, 7, 238, 2007.
72. Malard, L.M., Pimenta, M.A., Dresselhaus, G., Dresselhaus, M.S., Raman spectroscopy in graphene. *Phys. Rep.*, 473, 51, 2009.
73. Shubha, B.S., Praveen, B.M., Bhat, V., Graphene calculation of specific surface area. *Int. J. Appl. Eng. Res.*, 7, 2581, 2023.

74. Sultanov, F.R., Daulbayev, C., Bakbolat, B., Mansurov, Z.A., Urazgaliyeva, A.A., Ebrahim, R., Pei, S.S., Huang, K.P., Microwave□enhanced chemical vapor deposition graphene nanoplatelets□derived 3D porous materials for oil/water separation. *Carbon Lett.*, 30, 81, 2020.
75. Sahoo, P.C., Dash, T., Mallick, S.C., Biswal, S.K., Reduced graphene oxide synthesis by dry planetary ball milling followed by arc plasma treatment of high pure graphite. *J. Mater. Res.*, 114, 299, 2023.
76. Zhao, M.Q., Zhang, Q., Huang, J.Q., Tian, G.L., Nie, J.Q., Peng, H.J., Wei, F., Unstacked double-layer templated graphene for high-rate lithium–sulphur batteries. *Nat.Commun.*, 5, 1, 2014.
77. Esmaeili, A. and Entezari, M.H., Facile and fast synthesis of graphene oxide nanosheets via bath ultrasonic irradiation. *J. Colloid Interf. Sci.*, 432, 19, 2014.
78. Huang, H., Chen, P.W., Zhang, X.T., Lu, Y., Zhan, W.C., Edge-to-edge assembledgraphene oxide aerogels with outstanding mechanical performance and superhigh chemical activity. *Small*, 9, 1397, 2013.
79. Montes-Navajas, P., Asenjo, N.G., Santamaría, R., Menéndez, R., Corma, A., García, H., Surface area measurement of graphene oxide in aqueous solutions. *Langmuir.*, 29, 13443, 2013.
80. Wang, H. and Hu, Y.H., Effect of oxygen content on structures of graphite oxides. *Ind. Eng. Chem. Res.*, 50, 6132, 2011.
81. Silva Filho, J.C., Venancio, E.C., Silva, S.C., Takiishi, H., Martinez, L.G., Antunes, R.A., A thermal method for obtention of 2 to 3 reduced graphene oxide layers from graphene oxide. *SN Appl. Sci.*, 2, 1450, 2020.
82. Rao, X., Zhang, L., Li, B., Zeng, X., Xiao, W., Lou, Y., Xie, H., Yan, H., Yib, Z., Zhong, S., Preparation of artificial graphite coated with sodium alginate as a negative electrode material for lithium-ion battery study and its lithium storage properties. *Mater. Adv.*, 3, 8958, 2022.
83. Fouda, A.N., Duraia, E.S.M., Almaqwashi, A.A., Facile and scalable green synthesis of N-doped graphene/CNTs nanocomposites via ball milling. *Eng. Phys. Mathematics.*, 12, 2090, 2020.
84. Marinho, B., Ghislandi, M., Tkalya, E., Koning, C.E., With, G.D., lectrical conductivity of compacts of graphene, multi-wall carbon nanotubes, carbon black, and graphite powder. *Powder Technol.*, 221, 351, 2012.
85. Lim, S., Park, H., Yammamoto, G., Lee, C., Suk, J.W., Measurements of the electrical conductivity of monolayer graphene flakes using conductive atomic force microscopy. *MDPI.*, 11, 2575, 2021.
86. Ould NE, M.L., Boujnah, M., Benyoussef, A., El Kenz, A., Electronic and electrical conductivity of AB and AA-stacked bilayer graphene with tunable layer separation. *Springer*, 30, 1263, 2011.
87. Pirzado, A.A., Dalmas, G., Dinh, L.N., Komissarov, I., Normand, F.L., Janowska, I., The electrical property of large few layer graphene flakes obtained by microwaves assisted exfoliation of expanded graphite. *Curr. Med. Chem.*, 3, 139, 2016.

88. Murata, H., Nakajima, Y., Saitoh, N., Yoshizawa, N., Suemasu, T., Toko, K., High- Electrical-Conductivity Multilayer Graphene Formed by Layer Exchange with Controlled Thickness and Interlayer. *Sci. Rep.*, 9, 4068, 2019.
89. Jafar, E., Kashif, M., Sahari, S.K., Ngaini, Z., Study on morphological, optical and electrical properties of graphene oxide (GO) and reduced graphene oxide (rGO). *Scientific.net.*, 917, 112, 2018.
90. Shin, H.J., Kim, K.K., Benayad, A., Yoon, S.M., Park, H.K., Jung, I.S., Jin, M.H., Jeong, H.K., Kim, J.M., Choi, J.Y., Lee, Y.H., Efficient reduction of graphite oxide by sodium borohydride and its effect on electrical conductance. *Adv. Funct. Mater.*, 19, 1987, 2009.
91. Pei, S., Zhao, J., Du, J., Ren, W., Cheng, H.M., Direct reduction of graphene oxide films into highly conductive and flexible graphene films by hydrohalic acids. *Carbon*, 48, 4466, 2010.
92. Er, E. and Cellikan, H., An efficient way to reduce graphene oxide by water elimination using phosphoric acid. *RSC Adv.*, 4, 29173, 2014.
93. Mohan, V.B., Jayaraman, K., Bhattacharyya, D., Brunauer–Emmett–Teller (BET) specific surface area analysis of different graphene materials: A comparison to their structural regularity and electrical properties. *Solid State Commun.*, 320, 114004, 2020.
94. Zhang, R., Yu, X., Yang, Q., Cui, G., Li, Z., The role of graphene in anti-corrosion coatings: A review. *Constr. Build Mater.*, 294, 123613, 2021.
95. Zhou, S., Yao, W., Wang, Z., Ma, L., Lu, Z., Hou, C., The first-principles calculations to explore the mechanism of oxygen diffusion on vacancy defective graphene in marine environment. *Appl. Surf. Sci.*, 525, 146585, 2020.
96. Ren, S., Cui, M., Li, W., Pu, J., Xue, Q., Wang, L., N-doping of graphene: toward long-term corrosion protection of Cu. *J. Mater. Chem. A.*, 6, 24136, 2018.
97. Zhu, Q., Li, E., Liu, X., Song, W., Li, Y., Wang, X., Liua, C., Epoxy coating with *in-situ* synthesis of polypyrrole functionalized graphene oxide for enhanced anticorrosive performance. *Prog. Org. Coat.*, 140, 1, 2020.
98. Li, X., Cai, W., An, J., Kim, S., Nah, J., Yang, D., Piner, R., Velamakanni, A., Jung, I., Tutuc, E., Large-area synthesis of high-quality and uniform graphene films on copper foils. *Science.*, 324, 1312, 2009.
99. Maurya, R., Siddiqui, A.R., Katiyar, P.K., Balani, K., Mechanical, tribological and anti-corrosive properties of polyaniline/graphene coated Mg-9Li-7Al-1Sn and Mg-9Li-5Al-3Sn-1Zn alloys. *J. Mater. Sci. Technol.*, 35, 1767, 2019.
100. Zhou, S., Wu, Y., Zhao, W., Yu, J., Jiang, F., Ma, L., Comparative corrosion resistance of graphene sheets with different structures in waterborne epoxy coatings. *Colloid. Surface A.*, 556, 273, 2018.
101. Mahato, N. and Cho, M.H., Graphene integrated polyaniline nanostructured composite coating for protecting steels from corrosion: synthesis, characterization, and protection mechanism of the coating material in acidic environment. *Constr. Build. Mater.*, 115, 618, 2016.

102. Jiulong, S., Wanwan, C., Ning, W., Lin, G., Weihua, L., Progress of boron nitride nanosheets used for heavy-duty anti-corrosive coatings. *ActaChim. Sin.*, 78, 1139, 2020.
103. Lin, G., Jiheng, D., Haibin, Y., Research in graphene-based anticorrosion coatings. *Prog. Chem.*, 28, 737, 2016.
104. Sun, J., Tang, X., Meng, G., Gu, L., Silane functionalized plasma-treated boron nitride nanosheets for anticorrosive reinforcement of waterborne epoxy coatings. *Prog. Org. Coat.*, 167, 106831, 2022.
105. Cheng, L., Wu, H., Li, J., Zhao, H., Wang, L., Polydopamine modified ultra-thin hydroxyapatite nanosheets for anti-corrosion reinforcement in polymeric coatings. *Corros. Sci.*, 178, 109064, 2021.
106. Chen, J. and Zhao, W., Silk fibroin-$Ti_3C_2T_X$ hybrid nanofiller enhance corrosion protection for waterborne epoxy coatings under deep sea environment. *Chem. Eng. J.*, 423, 130195, 2021.
107. Zhu, X., Yan, Q., Cheng, L., Wu, H., Zhao, H., Wang, L., Enhancing the mechanical and tribological properties of epoxy composites *via* incorporation of reactive bio-based epoxy functionalized graphene oxide. *Chem. Eng. J.*, 389, 124435, 2020.
108. Huang, H., Sheng, X., Tian, Y., Zhang, L., Chen, Y., Zhang, X., Two-dimensional nanomaterials for anticorrosive polymeric coatings: a review. *Ind. Eng. Chem. Res.*, 59, 15424, 2020.
109. Balani, K., Harimkar, S.P., Keshri, A., Chen, Y., Dahotre, N.B., Agarwal, A., Multiscale wear of plasma-sprayed carbon-nanotube-reinforced aluminum oxide nanocomposite coating. *Acta Mater.*, 56, 5984, 2008.
110. Keshri, A.K. and Agarwal, A., Wear behavior of plasma-sprayed carbon nanotubereinforced aluminum oxide coating in marine and high-temperature environments. *J. Therm. Spray Technol.*, 20, 1217, 2012.
111. Liu, L., Ryu, S., Tomasik, M.R., Stolyarova, E., Jung, N., Hybertsen, M.S., Steigerwald, M.L., Brus, L.E., Flynn, G.W., Graphene oxidation: thickness-dependent etching and strong chemical doping. *Nano Lett.*, 8, 1965, 2008.
112. Rayes, M.M., Abdo, H.S., Khalil, K.A., Erosion-corrosion of cermet coating. *Int.J. Electrochem. Sci.*, 8, 117, 2013.
113. Bunch, J.S., McEuen, L.P., Verbridge, S.S., Alden, J.S., Van der Zande, A.M., Parpia, J.M., Craighead, H.G., Impermeable atomic membranes from graphene sheets. *Nano Lett.*, 8, 2008, 2458.
114. Refait, P., Jeannin, M., Sabot, R., Antony, H., Pineau, S., Corrosion and cathodic protection of carbon steel in the tidal zone: Products, mechanisms and kinetics. *Corros. Sci.*, 90, 375, 2015.
115. Ituen, E., James, A., Akaranta, O., Sun, S., Eco-friendly corrosion inhibitor from Pennisetum purpureum biomass and synergistic intensifiers for mild steel. *Chin. J. Chem. Eng.*, 24, 1442, 2016.

116. Wanees, S.A., Radwan, A.B., Alsharif, M.A., Haleem, S.M., Initiation and inhibition of pitting corrosion on reinforcing steel under natural corrosion conditions. *Mater. Chem. Phys.*, 190, 79, 2017.
117. Mu, J., Gao, F., Cui, G., Wang, S., Tang, S., Li, Z., A comprehensive review of anticorrosive graphene-composite coatings. *Prog. Org. Coat.*, 157, 106321, 2021.
118. Ding, R., Chen, S., Lv, J., Zhang, W., Zhao, X., Liu, J., Wang, X., Gui, T., Li, B., Tang, Y., Li, W., Study on graphene modified organic anti-corrosion coatings: A comprehensive review. *J. Alloys Compd.*, 806, 611, 2019.
119. Ding, R., Li, W., Wang, X., Gui, T., Li, B., Han, P., Tian, H., Liu, A., Wang, X., Liu, X., Gao, X., Wang, W., Song, L., A brief review of corrosion protective films and coatings based on graphene and graphene oxide. *J. Alloys Compd.*, 764, 1039, 2018.
120. Singh Raman, R.K. and Tiwari, A., Graphene: the thinnest known coating for corrosion protection. *JOM*, 66, 637, 2014.
121. Schriver, M., Regan, W., Gannett, W.J., Zaniewski, A.M., Crommie, M.F., Zettl, A., Graphene as a long-term metal oxidation barrier: worse than nothing. *ACS Nano*, 7, 5763, 2013.
122. Sun, W., Yang, Y., Yang, Z., Wang, L., Wang, J., Xu, D., Liu, G., Review on the corrosion-promotion activity of graphene and its inhibition. *J. Mater. Sci. Technol.*, 91,278, 2021.
123. Chen, S., Brown, L., Levendorf, M., Cai, W., Ju, S.Y., Edgeworth, J., Li, X., Magnuson, C.W., Velamakanni, A., Piner, R.D., Kang, J., Park, J., Ruoff, R.S., Oxidation resistance of graphene-coated Cu and Cu/Ni alloy. *ACS Nano*, 5, 1321, 2011.
124. Ye, X., Long, J., Lin, Z., Zhang, H., Zhu, H., Zhong, M., Direct laser fabrication of large-area and patterned graphene at room temperature. *Carbon*, 68, 784, 2014.
125. Nayak, P.K., Hsu, C.J., Wang, S.C., Sung, J.C., Huang, J.L., Graphene coated Ni films: a protective coating. *Thin Solid Films*, 529, 312, 2013.
126. Kiran, N.U., Dey, S., Singh, B.P., Besra, L., Graphene coating on copper by electrophoretic deposition for corrosion prevention. *Coatings*, 7, 214, 2017.
127. Pu, N.W., Shi, G.N., Liu, Y.M., Sun, X., Chang, J.K., Sun, C.L., Ger, M.D., Chen, C.Y., Wang, P.C., Peng, Y.Y., Lawes, S., Wu, C.H., Graphene grown on stainless steel as a high-performance and ecofriendly anticorrosion coating for polymer electrolyte membrane fuel cell bipolar plates. *J. Power Sources*, 282, 248, 2015.
128. Giovannetti, G., Khomyakov, P.A., Brocks, G., Karpan, V.M., van den Brink, J., Kelly, P.J., Doping graphene with metal contacts. *Phys. Rev. Lett.*, 101, 026803, 2008.
129. Zhou, F., Li, Z., Shenoy, G.J., Li, L., Liu, H., Enhanced room-temperature corrosion of copper in the presence of graphene. *ACS Nano*, 7, 6939, 2013.

130. Wlasny, I., Dabrowski, P., Rogala, M., Kowalczyk, P.J., Pasternak, I., Strupinski, W., Baranowski, J.M., Klusek, Z., Role of graphene defects in corrosion of graphene-coated Cu(111) Surface. *Appl. Phys. Lett.*, 102, 111601, 2013.
131. Zhang, Y.H., Zhang, H.R., Wang, B., Chen, Z.Y., Zhang, Y.Q., Wang, B., Sui, Y.P., Zhu, B., Tang, C.M., Li, X.L., Xie, X.M., Yu, G.H., Jin, Z., Liu, X.Y., Role of wrinkles in the corrosion of graphene domain-coated Cu surfaces. *Appl. Phys. Lett.*, 104, 143110, 2014.
133. Gergely, A. and Kristof, T., Corrosion protection with ultrathin graphene coatings: a review. *Hung. J. Ind. Chem.*, 41, 83, 2013.
134. Jin, F., Li, X., Park, S., Synthesis and application of epoxy resins: a review. *J. Ind. Eng. Chem.*, 29, 1, 2015.
135. Meng, F., Liu, L., Tian, W., Wu, H., Li, Y., Zhang, T., Wang, F., The influence of the chemically bonded interface between fillers and binder on the failure behaviour of an epoxy coating under marine alternating hydrostatic pressure. *Corros. Sci.*, 101, 139, 2015.
136. Shi, C., Shao, Y., Wang, Y., Meng, G., Liu, B., Influence of submicron-sheet zinc phosphate synthesised by sol–gel method on anticorrosion of epoxy coating. *Prog. Org. Coat.*, 117, 102, 2018.
137. Erdmenger, T., Guerrero-Sanchez, C., Vitz, J., Hoogenboom, R., Schubert, U., Recent developments in the utilization of green solvents in polymer chemistry. *Chem. Soc. Rev.*, 39, 3317, 2010.
138. Wang, N., Cheng, K., Wu, H., Wang, C., Wang, Q., Wang, F., Effect of nano-sized mesoporous silica MCM-41 and MMT on corrosion properties of epoxy coating. *Prog. Org. Coat.*, 75, 386, 2012.
139. Dhokea, S., Khanna, A., Jai Mangal Sinha, T., Effect of nano-ZnO particles on the corrosion behavior of alkyd-based waterborne coatings. *Prog. Org. Coat.*, 64, 371, 2009.
140. Wang, N., Fu, W., Zhang, J., Li, X., Fang, Q., Corrosion performance of waterborne epoxy coatings containing polyethylenimine treated mesoporous-TiO_2 nanoparticles on mild steel. *Prog. Org. Coat.*, 89, 114, 2015.
141. Behzadnasab, M., Mirabedini, S., Kabiri, K., Jamali, S., Corrosion performance of epoxy coatings containing silane treated ZrO2 nanoparticles on mild steel in 3.5% NaCl solution. *Corros. Sci.*, 53, 89, 2011.
142. Yang, S., Zhu, S., Hong, R., Graphene oxide/polyaniline nanocomposites used in anticorrosive coatings for environmental protection. *Coatings*, 10, 1, 2020.
143. Yu, Z., Di, H., Ma, Y., He, Y., Liang, L., Lv, L., Ran, X., Pan, Y., Luo, Z., Preparation of graphene oxide modified by titanium dioxide to enhance the anti-corrosion performance of epoxy coatings. *Surf. Coat. Technol.*, 276, 471, 2015.
144. Qiu, S., Li, W., Zheng, W., Zhao, H., Wang, L., Synergistic effect of polypyrrole-intercalated graphene for enhanced corrosion protection of aqueous coating in 3.5% NaCl solution. *ACS Appl. Mater. Interfaces*, 39, 34294, 2017.

145. Pourhashem, S., Vaezi, M.R., Rashidi, A., Investigating the effect of SiO2-graphene oxide hybrid as inorganic nanofiller on corrosion protection properties of epoxy coatings. *Surf. Coat. Technol.*, 311, 282, 2017.
146. Sodani, K.A.A.A., Maslehuddin, M., Amoudi, O.S.B.A., Saleh, T.A., Shameem, M., Performance of corrosion inhibitors in cracked and uncracked silica fume cement concrete beams. *Eur. J. Environ. Civ. Eng.*, 24, 1573, 2018.
147. Alamry, K.A., Hussein, M.A., Harunacd, A.M.K., Salehc, T.A., The inhibition performance of a novel benzenesulfonamide-based benzoxazine compound in the corrosion of X60 carbon steel in an acidizing environment. *RSC Adv.*, 11, 7078, 2021.
148. Haruna, K., Saleh, T.A., Quraishi, M.A., Expired metformin drug as green corrosion inhibitor for simulated oil/gas well acidizing environment. *J. Mol. Liq.*, 315, 113716, 2020.
149. Suleimana, R.K., Kumara, A.M., Adesinaa, A.Y., Badourb, F.A.A., Melianic, M.H., Saleh, T.A., Hybrid organosilicon-metal oxide composites and their corrosion protection performance for mild steel in 3.5% NaCl solution. *Corros. Sci.*, 169, 108637, 2020.
150. Ayra, O.R., Tahchieva, A.B., Lopez, V.S., Isern, N.L., Superhydrophobic PDMS coated 304 stainless-steel mesh for the removal of HDPE microplastics. *Prog. Org. Coating*, 170, 1, 2022.
151. Dash, T., Rout, T.K., Palei, B.B., Bajpai, S., Kundu, S., Bhagat, A.N., Satpathy, B.K., Biswal, S.K., Rajput, A., Sahu, A.K., Biswal, S.K., Synthesis of α Al_2O_3–graphene composite: a novel product to provide multi functionalities on steel strip surface. *SN Appl. Sci.*, 2, 1, 2020.
152. Awang, M. and Khalili, A.A., Pedapati, S.Rao., A Review: thin protective coating for wear protection in high-temperature application. *Metals*, 10, 1, 2020.
153. Umashanker, L., Bharathesh, T.P., Saravanan, R., Ceramic protective based coatings for low carbon steel substrate – a Review. *Int. J. Electr. Syst. Control*, 10, 26080, 2020.
154. Li, S., Song, G., Zhang, Y., Fu, Q., Pan, C., Graphene-reinforced Zn–Ni alloy composite coating on iron substrates by pulsed reverse electrodeposition and its high corrosion resistance. *ACS Omega*, 6, 13728, 2021.
155. Torres, B., Garrido, M.A., Rico, A., Rodrigo, P., Campo, M., Rams, J., Wear behavior of thermal spray Al/SiCp Coatings. *Wear*, 268, 828, 2010.
156. Park, J.H. and Park, J.M., Electrophoretic deposition of graphene oxide on mild carbon steel for anti-corrosion application. *Surf. Coat. Technol.*, 254, 167, 2014.
157. Kirkland, N.T., Schiller, T., Medhekar, N., Birbilis, N., Exploring graphene as a corrosion protection barrier. *Corros. Sci.*, 56, 1, 2012.
158. Debasish, D., Bajpai, S., Gochhayat, S., Dash, T., Pati, A.R., Patra, P.K., Bhagat, A.N., Singh, S.K., Rout, T.K., Plasma processing of Fe-P/rGO powder for making robust wear resistance and anti-corrosion coating over mild steel. IOP Publishing., 9, 1, 2022.

159. Haruna, K., Alhems, L.M., Saleh, T.A., Graphene oxide grafted with dopamine as an efficient corrosion inhibitor for oil well acidizing environments. *Surface. Interfac.*, 24, 1, 2021.
160. Saleh, T.A., Haruna, K., Mohammed, A.R.I., Octanoate grafted graphene oxide as an effective inhibitor against oil well acidizing corrosion. *J. Mol. Liq.*, 325, 115060, 2021.
161. Haruna, K. and Saleh, T.A., N, N′ -Bis-(2-aminoethyl)piperazine functionalized graphene oxide (NAEP-GO) as an effective green corrosion inhibitor for simulated acidizing environment. *J. Environ. Chem.*, 9, 1, 2021.
162. Hamidon, T.S., Zakaria, T.P.Y.F.A., Hussin, M.H., Potential of zinc based-graphene oxide composite coatings on mild steel in acidic solution. *J. Indian Chem. Soc.*, 98, 100243, 2021.
163. Ye, X., Lin, Z., Zhang, H., Zhu, H., Liu, Z., Zhong, M., Protecting carbon steel from corrosion by laser *in situ* grown graphene films. *Carbon.*, 94, 326, 2015.
164. Quezada-Renterıa, J.A., Chazaro-Ruiz, L.F., RangelMendez, J.R., Synthesis of reduced graphene oxide (rGO) films onto carbon steel by cathodic electrophoretic deposition: anticorrosive coating. *Carbon*, 122, 266, 2017.
165. Prasai, D., Tuberquia, J.C., Harl, R.R., Jennings, G.K., Bolotin, K.I., Graphene: corrosion-inhibiting coating. *ACS Nano*, 6, 1102, 2012.
166. Li, J., Cui, J., Yang, J., Ma, Y., Qiu, H., Yang, J., Silanized graphene oxide reinforced organofunctionalsilane composite coatings for corrosion protection. *Prog. Org. Coat.*, 99, 443, 2016.
167. Haruna, K., Saleh, T.A., Obot, I.B., Umoren, S.A., Cyclodextrin-based functionalized graphene oxide as an effective corrosion inhibitor for carbon steel in acidic environment. *Prog. Org. Coat.*, 128, 157, 2019.
168. Naser, A.A., Al-sawaad, H.Z., Al-mubarak, A.S., Novel graphene oxide functionalization by urea and thiourea, and their applications as anticorrosive agents for carbon steel alloy in acidic medium. *J. Mater. Environ. Sci.*, 11, 404, 2020.
169. Ansari, K.R., Chauhan, D.S., Quraishi, M.A., Saleh, T.A., Bis(2-aminoethyl) aminemodified graphene oxide nanoemulsion for carbon steel protection in 15% HCl: Effect of temperature and synergism with iodide ions. *J. Colloid Interface Sci.*, 564, 124, 2020.
170. Gupta, R.K., Malviya, M., Ansari, K.R., Lgaz, H., Chauhan, D.S., Quraishi, M.A., Functionalized graphene oxide as a new generation corrosion inhibitor for industrial pickling process: DFT and experimental approach. *Mater. Chem. Phys.*, 236, 121727, 2019.
171. Gupta, R., Malviya, M., Verma, C., Quraishi, M., Aminoazobenzene and diaminoazobenzene functionalized graphene oxides as novel class of corrosion inhibitors for mild steel: experimental and DFT studies. *Mater. Chem. Phys.*, 198, 360, 2017.
172. Gupta, R.K., Malviya, M., Verma, C., Gupta, N.K., Quraishi, M.A., Pyridine-based functionalized graphene oxides as a new class of corrosion inhibitors for mild steel: an experimental and DFT approach. *RSC Adv.*, 7, 39063, 2017.

173. Raza, M.A., Rehman, Z.U., Ghauri, F.A., Ahmad, A., Ahmad, R., Raffi, M., Corrosion study of electrophoretically deposited graphene oxide coatings on copper metal. *Thin Sol. Fil.*, 620, 150, 2016.
174. He, W., Zhu, L., Chen, H., Nan, H., Li, W., Liu, H., Wang, W., Electrophoretic deposition of graphene oxide as a corrosion inhibitor for sintered NdFeB. *App. Sur. Sci.*, 279, 416, 2013.
175. Park, J.H. and Park, J.M., Electrophoretic deposition of graphene oxide on mild carbon steel for anticorrosion application. *Surf. Coat. Technol.*, 26, 321, 2014.
176. Mohapatra, S.S., Dash, T., Patnaik, T.K., Biswal, S.K., Improvement of hardness and anti-corrosion behavior of mild steel by plasma sprayed coating of Al2O3/GO composite. *J. Indian Chem. Soc.*, 100, 1, 2023.
177. Ding, J., Shi, S., Yu, H., Study on modification of lignin as dispersant of aqueous graphene suspension and corrosion performance in waterborne G/epoxy coating. *I.J.A.E.R.S.*, 3, 101, 2016.
178. Yadav, M. and Maurya, A.K., Thermal, mechanical and water barrier properties of graphene oxide/polyvinyl alcohol/polyol composite films. *J. Indian Chem. Soc.*, 99, 100456, 2022.
179. Saleh, T.A., Haruna, K., Mohammed, A.R.I., Octanoate grafted graphene oxide as an effective inhibitor against oil well acidizing corrosion. *J. Mol. Liq.*, 325, 115060, 2021.
180. Ahmad, I., Cao, H.Z., Chen, H.H., Zhao, H., Kennedy, A., Zhu, Y.Q., Carbon toughened aluminium oxide nanocomposite. *J. Eur. Ceram. Soc.*, 30, 865, 2010.
181. Choudhary, R.K., Kain, V., Hubli, R.C., Formation of alumina aluminide coatings on ferritic-martensitic T91 steel. *J. Min. Metall. B Metall.*, 50, 165, 2014.
182. Stolyarov, V.V., Misochenko, A.A., Zholnin, A.G., Grigiriev, E.G., Klyatskina, E.A., Structure and properties of Al_2O_3/graphene nanocomposite processed by spark plasma sintering. *IOP Conf. Ser. Mater. Sci. Eng.*, 218, 1, 2017.
183. Dash, T., Sahoo, G.K., Palei, B.B., Rout, T.K., Graphene–Alumina Composite: The Advanced Coating Material for Developing Harder Surface of Steel Strip Substrate, in: *Advances in Materials Processing and Manufacturing Applications. iCADMA 2020. Lecture Notes in Mechanical Engineering*, A. Patnaik, E. Kozeschnik, V. Kukshal (Eds.), Springer, Singapore, 2021, 2020, https://doi.org/10.1007/978-981-16-0909-1_45.
184. Yina, H., Daia, Q., Haob, X., Huanga, W., Wanga, X., Preparation and tribological properties of graphene oxide doped alumina composite coatings. *Surf. Coat. Tech.*, 352, 411, 2018.
185. Hentour, K., Marsal, A., Turq, V., Weibel, A., Ansart, F., Sobrino, J.M., Chen, Y.M., Garcia, J., Franc, P., Cardey, O., Laurent, C., Carbon nanotube/alumina and graphite/alumina composite coatings on stainless steel for tribological applications. *Mater. Today*, 8, 118, 2016.

9

Bioinspired Strategies for Corrosion Protection and Antifouling Coatings

K. R. C. Soma Raju[1], Aarti Gautam[1,2], Ramay Patra[1,2], K. Srinivasa Rao[1], K.V. Gobi[2] and R. Subasri[1]*

[1]*Centre for Sol-Gel Coatings, International Advanced Research Centre for Powder Metallurgy and New Materials (ARCI), Balapur, Hyderabad, India*
[2]*Department of Chemistry, National Institute of Technology, Warangal, Telangana, India*

Abstract

Steel is a ubiquitous material in load-bearing structures and vehicles such as automobile and ship components worldwide, but its susceptibility to corrosion poses significant challenges. Metal loss in offshore conditions is found to be more than that in onshore conditions. The sea has many contaminants such as microorganisms, bacteria, and parasites such as barnacles. They colonize surfaces, form biofilms, and produce by-products such as organic acids, hydrogen sulfide, and slime. Additionally, the turbulent boundary layer created by barnacles and slime along the mobile vessel body increases drag, prevents ship from clean cutting of water and lead to increased fuel consumption. Therefore, offshore surfaces require both antifouling and anticorrosion coatings for better performance and improved service life.

Conventionally, organic coatings broadly called paints are being used for protection against corrosion which act as a barrier between the metal substrate and the corrosion media. Among various classes of coatings, organic-inorganic hybrid sol-gel coatings are very effective, economical, environmentally friendly, flexible, and versatile to obtain corrosion protection. The addition of corrosion inhibitors into the sol-gel matrix to further enhance resistance is a widely researched topic. Incorporating corrosion inhibitors directly into coating formulations can potentially trigger undesirable and premature consumption. Thus, encapsulation of the corrosion inhibitor becomes a necessary approach to address the issue. Further, taking cues from nature and the self-healing ability of the human body,

**Corresponding author*: subasri@arci.res.in

Hari Murthy, Vinay Jha Pillai, Kukatlapalli Pradeep Kumar and Matthew Cowan (eds.) Novel Anti-Corrosion and Anti-Fouling Coatings and Thin Films, (251–286) © 2024 Scrivener Publishing LLC

incorporating encapsulated corrosion inhibitor into the coatings can provide a self-healing effect, enhancing the corrosion protection and prolonging the life span of metal.

Moreover, low surface free energy thin films derived from sol-gel organic and inorganic hybrids render the coated surfaces hydrophobic, preventing contaminants and bacterial adhesion thereby serving as an effective antifouling coatings for many practical applications. Since the traditional method of preparing antifouling coating is not appreciated due to its cytotoxicity, researchers have now completely focused on nature-inspired antifouling coatings. Organic-inorganic nanocomposite coatings help in implementing ideas derived from nature using the concepts of superhydrophobic coating and liquid-infused omniphobic slippery coating for imparting antifouling functionality. For instance, the lotus leaf-inspired superhydrophobic surface with micro/nanoscale roughness effectively repels microorganisms. Alternatively, eco-friendly biocides can be integrated into coatings for direct killing of attached microorganisms.

Hence, a sol-gel coating protects the coated surface not only from corrosion, but also from fouling due to the surrounding medium. This chapter will cover recent advancements in biomimetic antifouling and anticorrosion coatings with a special emphasis on sol-gel technique.

Keywords: Bioinspired surfaces, sol-gel coatings, self-healing, anticorrosion, superhydrophobic, antifouling

9.1 Introduction

Fouling is a property that refers to the unwanted accumulation and growth of substances on surfaces exposed to a fluid medium, particularly freshwater or seawater. It is a common phenomenon that occurs in various industrial and natural systems, ranging from pipes and heat exchangers to ship hulls and submerged structures. The substances that cause fouling include biological organisms such as algae, barnacles, mussels, and bacteria, as well as non-living matter such as sediment, minerals, scale, slime, and debris. These substances can attach to surfaces and form layers or deposits that alter the surface properties and consequently hinder fluid flow, reduce heat transfer efficiency, increase pressure drop, promote corrosion, and create favorable environments for further fouling. Fouling can have significant negative effects on the performance, efficiency, and life span of stationary equipment and mobile transport systems such as ships and other structures; it can lead to increased fuel consumption in the case of ships. Ships include naval ships and merchant ships such as container ships, tankers, cargo ships, bulk carriers, and cruise ships, and hence, the problem of fouling is quite significant and needs immediate attention. It was found that an unprotected vessel bottom can gather 150 kg fouling/m^2 in a 6-month exposure to sea. This could be as high as 6,000 tonnes of fouling for a crude carrier with 40,000 m^2 underwater surface area. The fuel consumption can easily go up by 50%, and hence, a good antifouling strategy can result in huge fuel savings, extended dry-docking, and increased vessel availability [1]. Fouling can start within 48 hours of contact with water depending on its temperature and nearly 80,000 tonnes of antifouling paints are being used all over the world annually to mitigate the same [2].

9.1.1 Types of Fouling Based on Factors Contributing to Fouling

The factors contributing to fouling can be (a) biological, (b) physical, (c) chemical, or a combination thereof such as corrosion fouling. Biological factors involve the settlement and growth of organisms on surfaces, often facilitated by adhesive secretions or biofilm formation. Marine biological organisms, such as barnacles, algae, mussels, slime, and bacteria can attach and grow on submerged surfaces. These organisms produce adhesives or secrete substances that facilitate attachment and colonization. They can form biofilms, which provide a favorable environment for other organisms to settle and grow.

A schematic representation of the biofouling process from the time of submersion of an article to the matured colony formation is shown below in Figure 9.1. Biofouling of a submerged surface is a very dynamic process as presented in Figure 9.1 and occurs in four phases over a period of time. Phase 1: Formation of a conditioning layer by absorbing organic molecules such as proteins, carbohydrates, and glycoproteins from the water body. Phase 2: Initiation of colonization by the bacteria and diatoms forming a biofilm that contains sessile microorganisms, i.e., immobile barnacle and an extracellular polysaccharide matrix. Biofilm helps in enhanced adhesion, protection to the microorganism, and controlled diffusion of nutrients to the colony of microorganisms. Phases 1 and 2 come under microfouling where bacterial growth and 'slime' formation due to the fluids they secret dominate within hours of the structure that comes in contact with the water body. Phase 3: Soft macrofouling by the algae, protozoa, diatoms,

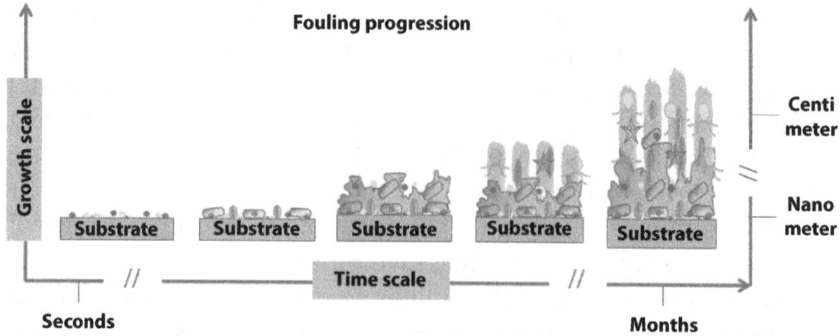

Figure 9.1 Schematic representation of the biofouling process from the time of submersion of an article to the matured colony formation.

and invertebrates colonizes the surface. Phase 4: Hard macrofouling by the tube worms and barnacles [3]. Phases 3 and 4 come under macrofouling.

Physical factors include sedimentation, deposition of suspended particles, and flow patterns that promote accumulation. In this case, particles, dust, dirt, airborne pollutants, or suspended matter settle and accumulate on surfaces. Turbulent flow patterns or stagnant regions in fluid systems can promote the deposition of particles or organisms onto surfaces. It can occur in various industrial processes, air filters, heat exchangers, and HVAC systems, where particles are carried by air or fluids.

Chemical factors encompass reactions and interactions between surfaces and the surrounding fluid, leading to mineral scaling or chemical reactions with dissolved substances. For example, mineral scaling can occur when dissolved minerals in water precipitate and form deposits on surfaces, leading to fouling. It commonly occurs in water systems, such as pipes, heat exchangers, and boilers, where water is heated or circulated. Chemical reactions with dissolved gasses or substances in the fluid medium can also contribute to fouling.

Corrosion fouling is the deposition of corrosion products or rust on surfaces. It can occur in environments where metal surfaces are exposed to corrosive substances, leading to the formation of unwanted deposits that can impair functionality or degrade performance. Marine corrosion is one of the severe threats and a big challenge for marine equipment in both inshore and offshore conditions. The aggressive saline conditions accelerate the metal surface corrosion causing breakdown or failures, which results in huge ocean economic losses. Organic fouling involves the accumulation of organic compounds on surfaces. This can include oils, greases, proteins, or other organic substances that adhere to surfaces and form layers or deposits.

The surface properties of materials play an important role in fouling. Surfaces with high surface energy tend to promote the adsorption and attachment of substances, while surfaces with low surface energy or specific textures can discourage fouling by reducing attachment and making it easier for fouling organisms or particles to be dislodged. Environmental factors, such as temperature, light, nutrient availability, and pH can also influence fouling. Different organisms thrive under specific environmental conditions, and hence, variations in these conditions can affect fouling rates.

Fouling can be influenced by operational factors, including flow rates, residence time, and fluid chemistry. Improper maintenance or operating conditions that promote stagnant zones or accumulation of debris can accelerate fouling. Therefore, the fouling rate will be very high on a stationary object when compared to that on a moving object. Understanding these factors can help in developing effective, antifouling strategies, and materials to prevent or minimize fouling in various applications.

It is important to note that these types of fouling can often occur simultaneously or in combination, and their severity and impact depend on the specific application and environment. Hence, different antifouling strategies as given below may be required to effectively address each type of fouling.

9.1.2 Methods to Prevent Fouling

The antifouling property refers to the ability of a material or surface to resist the attachment and growth of unwanted substances, such as organisms, biofilms, or deposits. There are different methods to prevent fouling. The first and most important yet simple method is to plan and regularly clean the surfaces that are prone to fouling. Any organic matter or debris that can attract marine organisms must be removed. This plan should include inspection schedules, cleaning protocols, and the use of appropriate antifouling measures. Regularly monitoring the surrounding environment for signs of fouling organisms or unusual bioactivity and early detection can help in taking prompt preventive measures. Yet, another easy method to prevent fouling is to modify the design or shape of structures to promote smooth water flow, reducing areas where fouling organisms can settle and grow. It is desirable in many applications where the accumulation of these substances can negatively impact performance, efficiency, or aesthetics.

However, the most widely used strategy to prevent fouling is through surface modification using antifouling coatings. Antifouling coatings contain chemicals or biocides that discourage the attachment and growth of marine organisms. Materials with low surface energy, such as certain polymers or coatings, can make it difficult for fouling organisms or particles to adhere. Low surface energy reduces the attractive forces between the surface and the foulants, making it easier for them to be dislodged or washed away. Teflon coatings and silica-based nanotechnology coatings are finding ground for low surface free energy coatings. A survey published by Markets and Markets Research Pvt. Ltd. estimated the antifouling coating market at U.S. $9.22 billion up to the year 2021, which is continuously growing [4].

In terms of published literature, a review of the Science Direct, Web of Science and Scopus reveals that there has been a rapid growth in the research in low or high surface energy coatings, antifouling, and corrosion protection coatings from the year 2000 onward. A total of 2257 publications were recorded in the Science Direct database, the majority of the publications were research articles followed by review articles, book chapters, conference abstracts, and others. Among them, about 935 articles were published in the domain of material science, 760 of them in chemical engineering, and 562 of them were published in environmental science–related subjects. It can also be seen that there is a systematic increasing trend in the publication of articles related to the field of antifouling coatings. The choice of antifouling strategy depends on the specific application and environmental conditions.

One of the reasons for fouling is corrosion, which can also be prevented by the use of coatings. Corrosion is a global challenge that has a detrimental impact on the service and safety life of metal infrastructure as well as equipment resulting in huge economic loss [5]. Application of protective/barrier coating on the metal surface is the most common strategy to prevent the corrosion process which includes polymer coatings [6], sol-gel coatings [7], and galvanization techniques [8]. Recent years have seen a rise in the importance of bioinspired structural and functional design of metal surfaces for overcoming the corrosion issue [9]. These nanostructures provide answers to many challenges faced by humans to prevent marine corrosion. Researchers have now designed and developed coatings, which have the tendency to mimic these nanostructures and thereby prevent corrosion of the metal surfaces. We may consider the same publication trend of antifouling to even corrosion studies as well. All the coatings that are applied for antifouling also can act as barriers against corrosion but all corrosion protection coatings need not be antifouling coatings.

This chapter will further discuss, in detail, recent developments in the area of bioinspired strategies to achieve both the antifouling and corrosion protection properties simultaneously through coatings, with special emphasis on sol-gel derived coatings.

9.1.3 Recent Developments in the Area of Antifouling Coatings

Historically, sailors used lime, arsenical, mercurial compounds, and *dichlorodiphenyltrichloroethane* (DDT1) as antifouling material in the ascending order of use. In the 1960s, the first effective antifouling paint for marine applications was introduced. Paint mixed with biocide TBT (tributyltin) was highly effective in preventing the attachment of marine organisms to ship hulls. TBT-based antifouling paints became widely used in the following decades due to their efficacy.

However, concerns about the environmental impact of TBT, particularly its toxicity to marine life due to leaching and subsequent killing of marine life such as the barnacles, led to its gradual phase-out and ban in many countries starting in the late 1980s. Moreover, the release of the biocide was uncontrolled and hence was consumed within a maximum time span of 2 years. By the end of the 1990s, the Marine Environment Protection Committee (MEPC) tasked by the International Marine Organization IMO raised the issue of toxicity in the use of TBT and many Western, European, and Asian major economies started prohibiting the use of TBT in antifouling paints. This prompted the search for alternative antifouling coatings that would be effective while being more environmentally friendly.

Since the phase out of TBT, significant research and development efforts have focused on the development of new-generation antifouling coatings. These coatings often utilize alternative biocides, such as copper compounds (95% of total antifouling paints in the United Kingdom), zinc-based materials, or organic biocides. Additionally, bioinspired approaches that mimic natural mechanisms of fouling prevention, such as surface texturing or low surface energy coatings, have gained attention.

The advancement of antifouling coatings continues to this day, with ongoing research into novel materials, coating formulations, and application techniques. It is worth noting that while antifouling properties can be highly effective, they may not be permanent. Over time, the effectiveness of antifouling materials or coatings may diminish due to wear, degradation, or the development of resistant fouling organisms. Hence, regular inspection, maintenance, and, if necessary, reapplication of antifouling measures are essential to ensure continued protection against fouling. The best solution is to develop coatings that offer effective fouling prevention while minimizing environmental impact and maintaining long-term durability.

9.1.4 Bioinspired Strategies for the Development of Functional Coatings

Biocide-free antifouling coatings that can reproduce the physicochemical properties of the naturally available surfaces such as the "lotus leaf effect" and "Salvinia effect," can greatly help in achieving an environment-friendly antifouling property for submersible structures [10]. Learning from nature is always a first priority when exploring any new ideas. Nature has served as a wonderful source of inspiration and has provided significant breakthroughs for solving technological problems in creative ways. The structural design which is biologically inspired or derived from nature is known as biomimetic and has aroused a lot of curiosity in researchers [11–14]. The primary source of inspiration for developing biomimetic materials is mainly from plants, marine organisms, and mammals, which include lotus leaves, nacre, mimosa, pitcher plants, and mussels. The comprehension of the intricate architecture of biological structures and their proper imitation has helped material scientists to develop new materials with extraordinary performance. Figure 9.2 presents examples of bioinspiration from nature that have led to exciting developments in the field of antifouling and corrosion protection coatings.

Figure 9.2 Examples of bioinspiration from nature that have led to exciting developments in the field of antifouling and corrosion protection coatings (source: no copyright images from public domain).

Published literature shows that the surface roughness parameter (R_a) of the various aquatic animals varies from 100 to 500 μm, 1 to 10 μm, and, in a few cases, reaches as low as the nanometer scale to exhibit antifouling properties. Similarly, creating micro- or nanoscale textures on a surface can deter fouling by reducing the contact area and altering the flow dynamics. These textures can disrupt the attachment of fouling organisms or prevent the formation of biofilms by minimizing available attachment sites [15]. This can involve incorporating specific chemical groups or coatings that deter attachment or growth [16].

The most popular types and frequently used bioinspired special surfaces to reduce corrosion and fouling are superhydrophobic coatings [17], slippery liquid-infused porous surfaces (SLIP) [18], and self-healing coatings [19]. We need to pay close attention to the two essential criteria in order to create a superhydrophobic surface. These are (i) hierarchical micro/nanostructure and (ii) low surface free energy [8]. It is also important to note that a superhydrophobic surface cannot be created by focusing on just one factor, such as either low surface energy or micro/nanoscale surface architecture. However, the combinatorial approach as presented in Figure 9.3 will always produce a superhydrophobic surface with a low rolling angle having multiple applications [20]. With time, researchers have shifted their attention from the lotus leaf to another intriguing plant called the pitcher plant, which has a microscale roughness architecture on the surface along with a nanoscale wax covering on it [21].

Inspired by the pitcher plant, Wong *et al.* [22] pioneered the concept of SLIPs by infusing perfluorinated fluids in epoxy resin–based nanostructured surfaces. The lubricant injected into the micro-nano architecture of SLIPs as shown in Figure 9.4, creates a continuous, dynamic liquid film on the substrate that serves as a barrier layer and offers superior corrosion prevention [21]. Fouling is discouraged due to its slippery surface.

An alternative of the process is a slippery liquid-infused surface (SLIS) wherein a slippery low friction surface was fabricated by irradiating the

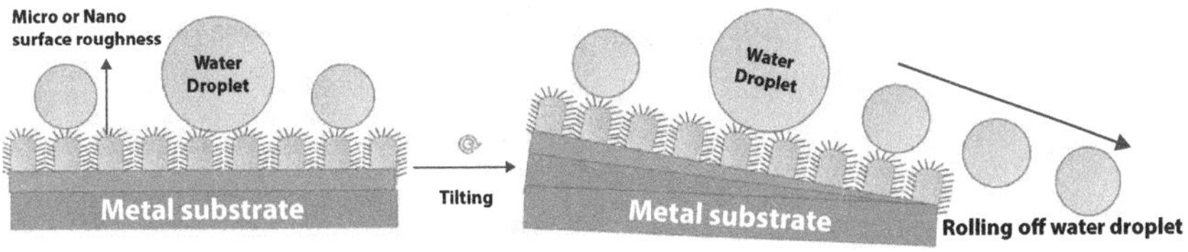

Figure 9.3 Schematic representation of the hierarchical roughness responsible for superhydrophobic coating (redrawn with permission from [17]).

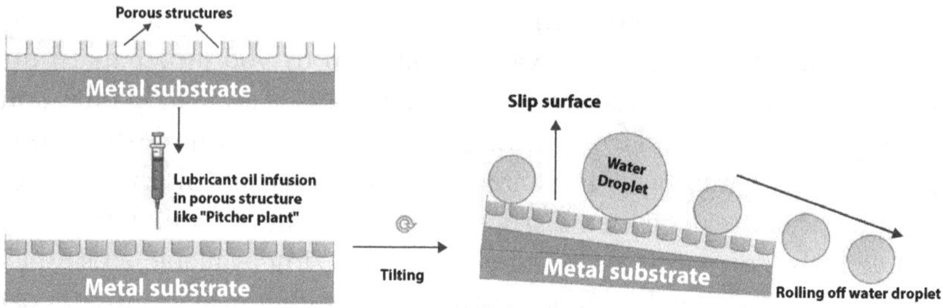

Figure 9.4 Schematic representation of the SLIPS (redrawn with permission from [18]).

methyl oleate (MO)–infused PDMS film. Such films too exhibited outstanding antifouling properties against fouling medium [23].

Researchers have also been intrigued by the ability of the human skin to self-heal and inspired by it; they have developed a self-healing coating as shown in Figure 9.5 for anticorrosion applications. For example, White *et al.* reported a polymeric material having self-healing properties. The healing agent was incorporated into a nanocontainer, which releases upon crack intrusion and thus heals the damaged area or crack [24]. Recently, the emergency response closure (ERC) feature of mimosa leaves also fascinated researchers to develop a smart anticorrosion coating. Taking inspiration from the ERC of mimosa leaves, Ding *et al.* [25] developed a smart epoxy coating on carbon steel using terpyridine derivative (TDD)–functionalized graphene oxide which has the ability to close water diffusion pathways in response to the Fe^{2+} production during the initiation of corrosion. Because of the reaction between the TDD and Fe^{2+} ions, the functionalized graphene oxide (tGO) sheets start to form a closed self-assembled 3D structure to prevent corrosion of metal surfaces by acting as a barrier layer to aggressive corrosive species. The anticorrosion behavior was confirmed *via* potentiodynamic polarization studies and electrochemical impedance spectroscopy [25].

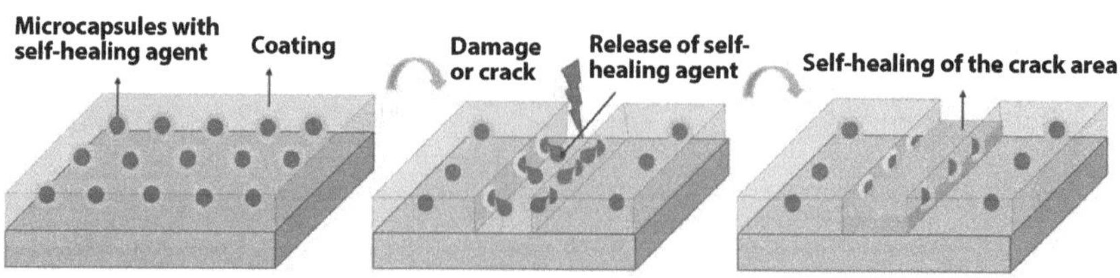

Figure 9.5 Schematic representation of the self-healing mechanism for protection against corrosion (redrawn with permission from [19]).

Inspired by the nacre and mussel structure, a new strategy of brick and mortar structure with enhanced diffusion pathway was proposed recently. Such structure as presented in Figure 9.6 has exhibited excellent corrosion protection [26]. Firstly, epoxy coating on Q235 mild steel (MS), was applied followed by deposition of nacre-mimicking graphene oxide layers *via* mussel-inspired dopamine method to obtain a final coating of epoxy-(graphene-dopamine). This process was repeated to achieve the desired thickness of epoxy-(graphene-dopamine)$_n$. Dopamine acted as a binder, enhancing the interface interaction between epoxy and graphene. Additionally, it served as a coating matrix, contributing to slowing down the corrosion process. Moreover, having a sandwich-like coating design, where the graphene is aligned parallel to the substrate, imparts a shielding effect and thus reduces galvanic corrosion by increasing the time and distance of direct contact with corrosive species. The composite coating thus obtained shows remarkable anticorrosion properties for 90 days of immersion in 3.5 wt% NaCl solution [26].

There has always been a concern that epoxy coatings peel from substrates when exposed to a strong salt spray or corrosive environment. Tian *et al.*, drew inspiration from a tree frog that exhibits strong wet adhesion due to the presence of a microstructure of its toe pads and developed a corrosion resistant self-healing coating, wherein coating material could be integrated with the interface due to the presence of bioinspired structures, that resulted in enhanced adhesion strength [27]. Bioinspired interfacial structure design was fabricated using a laser based manufacturing machine. Calcium alginate nanocontainer loaded with Tung oil as a corrosion inhibitor was incorporated in the epoxy coating and deposited on steel substrate. The adhesion of the modified coating was increased by three times when compared to only epoxy coating without compromising the corrosion inhibition property.

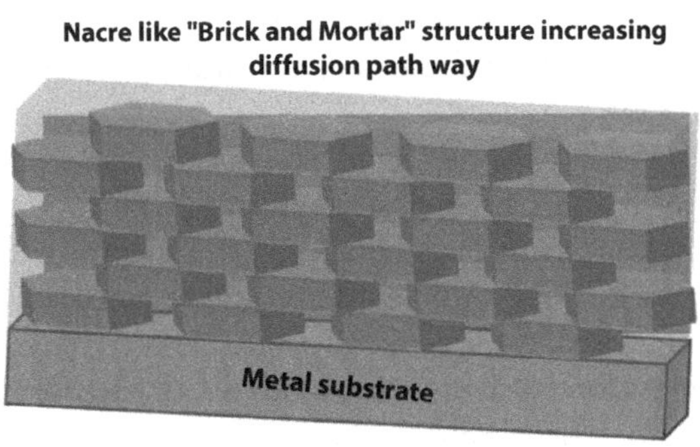

Figure 9.6 Schematic representation of nacre-like "brick and mortar" structure in the coating.

9.1.5 Methods to Generate Bioinspired Surfaces with Emphasis on Sol-Gel Technology

Various manufacturing methods, as presented below, were used for the successful reproduction of micro- and nanoscale features on the coated surfaces [28], thereby mimicking the bioinspired surfaces.

a. Photolithography: It involves the laborious process of printing, etching, and transfer of layers using a beam of photons. It is used for the production of field effect transistors and integrated circuits [29–31].
b. Electron beam lithography and ion beam lithography: A focussed electron or ion beam is used for direct printing or scanning a thin layer of organic film pattern and transfer to a substrate surface [32–36].
c. Electrochemical micromachining: This method is an electrolysis-based electrochemical surface texturing on an electrically conductive substrate. The process is amenable to automation and large-area processing [37–39].
d. Atomic layer deposition: Films can be deposited through a vapor deposition process with high precision and uniformity at the atomic scale through chemical surface reactions. Suitable to deposit films on high aspect ratio models such as jewelry [40].
e. Soft lithography and micro-contact printing: PDMS polymeric films with nano and micro features can be printed on the substrate surface using a stamping technique [41–45].
f. Hot embossing: Thermoplastics are used to replicate the micropatterns in softened condition with minimal stress [46–48].
g. 3D printing: Economical and fast additive prototyping process to create three-dimensional features on substrate surface [49, 50].
h. Picosecond laser texturing: A laser beam is used to remove material at a microscopic scale creating textures on the substrate surface using laser ablation technique [51, 52].
i. Sol-gel technology: Ambient temperature-cured sol-gel coatings can create surface micro/nanoscale hierarchical microstructures along with organic ligands responsible for imparting superhydrophobic properties [2].

All of the above methods were developed concurrently with unique features, but different limitations such as difficulty to adopt for large-area processing, requiring to process in evacuated chambers and lack of precision control over processing conditions, have posed restrictions to consider some of them for industrial application. Hence, among all the coating types, hybrid sol-gel coatings are more widely discussed as they are the best replacement for toxic coatings. The sol-gel process involves the conversion of a liquid precursor solution (sol) into a solid gel network through hydrolysis and condensation reactions. Hybrid sol-gel coatings are conventional and easy to synthesize. This process allows for the deposition of thin films or coatings on various surfaces, including metals, glass, ceramics, and polymers, and on large areas. Sol-gel–based antifouling coatings can be prepared using various precursors such as organically modified silanes, metal alkoxides, and organic compounds that undergo hydrolysis and condensation reactions to form a stable gel network.

These coatings are quite compatible with paints, eco-friendly, and cost-effective in nature. Modification of sol-gel coating becomes easier because of the presence of colloidal particles in them, acting as a cavity for loading corrosion inhibitors or antifouling agents to enhance their anticorrosion and antifouling properties [53]. High surface energy results in hydrophilic properties, while low surface free energy results in hydrophobic properties. Both conditions can be used as self-cleaning and antifouling surfaces [54]. However, hydrophobic condition along with surface micro- and nanoscale geometry, provides hierarchical surface roughness and helps in repelling water. This, coupled with the slippery nature of the coating, prevents the adherence of any foreign body to the coated surface [55].

Certain surfaces possess self-cleaning properties, enabling them to shed or remove fouling substances through natural forces such as water flow, wind, or vibrations. These surfaces are designed to prevent fouling by promoting the easy removal of attached organisms or debris. Polydimethylsiloxane (PDMS) is one of the well-known silicone elastomers that reduces surface tension to result in a hydrophobic surface of the substrate to which it has been applied and keeps the surface free from adhesion and growth of microbes [56].

Some materials can also be used to release bioactive compounds that inhibit the attachment or growth of fouling organisms. Copper- and zinc-based compounds such as copper or copper oxide nanoparticles, copper thiocyanate and zinc oxide, and zinc pyrithione may have antimicrobial, antifungal, or antifouling properties for effectively preventing or reducing fouling. Silane coupling agents such as 3-aminopropyltrimethoxysilane (APTMS), 3-aminopropyltriethoxysilane (APTES),

3-glycidyloxypropyltrimethoxysilane (GPTMS), and conductive polymers polyaniline (PANI) and its derivatives are used to enhance compatibility of nanofillers in the polymer matrix coatings [57].

9.1.5.1 *Bioinspired, Sol-Gel–Based Antifouling Coatings*

Tetraethyl orthosilicate (TEOS) is one of the most commonly used precursors for sol-gel coatings. It is a silicon alkoxide that undergoes hydrolysis and condensation reactions to form a silica-based network. TEOS provides good adhesion and durability to the coating and forms hybrids when added with polysaccharides such as chitosan, hyaluronic acid, alginate, chondroitin-6-sulfate, and heparin. TEOS-based sols with anionic polysaccharides such as alginate or heparin showed promising antifouling performance [54]. Methyltrimethoxysilane (MTMS) is another commonly used silane precursor for sol-gel coatings. It is a silicon alkoxide that can be used as a co-precursor with TEOS or as a standalone precursor. MTMS helps in attaching the hydrophobic methyl groups to the hierarchical surface roughness created by the cellulose nanocrystal (CNC)–dispersed SiO_2 nanoparticulate sols. MTMS-based coatings offer enhanced hydrophobicity and fouling resistance [58].

Octyltriethoxysilane (OTES) is an alkyl silane precursor that can be used to introduce hydrophobicity to TEOS-based organically modified silicate (ORMOSIL) hybrid sol-gel coatings. It forms a hydrophobic surface due to a long-chain hydroxyl group that resists fouling by preventing the adhesion of fouling organisms. Sol synthesized with TEOS, n-octadecyltrimethoxysilane C18, and n-octyltriethoxysilane C8 can exhibit enhanced antifouling properties when nanoscale roughness is created on the coated surface by adding tridecafluoro-1,1,2,2-tetrahydrooctyl-triethoxysilane and incorporating nanoscale inorganic particles in the sol [2]. Zirconium alkoxides such as zirconium (IV) propoxide or zirconium (IV) n-butoxide can be used as precursors for sol-gel coatings. These precursors provide excellent antifouling properties due to the release of metal ions, which are toxic to fouling organisms. CNCs were used as nanofillers in a hybrid sol synthesized using tetrapropyl zirconate, 3-glycidyloxypropyltrimethoxysilane, and MTMS. Such biomimetic coating resulted in micro/nanoscale features on the coated surface with the best antifouling properties [33]. Zirconium–n-propoxide and hexamethyldisilazane (HMDS) were used to develop transparent, antireflective coating with self-cleaning properties. The water contact angle (WCA) could be increased to 101° while maintaining a transmittance of 91.2% [59]. The antifouling effect multiplies when a suitable biocide is added to the coating. Titanium alkoxides, such as titanium (IV) isopropoxide or titanium(IV) n-butoxide, are commonly used precursors for sol-gel coatings. Titanium-based coatings offer good antifouling properties due to their biocidal effects and photoinduced photocatalytic oxidation. TiO_2 disinfects surfaces and exhibits self-cleaning capabilities. SiO_2 core TiO_2 shell nanoparticles were functionalized using MTMS and used in polyurethane-based paint to impart hydrophobicity (143.8° WCA) and antimicrobial dual functionality in a single coating [60]. Inorganic biocides such as Ag, Cu, and Fe_2O_3 nanoparticles were being used to further enhance the antimicrobial and antifouling properties of titania-based sol-gel coatings [61].

Fluorine-free PDMS-based precursors can be incorporated into sol-gel coatings. PDMS derivatives impart hydrophobicity and low surface energy, preventing the adhesion of fouling organisms. Incorporation of 2.5 wt% nanoparticles such as zinc oxide and multi-walled carbon nanotube MWCNT enhanced their contact angle to 156° attaining superhydrophobic property while maintaining excellent long-term antifouling properties and attained 95% efficiency in separation of water-oil mixtures with excellent reusability [62]. It was found that hydrophobic compositions prepared with 50/50 mole ratio n-octyltriethoxysilane (C8-TEOS), tetramethylorthosilane (TMOS); 45/55 mole ratio n-propyltrimethoxysilane (C3-TMOS), TMOS; 2/98 mole ratio bis[3-(trimethoxysilyl)propyl]-ethylenediamine (enTMOS), tetraethyl orthosilicate (TEOS); 50/50 mole ratio 3,3,3-trifluoropropyltrimethoxysilane (TFP-TMOS), TMOS, inhibited fouling of zoospores. Sol with TEOS/TMOS exhibited higher hydrophobicity and low critical surface tension and could able to reduce the adhesion and higher release of Ulva sporeling biomass and juvenile of tropical *B. amphitrite* barnacles [63]. These are just a few examples of precursors used in sol-gel–based antifouling coatings. The choice of precursors depends on the specific requirements of the coating, such as desired properties, substrate compatibility, and environmental considerations.

Specific additives or functional groups can be incorporated into the precursor solution to impart antifouling properties to the sol-gel coating. These additives may include biocides, antimicrobial agents, or low surface energy compounds that discourage the attachment of fouling organisms or substances. Metal and metal oxide nanoparticles such as Ag, ZnO, Cu, CuO, and Fe_2O_3 were extensively used as biocides for antifouling and antimicrobial applications [61]. Both Cu and Zn are known as good biocides but Cu is more toxic to aquatic organisms as it builds in marine habitats, when compared to Zn/ZnO or zinc pyrithione. Different nanoparticles not only increase the dose of metal to the organism that comes in contact with them, but also dissolve to a different extent in water and release toxic metal ions. Some of the nanoparticles such as TiO_2 and Bi_2WO_6 can be used as visible light photo-catalysts in the coating for enhanced antifouling application. Visible light catalysis of Bi_2WO_6 in the coating continuously releases hydrogen peroxide H_2O_2. Hydroxyl radicals from H_2O_2 degrade the adsorbed fouling species and result in excellent antifouling properties [16]. Performance efficiency and toxin release characteristic studies show that nanoparticulate metal biocide coatings did not exhibit any benefit over micro-sized particles in long-term use. But, the coating matrix plays an important role in leaching and affecting the ability of biocide on the antifouling properties [64].

Polyaniline (PANI) is one of the commonly used conductive polymer additives during sol synthesis whenever electrical conductivity is required for the coatings. It has dual functionalities and imparts corrosion protection as well as antifouling properties due to its redox activity and charge transfer ability. Moreover, it interacts with bacterial cell walls and kills it due to oxidative stress [57]. The surface of the substrate can be modified during the sol-gel process to enhance the antifouling properties. This can be achieved by incorporating hydrophobic precursors such as n-octadecyltrimethoxysilane C18 and n-octyltriethoxysilane C8 or hydrophilic compounds such as hyaluronic acid, alginate, and heparin which can deter the attachment of fouling organisms to substances [16, 54]

Cross-linking agents can be added to the sol-gel precursor solution to enhance the mechanical strength and durability of the resulting coating. This helps in maintaining the integrity of the antifouling coating over extended periods of use. Ethylene glycol and polyethylene glycol, diethylenetriamine (DETA), ethylenediamine (EDA), and N-aminoethylpiperazine (AEP) have been investigated for their network forming and density enhancing ability, which are found to be important functions in enhancing corrosion protection of coatings. Reactive sites of the cross-linking agent were found to be the most important factor in obtaining adherent and durable coating [65, 66]. The presence of (3-aminopropyl)triethoxysilane (APTES) precursor as a cross-linking agent in the coatings proved to enhance self-healing ability and enhanced corrosion protection by promoting adhesion of coating to the AA2021-T3 substrate [67].

Sol-gel coatings can be applied onto surfaces through various techniques, including dip-coating, spin-coating, spraying, or roll-to-roll coating, depending on the size and shape of the substrate. The performance of sol-gel–based antifouling coatings can be evaluated through standardized tests such as biofouling stability assay in aqueous media, laboratory diatom attachment and removal assay, WCA measurements, and durability assessments under simulated operating conditions [54]. Immersion in seawater for a few months, periodic withdrawal, and visual examination at regular intervals are also being carried out as a standard test by the researchers [68].

It is worth noting that the effectiveness and durability of sol-gel–based antifouling coatings can vary depending on factors such as the specific formulation, substrate material, environmental conditions, and type of fouling organisms or substances being targeted. Therefore, careful selection of materials and optimization of the coating process are crucial for achieving desired antifouling properties.

Characteristics of a Good Biocide in Antifouling Coatings

a. Broad spectrum activity and high performance such that more bacterial species can be controlled
b. Low water solubility so that leaching can be negligible
c. Compatible with paint raw materials for enhanced adhesion
d. Low toxicity such that it is not hazardous to the mammalian and non-mammalian life and ultimately ending up in the human food chain
e. Non-persistent to the environment such that its life cycle is minimal and finally,
f. Affordable price

Table 9.1 summarizes different combinations of precursor and active materials reported for the generation of sol-gel–based antifouling coatings.

9.1.5.2 Bioinspired Sol-Gel–Based Anticorrosion Coatings

Protection of metal surfaces in marine environments and bioinspired corrosion protection coatings based on water repellence, i.e., superhydrophobic, are the most widely researched topics. Superhydrophobic coatings are the ones that have the tendency to repel the water droplet from the surface by an angle of more than 150°. To obtain such a coating, nano- and microstructures are introduced in the coating so that they have water-repelling action. The best example from nature is the lotus leaf, though submerged in water continuously, it has the tendency to roll off the water droplet due to its waxy surface. For example, Li *et al.* reported a one-step electrodeposition process to fabricate a bioinspired superhydrophobic surface on a copper substrate that exhibits potential corrosion protection. They have used the mixture of myristic acid and $CaCl_2$ solution in ethanol for electrodeposition to develop a Ca–myristic acid complex coating with a micro/nanostructure [75].

Table 9.1 Summary of different precursor and active materials used for the generation of sol-gel–based antifouling coatings.

S. no	Reference no.	Strategy	Raw materials	Functional material used	Substrate	Coating system
1	[2] Detty *et al.*	Hydrophobic sol-gel coating	n-Octyltriethoxysilane (C8) and tetraethyl orthosilicate (TEOS) in 1:1	C8	Fiberglass boat hull	Brush coating
2	[16] Scurria *et al.*	Hydrophobic sol-gel coating with biocide	TEOS, C18, C8, and Bi_2WO_6	C18, C8, and Bi_2WO_6	Glass and steel grades used for ships	Spin coating
3	[28] Richards *et al.*	Hydrophobic coating with ORMOSILs by sol-gel process	TEOS, DMDEOS, ETEOS, and TMES	DMDEOS	Glass	Spin coating
4	[54] Wanka *et al.*	Hydrophilic sol-gel coating	TEOS and polysaccharides	Polysaccharides	Glass	Spin coating
5	[58] Duan *et al.*	Superhydrophobic sol-gel coating with cellulose nanocrystals	TPOZ, GPTMS, MTMS, and SiO_2 nanoparticles	TPOZ	Glass slide	Spin coating
6	[59] Dogra *et al.*	Sol-gel hydrophobic coating	Zr–n-propoxide, HMDS	HMDS	Glass	Spin coating

(Continued)

Table 9.1 Summary of different precursor and active materials used for the generation of sol-gel–based antifouling coatings. (*Continued*)

S. no	Reference no.	Strategy	Raw materials	Functional material used	Substrate	Coating system
7	[62] Subodh *et al.*	Sol-gel superhydrophobic coating	PDMS-modified MWCNTs/ZnO composite	PDMS	Cu mesh, filter paper, aluminum, and glass	Dip coating
8	[63] Tang *et al.*	Sol-gel hydrophobic coating	C8-TEOS/TMOS	C8-TEOS	Glass	Spin coating
9	[69] Peng *et al.*	Nanofiller addition or chemical modification of PDMS	PDMS with CNT/graphene oxide/TiO2 or PDMS with epoxy/urethane/urea	PDMS	Steels	Brush coating
10	[70] Yue *et al.*	PDMS epoxy hybrid coating	Epoxy resin (E-44), aminopropyl terminated polydimethyl siloxane (NH2-PDMS), hexamethylene diisocyanate, and aminated UiO-66 nanoparticles	PDMS	TC4 titanium alloy	Brush coating

(*Continued*)

Table 9.1 Summary of different precursor and active materials used for the generation of sol-gel–based antifouling coatings. (*Continued*)

S. no	Reference no.	Strategy	Raw materials	Functional material used	Substrate	Coating system
11	[71] Wenhuan *et al.*	Solid slippery surface prepared by sol-gel process	TEOS, H_2SO_4, and PDMS	PDMS grafting	Glass	Dip coating
12	[72] Madasamy *et al.*	Membrane synthesis	Chitosan and ZnO	Chitosan/ZnO nanocomposite	Standalone film	Casting
13	[73] Mohamed *et al.*	Super hydrophobic coating with inorganic biocide	PDMS + graphene oxide/reduced graphene oxide and boehmite nanorods	PDMS	Graphene film and mild steel	Dip coating
14	[74] Ting *et al.*	Superhydrophobic sol-gel coating	PFDTS, APTES, TEOS, and ZnO particles	PFDTS	Glass	Dip coating

Recent research work carried out by Zhang and his coworkers reported the use of water-based ZrO_2-SiO_2 sol-gel coatings for corrosion protection of carbon steel. They have used Ce^{3+} ions as the corrosion inhibitor to load them into a self-shaping endow clay called bentonite. This layered structure of bentonite clay shows a delay in the corrosion process by entrapping the chloride ions, and furthermore, the sustained release of the corrosion inhibitor increases the corrosion resistance of the coating by many folds. The coating thus prepared has dual functionality, i.e., water repellence with WCA of 115° as well as corrosion resistance [76].

Similar work was reported by England *et al.*, where they have reported the development of multilayered hydrophobic surfaces on copper substrates. The prepared coatings were transparent with lamellar-like structures having excellent hydrophobicity and corrosion resistance properties. The hydrophobic surfaces synthesized have a unique behavior of regeneration by delaminating the topmost layer after photochemical degradation. Copper substrates even after exposure to salt water for 24 hours show no sign of corrosion and retain their lustrous properties too [77].

Liu *et al.* in their work, have reported the use of sodium molybdate dihydrate as a corrosion inhibitor intercalated in between the layered double hydroxide (LDH). The LDHs were pre-treated with hydrophobic silane 3-aminopropyltriethoxysilane (ATES) to impart water-repelling properties. For multilayer coating synthesis, the pre-treated and inhibitor-modified LDH layers were sequentially deposited with polyvinyl butyral (PVB) layers. The as-synthesized multilayered hybrid coating deposited on MS has a compact nacre-like structure which imparts an excellent barrier effect. The active corrosion inhibition was attributed to the stimuli-responsive release of corrosion inhibitors [78].

In another work, Zhao *et al.* [79] created a superhydrophobic surface on a stainless steel substrate for anticorrosion performance, taking inspiration from the "lotus effect". The substrates were first etched with a solution of $FeCl_3$ to create a micron-sized pore structure, and then with a solution of 1M HCl to create a nanosized pit. In order to impart low surface energy, the substrates were treated with a solution of 1H, 1H, 2H, and 2H-perfluorodecyltriethoxysilane (PFDTES). Electrochemical studies revealed that the polarization resistance of the modified substrate is about 28 times higher than that of the bare stainless steel substrate.

Miller *et al.* developed an anticorrosive coating on a copper substrate utilizing the superhydrophobic concept [80]. They could achieve corrosion inhibition efficiency as high as 98.82% by one-step electrodeposition method from the electrolytic solution containing cobalt nitrate, myristic acid, and ethanol. The as-fabricated superhydrophobic surfaces have exceptional antifouling as well as self-cleaning properties.

Taking this point into consideration, Zheng *et al.* have fabricated hydrophobic graphene on Al alloy by the chemical vapor deposition (CVD) method. It was observed that the CVD-grown graphene-coated Al alloy has shown WCA of approximately 130° which helps in achieving hydrophobicity as well as anticorrosion property due to its barrier effect. Therefore, such dual-functional coating can be helpful for engineering materials used in marine applications [81]. Zhang *et al.* have reported the synthesis of superhydrophobic coating on Al alloy, AA5083 by a simple and one-step hydrothermal process. To achieve this, they have used pentadecafluorooctanoic acid to create microstructure roughness on the substrate, thereby increasing water repellent property. Due to the presence of flower-like structures and due to the presence of carbon-fluoride groups, the interfacial contact area decreases the rate of diffusion and, hence, the surface exhibits excellent self-cleaning and anticorrosion properties [82]. Another research group has developed a novel slippery zinc phosphate coating on MS. Initially, they dipped the MS substrates in the zinc phosphate bath followed by surface modification using fluoroalkylsilane (FAS). To this modified MS surface, the Krytox 100 lubricant was infused by using a spin-coater. The coating thus obtained was abbreviated as SZP (slippery zinc phosphate coating). The WCA of the FAS-modified MS was 156.6° illustrating its excellent superhydrophobicity. After infusing the liquid, the WCA decreased to 122.8°. It was observed that even after a decrease in hydrophobicity, the anticorrosion property of the MS was excellent as the corrosion resistance of this coating has increased by five orders when compared to only modified zinc phosphate coating. Therefore, developing such a coating will be helpful in corrosion retarding process [83].

Ji *et al.* have reported the use of epoxy thermoset composite coatings based on micro- and nanoscaled hierarchical structures synthesized using photolithography and colloidal lithography, respectively. Initially, microscaled cylinder-like structures were fabricated on cold-rolled steel (CRS) *via* the photolithographic method followed by the fabrication of small nano-tip–sized structures using the colloidal lithography method. The final coating deposition was carried out by PDMS which serves as a soft negative template. The WCA of the surface-modified epoxy thermoset was 160° suggesting that the surface is superhydrophobic in nature which is much higher than the only epoxy thermoset coating with the structure of natural *Xanthosoma sagittifolium* leaf, which has a WCA of 123°. The anticorrosion studies carried out with and without surface-modified structures on CRS confirm that the surface with micro- and nanoscaled tips shows better corrosion resistance with a corrosion current of 57.8 $\mu A/cm^2$ when compared to CRS without any surface modifications having a corrosion current of 310.5 μA/cm^2 [84].

Ouyang's group introduced a novel coating that has self-healing ability and corrosion resistance. Their work was inspired by natural fruit pomegranate and nepenthes pitcher plants. The walls of pomegranate restrict the rotting of whole fruit because of the presence of a wall-like structure in it. Similarly, they have designed a coating system where they have built a $Ce(OH)_3$ nanowall structure with partitions on carbon steel with an electrodeposition method. This design helps in corrosion inhibition in the horizontal direction. For self-healing purposes, they have used nanocomposite liquid-infused surfaces filled with oleic acid and dodecanethiol in vertical direction, which are inspired from the nepenthes pitcher plant consisting of porous Fe_3O_4 nanoparticles. As coated carbon steel, even after 600-hour immersion in 3.5 wt% NaCl solution, shows a high impedance value of 1.19×10^8 ohm·cm^2 when compared to single-layer diffused coatings on carbon steel. Therefore, this dual biomimetic offers a versatile strategy to inhibit corrosion in both directions [85].

Chen *et al.*, for the first time, have reported the use of underwater superoleophilic and antibiofouling coating on glass. The coating has self-repairing properties because of the presence of hierarchical microgel spheres and due to the 3D hydrophilic polymeric chains. The self-repairing capability of the coating bio-mimics the healing mechanism of whale and shark skin. Due to the presence of micro and nano-surface structures, the coating possesses oil-repellent and biofouling properties in adverse conditions such as highly acidic, basic, or salty environments [86].

Aluminum and its alloys undergo corrosion in marine environments; therefore, recent research carried out by Jiang and his coworkers focuses on the development of anticorrosion surfaces using a liquid-infused method. Here, they have created a porous structure on Al alloy *via* laser ablation method which can store the lubricant oil for self-healing behavior. These structures were then covered by LDH films to preserve the lubricant oil for an extended period of time. Electrochemical studies after immersion in 3.5 wt% NaCl solution reveal that the current density of the sample with these micro- and nano-surfaces was the least with a value of 0.02 μA/cm^2 even after 48 hours of exposure when compared to untreated Al alloy substrate having higher corrosion current of 0.25 μA/cm^2 [87].

9.1.5.2.1 Self-Healing Coatings for Corrosion Protection

Inspired by the self-healing mechanism from nature, Xu *et al.* developed poly(urea-urethane)-graphite carbon nitride nanosheets elastomeric composite coating on Al 2024 alloy surface which exhibited better anticorrosion performance and self-healing function compared to bare substrate [88]. Multiple hydrogen bonds derived from urea, urethane, and amide groups present in the coated material provided autonomous self-healing at room temperature.

While developing a smart anticorrosion coating, health and environmental risks also should be considered. Therefore, eco-friendly coatings have captured the attention of researchers in view of the minimal use of volatile compounds and low pollutant features. In this regard, Deng *et al.* [89] developed graphene/lacquer anticorrosion coating on tin plate and copper substrate. Graphene was modified by an aqueous stabilizer, lignin tripolymer, which was then added to raw lacquer. Fabricated coating can be dried under natural conditions which provides excellent corrosion protection.

Chen's research group has focused on designing self-healing corrosion-resistant coating on steel. Water-born epoxy coating has been used for this purpose with an adhesion promoter, polydopamine (PDA), a glue on mussels. Benzotriazole (BTA) has been used as a corrosion inhibitor-loaded into a halloysite nanotube as the nanocontainer. Loaded HNTs were coated or wrapped with PDA called PBH, which is, in turn, again coupled with graphene oxide sheets. Epoxy coating in combination with loaded GO@PBH acts as a strong barrier providing excellent anticorrosion performance. Corrosion inhibition using BTA as a corrosion inhibitor is achieved when the coating is subjected to an acidic environment which enables its release and helps in forming a complex between catechol of PDA and Fe^{3+} ions. Electrochemical studies reveal that the corrosion current density was least for GO@PBH epoxy coating (5.2×10^{-11} A/cm^2) when compared to pure epoxy coating [90].

Recent research work carried out by Gautam *et al.* has reported the optimization studies based on BTA loading into halloysite nanotubes followed by dispersion into a sol-gel matrix for investigating the corrosion behavior of mild steel. For the controlled release of BTA from HNT, the tube ends were capped using urea-formaldehyde microcapsules (UFMCs). The use of polymeric microcapsules in the coating matrix again increases the coating durability by increasing the thickness of the coating which increases the diffusion path of the corrosive media, which is a 3.5 wt% NaCl solution. Electrochemical studies confirm that the coating with capped inhibitor-loaded HNTs possesses higher corrosion resistance of 10^8 Ω·cm^2 which has decreased by two orders, i.e., 10^6 Ω·cm^2 for uncapped inhibitor-loaded HNTs coating on MS [91].

Manasa *et al.* have reported the use of HNT loaded with cationic corrosion inhibitors, i.e., Ce^{3+}/Zr^{4+}, and mixed into sol-gel silica matrix for coatings on aluminum alloy A356.0. Coatings generated were then characterized for their self-healing action using electrochemical studies. On exposure to corrosive media such as 3.5 wt% NaCl solution, the inhibitors release into the scribe area thereby self-healing the damaged part. Electrochemical studies confirm that the coatings with inhibitor-loaded HNT impart better corrosion resistance and self-healing properties when compared to uncoated substrates [92]. Similar work was carried out in the same research lab by Adsul *et al.* using magnesium alloy as the metal substrate. The self-healing property and corrosion resistance behavior were investigated using weight loss experiments, potentiodynamic polarization, and electrochemical impedance spectroscopy. The scanning vibrating electrode technique (SVET) confirmed the self-healing mechanism of AZ91D substrates by the release of cationic corrosion inhibitors into the scribe area [93].

Table 9.2 summarizes the anticorrosion coatings generated on a variety of substrates by different coating techniques and reported in the literature.

9.1.6 About Antifouling Paint Particles (APP)

Antifouling paints are applied as protective coatings on ship and boat hulls to avoid fouling settlement by marine organisms and to prevent various navigational problems. After a ship is launched, biocides in the paints slowly leach out as seawater penetrates the surface of the paints. Due to the continuous leaching of biocides, the biocide in the antifouling paint becomes depleted or exhausted, initiating the growth of the foulants. Periodic maintenance or cleaning of the hull surface is required, depending on the conditions of the antifouling paints on the ship hull. High-pressure water blasting generates wastewater containing paint residues that are contaminated with biocides and metals. The contaminated wastewater can be introduced into the surrounding marine environment without any treatment and the impact of wastewater release on marine environments can be detrimental. Both dry dock abrasive water cleaning and the in-water cleaning process may pose adverse effects on the marine environment when the waste is not entirely captured or treated appropriately. Few studies have assessed the environmental risk associated with the release of wastewater containing these paint substances called antifouling paint particles (APP) [94–99]. Hence, even though an environmentally friendly biocide can be incorporated into the paint, if the paints used are not environmentally friendly, the impact caused

Table 9.2 Summary of substrates, coatings with different functionality, and coating techniques used by various researchers.

Sr. no.	Substrate	Strategy/functionality	Remarks	References
1	Copper	Superhydrophobicity	Electrodeposition method with myristic acid and $CaCl_2$ solution in ethanol	[75]
2	Carbon steel	Superhydrophobicity	Ce^{3+}ions intercalated in bentonite nanoclay–based ZrO_2-SiO_2 sol-gel coating	[76]
3	Copper	Superhydrophobicity	Lamellar hydrophobic surfaces	[77]
4	Mild steel	Superhydrophobicity	ATES-treated LDH loaded with sodium molybdate followed by PVB deposition	[78]
5	Stainless steel	Superhydrophobicity	Etching with $FeCl_3$ and 1M HCl, PFDTES for low surface energy	[79]
6	Copper	Superhydrophobicity	Electrodeposition with cobalt nitrate, myristic acid, and ethanol	[80]
7	Al alloy	Superhydrophobicity	Graphene deposition using the CVD method	[81]
8	Al alloy	Superhydrophobicity	Hydrothermal method using pentadecafluorooctanoic acid	[82]
9	Mild steel	SLIPs	Zinc phosphating along with FAS	[83]
10	Cold-rolled steel	Superhydrophobicity	Epoxy thermoset coating along with photolithography and colloidal lithography	[84]

(Continued)

Table 9.2 Summary of substrates, coatings with different functionality, and coating techniques used by various researchers. (*Continued*)

Sr. no.	Substrate	Strategy/functionality	Remarks	References
11	Carbon steel	SLIPs and self-healing	$Ce(OH)_3$ nanowall structure, oleic acid, and dodecanethiol	[85]
12	Glass	Oleophilic and antifouling	Self-repairing microgel spheres and polymeric chains	[86]
13	Al alloy	SLIPs	Laser ablation, layered double hydroxide films to preserve lubricant oil	[87]
14	Al alloy	Self-healing	Poly(urea-urethane)-graphite carbon nitride coating	[88]
15	Copper and tin plate	Self-healing	Lignin tripolymer–modified graphene/lacquer coating	[89]
16	Steel	Self-healing	Calcium alginate nanocontainers loaded with Tung oil in epoxy coating	[27]
17	Mild steel	Self-healing	Nacre-mimicking graphene oxide deposited *via* mussel-inspired dopamine method	[26]
18	Steel	Self-healing	BTA-loaded HNT wrapped with polydopamine and coupled with GO sheets using epoxy coating	[90]
19	Mild steel	Self-healing	Capped BTA-loaded HNT-based sol-gel coatings	[91]
20	Al alloy	Self-healing	Ce^{3+}/Zr^{4}-loaded HNT-based sol-gel coatings	[92]
21	Mg alloy	Self-healing	Ce^{3+}/Zr^{4+}-loaded HNT-based sol-gel coatings	[93]

due to the antifouling paint particles will be very severe. This indicates the immediate need to focus more on developing an environment-friendly paint matrix into which an environment-friendly biocide can be introduced for generating an antifouling surface. To achieve long-lasting antifouling properties of amphoteric materials, self-healing and self-renewal of amphoteric surfaces are a promising antifouling strategy.

9.2 Conclusions

Bioinspired multifunctional coatings that mimic complex geometrics of shark skins and lotus leaves, inhibiting the biofouling, are well studied and practically implemented to impart antifouling properties to the coated surface. These coatings first act as simple barriers for corrosive species and protect the coated surface from corrosion. Secondly, they naturally develop hydrophobic properties due to the nano scale geometry, hierarchical surface roughness, and surface modification, which enable them to protect the surface by repelling water bodies upon contact.

Acknowledgement

The authors gratefully acknowledge Director, ARCI, Hyderabad for the constant encouragement provided throughout the course of writing this chapter.

References

1. Qureshi, A. and Ramaswamy, R., Evolution of Antifouling Paints. *Int. J. Sci. Res.*, 6, 2319, 2017.
2. Detty, M., Ciriminna, R., Bright, F., Pagliaro, M., Environmentally benign sol-gel antifouling, and foul-releasing coatings. *Acc. Chem. Res.*, 2, 678, 2014.
3. Kirschner, M. and Brennan, A., Bio-inspired antifouling strategies. *Annu. Rev. Mater. Res.*, 42, 211, 2012.
4. https://www.alliedmarketresearch.com/global-antifouling-paints-and-coatings-markets-A01264 published in Jul 2022 by Markets and Markets Research Pvt. Ltd.
5. Li, X., Zhang, D., Liu, Z., Du, C., Dong, C., Material Science: Share corrosion data. *Nature*, 527, 441, 2015.

6. Karteek, B., Ramyakrishna, P., Ravi, A., Rajender, B., Sashivinay Kumar, G., The role of anticorrosive polymer coatings for the protection of metallic surface. *Corros. Rev.*, 39, 547, 2021.
7. Zheludkevich, M., Miranda, I., Ferreira, M., Sol-gel coatings for corrosion protection of metals. *J. Mater. Chem.*, 15, 5099, 2005.
8. Revamma, B., Vijayan, A., Basheer, R., Arunima, S., Sha, M. *et al.*, Zn Wetted CeO_2 based composite galvanization: An effective route to combat biofouling. *ACS Appl. Bio. Mater.*, 2, 3774, 2019.
9. Zhang, L., Tan, Z., Zhang, C., Tang, J., Yao, C., You, X., Hao, B., Research on metal corrosion resistant bio inspired special wetting surface based on laser texturing technology: A review. *Micromachines*, 13, 1431, 2022.
10. Sullivan, T. and Callaghan, I., Recent developments in biomimetic antifouling materials: a review. *Biomimetics*, 5, 58, 2020.
11. George, J., Vijayan, P., Hoang, A., Kalarikkal, N., Nguyentri, P., Thomas, S., Recent advances in bio-inspired multifunctional coatings for corrosion protection. *Prog. Org. Coat.*, 168, 106858, 2022.
12. Vijayan, P. and Pugila, D., Biomimetic multifunctional materials: a review. *Emerging. Mater.*, 2, 391–415, 2019.
13. Rudnik, E. and Chat, K., A brief review on bioinspired superhydrophobic electrodeposited nickel coatings. *Trans. Inst. Met*, 96, 185, 2018.
14. Wang, J., Shao, R., Zhang, Y., Guo, L., Jiang, H., Biomimetic graphene surfaces with superhydrophobicity and iridescence. *Chem. Asian J.*, 7, 301, 2012.
15. Malshe, A., Bapat, S., Rajurkar, K., Haitjema, H., Bio-inspired textures for functional applications. *CIRP Ann. Manuf.*, 67, 627, 2018.
16. Scurria, A., Scolaro, C., Sfameni, S., Carlo, G., Pagliaro, M., Visco, A., Ciriminna, R., Towards aquasun practical utilization: Strong adhesion and lack of ecotoxicity of solar-driven antifouling sol-gel coating. *Prog. Org. Coat.*, 165, 106771, 2022.
17. Zheng, Q. and Lu, C., Size Effects of Surface Roughness to Superhydrophobicity. *Proc. IUTAM.*, 10, 462, 2014.
18. Feat, A., Federle, W., Kamperman, M., Gucht, J., Coatings preventing insect adhesion: An overview. *Prog. Org. Coat.*, 134, 349, 2019.
19. Lan, P., Nunez, E., Polycarpou, A., Advanced polymeric coating and their application: Green tribology, in: *Encyclopedia of Renewable and Sustainable Materials*, p. 345, Elsevier, 2020.
20. Yang, Z., Tian, Y., Zhao, Y., Yang, C., Study on the fabrication of super-hydrophobic surface on inconel alloy via nanosecond laser ablation. *Materials*, 12, 278, 2019.
21. Yan, W., Xue, S., Xiang, B., Xhao, X., Zhang, W., Mu, P., Li, J., Resent advances of slippery liquid infused porous surfaces with anti-corrosion. *Chem. Commun.*, 59, 2182, 2023.
22. Wong, T., Kang, S., Tang, S., Smythe, E., Hatton, B., Grinthal, A., Aizenberg, J., Bioinspired self-reparing slippery surfaces with pressure-stable omniphobicity. *Nature*, 477, 443, 2011.

23. Yang, H., Qin, L., Wang, F., Mawignon, F., Dong, M., Wu, Y., Zhang, Y., Ma, Z., A facile method to fabricate the durable and self-protective coating for marine applications. *Surf. Coat. Technol.*, 452, 129124, 2023.
24. White, S.R., Sottos, N.R., Geubelle, P.H., Moore, S.R., Kessler, M.R., Sriram, S.R., Brown, E.N., Viswanathan, S., Autonomic healing of polymer composites. *Nature.*, 409, 794, 2001.
25. Ding, J., Zhao, H., Shao, Z., Yu, H., Bioinspired Smart Anticorrosive Coatings with an Emergency-Response Closing Function. *ACS Appl. Mater. Interfaces.*, 11, 42646, 2019.
26. Zhu, X., Zhao, H., Wang, L., Xue, Q., Bioinspired ultrathin graphene nanosheets sandwiched between epoxy layers for high performance of anticorrosion coatings. *J. Chem. Eng.*, 410, 128301, 2021.
27. Tian, H., Zhan, Y., Tian, L., Sun, J., Corrosion resistance self-healing coating with bioinspired interfacial structure. *Prog. Org. Coat.*, 174, 107303, 2023.
28. Richards, C., Slaimi, A., Connor, N., Barrett, A., Kwiatkowska, S., Regan, F., Bio-inspired surface texture modification as a viable feature of future aquatic antifouling strategies: A Review. *Int. J. Mol. Sci.*, 21, 5063, 2020.
29. Ermis, M., Antmen, E., Hasirci, V., Micro and nanofabrication methods to control cell-substrate interactions and cell behavior: A review from the tissue engineering perspective. *Bioact. Mater*, 3, 355, 2018.
30. Cirelli, R.A., Watson, G.P., Nalamasu, O., *Encyclopedia of Materials: Science and Technology*, 2nd ed, p. 6441, Elsevier Ltd, Amsterdam, The Netherlands, 2001.
31. Murad, R. and Xichun, H., Promising lithography techniques for next-generation logic devices. *Nanomanuf. Metrol.*, 1, 67, 2018.
32. Watt, F., Bettiol, A.A., Van Kan, J.A., Teo, E.J., Breese, M.B.H., Ion beam lithography and nanofabrication: A review. *Int. J. Nanosci.*, 4, 269, 2005.
33. Tocce, E.J., Liliensiek, S.J., Wilson, M.J., Yanez-Soto, B., Nealey, P.F., Murphy, C.J., *Comprehensive Biomaterials*, p. 527, Elsevier Ltd, Amsterdam, The Netherlands, 2011.
34. Charlton, M.D.B., Photonic Crystal Nitride LEDs, in: *Nitride Semiconductor Light-Emitting Diodes (LEDs)*, 2nd ed, p. 327, Elsevier Ltd, Amsterdam, The Netherlands, 2018.
35. Baglin, J.E.E., Ion beam nanoscale fabrication and lithography-A review. *Appl. Surf. Sci.*, 258, 4103, 2012.
36. Martinex-Chapa, S.O., Salazar, A., Madou, M.J., Two-Photon Polymerization as a Component of Desktop Integrated Manufacturing Platforms, in: *Three-Dimensional Microfabrication Using Two-Photon Polymerization*, p. 374, Elsevier Ltd, Amsterdam, The Netherlands, 2016.
37. Dharmalingam, S. and Bhaskaran, B., Need & Overview of electrochemical micro machining. *Int. J. ChemTech, Res.*, 10, 35, 2017.
38. Bhattacharyya, B.Ã., Munda, J., Malapati, M., Advancement in electrochemical micro-machining. *Int. J. Mach.Tools Manuf.*, 44, 1577, 2004.

39. Hao, X., Wang, L., Wang, Q., Guo, F., Tang, Y., Ding, Y., Lu, B., Surface micro-texturing of metallic cylindrical surface with proximity rolling-exposure lithography and electrochemical micromachining. *Appl. Surf. Sci.*, 257, 8906, 2011.
40. Steck, J.G., Afshar-Mohajer, M., Sun, Q., Meng, X., Zou, M., Fabrication and tribological characterization of deformation-resistant nano-textured surfaces produced by two-photon lithography and atomic layer deposition. *Tribol. Int.*, 132, 75, 2019.
41. Rogers, J.A. and Nuzzo, R.G., Recent progress in soft lithography. *Mater. Today*, 8, 50, 2005.
42. Qin, D., Xia, Y., Whitesides, G.M., Soft lithography for micro- and nanoscale patterning. *Nat. Protoc.*, 5, 491, 2010.
43. Zhang, F., Fu, Y., Yu, X.-Y., *Physical Chemistry of Gas-Liquid Interfaces*, p. 245, Elsevier Ltd, Amsterdam, The Netherlands, 2018.
44. Fink, J.K., *Reactive Polymers Fundamentals and Applications; A Concise Guide to Industrial Polymers*, p. 49, Elsevier Ltd, Amsterdam, The Netherlands, 2013.
45. Harvey, E. and Ghantasala, M., *Nanostructure Control of Materials*, p. 303, Elsevier Ltd, Amsterdam, The Netherlands, 2006.
46. Chen, Y., Applications of nanoimprint lithography/hot embossing: A review. *Appl. Phys. A Mater. Sci. Process*, 121, 451, 2015.
47. Peng, L., Deng, Y., Yi, P., Lai, X., Micro hot embossing of thermoplastic polymers: A review. *J. Micromech. Microeng.*, 24, 13001, 2014.
48. Sequeiros, E.W., Emadinia, O., Vieira, M.T., Vieira, M.F., Development of metal powder hot embossing: A new method for micromanufacturing. *Metals*, 10, 388, 2020.
49. Lee, J., An, J., Chua, C.K., Fundamentals and applications of 3D Printing for novel materials. *Appl. Mater. Today*, 7, 120, 2017.
50. Richards, C., Barrett, A., Maguire, I., Kwiatkowska, S., Regan, F., Marine Inspired textured materials for reduction of biofouling on surfaces, in: *Proceedings of the IEEE Conference Proceedings, OCEANS 2019*, Marseille, France, 17–20 June 2019, p. 1.
51. Xian, J., Wang, X., Fu, X., Zhang, Z., Liu, L., Kang, M.A., Simple model to predict machined depth and surface profile for picosecond laser surface texturing. *Appl. Sci.*, 8, 2111, 2018.
52. Whitehead, K.A., Colligon, J., Verran, J., Retention of microbial cells in substratum surface features of micrometer and sub-micrometer dimensions. *Colloids Surf. B.*, 41, 129, 2005.
53. Garcia, A., Casas, A., Morales, A., The effect of the organosilane content on the barrier features of sol-gel anticorrosive coatings applied on carbon steel. *Prog. Org. Coat.*, 139, 105418, 2020.
54. Wanka, R., Koc, J., Clarke, J., Hunsucker, K., Swain, G., Aldred, N., Finlay, J., Clare, A., Rosenhahn, A., Sol–gel-based hybrid materials as antifouling and

fouling-release coatings for marine applications. *ACS Appl. Mater. Interfaces*, 12, 53286, 2020.

55. Vinod, B., Damodaran, N., Murthy, S., Bio-inspired strategies for designing antifouling biomaterials. *Biomater. Res.*, 20, 18, 2016.
56. Han, X., Wu, J., Zhang, X., Shi, J., Wei, J., Yang, Y., Feng, Y., The progress on antifouling organic coating: From biocide to biomimetic surface. *J. Mater. Sci. Technol.*, 61, 46, 2021.
57. Sepideh, P., Abdolvahab, S., Farhad, S., Elham, G., Xiaohong, J., Ziyang, Z., Xiaofan, Z., Majid, M., Jizhou, D., Alimorad, R., Baorong, H., Antifouling nanocomposite polymer coatings for marine applications: A review on experiments, mechanisms, and theoretical studies. *J. Mater. Sci. Technol.*, 118, 73, 2022.
58. Duan, Y., Wu, J., Qi, W., Su, R., Eco-friendly marine antifouling coating consisting of cellulose nanocrystals with bioinspired micromorphology. *Carbohydr. Polym.*, 304, 120504, 2023.
59. Dogra, V., Kishore, C., Mishra, A., Gaur, A., Verma, A., Sol-Gel preparation and wetting behaviour analysis of hydrophobic Zirconium based nano-coating: Implications for solar panel coating. *J. Chem. Eng.*, 15, 100507, 2023.
60. Verma, J., Gupta, A., Kumar, D., Steel protection by SiO_2/TiO_2 core-shell based hybrid nanocoating. *Prog. Org. Coat.*, 163, 106661, 2022.
61. Calabrese, C., Parola, V., Testa, M., Liotta, L., Antifouling and antimicrobial activity of Ag, Cu and Fe nanoparticles supported on silica and titania. *Inorganica Chim. Acta*, 529, 120636, 2022.
62. Subodh, B., Sumit, B., Bhim, S., Nakshatra, B.S., Multifunctional and fluorine-free superhydrophobic composite coating based on PDMS modified MWCNTs/ZnO with self-cleaning, oil-water separation, and flame retardant properties. *Colloids Surf. A*, 597, 124776, 2020.
63. Tang, Y., Finlay, J., Kowalke, G., Meyer, A., Bright, F., Callow, M., Callow, J., Dean Wendt, D., Detty, R., Hybrid xerogel films as novel coatings for antifouling and fouling release. *Biofouling*, 21, 59, 2005.
64. Miller, R., Adeleye, A., Page, H., Kui, L., Lenihan, H., Keller, A., Nano and traditional copper and zinc antifouling coatings: metal release and impact on marine sessile invertebrate communities. *J. Nanopart Res.*, 22, 129, 2020.
65. Jin, H., Tian, L., Binga, W., Zhao, J., Ren, L., Bioinspired marine antifouling coatings: Status, prospects, and future. *Prog. Mater. Sci.*, 124, 100889, 2022.
66. Vreugdenhil, A., Gelling, V., Woods, M., Schmelz, J., Enderson, B., The role of crosslinkers in epoxy–amine crosslinked silicon sol–gel barrier protection coatings. *Thin Solid Films*, 517, 538, 2008.
67. Lollivier, G., Gressier, M., Ansart, F., Aufray, M., Menu, M., Influence of hybrid sol-gel crosslinker on self-healing properties for multifunctional coatings. *Materials*, 14, 5382, 2021.
68. Mohan, M., Bhanu Prakash, T., Mukherjee, A., Antifouling paint formulations with ZnO and Fe_2O_3 nano-paints for marine applications. *Sustain. Chem. Pharm.*, 30, 100858, 2022.

69. Peng, H., Qingyi, X., Chunfeng, M., Guangzhao, Z., Silicone-Based Fouling-Release Coatings for Marine Antifouling. *Langmuir*, 36, 2170, 2020.
70. Yue, Z., Teng, X., Ji-Ming, H., A robust, room-temperature curable and molecular-level superhydrophobic coating with excellent antibacterial and antifouling properties. *Chem. Eng. J.*, 450, 136557, 2022.
71. Wenhuan, W., Jiawei, L., Peng, W., Junfei, O., Dun, Z., Fabrication of polydimethylsiloxane-attached solid slippery surface with high underwater transparency towards the antifouling of optical window for marine instruments. *J. Colloid Interface Sci.*, 623, 832, 2022.
72. Madasamy, M., Munusamy, T., Wen-Yi, Y., Jiunn-Fwu, L., Gurusamy, A., A versatile chitosan/ZnO nanocomposite with enhanced antimicrobial properties. *Int. J. Biol. Macromol.*, 80, 121, 2015.
73. Mohamed, S., Nesreen, A., Shimaa, A.H., Zhifeng, H., Jing, M., A comparative study between two novel silicone/graphene-based nanostructured surfaces for maritime antifouling. *J.Colloid Interface Sci.*, 606, 367, 2022.
74. Ting, S., Hongning, W., Qian, J., Ruoyu, C., Preparation of a transparent coating with superamphiphobic and antifouling properties. *Mater. Chem. Phys.*, 293, 126888, 2023.
75. Li, B., Miller, R., Zhang , H., Ouyang , Y., Qiu , R., Hu , S., Niu, H., A facile one-step route to fabricate bio-inspired superhydrophobic matrix with high corrosion inhibition to Cu metal. *Surf. Interfaces*, 25, 101189, 2021.
76. Zhang, N., Zhou, B., Liu, Y., Yang, B., Zheng, H., Wu, Y., Dual-functional anti-corrosion coatings with surface hydrophobicity and internal smart releasing Ce^{3+} loaded in bentonite on carbon steel. *J. Mater. Res. Technol.*, 22, 879, 2023.
77. England, M.W., Urata, C., Dunderdale, G.J., Yagihashi, M., Hozumi, A., Bio-Inspired Layered Hybrid Films Showing Long-Lasting Corrosion Resistance and Repeatable Regeneration of Surface Hydrophobicity. *J. Nanosci. Nanotechnol.*, 16, 9166, 2016.
78. Liu, A., Tian, H., Li, S., Ju, X., Yang, H., Sun, Y., Wang, L., Li, W., Bioinspired layered hybrid coatings with greatly enhanced barrier effect and active corrosion protection performance. *Prog. Org. Coat.*, 152, 106131, 2021.
79. Zhao, E., Li, Y., Gao, L., Yang, S., Ma, T., Anti-corrosion Properties of a Bioinspired superhydrophobic surface on Stainless Steel. *Int. J. Electrochem. Sci.*, 12, 9855, 2017.
80. Miller, H., Wei, Y., Ma, C., Li, L., Shao, J., Hu, S., Weamie, S., Bioinspired super-hydrophobic fractal array via a facile electrochemical route: preparation and corrosion inhibition for Cu. *RSC Adv.*, 12, 265, 2022.
81. Zheng, Z., Liu, Y., Bai, Y., Zhang, J., Han, Z., Ren, L., Fabrication of biomimetic hydrophobic patterned graphene surface with ecofriendly anti-corrosion properties for Al alloy. *Colloids Surf.*, 500, 64–71, 2016.
82. Zhang, B., Wang, J., Zhang, J., Bioinspired one step hydrothermal fabricated superhydrophobic aluminum alloy with favorable corrosion resistance. *Colloids Surf.*, 589, 124469, 2020.

83. Xiang, T., Zheng, S., Zhang, M., Sadig, H.R., Li, C., Bioinspired Slippery Zinc Phosphate Coating for Sustainable Corrosion Protection. *ACS Sustain. Chem. Eng.*, 6, 10960, 2018.
84. Ji, W.F., Ahmed, M.M., Luo, K.H., Chen, K.Y., Lee, Y.C., Yeh, J.M., Epoxy thermoset coatings with fine controllable hierarchical structures prepared from bio-inspired photo-/colloidal lithography technique for anticorrosion application. *Prog. Org. Coat.*, 152, 106132, 2021.
85. Ouyan, Y., Zhao, J., Qiu, R., Hu, S., Niu, H., Zhang, Y., Chen, M., Biomimetic partition structure infused by nano-compositing liquid to form bio-inspired self-healing surface for corrosion inhibition. *Colloids Surf. A*, 596, 124730, 2020.
86. Chen, K., Zhou, S., Wu, L., Self-Healing Underwater Superoleophobic and Antibiofouling Coatings Based on the Assembly of Hierarchical Microgel Spheres. *ACS Nano*, 10, 1386, 2016.
87. Jiang, S., Zhang, H., Song, K., Liu, X., Protection application of liquid-infused surface with self-healing via regional growth of layered double hydroxide films on aluminum alloy. *Colloids Surf., A*, 612, 125996, 2021.
88. Xu, J., Ye, S., Ding, C., Ling, T., Fu, J., Autonomous self-healing supramolecular elastomer reinforced and toughened by graphitic carbon nitride nanosheets tailored for smart anticorrosion coating applications. *J. Mater. Chem. A*, 6, 5887, 2018.
89. Deng, Y., Bai, W., Chen, J., Zhang, X., Wang, S., Lin, J., Xu, Y., Bio-inspired electrochemical corrosion coatings derived from graphene/natural lacquer composites. *RSC Adv.*, 7, 45034, 2017.
90. Chen, C., Xiao, G., He, Y., Zhong, F., Li, H., Wu, Y., Chen, J., Bio-inspired superior barrier self-healing coating: Self-assemble of graphene oxide and polydopamine-coated halloysite nanotubes for enhancing corrosion resistance of waterborne epoxy coating. *Prog. Org. Coat*, 139, 105402, 2020.
91. Gautam, A., Siva, T., Sathiyanarayanan, S., Gobi, K.V., Subasri, R., Capped inhibitor-loaded halloysite nanoclay-based self-healing silica coatings for corrosion protection of mild steel. *Ceram. Int.*, 48, 30151, 2022.
92. Manasa, S., Jyothirmayi, A., Siva, T., Sathiyanarayanan, S., Gobi, K.V., Subasri, R., Effect of inhibitor loading into nanocontainer additives of self-healing corrosion protection coatings on aluminum alloy A356.0. *J. Alloys Compd.*, 726, 969, 2017.
93. Adsul, S.H., Siva, T., Sathiyanarayanan, S., Sonawane, S.H., Subasri, R., Self-healing ability of nanoclay-based hybrid sol-gel coatings on magnesium alloy AZ91D. *Surf. Coat. Technol.*, 309, 609, 2017.
94. Sanye, S., Fiamma, A., Ítalo, B., Fabio, A., Lopes L., G., Are antifouling paint particles a continuous source of toxic chemicals to the marine environment? *J. Hazard. Mater.*, 330, 76, 2017.
95. Mahima, J. and Naveen, N., Boat paint and epoxy fragments - Leading contributors of microplastic pollution in surface waters of a protected Andaman bay. *Chemosphere*, 312, 137183, 2023.

96. Russell, G., Victor, H., Jaime, R., Ítalo, B., Maíra, C., Gilberto, F., Mexican paradise under threat: The impact of antifouling biocides along the Yucatán Peninsula. *J. Hazard. Mater*, 427, 128162, 2022.
97. Zhi, Y., Jee-Hyun, J., Andrew, L., Cheolho, Y., Dongju, S., Moonkoo, K., Seawater contamination associated with in-water cleaning of ship hulls and the potential risk to the marine environment. *Mar. Pollut. Bull.*, 171, 112694, 2021.
98. Oh, Y., Jung-Hoon, K., Sang, H., Won, J., Spatial distribution of microplastic in the surface waters along the coast of Korea. *Mar. Pollut. Bull.*, 155, 110729, 2020.
99. Zhenchun, L., Peng, L., Shaowei, C., Xiaoting, L., Yunwu, Y., Tianwei, L., Ye, W., Ning, T., Yunxue, L., Yaxin, G., Bioinspired marine antifouling coatings: Antifouling mechanisms, design strategies and application feasibility studies. *Eur. Polym. J.*, 190, 111997, 2023.

10

Implementation of Nanotechnology in Anticorrosion Material Development for Food Packaging

Aparna Ray Sarkar*, **Dwaipayan Sen and Pramita Sen**

Department of Chemical Engineering, Heritage Institute of Technology, Kolkata, West Bengal, India

Abstract

One of the primary concerns in any food packaging industry is the development of coating for food-bearing containers and packages to ensure freshness by restricting the staleness of food because of its contamination with atmospheric bacterial agents. There are three important parameters that primarily affect corrosion resistance—barrier properties (WVTR and OTR), adhesion, and electrochemical impedance. Nanocomposite (NP)-based anticorrosive material imparts antibiotic action to protect and preserve canned food as well as packaged food material. Several nanoparticles like zinc oxide nanocomposites (Zn NPs), chitosan, titanium dioxide nanocomposites (TiO_2 NPs), silver oxide nanocomposites (AgO NPs), nanoclay particles, nanocellolulose, and carbon nanotubes (CNTs) are applied as anticorrosive coating materials along with antibiotic properties. Sol-gel techniques are applied to synthesize the NPs required for food packages. Primarily, casting is a widely accepted technology for nanofilm production. Several properties such as deacetylation, solubility, self-healing, and quantitative toxicity play important roles in modulating NP activities. A sulfate-reducing bacterium, through genetically expressed proteins, forms biofilms on metal surfaces under various environments that later induce biocorrosion of the metal surfaces. ZnO NPs and Ag NPs have antimicrobial properties that could lead to protection against contamination, while ZnO NPs will enhance the preventive measures for both water vapor permeation and gas permeation.

Keywords: Packaging, nanoclay, biopolymer, CNT, chitosan, barrier properties, nanocomposite

**Corresponding author*: aparnaray.sarkar@heritageit.edu

Hari Murthy, Vinay Jha Pillai, Kukatlapalli Pradeep Kumar and Matthew Cowan (eds.) Novel Anti-Corrosion and Anti-Fouling Coatings and Thin Films, (287–322) © 2024 Scrivener Publishing LLC

10.1 Introduction

One of the biggest concerns among the food packaging industries is the preservation of food for a longer tenure in order to ensure the food safety within the consumers. Another prime concern in food packaging industries is the development of toxicity due to food contact with the coating. Therefore, coating material must possess atmospheric-resistant and water vapor–resistant properties along with no toxicity. Several associations like European Food Safety Authority (EFSA), Food Safety and Standard Authority of India (FSSAI) and Food and Drug Administration (FDA) have framed regulations to establish standards for the food packaging. These are some of the agencies that are thoroughly monitoring food safety and initiate the approval process for any material related to the food. Thus, the approval for the coating material for canned food packages is also one of the foremost criteria in order to implement it commercially.

A brief of such nanoparticles used as a coating in the food packaging industry is mentioned in Table 10.1, while a detailed discussion is given further in subsequent sections.

10.2 Synthesis of Nanoclay-Based Composites for Food Packaging

Nanoclay are optimized clay minerals comprising 1 nm thick alumina silicate layer surface and stacked with 10 nm sized multilayer stacks. This endows them with excellent aspect ratio and specific surface area as high as 657 m^2/g and remarkable exfoliation characteristics. Nanoclays, which are primarily nanoparticles of mineral silicate with layered structures, are mostly used for food packaging applications including montmorillonite (MMT) and halloysites.

10.2.1 Halloysite-Based Nanocomposites

Active food packaging materials are synthesized using halloysite nanotubes (HNTs). The unique tubular morphology, high aspect ratio, low cost, and environment friendliness make them ideal candidates for nanofillers in polymeric films. These nanofillers improve the thermal and mechanical properties, along with gas barriers of the films used in packaging for food [14]. In one of the studies, polylactic acid (PLA)/HNT-based bionanocomposite films were prepared by dissolving PLA in chloroform followed by stirring and sonication. This was followed by casting a definite

Table 10.1 Overview of function and application of nanomaterials used in food packaging.

Nanomaterial	Polymer matrix	Function	Application	References
Chitosan nanoparticles (CNPs)	Hydroxypropyl methylcellulose (HPMC) film with chitosan nanoparticles (CNPs)	Antimicrobial agent	Incorporated in the polylactic acid film to improve barrier properties. Furthermore, there will be an improvement in the antimicrobial properties because of the chelation effect and ionic interaction	[1–4]
Nanocellulose	(i) PVA/chitosan film with cellulose nanocrystal (CNC) (ii) Starch PVA film with nanocrystal (iii) Poly(3-hydroxybutyrate)/poly(caprolactone) blend film with cellulose nanocrystals (CNC)	Antifungal and antibacterial agents Gas/moisture barrier enhancer	Used in paper and composites industries because of its strength along with its biodegradability	[1, 5–7]

(Continued)

Table 10.1 Overview of function and application of nanomaterials used in food packaging. (*Continued*)

Nanomaterial	Polymer matrix	Function	Application	References
Silver nanoparticles (AgNPs)	(i) Starch PVA film containing AgNPs (ii) AgNPs encapsulated within polyacrylic acid conjugated with the cellulose/chitosan composite films (iii) AgNPs with PP	Antimicrobial agent	Ag-loaded starch PVA film–coated container keeps fruit, vegetable, and food fresh with three to four times longer shelf life	[1, 8]
Zinc oxide nanocomposites (ZnONPs)	(i) Polyurethane and chitosan with ZnO-NPs (ii) Starch PVA composite film with ZnO-NPs	Antimicrobial agent, barrier properties, and shelf-life enhancer	Integrated in food grip wrap film	[1, 9]
Titanium dioxide nanomaterials (TiO_2 NPs)	(i) TiO2-PVA/xylan composites (ii) Polylactic acid with TiO2 and AgNPs film	UV absorber Oxygen forger	Incorporated into plastic film for UV shielding performance of food package Blended with PVC in modified active packaging due to its low water vapor permeability	[1, 10, 11]

(*Continued*)

Table 10.1 Overview of function and application of nanomaterials used in food packaging. (*Continued*)

Nanomaterial	Polymer matrix	Function	Application	References
Nanoclay	(i) Polylactic acid with nanoclay (ii) Transglutaminase and montmorillonite nanoclay incorporated fish protein film	Water vapor transmission (WVTR) reducer, properties enhancer, and anticorrosive	Combined with resin and PET film enhanced mechanical properties and biodegradability Water vapor permeation (WVP), solubility, and tensile strength (TS) will enhance when combined with biodegradable polymers	[1, 10, 12]
Carbon nanotubes (CNTs)	(i) PVA film modified with ZnO-doped multi-walled carbon nanotubes (CNTs) (ii) Polylactic acid and poly(caprolactone) CNTs	Antifungal agent, filler, and barrier enhancer	Incorporated with PVA film to enhance mechanical, thermal, and barrier properties Enhance UV resistance, decrease glass transition temperature, and inhibit fungal growth	[13]

amount of the solution and drying for 72 hours after which the nanocomposite was ready for use [15]. HNT can be further used as a carrier for active antimicrobial agents and antioxidants [16]. Reduction-stabilization strategy with tannic acid (TA) was implemented for the *in situ* synthesis of the AgNPs template on HNT. Functionalization of HNT was carried out using (3-aminopropyl)triethoxysilane (APTES) followed by dispersion in a solvent by ultrasound. TA was added to the HNT dispersion, stirred, and centrifuged for carrying out TA adsorption on the surface of HNTs. A solution of silver nitrate was added next to the halloysite dispersion followed by centrifugation, washing, and drying at 60°C [17]. HNTs have been used as ethylene scavengers [16].

Nanocomposites of HNTs and polyethylene with ethylene scavenging and barrier properties have been explored by Tas and coworkers [18]. The nanocomposites were prepared by a process of extrusion at 650 rpm and 165°C to 185°C in a twin screw extruder containing mixtures of low-density polyethylene (LDPE) and HNT granules. Following this operation, nanocomposites were cooled, pelletized, and blown into 55- to 60-μm films in a single screw extruder at 150°C to 160°C at a screw speed of 80 rpm. Antimicrobial nanocomposite development using HNTs and essential oils is shown in Figure 10.1a. The combination of HNTs and biopolymers in forming antimicrobial nanocomposites is presented in Figure 10.1b.

10.2.2 Montmorillonite-Based Composites

One of the primary advantages of MMT-based nanoclays in food packaging is the miscibility of the hydrophilic MMT with hydrophilic polymers. This provides good thermomechanical properties and oxygen along with water vapor barrier properties. Furthermore, modification of MMT with either Ag^{+} or Cu^{2+} nanoparticles used to be done to prepare antibacterial nanofillers. Moreover, the modification of the large specific area for modified MMT with antioxidants and antimicrobial agents is primarily done toward preparing nanocarriers for smart packaging applications. One essential feature of layered clay mineral is the ability to carry out pillaring of metal oxides within the space between the layers of MMT producing nanosensors for food packaging applications. Organically modified MMT (OMMT) can be mixed with hydrophobic polymers improving their gas barrier properties [20]. In one of the studies [21], ethylene-co-vinyl acetate (EVA) was mixed with MMT in a co-rotating twin extruder using a temperature profile of 150°C to 190°C and 50-rpm screw speed. The

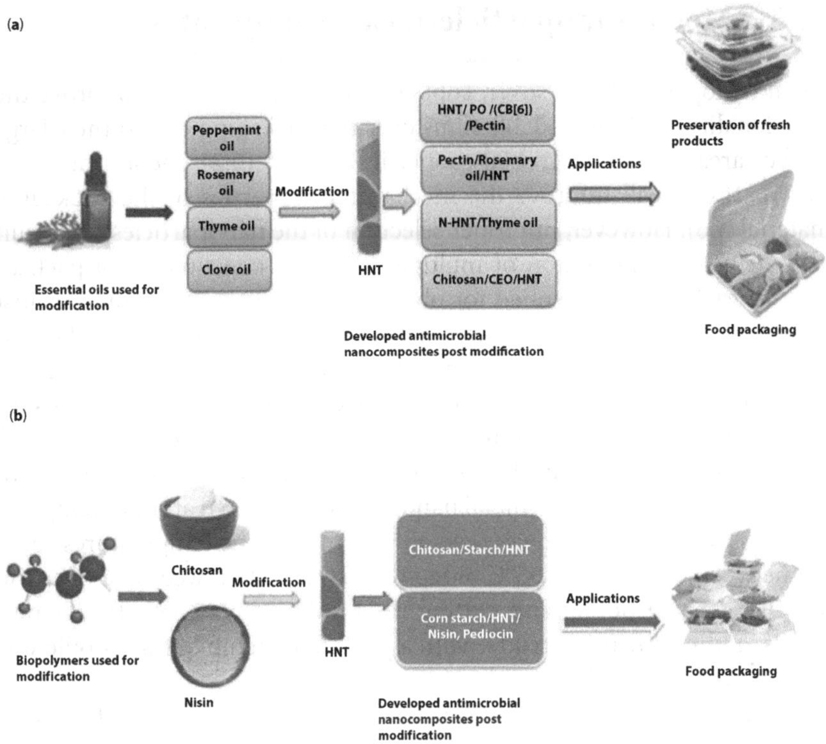

Figure 10.1 (a) HNT-based nanocomposites for food packaging with essential oil. (b) Nanocomposites of HNT and biopolymer for food packaging (reprinted with permission from [19]).

extrudate was cut into pellets, blended with LDPE in the extruder, and blended with LDPE in the twin extruder. Nanocomposite films prepared using sodium MMT as filler material for PLA were used as packaging material for processed meat products due to their ability to control lipid oxidation of processed meat products [22, 23]. The PLA/MMT composite films were prepared by melt processing technique. MMT in powder form and PLA in pellet form were mixed at 60-rpm speed and 180°C followed by compression molding for preparing a film 100 to 150 μm at 180°C at 5 MPa and 3 min and cooling for 10 min [22]. Other studies have prepared polypropylene/clay-based film-type nanocomposites by melt extrusion for food packaging applications [24, 25]. The studies have reported that the addition of clay reduced the oxygen transmission and water vapor transmission rate within the packaging material [24].

10.3 Metal Nanoparticle-Based Composites

Metal nanoparticles like silver, copper oxide, and zinc oxide improve the functionality of biodegradable films over the packaging due to their large surface area and antimicrobial properties (Figure 10.2). Incorporation of such particles will improve the mechanical properties of the packaging material [26]. However, improper selection of the nanoparticles can result in toxicity. A combination of multiple MNPs as constituents of packaging material is recommended for obtaining the best performance against microbes. The enhanced antimicrobial activity of biopolymer-based nanocomposites containing multiple metallic nanoparticles has been evaluated during the storage of fresh peaches and nectarines [27]. Nanocomposites were prepared using starch and poly(butylene adipate-co-terephthalate) (PBAT) after loading with silver and zinc oxide nanoparticles. A mixture of starch, PBAT, glycerol, and metal nanoparticles prepared at room temperature was extruded at a screw speed of 70 rpm with the temperatures varying between 110°C and 135°C along the extruder followed by a decrease to 110°C. After cooling and cutting the product from the extruder into pellets, a film-blowing extruder with a predefined temperature profile was used to prepare the film.

A number of studies have been reported on the synthesis of packaging material for meat products with the aid of zinc oxide nanoparticles [29]. Sodium hydroxide solution of strength 5 M was added dropwise for 20 min to an aqueous solution of $ZnCl_2$ heated to 90°C. Washing of the precipitate was followed by peptization with 2-propanol and centrifugation at 6000 rpm. Calcination of the resulting particles was carried out at 300°C for 6 hours followed by crushing to yield the ZnO nanoparticles [30]. These nanoparticles were used as nanofillers during the fabrication of chitosan-based films [30] and carboxymethyl cellulose-based films [31] as food packaging material. CMC was dissolved in an aqueous solution of glycerol at 70°C to which a suspension of ZnO nanoparticles was added by ultrasonication for 30 min. Subsequently, grape seed extract was added to the solution and mixed, and Teflon-coated glass plates were used to cast the solutions [31]. The films with excellent antioxidant properties and high mechanical and water vapor barrier characteristics were used as packaging material for beef. Chitosan-based nanocomposites with ZnO nanofillers having good antibacterial properties were prepared using a similar methodology for storing fresh poultry meat [32].

Silver nanoparticles have been used as prospective packaging material for food items, namely, fresh fruit, meat, and consumer products owing to

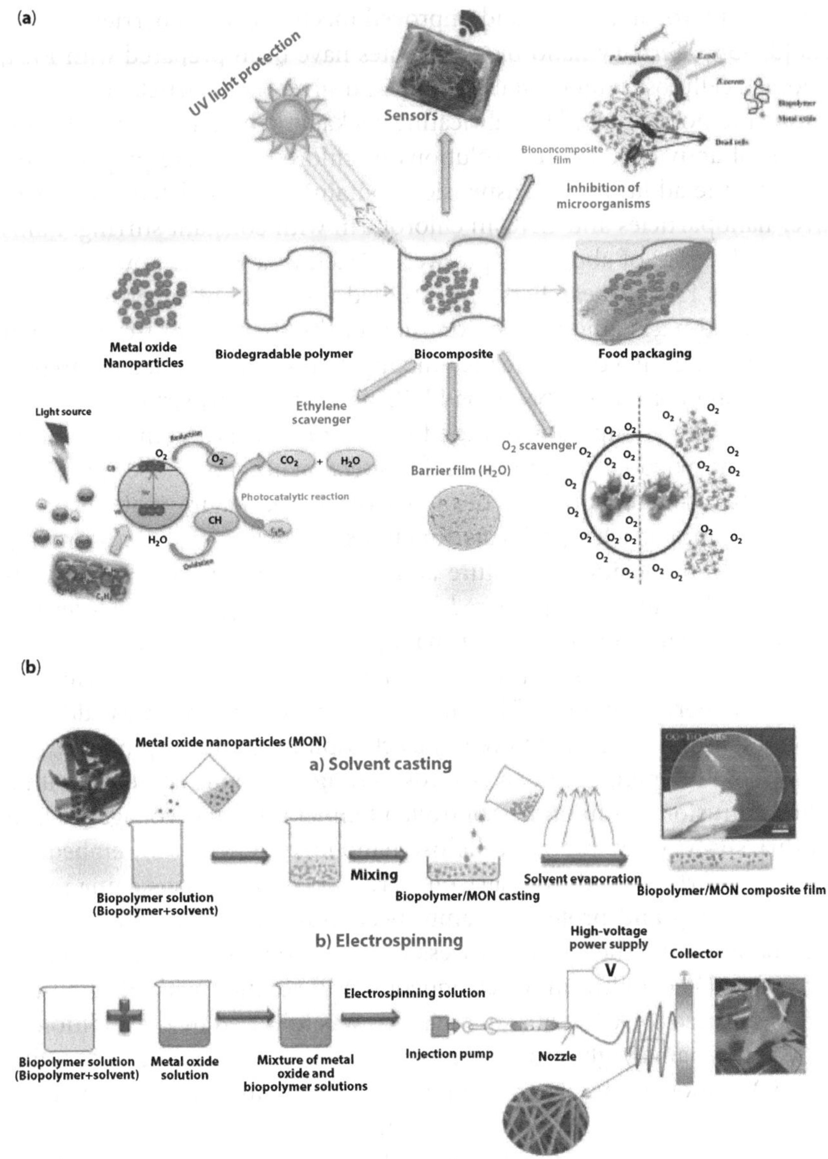

Figure 10.2 (a) Schematic representation of coating of metal nanoparticles incorporated biodegradable polymer during smart food packaging. (b) Schematic representation of biopolymer-metal oxide film synthesis through solvent casting and electrospinning (reprinted with permission from [28]).

their antimicrobial activity, and improved mechanical and barrier properties [33–35]. Ternary nano-biocomposites have been prepared with PLA, modified cellulose nanocrystals (CNC), and silver nanoparticles by solvent casting method and used for fabricating packaging material with high antimicrobial activity [34]. PLA solutions in chloroform were prepared followed by the addition of a suspension containing a calculated quantity of silver nanoparticles and CNC in chloroform with constant stirring. Films with thicknesses of about 250 μm were placed in a vacuum oven at 40°C for 2 weeks [34]. Researchers have studied the effect of silver nanocomposite-based packaging on the extension of the shelf life of fresh chicken stored at 4°C until 5 days. The melt mixing method was carried out to prepare the nanocomposites based on LDPE and silver nanoparticles in a twin screw extruder. Temperatures varied from 120°C to 170°C in the heating zones of the extruder with the spinning speed maintained at 60 rpm [36].

Multifunctional photocatalytically active food packaging material based on TiO_2 nanoparticles performs functions, namely, oxygen scavenging, ethylene decomposition, moisture and CO_2 absorption, and UV shielding [37, 38]. Nanofibers prepared by electrospinning the maize-derived biopolymer zein containing TiO_2 nanoparticles for achieving improved photocatalytic activities and thermal stability against ethylene during the storage of cherry tomatoes. TiO_2 at different concentrations was added to ethanol solutions of zein followed by electrospinning with a spinneret needle at room temperature in an electrospinning apparatus equipped with a syringe infusion pump under controlled temperature and voltage [39]. In another study, whey protein/cellulose nanofibers (CNF) with embedded TiO_2 nanoparticles and rosemary oil were used as active packaging material for storing and protecting lamb meat from microbial and chemical degradation [40]. The casting process was used to fabricate the whey protein isolate (WPI) films in which aqueous WPI solution with pH adjusted to 8 was heated to 90°C with stirring. This was followed by the addition of ingredients, namely, glycerol acting as a plasticizer, CNF acting as nanofillers, TiO_2 nanoparticles as an antimicrobial agent, and rosemary oil as an antioxidant. The solution was agitated using an ultrasonic homogenizer, vacuum-treated for 40 min, cast into plates, dried, and peeled off from the plates.

10.4 Synthesis of Biopolymer-Based Packaging Material

Bio-based materials for food packaging like bio-based nanomaterial, bio-based fibers, and composites are of primary concern nowadays. Bio-based polymers are biodegradable varieties of polymers produced by plants or microorganisms from renewable resources, namely, polysaccharides, wood, food and agricultural waste, and lignocelluloses [41]. Bio-based polymers are classified into four categories, namely, polymers extracted from biomass, biomass-derived synthetic polymers, micro-organism derived polymers and biodegradable polymers from petrochemical monomers [42]. Biopolymers that are derived from microorganisms including dextran, xanthan, alginate, cellulose, hyaluronic acid, polysaccharides, and polyhydroxyalkanoates (PHA). Bioplastic formulations based on poly(hydroxybutyrate-co-valerate) (PHBV)/tannin films have recently been used for smart food packaging applications [43]. PHBV/tannin films with a thickness of 80 μm were prepared by solvent casting technique in which an appropriate amount of tannin powder was primarily dissolved in formic acid followed by an addition of PHBV along with a continuous stirring at 80°C. The solution was dried using a fume hood and stored at room temperature. Chitosan-based films are currently in use for preparing active and intelligent packing material. The inclusion of additional additives enhances the mechanical, and barrier properties of the prepared films [44]. In one of the studies, the casting method uses chitosan films, which are prepared after dissolving chitosan powder in an aqueous solution of glacial acetic acid followed by the addition of glycerol and hydroalcoholic along with aqueous plant extracts to the solution. The resulting solutions were degassed and cast on polystyrene Petri dishes after oven-drying at 35°C for 48 hours. The dried films were peeled and stored for 72 hours in a desiccator for 72 hours before use [45].

Nowadays, proteins have emerged as prospective food packaging material owing to their gloss and transparency. They have reasonably good barrier properties for organic vapor and oxygen although possess selective permeability to gasses. One of the basic advantages of them is their ability to be altered physically, chemically, and enzymatically to suit them for specific applications [46]. Edible films/coatings are thin layers of material that can modify the molecular exchange between the food and the environment and also help in maintaining freshness, ease of transportation, storage, and display. They have the ability to control the deterioration of fish and meat and extend the shelf life of tropical fruits, berries, and vegetables [47]. Gelatin is a type of material that has strong water binding and gel formation ability and water vapor barrier properties. Gelatin dispersions were prepared by adding fish gelatin powder to distilled water followed by heating and the addition of glycerol. Aloe vera extraction was added in different concentrations to the gelatin dispersion with continuous stirring. The casting of the solution was further carried out to prepare the film [48].

In one of the studies, acetylated hemicellulose nanocellulose (AH-ACNC) films were modified with PCL films for preparing active packaging material for aqueous, fatty, acidic, and alcoholic food. AH-ACNC films are promising packaging materials owing to their commendable oxygen and gas barrier properties and mechanical strength. Mixtures of AH with different concentrations of ACNC were prepared by dissolving the above in DMSO, followed by heating in a water bath. The films were prepared by solvent casting followed by the addition of PCL coating on the films by dip coating technique. Oven drying of the prepared films at 25°C for 24 hours was carried out to prepare films with controlled thickness [49]. The methods of synthesis of different types of packaging material are listed in Table 10.2.

10.5 Structural Features of Different Nanocomposites

10.5.1 Nanoclay

Nanoclay is known as a major reinforcement filler for bio-based polymers manifesting enhanced barrier properties, modulus, creep resistance, and mechanical strength of the biopolymer. Furthermore, the tensile strength, water vapor permeability, and hydrophobic behavior of agar films are substantially influenced by nanoclay, when used as the filler material. Several nanomaterials are developed such as MMT nanoclay, nanofibrillated cellulose, and chitosan. An elaborative discussion on MMT has been discussed here in order to understand their primitive features in invoking the aforesaid properties.

Montmorillonite Structure

The MMT $[(Na,Ca)_{0.33}(Al,Mg)_2(Si_4O_{10})(OH)_2.nH_2O]$ is a prototypical clay (Figure 10.3) used in polymer nanocomposite (PNC) application. This clay is covered with highly anisotropic platelets having an average thickness of ~1 nm separated by a thin layer of water [52]. Each platelet contains aluminum, magnesium hydroxide, and silicon oxide in a sandwiched layer condition. It shows excellent cation exchange capacity along with the benefits of a large surface area and good swelling behavior [53]. The MMT has an outstanding ability as a filler medium in PNCs due to its structural characteristics. Few percentages of polymer matrix addition with MMT help to increase the polymer strength and the barrier properties.

Table 10.2 Different nanomaterial synthesis methods for food packaging.

Name of packaging material	Type of food for which the material is used	Method of preparation	Properties affected	References
Bio-nanocomposite films of potato starch/halloysite nanotubes	Fruits and vegetables	Casting of starch dispersion containing halloysite nanoparticles and sorbitol-glycerol plasticizer	Tensile strength increased by 25.4%, water vapor (WVP) and oxygen permeability (OP) decreased by 71.5% and 50.31% with 5% HNT	[50]
Bio-nanocomposite films of polylactic acid/halloysite nanotubes	Cherry tomato	Casting of PLA solution with different concentrations of HNT	Tensile strength increased 35%, WVP and OP decreased 28.6% and 50% with 3% HNT	[15]
Bio-nanocomposite film of poly(lactic acid) (PLA) and sodium montmorillonite (MMT)	Processed meat	Melt processing in which mixtures of MMT powder and PLA pellets undergo compression molding at 5 MPa	Water permeability decreases by 50% after 90 days, reduction in lipid oxidation	[22]

(Continued)

Table 10.2 Different nanomaterial synthesis methods for food packaging. (*Continued*)

Name of packaging material	Type of food for which the material is used	Method of preparation	Properties affected	References
Bio-nanocomposite films of starch/ poly(butylene adipate-co-terephthalate) (PBAT) doped with silver and zinc oxide nanoparticles	Peaches and nectarines	Extrusion of mixtures of starch, PBAT, glycerol, and nanoparticles blowing using twin screw extruder	Tensile strength increases by 23.2% and WVP decreases by 14.85% by adding silver nanoparticles and ZnO at a 3:2 ratio	[27]
Bio-nanocomposite films of carboxymethyl cellulose (CMC) with ZnO nanoparticles and grape seed extract	High-fat meat products	Solution casting method in which ZnO nanoparticles were added to a solution of CMC in glycerol followed by casting and drying	Tensile strength increases by 26.4% and WVP decreases by 19%	[31]
Bio-nanocomposite of low-density polyethylene (LDPE)/ Ag/TiO_2	Fresh chicken	Melt mixing and sol-gel method	Enhancing the shelf life of chicken using Ag TiO_2 nanoparticles compared to clay and TiO_2	[36]

(*Continued*)

Table 10.2 Different nanomaterial synthesis methods for food packaging. (*Continued*)

Name of packaging material	Type of food for which the material is used	Method of preparation	Properties affected	References
Bio-nanocomposite films of gelatin and Aloe vera gel	Active packaging material for different food items	Solution casting of gelatin and Aloe vera dispersions	Elongation at break percentage increased by 32% and WVP was insignificantly affected, radical scavenging activity increased from 65.78% to 74.76% with 9% Aloe gel	[48]

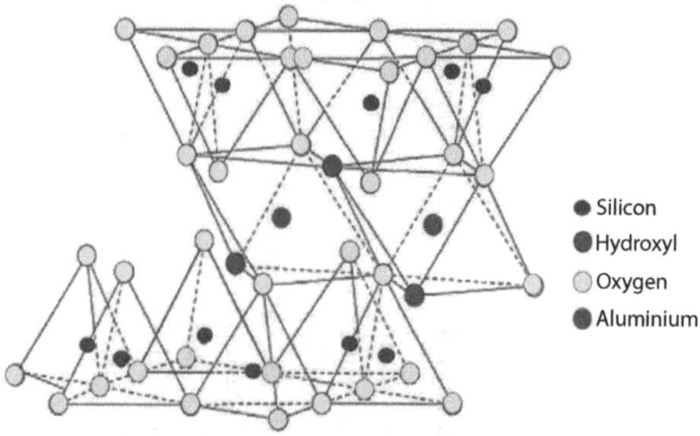

Figure 10.3 Structure of montmorillonite (reprinted with permission from [51]).

Characteristics of Nanoclay

Process conditions have an impact on the *delamination* and *dispersion* of clay additives within the matrix. Specimen elongated high shear rate and low shear rate influence the formation of nanocomposites [54]. Shear stress is given by Newtonian law (Equation 10.1)

$$\sigma = \eta\gamma \tag{10.1}$$

where, η = melt viscosity and γ= shear rate.

Shear stress has a substantial influence on nanocomposite processing. Higher shear stresses to platelets are generated due to the higher melt viscosity during melt mixing within the extruder breaking the organoclay particles into stacks of platelets or tactoids. Dispersion of nanoclay within the thermoset matrix depends on Gibb's free energy, where during complete exfoliation of clay in polymer matrix the free energy is given by Equation 10.2.

$$(\Delta G) = (\Delta G_{po}) + (\Delta G_{nc}) \tag{10.2}$$

where ΔG_{po} is the free energy contributed by the polymer matrix and ΔG_{nc} is the free energy contributed by the nanoclay.

However, the degree of exfoliation depends on the total enthalpy. The complete exfoliation happens during the heat released at the time of polymerization within the intergallary before the gel point is more than that of the Van-der Waals attractive forces between the interlayer (gallery). Exfoliated nanoclays assemble easily and are detached into single flakes. A moderate dispersion level of the clay into the polymer matrix is permissible due to the interpolated pattern of exfoliated nanoclay. This feature shows the best barrier properties of nanoclay composites.

10.5.2 Chitosan Nanoparticles (CNPs)

Chitosan-based nanocomposites have a significant impact on the enhancement of physical, mechanical, and thermal stability of polymers. The composite form also contributes to either an increase in the effective surface area or in developing good physicochemical properties. Chitosan nanocomposites such as chitosan metal/metal oxide (e.g., silver chitosan nanocomposites), chitosan carbon material, chitosan polymer mixture/copolymer, and chitosan clay composites are developed. The chitosan polymer mixture among them is discussed below.

Structure of Chitosan Polymer Matrix

Chitosan polymer matrix is developed in order to manifest application-specific properties. Figure 10.4 shows the chemical structure of chitosan. Polymer contained activated carboxyl groups, which are covalently linked with the primary amine groups of chitosan because of its high reactivity. Carbodiimide cross-linking reagents initiate the formation of an amide bond between amine and carboxyl groups [55, 56]. Chitosan-polymer hydrogels and N-alkyl carbamate linkages are formed due to either physical mixture or physical interaction between polymer and chitosan.

Characteristics of Chitosan

The degree of deacetylation (DD) is a chemical characteristic of chitosan determining the content of free amino groups ($-NH_2$) formed by the partial replacement of acetyl groups ($-C_2H_3O$) that results in a copolymer of N-acetyl-glucosamine and D-glucosamine. The solubility of chitosan in weak acids is high due to the presence of a high percentage of amino groups. Chitosan polymer has versatile applications in industry due to highly reactive amino groups. The DD value of chitosan film varies from

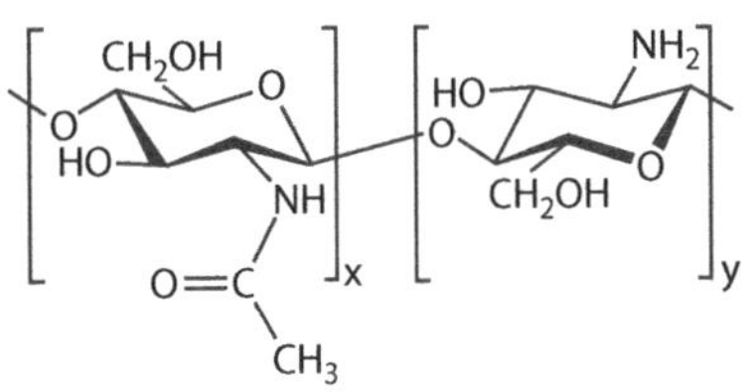

Figure 10.4 Structure of chitosan (reprinted with permission from [57]).

88.1% to 95.2% [58]. A high DD value gives better water barrier properties and enhances antimicrobial efficiency. The molecular weight (MW) of chitosan also affects the physicochemical and antimicrobial properties. With the increase in the MW of CH films, the water barrier property and tensile strength were also being enhanced. *Self-healing* is the ability of a material to either repair or heal damages automatically or with some external simulation independently. Self-healing properties of chitosan- and cerium (Ce)-based coatings protect aluminum alloy 2024 from corrosion [59], while 2-mercaptobenzothiazole (MBT) is an effective corrosion inhibitor with chitosan-based coating for aluminum alloys 2024.

10.5.3 Titanium Dioxide Nanomaterials (TiO_2 NPs)

TiO_2 NPs have been included in food packages as either a whitener or photocatalytic antibacterial agent since 1996. TiO_2 NPs block the UV radiations attributing to an enhancement in the chemical, mechanical, and barrier properties of the films. TiO_2 NPs are also used as a coloring agent in various processed foods due to their high index of refraction, brightness, and color stability [60].

Structure of TiO_2 Nanoparticles

Figure 10.5 shows the crystal structure of TiO_2 polymorphs. Among rutile and anatase crystal phases, rutile is considered the most stable bulk phase, while anatase is a more stable phase in the nanoscale because of its lower surface energy compared to rutile. TiO_2 prepared after mixing anatase and rutile with a specific ratio, called mixed-phase TiO_2, manifests extraordinary photoelectric properties and better photocatalytic performance compared to single-phase TiO_2 [61].

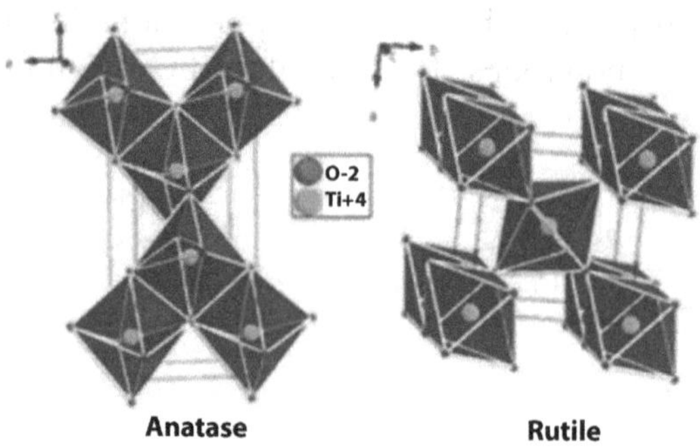

Figure 10.5 Crystal structure of TiO_2 polymorphs (reprinted with permission from [62]).

Characteristics of TiO_2 Nanoparticles

The unsaturated polyphospholipid components of amicrobial cell membranes are oxidized by TiO_2 NPs resulting in a *biocidal* effect. TiO_2 produces superoxide anions, hydrogen peroxide, and hydroxyl radicals that directly destroy microbial cell walls under sunlight or UV radiation (*antimicrobial* effects). TiO_2 NPs produce more reactive oxygen due to their large surface area which helps to damage the cells of the microorganisms. Moreover, chemical interactions between the polymeric matrix and treated nanoparticles show substantially high barrier properties. Chitosan is widely used in the food packaging industry although due to its weak mechanical and barrier properties nowadays, chitosan is not considered an alternative to synthetic plastics. These limitations can be improved by using chemical cross-linkers and physical reinforcement with nanofillers. Recently TiO_2 has been used as chemical reinforcing agents to prepare different composites. The incorporation of TiO2 improves the mechanical and barrier properties of TiO_2 chitosan–based film [59].

10.5.4 Silver Nanoparticles (Ag NPs)

The significant usage of silver nanoparticles (Ag NPs) is as an antimicrobial agent as coating over the food package to enhance the shelf life of food. It has antibacterial as well as antioxidant properties. AgNPs are slowly released into the microbes present on the surface of the food considered a hotspot for the microbes. Then, the cells of the microbes start damaging due to the toxicity characteristics of AgNPs. These activities limit microbial growth and prolong the shelf life of food. Hence, Ag NPs in the form of matrix within active packaging films and coatings are beneficial to food [63].

Structure of Silver Nanoparticles (AgNPs)

Figure 10.6 shows that the Ag NP surfaces contain silver ions (Ag^+) and deliver them to the interior of the microorganisms causing DNA damage [65]. Ag NPs are ineffective in reducing the growth rate of Ag^+-resistant *E. coli* strains [66]. Compared to $AgNO_3$, Ag NPs are more toxic to algae. Ag NPs' surfaces effectively catalyze the formation of free radicals within the bacterial cells followed by cell death through oxidative stress. Ag NPs can also bind with membrane proteins and form a pit which causes other morphological changes. They react with the phosphorous groups of DNA [67] resulting a damage to DNA along with other cellular components. Furthermore, silver chloride (silver salts), silver sulfide, and silver nitrate (silver ion–based materials) have higher antimicrobial activity.

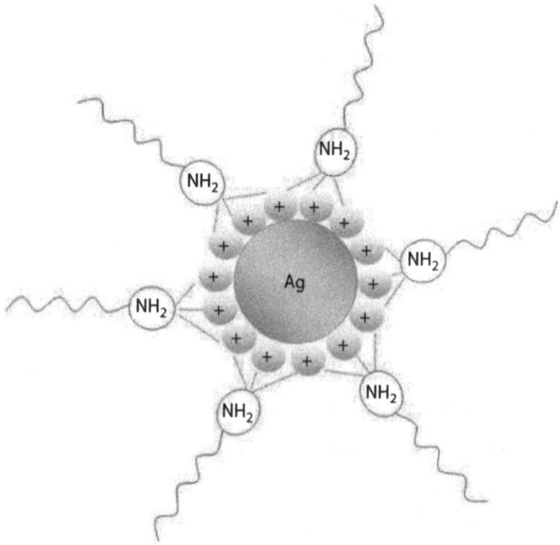

Figure 10.6 Structure of silver nanoparticles (reprinted with permission from [64]).

Characteristics of Silver Nanoparticles (AgNPs)
Quantitative toxicity depends on numerous factors. The toxicity is inversely proportional to nanoparticle diameter. The toxicity significantly increases with the decrease in nanoparticle diameter [68]. Smaller Ag NPs have relatively larger surface area resulting in higher protein binding efficiencies. The released Ag^+ passes easily into the bacterial membrane pores. The toxicity also depends on the nanoparticle shapes. Triangular Ag NPs show better bactericidal properties compared to spherical and rod-shaped particles [69]. Moreover, the surface charge, solubility, surface coating, and degree of agglomeration affect the antimicrobial properties of Ag NPs.

10.5.5 Zinc Oxide Nanocomposite (ZnO NPs)

Affordable ZnO NPs are enormously used in the food packaging industry because of their excellent antimicrobial properties, UV blocking, and less toxicity to living beings compared to Ag NPs. The matrix of ZnO NPs with polymeric films has excellent mechanical strength and durability due to enhanced dispersion of ZnO NPs within increased tortuosity of PLA [70].

Structure and Characteristics of ZnO NPs
The antibacterial activities of ZnO NPs start after penetration through any microbial cell structure. The antibacterial performances depend on the

concentration of ZnO. The presence of ZnO NPS in food packages prolongs the shelf life of food by controlling the pH change [71]. ZnO can exert a significant amount of hydrogen peroxide under visible light and UV radiation which causes oxidative stress in bacterial cells. The "cell damage" efficiency increases with decreasing nanoparticle size thus preserving the food for a longer period of time.

10.5.6 Carbon Nanotubes (CNTs)

The matrix of CNTs and polymers (PVOH, PP, nylon, and PLA) are developed for the food packaging industry as an antimicrobial agent and smart sensor. The CNTs have the ability to puncture the membrane cell which causes irreversible microbial cell damage. CNTs are considered one of the promising candidates as conductive additives due to their excellent electrical conductivity and unique length-to-diameter ratio.

Structure and Characteristics of Carbon Nanotubes (CNTs)
CNTs are categorized into single-walled CNTs and multi-walled CNTs. Both types of nanotubes provide high elastic modulus and also high tensile strength during their incorporation within the polymer matrix. The long-chain and thin SWCNTs damage the *E. coli* bacterial membrane [72–74].

10.6 Application of Different Nanomaterials in Canned Food Packaging

So far, several preventive nonhazardous materials have been used as anti-corrosive agents, which are biodegradable leading to less hazardous to the environment. Biopolymers are those materials that are found to be mostly used as coating material for food packaging. Biopolymers are considered renewable packaging materials, an alternative to petroleum-based polymers. Polysaccharides, lipids, proteins, PLA, PHB, PVA, PCL, poly(butylenes succinate-co-adipate), and their blends are used as the coating material in case during food packaging [33]. However, the use of biopolymers is constrained because of their handling issues with their high MW, viscosity, hydrophobicity, and melting nature. Mixing polymers with additives like silver NPs, nano-zinc oxide, titanium dioxide NPs, and titanium nitride NPs becomes advantageous because of their different antibacterial property, capacity to modulate barrier properties, and capacity to increase mechanical strength. However, the major concern with the inclusion of NPs is their toxicity. For example, nano-zinc oxide becomes more advantageous compared to silver NPs because of their low toxicity [33]. The properties of the polymer nanomaterials with respect to the traditional composites are attributed to the dispersion and distribution of nanoparticles within the matrix along with their interfacial adhesion. Such reactivity, which in turn somehow being affected during the processing, also affects the structure and morphology of the polymer nanocomposites [75]. Silvestre *et al.* [12] conducted a very thorough investigative study on this structural property of the filler material clay to understand how far they impart enhanced barrier properties. According to them, primarily there are two structural forms of clay nanocomposites. One is called intercalated and the other is known as exfoliated, where the individual silicate layers are suspended within the polymer matrix attributing to a substantial penetration of the polymer in between the silicate layers that eventually increases the spacing between the layers may be extended up to 10 nm or more (Figure 10.7) [12]. Therefore, they form a homogeneous well-dispersed polymer matrix with the suspended silicate layers in its exfoliated structural form. Such a well dispersion within the matrix clay provides a good barrier toward the gas barrier and also being useful with the co-extrusion process. With the enhanced barrier property, this nanocomposite has the potential to be

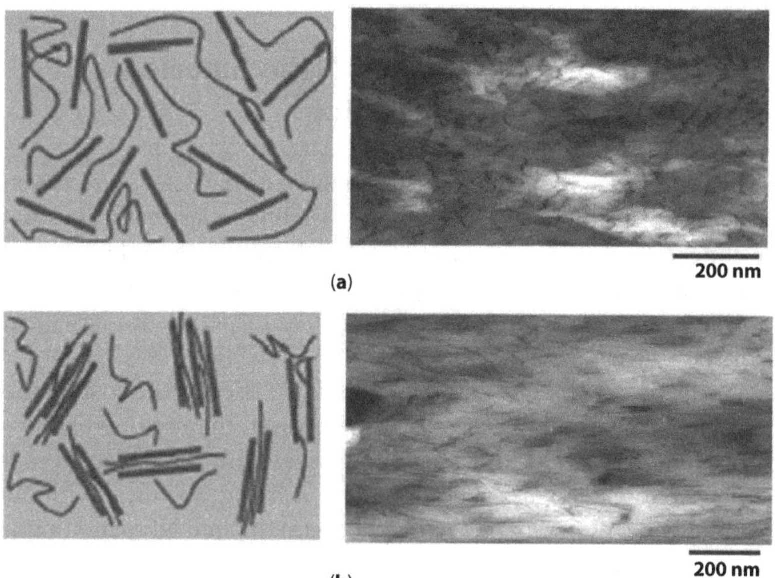

Figure 10.7 (a): Exfoliated polymer matrix. (b) Intercalated polymer matrix (reprinted with permission from [12]).

used as a coating material for the preservation of processed meat, cheese, confectionery, cereals, and boil-in-the-bag foods. Furthermore, with a good co-extrusion capacity, it has wide applications in fruit juice, dairy, beer, and carbonated drink packaging industries.

Another typical property of the nanocomposite coating is its antimicrobial nature increasing with the decrease in the particle size. Zinc oxide nanoparticles, which are nontoxic in nature, inhibit the respiratory enzymes within the bacterial species through the release of Zn^{2+} ions leading to the formation of reactive oxygen species (ROS) and this further generates oxidative stress on the bacterial cell resulting in the death of the cell [76]. However, not restricted to antimicrobial properties, zinc oxide nanoparticles on mixing with polymers also increase the mechanical strength. In a study [77], it was observed that with the addition of zinc oxide nanoparticles with ethylene vinyl alcohol (EVOH) copolymer, the mechanical strength increased due to dense cross-linking between the polymer matrix and zinc oxide. According to them, the barrier properties also increased along with the addition of nano-zinc oxide. As discussed previously, the barrier properties depend on the uniform distribution of the nanoparticles within the polymer matrix that reduces the diffusion through it. In addition to zinc oxide nanoparticles, the particles are uniformly dispersed within the matrix, and also due to dense cross-linking, it creates a tortuous path within the composite that reduces water vapor permeation along with gas permeation. There was around a 49% to 67% decrease in water vapor permeation rate (WVPR), while a 97% decrease was found in gas permeation rate with a 2% nano-zinc oxide–EVOH composite [76]. However, it is a difficult task to address a common food packaging coating due to the variety of food materials requiring various postprocessing treatments that may introduce additional intricacy with the management of coating inside food containers. Ca^{2+} scaling is one of the primary issues with any of the containers, even with the steel one. Especially, this tendency becomes aggravated, when the food is being treated with any calcium salt solution. For example, Broccoli (*Brassica oleracea* var. *italica*) is in the family of Cruciferae, which is used as a vegetable for its nutritious value. Storage is always a big problem for these vegetables, especially storing them in a cold condition after harvest. In a calcium chloride solution, dipping vegetables inhibits cell-degrading enzymes and hardening of cell walls takes place, leading to a proper storage of broccoli [77]. Therefore, food containers here are much more prone to Ca^{+2} scaling. Silver nanoparticles possess antiscaling property along with antibacterial and anticorrosion properties.

The adsorption of the nanoparticles on the scale crystals restricts their further growth. Silver nanoparticles also form complexes with Ca^{2+} ions and thus reduce Ca^{2+} ions' affinity for the scale-forming anions (CO_3^{2-} or SO_4^{2-}). The other possible mechanism that might prevail is due to the selective adsorption of silver nanoparticles on the steel surface manifesting the blocking of the sites where Ca^{2+} scales might have settled [78]. It was seen that with around 1000 ppm of silver nanoparticles within the composite, the scaling inhibition efficiency increases by almost 25% at 30°C. However, with the increase in temperature, inhibitor binding on the substrate material decreases. Furthermore, silver nanoparticle–based biocomposite packaging is the most promising form of active food packaging imparting extended shelf life of foods after reducing the risk of pathogen contamination. Table 10.3 [79] shows the potential composite materials with silver nanoparticles for food packaging showing the inhibition of different pathogens. These microbial contamination leads to a corrosion of the container, called biocorrosion. This type of corrosion happens when three Ms microbes, media, and metal will be in the same place and the formation of biofilm happens [80]. Steel containers, which are frequently used in beverage industries, are prone to sulfate-reducing bacteria (SRB) that create rupture over the passive layer (consists of both chromium and ferric oxide) leading to the formation of pits. A sulfate-reducing bacterium, through genetically expressed proteins, forms biofilms on metal surfaces under various environments that later induce biocorrosion of the metal surfaces [81]. However, biocorrosion is restricted not only to metals but also to plastic materials [82] and papers [83]. Primarily, aerobic microorganisms through enzymatic degradation of synthetic plastics like polyolefins are of major concern in the case of the bicorrosion of plastics. Meanwhile, with papers, several microbial growths on the surface of the paper board generate volatile matter contributing to the formation of odors that finally taint the food inside the packets. Therefore, it is preferable that the anticorrosion coating will also have antimicrobial properties in order to restrict corrosion as observed with many nanoparticles.

So far, from the aforesaid discussion, it is much evident that corrosion and microbial attack are somehow interlinked leading to a selection of coating material that could have preventive properties for both the occurrences. ZnO NPs and Ag NPs have antimicrobial properties that could lead to protection against contamination, while ZnO NPs will enhance the preventive measures for both water vapor permeation and gas permeation. However, apart from biocorrosion, the other intricacy is normal corrosion because of electrochemical reactions, especially with the metallic container as in the case of the beverage cans, when the passive layer gets ruptured.

Table 10.3 Different types of the nanocomposite polymer matrix for food packaging.

Polymer matrix	Tested food	Tested microorganisms
LDPE with Ag and ZnO NPs	Orange juice	Yeast and total aerobic bacteria
LDPE with Ag NPs	Barberry	Aerobic bacteria
LDPE with Ag NPs, TiO_2, and kaolin	Strawberry	-
PVC with Ag NPs	Minced beef	Mesophilic bacteria, *S. Aureus, and E. coli*
EVOH with Ag NPs	Chicken, pork, cheese, lettuce, apples, peels, and eggshells	*Salmonella* spp. and *L. monocytogenes*
Polyethylene with Ag and TiO_2 NPs	Fresh apples, white slice bread, fresh carrots, soft cheese, atmosphere packaging milk powder, and fresh orange juice	*Penicillium*, *Lactobacillus*, *S. aureus*, *Coliforms*, *E. coli*, and *Listeria*
Cellulose with Ag NPs	Beef meat and fresh-cut melon	Aerobic bacteria, lactic acid bacteria, *Pseudomonas* sp., *Enterobacteriaceae*, Mesophilic aerobic bacteria, psychrotrophic bacteria, yeasts, and molds

LDPE: Low-density polyethylene, Ag NPs: Silver nanoparticles; ZnO NPs: Zinc oxide nanoparticles; TiO_2 NPs: Titanium dioxide nanoparticles

TiO_2 NPs have the ability to improve the preventive features against corrosion along with the enhancement of mechanical properties of nanocomposite coatings through proper dispersion within the polymer matrix. The well dispersion also enhances the barrier properties, while further non-toxic TiO_2 shows its potential against microbial contamination. In a study by Kumar *et al.* [84], TiO_2 NPs in epoxy coating show good preventive measures for corrosion. Furthermore, they had shown that graphene in the composite enhances its protection capacity against corrosion. However, when the question arises with the coating for food packaging containers, the toxicity of either the polymer or the nanoparticle is of primary interest. Not only even graphene but also oxides of graphene with epoxy nanocomposite coating act as a corrosion protector [85]. Furthermore, in the case of packaging based on nanocomposites, which is much more relevant to food packaging, graphene senses any microbial changes occurring in the product package in order to identify specific food-borne pathogens or gasses. Especially, graphene oxide is one of the promising packaging materials within the graphene family because of its relevant characteristics, like high barrier properties, large specific surface area, antibacterial and antifungal antioxidants, and biocompatibility [86].

Epoxy resin coating is a common form of coating material within the food container where it separates food in contact with the metal surface. However, sole coating cannot increase the barrier properties specifically water vapor permeation restriction. On the contrary, CNT-mixed epoxy nanocomposite not only increases the barrier properties but also restricts the anticorrosiveness of the coating by restricting the water permeation. The volume fraction and the diffusivity of water vapor through coating are given by Equations 10.3 and 10.4 [87].

$$V_t = \frac{100\log\left({C_t}/{C_0}\right)}{\log \varepsilon_{water}} \tag{10.3}$$

$$D = \frac{L^2\pi[slope]^2}{4} \tag{10.4}$$

Here, V_t is the volume fraction of water absorbed within the coating at any time 't'; D is the diffusion coefficient of water vapor through the coating; C_t is the capacitance of an organic coating at saturation time 't'; C_0 is the capacitance of an organic coating at initial time 't=0'; C_c is the capacitance of an organic coating at any time 't'; and ε_{water} is the dielectric constant of water at a certain temperature.

The slope in Equation 10.4 is determined from the graph between $\log(C_c)$ and $t^{1/2}$ as shown in Figure 10.8 [87]. As seen from the figure, the water vapor permeation increases at the very initial stage, while it becomes constant after a certain time attributing to a saturation level. Furthermore, it is also assessed that 0.5% CNT-mixed nanocomposite shows minimum diffusion resulting minimum volume fraction of water absorbed within the coating

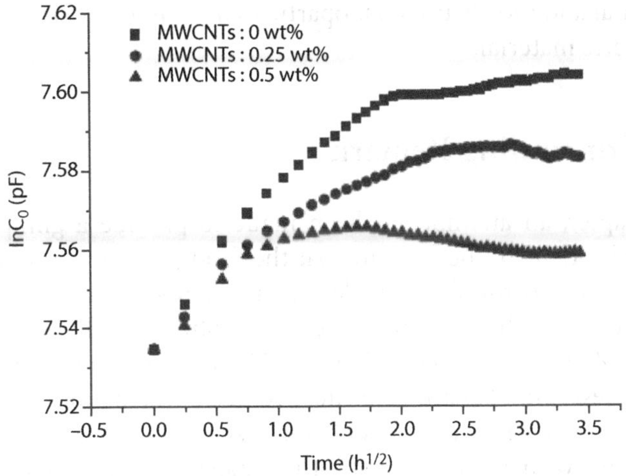

Figure 10.8 $\ln C_C$ versus $t^{1/2}$ curve for epoxy-coated carbon steel with different MWCNT content (reprinted with permission from [87]).

material. Thus, an increase in the barrier properties restricts the water contact with the metal and henceforth corrosion of the metal. Williams *et al.* [88] made a study on the use of allyl isothiocyanate and CNT nanocomposites to understand the preservation of shredded and cooked meat. They found that more than 28% of allyl isothiocyanate concentration along with more than 0.02% of CNT concentration possesses potential microbial reduction capacity attributing to 40-day preservation. Such antimicrobial properties will also lead to a reduction in biocorrosion. However, the nanocomposite is not only limited to either inorganic or graphene families but also fabricated with the mixing of polymers with nano-biopolymers such as nanocellulose. As discussed previously, the primary attributes of food packaging coating are anticorrosion, antimicrobial, and enhanced barrier properties. In a study [89], it was shown that the epoxy coating cannot restrict the permeation of corrosive agent through penetration within it, while 1 wt% and 1.5 wt% nanocellulose-mixed epoxy coating can resist such penetration and henceforth the corrosion of the substrate.

Hence, with the aforesaid discussion, it is much understandable that different nanoparticles along with the polymers in composite form can prevent the corrosion of the food package along with the prevention of the water and gas permeation toward their contact with the surface. However, the intricacy lies with the selection of the nanoparticles as they might impart toxicity to the food. In the above discussion, the applicability of nanoparticles as anticorrosive agents is thus restricted with the conventional and mostly used nanoparticles in composite form along with the polymeric material.

10.7 Concluding Remark

In this chapter, an attempt has been made to provide a glimpse of the nanoparticles that can be used to coat the food package without imparting any toxicity to the food inside. Several nanoparticles like nanoclay, Ag, TiO_2, chitosan-based nanoparticles, nanotubes, and composite with synthetic polymer have been elaborated. Not only elaboration is confined to the synthesis but also how the effectiveness of the nanomaterial is correlated with their structure is also being discussed. The primary concern with these nanoparticles is to restrict the staleness of the food promoted due to their aerial contamination that happens after the corrosion of the food can. These corrosions might be enhanced with the uncoated food package either because of low resistant properties or due to biocorrosion. Hence, the coating material must possess barrier properties along with antimicrobial properties although the selection of coating material is application specific.

References

1. Ashfaq, A., Khursheed, N., Fatima, S., Anjum, Z., Younis, K., Application of nanotechnology in food packaging: Pros and Cons. *J. Agric. Food Res.*, 7, 100270, 2022.
2. de Moura, M.R., Aouada, F.A., Avena-Bustillos, R.J., McHugh, T.H., Mattoso, L.H.C., Improved barrier and mechanical properties of novel hydroxypropyl methylcellulose edible films with chitosan/tripolyphosphate nanoparticles. *J. Food Eng.*, 92, 4, 448–453, 2009.
3. Divya, K. and Jisha, M.S., Chitosan nanoparticles preparation and applications. *Environ. Chem. Lett.*, 16, 1, 101–112, 2018.
4. Yanat, M. and Schroën, K., Preparation methods and applications of chitosan nanoparticles; with an outlook toward reinforcement of biodegradable packaging. *React. Funct. Polym.*, 161, 104849, 2021.
5. Arrieta, M.P., Fortunati, E., Burgos, N., Peltzer, M.A., López, J., Peponi, L., Chapter 7 - Nanocellulose-Based Polymeric Blends for Food Packaging Applications, in: *Multifunctional Polymeric Nanocomposites Based on Cellulosic Reinforcements*, D. Puglia, E. Fortunati, J.M. Kenny (Eds.), pp. 205–252, William Andrew Publishing, 2016.

6. Ferrer, A., Pal, L., Hubbe, M., Nanocellulose in packaging: Advances in barrier layer technologies. *Ind. Crops Prod.*, 95, 574–582, 2017.
7. Naseer, B., Srivastava, G., Qadri, O.S., Faridi, S.A., Islam, R.U., Younis, K., Importance and health hazards of nanoparticles used in the food industry. 7, 6, 623–641, 2018.
8. Cano, A., Cháfer, M., Chiralt, A., González-Martínez, C., Development and characterization of active films based on starch-PVA, containing silver nanoparticles. *Food Packag. Shelf Life*, 10, 16–24, 2016.
9. K, S.S., I., M.P., R., G.R., Mahua oil-based polyurethane/chitosan/nano ZnO composite films for biodegradable food packaging applications. *Int. J. Biol. Macromolecules*, 124, 163–174, 2019.
10. Blasco, C. and Picó, Y., Determining nanomaterials in food. *TrAC Trend. Anal. Chem.*, 30, 1, 84–99, 2011.
11. Peiró, A.M., Colombo, C., Doyle, G., Nelson, J., Mills, A., Durrant, J.R., Photochemical Reduction of Oxygen Adsorbed to Nanocrystalline TiO2 Films: A Transient Absorption and Oxygen Scavenging Study of Different TiO2 Preparations. *J. Phys. Chem. B*, 110, 46, 23255–23263, 2006.
12. Silvestre, C., Duraccio, D., Cimmino, S., Food packaging based on polymer nanomaterials. *Prog. Polym. Sci.*, 36, 12, 1766–1782, 2011.
13. Cui, R., Jiang, K., Yuan, M., Cao, J., Li, L., Tang, Z., Qin, Y., Antimicrobial film based on polylactic acid and carbon nanotube for controlled cinnamaldehyde release. *J. Mater. Res. Technol.*, 9, 5, 10130–10138, 2020.
14. Li, Q., Ren, T., Perkins, P., Hu, X., Wang, X., Applications of halloysite nanotubes in food packaging for improving film performance and food preservation. *Food Control*, 124, 107876, 2021.
15. Risyon, N.P., Othman, S.H., Basha, R.K., Talib, R.A., Characterization of polylactic acid/halloysite nanotubes bionanocomposite films for food packaging. *Food Packag. Shelf Life*, 23, 100450, 2020.
16. Kumar, L., Deshmukh, R.K., Hakim, L., Gaikwad, K.K., Halloysite Nanotube as a Functional Material for Active Food Packaging Application: A Review. *Food Bioproc. Technol.*, 2023.
17. Stavitskaya, A., Shakhbazova, C., Cherednichenko, Y., Nigamatzyanova, L., Fakhrullina, G., Khaertdinov, N., Kuralbayeva, G., Filimonova, A., Vinokurov, V., Fakhrullin, R., Antibacterial properties and *in vivo* studies of tannic acid-stabilized silver–halloysite nanomaterials. *Clay Miner.*, 55, 2, 112–119, 2020.
18. Tas, C.E., Hendessi, S., Baysal, M., Unal, S., Cebeci, F.C., Menceloglu, Y.Z., Unal, H., Halloysite Nanotubes/Polyethylene Nanocomposites for Active Food Packaging Materials with Ethylene Scavenging and Gas Barrier Properties. *Food Bioproc. Technol.*, 10, 4, 789–798, 2017.
19. Saadat, S., Pandey, G., Tharmavaram, M., Braganza, V., Rawtani, D., Nano-interfacial decoration of Halloysite Nanotubes for the development of antimicrobial nanocomposites. *Adv. Colloid Interface Sci.*, 275, 102063, 2020.

20. Giannakas, A.E. and Leontiou, A.A., Montmorillonite Composite Materials and Food Packaging, in: *Composites Materials for Food Packaging*, pp. 1–71, 2018.
21. Hosseinkhanli, H., Sharif, A., Aalaie,J., Khalkhali, T., Akhlaghi, S., Oxygen permeability and the mechanical and thermal properties of (Low-density Polyethylene)/Poly(ethylene co-vinyl acetate)/Organoclay blown film nanocomposites. *J. Vinyl Addit. Techn.*, 19,132–139, 2013.
22. Vilarinho, F., Andrade, M., Buonocore, G.G., Stanzione, M., Vaz, M.F., Sanches Silva, A., Monitoring lipid oxidation in a processed meat product packaged with nanocomposite poly(lactic acid) film. *Eur. Polym. J.*, 98, 362–367, 2018.
23. Vilarinho, F., Vaz, M.F., Silva, A.S., The Use of Montmorillonite (MMT) in Food Nanocomposites: Methods of Incorporation, Characterization of MMT/Polymer Nanocomposites and Main Consequences in the Properties. *Recent Pat. Food, Nutr. Agric..*, 11, 1, 13–26, 2020.
24. Ayhan, Z., Cimmino, S., Esturk, O., Duraccio, D., Pezzuto, M., Silvestre, C., Development of Films of Novel Polypropylene based Nanomaterials for Food Packaging Application. *Packag. Technol. Sci.*, 28, 7, 589–602, 2015.
25. Khalaj, M.-J., Ahmadi, H., Lesankhosh, R., Khalaj, G., Study of physical and mechanical properties of polypropylene nanocomposites for food packaging application: Nano-clay modified with iron nanoparticles. *Trends Food Sci. Technol.*, 51, 41–48, 2016.
26. Kumar, A., Kumar, A., Vats, C., Sangwan, P., Kumar, V., Abhineet, Chauhan, P., Chauhan, R.S., Chaudhary, K., Recent insights into metallic nanoparticles in shelf-life extension of agrifoods: Properties, green synthesis, and major applications. *Front. Sustain. Food Syst.*, 6, 2022.
27. Zhai, X., Zhou, S., Zhang, R., Wang, W., Hou, H., Antimicrobial starch/poly(butylene adipate-co-terephthalate) nanocomposite films loaded with a combination of silver and zinc oxide nanoparticles for food packaging. *Int. J. Biol. Macromol.*, 206, 298–305, 2022.
28. Nikolic, M.V., Vasiljevic, Z.Z., Auger, S., Vidic, J., Metal oxide nanoparticles for safe active and intelligent food packaging. *Trends Food Sci. Technol.*, 116, 655–668, 2021.
29. Smaoui, S., Chérif, I., Ben Hlima, H., Khan, M.U., Rebezov, M., Thiruvengadam, M., Sarkar, T., Shariati, M.A., Lorenzo, J.M., Zinc oxide nanoparticles in meat packaging: A systematic review of recent literature. *Food Packag. Shelf Life*, 36, 101045, 2023.
30. Priyadarshi, R. and Negi, Y.S., Effect of Varying Filler Concentration on Zinc Oxide Nanoparticle Embedded Chitosan Films as Potential Food Packaging Material. *J. Polym. Environ.*, 25, 4, 1087–1098, 2017.
31. Priyadarshi, R., Kim, S.-M., Rhim, J.-W., Carboxymethyl cellulose-based multifunctional film combined with zinc oxide nanoparticles and grape seed extract for the preservation of high-fat meat products. *Sustain. Mater. Technol.*, 29, e00325, 2021.

32. Souza, V.G., Rodrigues, C., Valente, S., Pimenta, C., Pires, J.R., Alves, M.M., Santos, C.F., Coelhoso, I.M., Fernando, A.L., Eco-Friendly ZnO/Chitosan Bionanocomposites Films for Packaging of Fresh Poultry Meat. *Coatings*, 10, 2020, doi: 10.3390/coatings10020110.
33. Chaudhary, P., Fatima, F., Kumar, A., Relevance of Nanomaterials in Food Packaging and its Advanced Future Prospects. *J. Inorg. Organomet. Polym. Mater.*, 30, 12, 5180–5192, 2020.
34. Fortunati, E., Rinaldi, S., Peltzer, M., Bloise, N., Visai, L., Armentano, I., Jiménez, A., Latterini, L., Kenny, J.M., Nano-biocomposite films with modified cellulose nanocrystals and synthesized silver nanoparticles. *Carbohydr. Polym.*, 101, 1122–1133, 2014.
35. Istiqola, A. and Syafiuddin, A., A review of silver nanoparticles in food packaging technologies: Regulation, methods, properties, migration, and future challenges. *J. Chin. Chem. Soc.*, 67, 11, 1942–1956, 2020.
36. Lotfi, S., Ahari, H., Sahraeyan, R., The effect of silver nanocomposite packaging based on melt mixing and sol–gel methods on shelf life extension of fresh chicken stored at 4 °C. *J. Food Saf.*, 39, 3, e12625, 2019.
37. Alizadeh Sani, M., Maleki, M., Eghbaljoo-Gharehgheshlaghi, H., Khezerlou, A., Mohammadian, E., Liu, Q., Jafari, S.M., Titanium dioxide nanoparticles as multifunctional surface-active materials for smart/active nanocomposite packaging films. *Adv. Colloid Interface Sci.*, 300, 102593, 2022.
38. Zhang, W. and Rhim, J.-W., Titanium dioxide (TiO2) for the manufacture of multifunctional active food packaging films. *Food Packag. Shelf Life*, 31, 100806, 2022.
39. Böhmer-Maas, B.W., Fonseca, L.M., Otero, D.M., da Rosa Zavareze, E., Zambiazi, R.C., Photocatalytic zein-TiO2 nanofibers as ethylene absorbers for storage of cherry tomatoes. *Food Packag. Shelf Life*, 24, 100508, 2020.
40. Alizadeh-Sani, M., Mohammadian, E., McClements, D.J., Eco-friendly active packaging consisting of nanostructured biopolymer matrix reinforced with TiO2 and essential oil: Application for preservation of refrigerated meat. *Food Chem.*, 322, 126782, 2020.
41. Agarwal, A., Shaida, B., Rastogi, M., Singh, N.B., Food Packaging Materials with Special Reference to Biopolymers-Properties and Applications. *Chem. Afr.*, 6, 1, 117–144, 2023.
42. Wang, J., Euring, M., Ostendorf, K., Zhang, K., Biobased materials for food packaging. *J. Bioresour. Bioprod.*, 7, 1, 1–13, 2022.
43. Ferri, M., Papchenko, K., Degli Esposti, M., Tondi, G., De Angelis, M.G., Morselli, D., Fabbri, P., Fully Biobased Polyhydroxyalkanoate/Tannin Films as Multifunctional Materials for Smart Food Packaging Applications. *ACS Appl. Mater. Interfaces*, 15, 23, 28594–28605, 2023.
44. Flórez, M., Guerra-Rodríguez, E., Cazón, P., Vázquez, M., Chitosan for food packaging: Recent advances in active and intelligent films. *Food Hydrocoll.*, 124, 107328, 2022.

45. Moalla, S., Ammar, I., Fauconnier, M.-L., Danthine, S., Blecker, C., Besbes, S., Attia, H., Development and characterization of chitosan films carrying Artemisia campestris antioxidants for potential use as active food packaging materials. *Int. J. Biol. Macromol.*, 183, 254–266, 2021.
46. Gómez-Estaca, J., Gavara, R., Catalá, R., Hernández-Muñoz, P., The Potential of Proteins for Producing Food Packaging Materials: A Review. *Packag. Technol. Sci.*, 29, 4-5, 203–224, 2016.
47. Lu, Y., Luo, Q., Chu, Y., Tao, N., Deng, S., Wang, L., Li, L., Application of Gelatin in Food Packaging: A Review. *Polymers*, 14, 2022, doi: 10.3390/polym14030436.
48. Sui Chin, S., Han Lyn, F., Nur Hanani, Z.A., Effect of Aloe vera (Aloe barbadensis Miller) gel on the physical and functional properties of fish gelatin films as active packaging. *Food Packag. Shelf Life*, 12, 128–134, 2017.
49. Mugwagwa, L.R. and Chimphango, A.F.A., Enhancing the functional properties of acetylated hemicellulose films for active food packaging using acetylated nanocellulose reinforcement and polycaprolactone coating. *Food Packag. Shelf Life*, 24, 100481, 2020.
50. Sadegh-Hassani, F. and Nafchi, A.M., Preparation and characterization of bionanocomposite films based on potato starch/halloysite nanoclay. *Int. J. Biol. Macromol.*, 67, 458–462, 2014.
51. Mustapha, S., Ndamitso, M.M., Abdulkareem, A.S., Tijani, J.O., Shuaib, D.T., Ajala, A.O., Mohammed, A.K., Application of TiO2 and ZnO nanoparticles immobilized on clay in wastewater treatment: a review. *Appl. Water Sci.*, 10, 1, 49, 2020.
52. Duncan, T.V., Applications of nanotechnology in food packaging and food safety: Barrier materials, antimicrobials and sensors. *J. Colloid Interface Sci.*, 363, 1, 1–24, 2011.
53. Bumbudsanpharoke, N., Choi, J., Ko, S., Applications of Nanomaterials in Food Packaging. 1533–4880.
54. Okamoto, M., Nam, P.H., Maiti, P., Kotaka, T., Hasegawa, N., Usuki, A., A House of Cards Structure in Polypropylene/Clay Nanocomposites under Elongational Flow. *Nano Lett.*, 1, 6, 295–298, 2001.
55. Pandiyaraj, K.N., Ramkumar, M.C., Arun Kumar, A., Padmanabhan, P.V.A., Pichumani, M., Bendavid, A., Cools, P., De Geyter, N., Morent, R., Kumar, V., Gopinath, P., Su, P.-G., Deshmukh, R.R., Evaluation of surface properties of low density polyethylene (LDPE) films tailored by atmospheric pressure non-thermal plasma (APNTP) assisted co-polymerization and immobilization of chitosan for improvement of antifouling properties. *Mater. Sci. Eng. C*, 94, 150–160, 2019.
56. Papadimitriou, L., Kaliva, M., Vamvakaki, M., Chatzinikolaidou, M., Immunomodulatory Potential of Chitosan-graft-poly(ε-caprolactone) Copolymers toward the Polarization of Bone-Marrow-Derived Macrophages. *ACS Biomater. Sci. Eng.*, 3, 7, 1341–1349, 2017.
57. Ibrahim, H.M. and Zairy, E.M.R.E.-., Chitosan as a Biomaterial — Structure, Properties, and Electrospun Nanofibers, in: *Concepts, Compounds and the*

Alternatives of Antibacterials, B. Varaprasad (Ed.), p. Ch. 4, IntechOpen, Rijeka, 2015.

58. Li, X. and Xia, W., Effects of concentration, degree of deacetylation and molecular weight on emulsifying properties of chitosan. *Int. J. Biol. Macromol.*, 48, 5, 768–772, 20110.
59. Bahal, M., Kaur, N., Sharotri, N., Sud, D., Investigations on Amphoteric Chitosan/TiO2 Bionanocomposites for Application in Visible Light Induced Photocatalytic Degradation. *Adv. Polym. Technol.*, 2019, 2345631, 2019.
60. Baranowska-Wójcik, E., Szwajgier, D., Oleszczuk, P., Winiarska-Mieczan, A., Effects of Titanium Dioxide Nanoparticles Exposure on Human Health-a Review. *Biol. Trace Elem. Res.*, 193, 1, 118–129, 2020.
61. Ye, M., Gong, J., Lai, Y., Lin, C., Lin, Z., High-Efficiency Photoelectrocatalytic Hydrogen Generation Enabled by Palladium Quantum Dots-Sensitized TiO2 Nanotube Arrays. *J. Am. Chem. Soc.*, 134, 38, 15720–15723, 2012.
62. Imanova, G.T., Crystal Structure of TiO2 Nanoparticles, in: *3rd International Scientific and Practical Internet Conference "Integration of Education, Science and Business in Modern Environment: Winter Debates*, Dnipro, Ukraine, 2022.
63. Kraśniewska, K., Galus, S., Gniewosz, M., Biopolymers-Based Materials Containing Silver Nanoparticles as Active Packaging for Food Applications-A Review. *Int. J. Mol. Sci.*, 21, 698, 2020.
64. Humbatova, S.F., Zeynalov, N.A., Taghiyev, D.B., Tapdiqov, S.Z., Mammedova, S.M., Chitosan Polymer Composite Material Containing of Silver Nanoparticle. *Dig. J. Nanomater. Biostruct.*, 11, 1, 39–44, 2016.
65. de Azeredo, H.M.C., Antimicrobial nanostructures in food packaging. *Trends Food Sci. Technol.*, 30, 1, 56–69, 2013.
66. Lok, C.N., Ho, Cm, Chen, R., Q.-Y. He, Yu, W.-Y., Sun, H., Tam, P.K.H., Chiu, J.F., Che, C.M., Silver nanoparticles: partial oxidation and antibacterial activities. 0949–8257.
67. Sondi, I. and Salopek-Sondi, B., Silver nanoparticles as antimicrobial agent: a case study on E. coli as a model for Gram-negative bacteria. *J. Colloid Interface Sci.*, 275, 1, 177–182, 2004.
68. Jeong, S.H., Yeo, S.Y., Yi, S.C., The effect of filler particle size on the antibacterial properties of compounded polymer/silver fibers. *J. Mater. Sci.*, 40, 20, 5407–5411, 2005.
69. Morones, J.R., Elechiguerra, A.J.L., Camacho, A., Holt, K., Kouri, J.B., Ramírez, J.T., Yacaman, M.J., The bactericidal effect of silver nanoparticles. *Nanotechnology*, 16, 2346, 2005.
70. Pantani, R., Gorrasi, G., Vigliotta, G., Murariu, M., Dubois, P., PLA-ZnO nanocomposite films: Water vapor barrier properties and specific end-use characteristics. *Eur. Polym. J.*, 49, 11, 3471–3482, 2013.
71. Pirsa, S. and Shamusi, T., Intelligent and active packaging of chicken thigh meat by conducting nano structure cellulose-polypyrrole-ZnO film. *Mater. Sci. Eng.C*, 102, 798–809, 2019.

72. Liu, S., Ng, A.K., Xu, R., Wei, J., Tan, C.M., Yang, Y., Chen, Y., Antibacterial action of dispersed single-walled carbon nanotubes on Escherichia coli and Bacillus subtilis investigated by atomic force microscopy. *Nanoscale*, 2, 12, 2744–2750, 2010.
73. Liu, S., Wei, L., Hao, L., Fang, N., Chang, M.W., Xu, R., Yang, Y., Chen, Y., Sharper and Faster "Nano Darts" Kill More Bacteria: A Study of Antibacterial Activity of Individually Dispersed Pristine Single-Walled Carbon Nanotube. *ACS Nano*, 3, 12, 3891–3902, 2009.
74. Song, J. and Jang, J., Antimicrobial polymer nanostructures: Synthetic route, mechanism of action and perspective. *Adv. Colloid Interface Sci.*, 203, 37–50, 2014.
75. Silvestre, C., Pezzuto, M., Cimmino, S., Duraccio, D., Polymer Nanomaterials for Food Packaging: Current Issues and Future Trends, in: *Ecosustainable Polymer Nanomaterials for Food Packaging*, C. Silvestre and S. Cimmino (Eds.), pp. 1–26, CRC Press, 2013.
76. Kim, I., Viswanathan, K., Kasi, G., Thanakkasaranee, S., Sadeghi, K., Seo, J., ZnO Nanostructures in Active Antibacterial Food Packaging: Preparation Methods, Antimicrobial Mechanisms, Safety Issues, Future Prospects, and Challenges. *Food Rev. Int.*, 38, 4, 537–565, 2022.
77. Parmar, F., Panchal, B., Macwan, S., Patel, S., Effect of packaging materials and calcium chloride on storage and shelf life of broccoli under cold storage condition. *Pharma Innov. J.*, 11, 9S, 2278–2284, 2022.
78. Ituen, E., Yuanhua, L., Verma, C., Alfantazi, A., Akaranta, O., Ebenso, E.E., Multifunctional silver nanocomposite: A potential material for antiscaling, antimicrobial and anticorrosive applications. *JCIS Open*, 3, 100012, 2021.
79. Carbone, M., Donia, D.T., Sabbatella, G., Antiochia, R., Silver nanoparticles in polymeric matrices for fresh food packaging. *J. King Saud Univ. - Sci.*, 28, 4, 273–279, 2016.
80. Lavanya, M., A Brief Insight into Microbial Corrosion and its Mitigation with Eco-friendly Inhibitors. *J. Bio Tribo Corros.*, 7, 3, 125, 2021.
81. Tripathi, A.K., Thakur, P., Saxena, P., Rauniyar, S., Gopalakrishnan, V., Singh, R.N., Gadhamshetty, V., Gnimpieba, E.Z., Jasthi, B.K., Sani, R.K., Gene Sets and Mechanisms of Sulfate-Reducing Bacteria Biofilm Formation and Quorum Sensing With Impact on Corrosion. *Front. Microbiol.*, 12, 2021.
82. Plakunov, V.K., Gannesen, A.V., Mart'yanov, S.V., Zhurina, M.V., Biocorrosion of Synthetic Plastics: Degradation Mechanisms and Methods of Protection. *Microbiology*, 89, 6, 647–659, 2020.
83. Kaushal, M., Sharma, G., Mittal, S., Controlling of Odour Issues in Packaging Industries. *Indian Pulp Paper Tech. Assoc.*, 30, 89–91, 2018.
84. Kumar, A., Anant, R., Kumar, K., Chauhan, S.S., Kumar, S., Kumar, R., Anticorrosive and electromagnetic shielding response of a graphene/TiO2–epoxy nanocomposite with enhanced mechanical properties. *RSC Adv.*, 6, 114, 113405–113414, 2016.

85. Harsha, V.S.S., Sharma, A., Tambe, P., Graphene oxide reinforced epoxy nanocomposites coatings for corrosion protection: a review. *J. Physics: Conf. Ser.*, 2225, 1, 012002, 2022.
86. Rossa, V.A.-O., Monteiro Ferreira, L.A.-O., da Costa Vasconcelos, S.A.-O., Tai Shimabukuro, E.T., Gomes da Costa Madriaga, V., Carvalho, A.P., Castellã Pergher, S.B., de Carvalho da Silva, F.A.-O., Ferreira, V.A.-O.X., Conte Junior, C.A.-O., de Melo Lima, T.A.-O., Nanocomposites based on the graphene family for food packaging: historical perspective, preparation methods, and properties. *RSC Adv.*, 12, 14084–14111, 2022.
87. Jeon, H., Park, J., Shon, M., Corrosion protection by epoxy coating containing multi-walled carbon nanotubes. *J. Ind. Eng. Chem.*, 19, 3, 849–853, 2013.
88. Williams, J.R., Rayburn, J.R., Cline, G.R., Sauterer, R., Friedman, M., Effect of allyl isothiocyanate on developmental toxicity in exposed Xenopus laevis embryos. *Toxicol. Rep.*, 2, 222–227, 2015.
89. Wonnie Ma, I.A., Shafaamri, A., Kasi, R., Zaini, F.N., Balakrishnan, V., Subramaniam, R., Arof, A.K., Anticorrosion properties of epoxy/nanocellulose nanocomposite coating. *BioResources*, 12, 2, 2912–2929, 2017.

11

Development and Characterization of Nanostructured Thin Films for Corrosion Control Applications

M. Geetha Devi[1]*, R. Senthilkumar[2] and Hebatallah Al Jabri[1]

[1]*Mechanical & Industrial Engineering Department, National University of Science & Technology, Muscat, Sultanate of Oman*
[2]*College of Engineering and Technology, University of Technology & Applied Sciences, Suhar, Sultanate of Oman*

Abstract
Corrosion is a massive problem all over the world due to the exposure of material surfaces to their neighboring environment and total control of corrosion is a challenging task. Most of the corrosion problems occur in the process industries and would be meritoriously addressed by suitable preventive measures. Conventional corrosion control measures, such as galvanization, cathodic protection, powder coating, and protective paints, shield the metal surface from deterioration. But, those techniques require regular maintenance. However, the majority of the approaches are either ineffective or expensive. By protecting the metal surface against corrosion by thin layer deposition using polymer-based coating at subatomic levels, the progress in nanoengineered techniques offers a beneficial and practical solution to such problems. The corrosion rates are governed by the period of contact between the metal surface and the protective coating. Due to the elevated maintenance cost of the corroded equipment in the process industries, it is obligatory to safeguard the device against corrosion. Hence, this chapter presents an overview of the corrosion problems encountered in process industries, the impact of corrosion in process equipment, and the role of nanotechnology in corrosion control. The development of nanostructured thin films using polymeric materials and nanoparticles for the corrosion inhibition studies of mild steel specimens is described. The utilization of titanium dioxide (TiO_2) nanoparticles and polyaniline (PANI) in the preparation of nanocomposite coating for corrosion

**Corresponding author*: geethadevi@nu.edu.om

Hari Murthy, Vinay Jha Pillai, Kukatlapalli Pradeep Kumar and Matthew Cowan (eds.) Novel Anti-Corrosion and Anti-Fouling Coatings and Thin Films, (323–352) © 2024 Scrivener Publishing LLC

inhibition studies is highlighted. Different techniques employed in the creation of nano–thin layers and their characterization techniques are illustrated. The thin nanolayer deposition process on mild steel pipe materials is carried out by dip coating technique with particular emphasis on stability studies. The developed nanocomposite thin films were tested for stability studies at various atmospheric conditions and the exposure time in order to establish optimum conditions. The nanoparticles and thin films used in the corrosion inhibition studies were subjected to a series of characterization techniques *viz* scanning electron microscopy (SEM), Fourier-transform infrared spectroscopy (FTIR), energy-dispersive x-ray analysis (EDX), X-ray diffraction (XRD), dynamic light scattering (DLS), and potentiostatic and polarization techniques. SEM was used to visualize the microstructural characterization and surface features of the preformed thin layer to assess the effectiveness of layer deposition. The significance of the selection of titanium dioxide (TiO_2)–PANI as the layer material is due to their large surface-to-volume ratio and stronger corrosion inhibition potential by shielding the metal surface in diverse conditions.

Keywords: Corrosion inhibitors, nanocoating, nanoparticles, nanocomposite, polyaniline, potentiodynamic test, scanning electron microscopy, thin films

11.1 Introduction

Corrosion is the deterioration or decay of materials due to the interaction with their surrounding environment. Corrosion is a serious issue affecting metal objects and damages the metal surface due to chemical and electrochemical reactions with oxygen, hydrogen, or hydroxide, which affects the economy and industry [1]. The life span and durability of metallic objects can be enhanced by protecting the material against corrosion. The rusting of metals occurs due to the formation and deposition of oxides or salt of the original metals by electrochemical corrosion. Corrosion may also occur in nonmetals such as ceramics and polymers. The adverse effects of corrosion are the reduction of mechanical strength, permeability to liquids and gasses, aesthetic appeal, and faster degradation rate leading to structural damage and the equipment will be at high risk. Various ways to control corrosion in metal industries are the use of appropriate corrosion inhibitors, the development of protective coatings, and the application of thin film technology to increase the life span of the metal surface [2–4]. Polymer coatings, organic resins, paints, epoxy resin, and conductive polymers are widely employed coating materials with high durability and improved performance to prevent the rusting process. Among these, epoxy resin is gaining attention owing to its attractive anticorrosion resistance, notable chemical stability, and better adhesive properties [5–7]. The appealing chemical and physical properties of polyaniline (PANI)-based coatings have been explored as a layered material on metal surfaces; however, the mechanical and adhesion properties are not attractive [8]. Organic polymers have a tendency to preclude the chemical reaction kinetics happening on the metallic surface and interfaces [9]. Metal substrates are coated with nonconducting polymers such as thin electropolymerized films of poly(1,3-diaminobenzene) (DAB) as well as conducting polymers like polypyrrole and PANI have been used as corrosion inhibitors on a larger scale [10]. The life span and performance of metallic surfaces can be improved using protective coatings as inhibitors [11]. The types of anti-corrosive coatings include epoxy, fluoropolymer, nanocomposite coatings, healing coatings, and hybrid coatings [12].

The transportation of oil and gas through the pipelines in the petrochemical industries experiences a destructive attack due to the corrosion of pipe material that causes a significant impact on the environment. The aqueous environments produce severe corrosion problems starting from the extraction stage to the oil processing stage. The pipeline networks in the petroleum sector will be damaged resulting in huge losses, and the management and control of corrosion lead to extra costs in terms of billions of dollars. The major challenges encountered by petroleum refinery operations include the physicochemical fundamentals of corrosion at various processing parts, mechanisms of corrosion, sources of corrosion, and contemporary mitigation techniques such as cathodic protection, engineering design, use of inhibitors, and metal surface coating. Many pollutants, some of which may bc sulfur and caustic in specific circumstances, are present throughout the transit of natural gas and crude oil. In the oil and gas processing industry, these impurities may include hydrogen sulfide, carbon dioxide (CO_2), or free water [13]. The interior surfaces of the oil and gas pipelines will interact with H_2S, CO_2, and free water which are at high corrosion risk.

The corrosion issues happening in the oil refineries and petrochemical industries have been less extensively studied by researchers, materials scientists, and corrosion engineers. A significant research gap was noticed in the development of intelligent, environmentally friendly, and high-temperature resistant corrosion inhibitors. Inadequate prediction models and insufficient data collection techniques during the processing of ammonium bisulfide (NH_4HS) lead to corrosion [14]. Hence, further research is required for the development of novel corrosion control techniques by introducing nanomaterials and super hydrophobic coatings with temperature-resistive properties (>100°C), for a feasible and comprehensive study of ammonium bisulfide, ammonium chloride, and hydrogen sulfide corrosion in oil refineries.

Corrosion is a severe problem in the construction and manufacturing sectors. Thus, a clear understanding of the various forms of corrosion and its causes is an essential topic of interest in recent times. Localized corrosion in the specific surface regions of oil and gas pipelines is a serious concern in oil refineries and may lead to pitting or stain corrosion thereby resulting in equipment failure. The majority of the corrosion prevention techniques are not effective and consume large amounts of inhibitors, which will result in heavy losses to the industry. Therefore, it is important to find cost-effective and long-lasting techniques to address the corrosion problems. Hence, the main focus of this chapter is the fabrication of thin films with nanometer precision using PANI and TiO_2 nanocomposite to appraise its corrosion inhibition effect on mild steel. The main reason for the selection of PANI-TiO_2 in this study is due to their nontoxic and anticorrosive nature with limited solubility of the film in aqueous media. The nanocomposite acts as a protective layer and inhibitor with improved passivation capacity compared to other types of coating. Hence, this chapter discusses the construction of PANI-TiO_2 nano–thin films for corrosion inhibition studies. Remarkable properties of anticorrosiveness and economic and environmental stability of PANI are considered for this study.

11.2 Various Forms of Corrosion in the Petroleum Industry

Various approaches are employed to protect the process equipment from corrosion including surface coating, inhibitors, cathodic protection, environmental alterations, choice of suitable metal and surface modifications, and plating [15].

11.2.1 Environmental Cracking or Caustic Embrittlement

Environmental cracking is the brittle fracture of steel or a ductile material due to the surrounding environmental conditions. Hydrogen attack, stress corrosion cracking (SCC), fatigue, and fretting corrosion are characteristics of this type of corrosion, which may impose substantial damage to the metal surface. Various strains generated through heat treatment processes, residual cold work, or welding might lead to sudden cracking.

11.2.2 Fretting Corrosion

Fretting corrosion occurs when two surfaces are continuously rubbed with each other. Corrosion and fretting happen simultaneously when materials are put under strain on a regular basis. Vibration and slip between materials under varying amounts of load will result in fretting corrosion.

11.2.3 High-Temperature Corrosion

As the name suggests, this type of corrosion might occur at extremely high temperatures (often exceeding 400°C). High-temperature corrosion is due to chemical degradation caused by gasses or molten metals. This particular type of corrosion occurs in gas turbines, furnaces, and other machinery parts in which hot gasses including certain contaminants are involved in the processing.

11.2.4 Pitting Corrosion

One type of localized corrosion that results in occasional tiny cavities or holes developed on the surface of metals in a restricted area is known as pitting corrosion [16–18]. The substrate is more inclined to corrosion when the barrier oxide layer formed on the surface begins to degrade. Pitting corrosion is an electrochemical process in which the cathode stands in for the remaining passive layer and the anode represents the pits. Pitting corrosion usually happens in the presence of a destructive anion such as chloride ion (Cl^-). The severity of pitting corrosion can vary subject to the logarithm of the bulk Cl^- concentration [19].

11.2.5 Crevice Corrosion

If an insulated metal surface has limited access to the surrounding environment, it may lead to crevice corrosion due to the presence of chloride ions and hence results in crevice corrosion. Electrochemical corrosion cells will be created when unprotected surfaces are exposed to an environment with an oxygen level lower than the surrounding [20] and are known as crevice corrosion, observed in the majority of oil pipelines [21].

11.2.6 Galvanic Corrosion

When two metals with dissimilar electrochemical actions come into electrical contact with one another, galvanic corrosion occurs. Corrosion on the active metal surface is driven by the passive metal. Environmental, geometrical, metallurgical factors, electrolyte characteristics, surface state, and reverse electrode potentials are the components that influence and contribute to this type of corrosion [22].

11.2.7 Stress Corrosion

Stress corrosion is also called fatigue corrosion. The breakdown of a metal as a result of stress and chemical attack is branded as SCC. SCC is a phenomenon linked to static tensile stress due to the surrounding environment. In some systems, a metallurgical state causes component failure as a result of the start and spread of a high aspect ratio crack. Fine cracks that can cause component failure and possibly structural failure occur as a result of SCC [23]. Chloride SCC (CSC) and sulfide SCC are the principal types of SCC. Steel specimens can experience SCC cracking in hydrogen sulfide environments causing severe problems in the oil drilling sector. However, in the presence of oxygen and chloride ions, under tensile stress, and at elevated temperature, the CSC develops in austenitic steels.

11.2.8 Intergranular Corrosion

Intergranular corrosion or localized corrosion develops at and near grain boundaries. As the grain boundary is very active, it has a tendency to attack the grain boundary. It is a special type of corrosion that is microstructurally impacted in which the 'region' of the alloy's grain boundaries is electrochemically separated from the nearby alloy microstructures. The binding force between crystal grains can be considerably weakened by Intergranular corrosion, and this would significantly reduce the mechanical strength of metals and alloys [24]. Intergranular corrosion occurs in nickel-base alloys, stainless steels, and aluminum/magnesium alloys [25].

11.2.9 Erosion Corrosion

This type of corrosion usually arises as a result of particle or relative fluid movement against the inner pipe surface. Erosion rates can rapidly increase as a result of fluid turbulence. Galvanic corrosion will occur if the region is anodic to the nearby pipe wall. The smooth fluid flow can be disrupted by improperly completed interior pipe surfaces or pits developing leading to turbulence of localized fluid. A rapid degradation rate might arise from this. An extremely serious pitting corrosion may result due to the interaction with cavitation, erosion, and corrosion under high pressure.

The use of wear-resistant materials can help avoid or decrease cavitation in conjunction with erosion corrosion with the support of proper design [26].

11.2.10 Sweet (CO_2) Corrosion

Sweet (CO_2) corrosion is a special type of corrosion phenomenon, particularly in the presence of carbon dioxide, and corrosion takes place in the inner regions of the oil pipelines. The corrosive nature of organic acids and hydrogen sulfide (H_2S) may also contribute to this form of corrosion. CO_2 corrosion is an electrochemical process that is commonly found in carbon steel.

11.2.11 Uniform Corrosion

This type of corrosion is the most frequent type of corrosive attack that develops an entire part of the material surface. Iron and steel specimens normally experience this type of corrosion when they are in contact with soil and water.

11.2.12 Microbiologically Influenced Corrosion (MIC)

Microbiologically influenced corrosion happens due to the existence and activity of different types of bacteria spread on the surface of the material in the form of biofilms. Most of the research studies dealing with bio-induced corrosion, biodegradation, MIC, and the microbial effect are often active. With the initiation of molecular microbiological systems in the oil and gas sector, MIC research underwent a revolution in the past decade. Oxygen deficiency, darkness, extreme salinity, acidic to basic pH, and extreme temperatures ranging from 290 K to 386 K are the favorable factors influencing MIC. Prediction and control of MIC are therefore challenging tasks due to a number of circumstances and constraints in estimating the magnitude of the influence of bacterial growth on corrosion [27–29].

11.3 Corrosion's Effects on Various Equipment Used in the Petrochemical Industry

11.3.1 Heat Exchanger

Heat exchangers are one of the major equipment that plays a crucial role in many process industries such as chemical and pharmaceutical, oil and gas, pulp and paper, and petroleum refining, etc. These industries depend heavily on heat exchangers since they are used as the main heat transfer equipment to exchange heat from one fluid to another [30]. Heat exchangers are constructed of metals that will corrode due to the transportation of these fluids. It may be exposed to a range of process conditions in each of these sectors. The direct cost of all forms of corrosion in the United States was estimated in 2022 to be $276 billion annually or 3.1% of the nation's gross domestic product [31].

11.3.1.1 Different Forms of Corrosion in Heat Exchangers

The corrosion process may take place in different forms depending on the type of material, climatic conditions, and operating circumstances. Heat exchangers are commonly employed to cool hydrocarbon gas using a coolant passing through the tubes [32]. The gas may escape due to galvanic corrosion between the titanium and carbon steel coating, overpressurizing the shell. The tubes and shell will be ripped off from the tube sheet, and the gas that emerges may result in flaring eventually leading to an explosion.

The tubes of vertical-type shell-and-tube heat exchangers need to be replaced every 2 to 5 years. Cooling water will be circulated around the tubes to cool the vapor and this may lead to heat exchanger damage (>70% damage for 159 out of 1410 tubes) and thereby develop brown and reddish rust deposition on the interior and exterior surfaces [33]. The corrosion will be very severe just below the periphery of the tube bundle. This is due to the corrosive nature of chlorides present in hot water which is circulated through the tubes. The vapor-liquid interface is developed by the buildup of non-condensable gasses and moves to the top of the shell region. The interior of the heat exchange tubes exhibits less uniform corrosion than the cooling water side. However, localized corrosion such as pitting corrosion and erosion corrosion will also progress on the damaged interface. Uniform corrosion is not the only type of corrosion seen on the interior surface of the tubes.

The tube bundles located at the central position of the shell-and-tube heat exchanger contain a mixture of condensate and brine, while the bottom and upper tubes received preferred fills of brine and steam, respectively [34]. Due to these circumstances, the sourness of the condensing water increases due to an increase in the concentration of CO_2 and H_2S gasses which are mostly acidic in nature. Swift growths of pits are experienced in one row of tubes due to the influence of high acidity. On top of that, precipitated arsenic from the brine in the pits enhances the corrosion rate due to galvanic effects.

Pitting-related failures are observed in plate-type heat exchangers employed in the textile industry cooling loops [35]. When the heat exchanger is opened for maintenance after 2000 hours of operation, it exhibits certain plates perforated by pits, which are better prevented by using 316-L stainless steel due to its corrosion-resistant potential. This may be due to the presence of excess chlorides in the cooling water that caused a local breakdown of the passive oxide coating.

Another type of corrosion problem is crevice corrosion takes place mostly in shell-and-tube heat exchangers. This crevice corrosion is seen in the spaces between the tube sheets and rolled tubes, open welds, and bolt holes. Additionally, intricate geometric patterns with sharp edges are vulnerable to crevice corrosion [36]. In oil coolers, crevice corrosion is found on the tube sheet [37]. However, the crevice is generated between the tube sheet and a deflector in the cooler's head, and not in between the tube sheet and the tubes. Stress corrosion cracking (SCC) and corrosion fatigue are frequently started at different stress concentrations [38]. After 48 hours of operational period at an oil refinery, the heat exchanger will fail due to the formation of polythionic acid from sulfate-bearing deposits and water vapor [39]. The interaction between the acid and internal forces in the bend results in cracks at the metal surface.

Erosion corrosion is a combination of corrosion and erosion, which is brought out by the relative movement of a highly corrosive liquid and a metal wall. When the fluid contains abrasive substances like sand particles, erosion corrosion will occur at the inlet of plate heat exchangers, which is more severe [40]. The inlet flow to the heat exchangers is similarly vulnerable to erosion corrosion, particularly on the tube bundle, and this phenom enon is sometimes referred to as inlet-tube corrosion [41].

On the welded areas of stainless steel (type 304), intergranular corrosion takes place frequently. Sensitization is the term used to describe the deterioration of the alloying element at the grain boundaries as a result of high-temperature operation. The corrosion resistance of these chromium-depleted areas is exhausted and the rest is the properties of the base metal [36]. Carbides are soluble at high temperatures and must acquire chromium from the adjoining metal in order to precipitate at the grain boundaries, thereby forming chromium carbides. Corrosion fractures may develop at the grain boundaries when the corroded metal is subjected to tensile stress and this phenomenon is named intergranular SCC [42].

Galvanic corrosion is frequently observed during the contact between the tubes and the baffles or tube sheets, delicate locations including the connections between the baffle and the shell in shell-and-tube heat exchangers [36]. Galvanic corrosion can occur at welded, brazed, or soldered seams in plate heat exchangers [40]. The collapse of a heat exchanger made up of carbon steel and titanium covering on an offshore area of gas production is a specific case of galvanic corrosion. In this circumstance, the carbon steel becomes rusted because it is anodic to the titanium [43].

Microbiologically induced corrosion is frequently observed in cooling water systems, where they slow down heat transfer rates and elevate pressure drops over the condenser [44]. A biofilm is created when these microbes adhere to a surface, which affects corrosion. Corrosion can be initiated by the microbes, which would not happen in the same environment in the absence of bacteria. Furthermore, the biofilm has the power to alter the kind of corrosion (for example, from uniform to localized corrosion) as well as the pace of corrosion. However, it is also conceivable that the biofilm does not influence corrosion. In a shell and tube heat exchange, corrosion tubercles are discovered on the shell, where the cooling water is circulated. While the majority of tubercles had a depth of 2 to 4 mm, some of them pierced the metal, which had an initial thickness of 6.2 mm [45]. According to studies, the metal is composed of a significant austenite phase and a smaller ferrite phase. At this ferrite phase, the surface is microscopically rough, which allows bacteria to adhere and eventually give rise to cells that concentrate oxygen. Local corrosion is caused by these cells, which also produce the tubercles.

11.3.2 Crude Distillation Overhead System

Although there is corrosion in every facet of the petrochemical industry, an overhead system of crude distillation is the focus area of this research study. Carbon steel, a metal that is naturally susceptible to corrosion, is employed as the material of construction in the crude distillation overhead system [46]. If metallic elements are in close contact with certain specific impurities present in crude oil such as acid, chlorides, and salty solutions, then corrosion initiates. Studies show that corrosion will always affect these metallic structures until they are protected [47]. The part of a refinery where crude oil is separated into various fractions is called the overhead system, which is referred to be the beginning of crude oil production in the petrochemical plant [48]. The fractions are separated based on their constituent boiling points and relative volatilities. The temperature can decrease below 20°C to distinguish the lighter materials from the heavier ones and can soar to 400°C. The overhead system in a refining operation is composed of a desalter unit, the top section of the distillation column, a furnace, a pre-flash drum, and a vacuum distillation part to make up the crude [49].

The presence of carbon dioxide, hydrogen sulfide, and chlorides in the crude distillation overhead system creates corrosion. Study shows that an acidic environment is a significant factor influencing the corrosion rate in the crude distillation overhead system [50]. Additionally, the extent of corrosion in the crude distillation unit is increased by the increased amounts of amines and ammonia. Sulfide and chlorides are dissolved in water to form corrosion in carbon steel during crude distillation. The crude distillation overhead system often experiences rapid failure due to a general corrosion attack that causes the metallic structure to lean down.

Pitting corrosion is more common when alloy steel is used to build the crude distillation overhead system, especially at temperatures around 338°C. When the processing temperature exceeds 310°C, the crude distillation overhead system also undergoes pitting corrosion, and this is often severe. Pitting corrosion occurs when corrosive salts come into contact with the crude distillation overhead system, seriously destroying the system. Owing to the presence of chloride and sulfide, pitting corrosion also happens in the crude distillation overhead system during the production process of crude oil [51].

In a crude distillation overhead system, intergranular corrosion occurs in the presence of sulfide, causing metallic fracture failure. The presence of hydrogen, amines, acids, and water is the reason for intergranular corrosion in crude distillation overhead systems [52]. Furthermore, the elements of oxygen and hydrochloric acid are the major causes of intergranular corrosion in crude distillation overhead systems.

Equimolar vaporization and condensation happen during the operation of the crude distillation overhead system. The bottom section of the distillation column is heated to a high temperature (extremely hot), and it cools off as it rises to the top, where it separates itself as an overhead product [53]. Due to their distinct boiling temperatures, the components with high boiling points will condense and travel toward the bottom while those with low boiling points rise at the top. Normally, crude oil is transported through the pipeline into the storage tank for the distillation process.

11.4 Conventional Corrosion Control Techniques

The solvents, natural gas, petroleum products, fuels, oil, and gas will affect the refinery equipment and are extremely susceptible to a range of corrosion phenomena due to a number of unified causes, which will turn into serious accidents. There are many ways to avoid or reduce corrosion risks, which include the selection of corrosion-resistant materials for appropriate function, using the right design principles, applying cathodic protection and anticorrosive chemical coatings, controlling technological parameters, and, most importantly, inspecting and monitoring the processes at each stage. The choice of corrosion control measures depends on the relationship between the corrosion-causing impurities in natural gas, chemical composition, and the process parameters of temperature, pressure, and flow rate regime. Corrosion is primarily influenced by the choice of metal, environmental characteristics and circumstances, and the border metal environment [54]. This falls into three categories: the treatment of the metal (material selection), environment (neutralization, removal of water, removal of CO_2 and H_2S, and removal of salts), and the boundary metal environment (injection of inhibitors, usage of coatings, and cathodic protection).

11.4.1 Engineering Design

Engineering design includes the use of standards, specifications, procedures, and reports for natural gas systems; the geometry of the equipment to make fewer crevices, control flow rate, prevent air infiltration, and avoid "dead legs"; and the selection of the right material of construction, protection methods, control, corrosion monitoring for designed period of service, choosing suitable coatings, corrosion inhibitors, biocides, anti-scaling agents, and cathodic protection.

11.4.2 Selection of Material for Construction

In natural gas technology, the choice of materials designates the application of metals and alloys, polymers, and composite materials (composites) appropriate to various situations. No material is flawless and impervious to all media and environmental influences. It is appropriate to refer to materials that are resistant to corrosion as stainless steels and nickel-based alloys. More than 90% of the materials used in natural gas systems are made of carbon steel. The API 5D standard is used for drilling pipes, the API 5CT standard for casings and tubes, and the API 5L standard for line pipes [55].

Design calculations for the usage of carbon steel specimens should include a particular corrosion allowance [56] and the appropriate heat treatment. In order to prevent internal corrosion, the application of surface coatings and cathodic protection must be combined with the use of carbon steel in natural gas systems. This also applies to equipment and structures that are underground or submerged.

The most commonly employed stainless steel alloy in natural gas systems is a type of 13Cr, which is also applied in seawater and wells. When the overall corrosion rate is less than 0.05 mm/year, it is feasible to choose a specific alloy that is acceptable for a given temperature, pressure, CO_2, H_2S, and NaCl concentration [57]. When varying amounts of H_2S and CO_2 are present, titanium alloys can be used in natural gas wells operated at high pressures and temperatures [58]. Cryogenic tanks used to store liquefied natural gas (LNG) at −162°C are made up of iron-nickel (9%) alloys.

Considering their use in natural gas systems, polymeric materials and composites have a specific position [59, 60]. These materials have substantially stronger chemical resistance when exposed to different gasses and solvents. Polymers have less tolerance towards high temperatures and inferior mechanical properties to metals. In natural gas systems, polymeric materials are utilized as seals, gaskets, rings, sumps, tanks, and pipelines that come into contact with freshwater, saltwater, cooling water, fire-fighting water, acidic and alkaline solutions, the atmosphere, and soil. Characteristics of polymeric materials are mechanical, thermal, electrical, and chemical resistance. The characteristics of composites and polymers are significantly impacted by temperature. Polymeric substances are susceptible to chemical breakdown when heated. It may become brittle and rigid as it cools. The coating materials chosen for the corrosion studies consist of polymeric and composite materials.

11.4.3 Coatings

The purpose of coating in metal is to protect the metal surface from any aggressive environment. The surface coating may be organic, inorganic, and metallic [58]. Paints, polymeric compounds, greases, and other paraffinic mixes are examples of organic coatings. Enamels, ceramics (including cement), and coating categories are utilized to safeguard a variety of buildings and equipment that are exposed to natural gas, the environment, water, dirt, and various technological advancements. The locations of natural gas distribution networks vary geographically. Therefore, it is crucial to develop coating standards that will safeguard buildings and machinery in a variety of circumstances, including those involving resistance to gas, fuel, water, aqueous solutions, environmental conditions, soil, high temperatures, and thermal insulation [61].

11.4.4 Cathodic Protection

The application of organic coatings to metallic structures is the prominent corrosion control technique, while cathodic protection is used as a supplement to safeguard the defects that always exist in organic coatings or manifest themselves during application and service. Electric current and electric potential can be used to prevent corrosion on metal surfaces thanks to the electrochemical process of metal corrosion in electrolyte solutions. Cathodic protection therefore functions only in electrolyte solutions and is ineffective in materials with high electrical resistivity that are unable to conduct electrical current, such as natural gas, air, oil, and fuels. Cathodic polarization can be achieved in practice in one of two ways: by connecting the primary metal to be protected to a less noble metal or by connecting to the negative pole of the external power source. The first approach relies on the practice of sacrificial anodes, which are used to convert metallic constructions into cathodes that do not corrode by being sacrificed and dissolved as anodes. The second cathodic protection relies on a connection to the rectifier's negative pole and the application of impressed electric current. This technique is also referred to as an active cathodic protection technique. Cathodic protection does not apply to media that is flowing through or being stored within, such as petrol, oil, gasoline, or water. It only applies to exterior surfaces that come into touch with the electrolyte. There are several standards for cathodic protection, testing procedures, and monitoring effectiveness [62, 63].

11.4.5 Corrosion Inhibitors

An essential element in the prevention and control of corrosion in natural gas systems is the application of corrosion inhibitors. The primary types of inhibitors employed are amines, which were applied in natural gas pipelines from the year 1940s through the 1980s as the exclusive method of corrosion inhibition. Those types of inhibitors were ineffective up to the 1980s when deep wells with high temperatures and pressures were discovered, at which point it was decided to utilize specific alloys.

Inhibitors work by adhering to metal surfaces and obstructing the entry of hostile compounds to those surfaces, which is the basis of their mechanism of action. Therefore, a typical increase in inhibitor concentration decreases the amount of metal exposed to hostile components. The majority of "filming amine" corrosion inhibitors used in gas systems create shielding on the surface of metal [64]. The components of crude oil have the ability to prevent corrosion on the inner surface of pipes when they transition into an aqueous phase [65, 66].

11.5 The Role of Nanotechnology in Corrosion Control

Nanotechnology is an emerging area of research with the potential to offer groundbreaking and revolutionary changes in the oil and gas sectors. The development and application of nanocoating have become an interesting area of interest in recent times. The attainment of corrosion control and ecological concerns are the strategic factor that requires significant consideration while formulating the surface coatings. Nanomaterials provide realistic solutions to safeguard oil pipelines from corrosion by protective coating in the form of polymeric layers at atomic levels [67–69]. The development of nanocoating over the surface of oil pipelines will enhance the life span and protect the surface from chemical and environmental attacks. The pipelines used to transport oil and gas from the refining operations are more susceptible to corrosion due to the presence of hydrocarbons and CO_2 [70, 71]. The conventional coating techniques include zirconium-based nanocoatings, chromium-based coatings, and polymer-based coatings. However, further development of corrosion inhibitors is needed to meet the current requirement of high-performance surface coatings by introducing the novel concept of nanotechnology. The problems encountered with the traditional coating methods can be addressed by nanotechnology-enabled coating. The attractive features of nanocoatings are enhanced surface area, fine particle diameter, and adjustable functional properties at nanoscale dimensions. The enhanced surface area offers excellent adhesion properties and hence increases the life of the coating. Additionally, the chemical, mechanical, and electrical properties will be augmented with nanosized materials making the surface coating harder, stronger, and better resistant to corrosion and wear. Nanoscale coating shows better corrosion resistance than microscale coating. The application of nanotechnology can help to develop smart coatings to withstand external factors such as stress, pH, heat, humidity, electromagnetic radiation, and coating deformation by discharging a regulated quantity of inhibitors in order to restore and cure faults and damages.

11.5.1 Types of Nanocoatings Employed in Corrosion Control

The protective coating will reduce or control corrosion with the help of a variety of coating materials. The corrosion rate can be minimized by the surface modification of metal by deposition of a protective layer by offering active or passive protection. The addition of chemical inhibitors to destructive environments results in active corrosion. Nanocoatings are suitable in eyeglasses, electronic devices, biomedical implants, and bone fixation materials. Nanostructured materials are widely employed as a coating layer in corrosion control due to the prospects of amalgamating materials with distinctive properties, such as wear resistance, increased hardness, attractive optical transparency, and toughness, making them scientifically attractive for various industrial applications.

11.5.1.1 Ceramic Nanocoatings

The common types of ceramic nanocoatings are alumina, tantalum oxide, titanium oxide, graphene-based nanocoatings, and zirconia. Titanium dioxide is preferred as anticorrosive nanocoating material due to its high refractive index, high abrasive and corrosion resistance, better photocatalytic activity, and attractive ultraviolet (UV) resistance. The coating methods include dip coating, atomic layer deposition, and sol-gel methods. Research studies show that nanocoating on stainless steel by sol-gel method exhibited better corrosion resistance.

11.5.1.2 Alumina Nanocoatings

Thin films of alumina exhibit better corrosion resistance with enhanced mechanical properties, and hence, they are preferred in surface passivation and gas diffusion barriers. The methods employed in the fabrication of alumina thin films are atomic layer deposition and sol-gel method. Better film density can be ensured by plasma-induced atomic-level layer deposition. Nanostructured alumina thin films of thickness around 10 nm are prepared by sol-gel process revealing greater corrosion resistance in the NaCl environment.

11.5.1.3 Tantalum Oxide Nanocoatings

These types of nanocoatings are preferred in nanosensors, capacitors, antireflective layers, microelectronics, and optical waveguides owing to their high dielectric strength, better chemical resistance, and high hardness. Tantalum oxide nanocoating exhibited a firm passive oxide layer on the substrate surface exposed to NaCl environments. The nature of the substrate and coating technique will influence the corrosion inhibition efficiency, and the rate of corrosion can be diminished by the application of tantalum oxide nanocoatings. Tantalum nitrate nanocoating showed a decrease in corrosion rate with an increase in temperature and acidity.

11.5.1.4 *Zirconium Oxide Nanocoatings*

These are a special type of ceramic nanomaterial applied as a coating material to prevent corrosion due to its attractive physicochemical characteristics such as improved chemical stability, better refractive index, exceptional temperature resistance, and low friction coefficient. The zirconia nanocoatings are applied in catalyst support, magnetic storage media, thermal barrier coatings, and optic devices. These types of nanocoatings are superior to chromium-based anticorrosive coatings. Zirconia-based nanocoatings can be fabricated by chemical vapor deposition, sol-gel, electrodeposition, and thermal spraying methods. The nanocoating will enhance the life span and adhesion properties.

11.5.1.5 *Graphene Oxide Nanocoatings*

Graphene oxide has emerged as a promising nanocoating for superior corrosion resistance and suppresses the growth of deposits and this type of coating presents good resistance to salt solutions. Graphene oxide nanocoating is an alternative type of coating for antibacterial activity. Multilayered graphene oxide nanocoating slows down the corrosion rates on nickel than on bare substrates. Graphene oxide nanocoating is preferred in corrosion control owing to the enhanced antibacterial activity of nickel-titanium shape memory alloy and wear resistance.

11.5.1.6 *Nanocomposite Coatings*

It is a form of anticorrosive coating. It consists of a matrix material amalgamated with polymers, ceramics, or metals and nanoparticles to improve the performance of nanocoating. Nanocomposite coatings are developed from polymers and nanoparticles with varying proportions to increase thermal resistance, strength, chemical resistance, improved stiffness, reduced thermal expansion, and reduced scratch resistance. Nanocomposites can be prepared by melt intercalation, *in situ* polymerization, and chemical processes such as emulsion polymerization and solution intercalation. The frequently employed nanoparticles in the preparation of nanocomposite film are silica, graphene oxide, alumina, carbon nanotubes, and chitosan, as well as polyvinyl chloride, polyvinyl acetate, and epoxy resins. PANI, polypyrrole, and polythiophene are the category of conductive polymers used in the fabrication of nanocomposite thin films. The hydrophobic nature of the graphene oxide and calcium carbonate nanoparticles will suppress the corrosion due to their extended surface area, thermal properties, and diffusion resistance. The incorporation of ZnO, Fe_2O_3, and Fe_3O_4 nanoparticles will protect the surface from abrasion resistance, scratch resistance, and UV protection, thus suppressing corrosion.

11.5.1.7 Metallic Nanocoating

Anticorrosive nanocoatings can be prepared from a variety of metals such as phosphorous (P), nickel (Ni), cadmium (Cd), zinc (Zn), tungsten (W), iron (Fe), copper (Cu), and cobalt (Co). This type of nanoscale coating shows excellent charge distribution on the surface of metals. The coatings on metallic surfaces are carried out by electrodeposition, sputtering, and multi-arc ion plating. The factors affecting the corrosion resistance of metallic nanocoatings are influenced by coating methodology, pH, grain size, nature of additives, coating thickness, composition, microstructural characteristics of nanocoating, etc. Anticorrosive thin layer coating can be fabricated using a variety of alloys such as Ni-P, Ni-W, and Ni-Zn, and this type of coating has received a great deal of attention as metallic anticorrosive coatings.

11.6 Application of Nanocomposite Thin Films in Corrosion Control

Polymer-based protective nanofillers and nanocoatings are broadly used as anticorrosive coatings due to their attractive properties of high chemical stability enhanced mechanical strength and excellent corrosion inhibition properties. The temperature and exposure time between the metal and chemical constituent are directly related to the rate of corrosion [72, 73]. This section focuses on the fabrication of nanoparticles and the development of nanocomposite thin film for corrosion inhibition studies. The nanoparticles and polymers used for the study are titanium dioxide and PANI, respectively.

11.6.1 Materials and Methods

Mild steel coupons (Product Code: 03-220-10) are received as a gift from one of the top petroleum companies in Oman, Petroleum Development Oman (PDO). The fabrication of TiO_2 nanoparticles was performed by the sol-gel method. Polyvinyl butyral, aniline blue (ANI), hydrochloric acid, and ammonium persulfate are purchased from Sigma-Aldrich. The mild steel coupon surface was polished to accomplish a mirror-bright finish followed by ultrasonication. The various characterization tools used are DLS, XRD, EDX, FTIR, and SEM. The surface topographical imaging was performed using an atomic force microscope (AFM XE-100 model) operated in noncontact mode. Potentiodynamic measurement using a potentiostat (GAMRY Interface 1010 series instrument) and weight loss method is employed to study the corrosion inhibition behaviors of PANI-TiO_2 composites. The PANI-TiO_2 composite powder was prepared by a polymerization reaction.

11.6.2 Synthesis of TiO_2 Nanoparticles and Preparation of Coating Solution

TiO_2 nanoparticles were synthesized by the sol-gel process [74] based on the hydrolysis and condensation reaction. The synthesis reaction was carried out at room temperature for 5 hours followed by centrifugal separation and drying at 110°C. The dried TiO_2 nanoparticles were ground into fine powder using a ball mill and powdered nanoparticles were subjected to further characterization tests and nanocomposite thin film assembly.

11.6.3 Fabrication of Nanocomposite and Thin Film Stability Studies

The composite coating solution was prepared by mixing PANI and TiO_2 under vigorous stirring for a duration of 20 hours followed by ultrasonication to obtain the required consistency. The resulting coating solution was utilized for the development of nanocomposite thin films by dip coating. The nanocomposite layers are deposited on the mild steel coupon surface by dipping the coupon in the coating solution for 2 min and then drying at ambient air. After drying, the resulting coated coupon was used for stability studies at varying environments by immersing in solutions of NaCl, H_2SO_4, and seawater for 24 hours.

11.6.4 Corrosion Inhibition Studies of PANI-TiO_2–Coated Specimen

The PANI-TiO_2–coated specimen was subjected to corrosion inhibition studies followed by atmospheric, potentiodynamic, and wet/dry tests to assess the corrosion performance. The efficiency of the study was observed from the graph of the amount of TiO_2 versus the corrosion rate.

11.7 Results and Discussion

The crystal structure and phase identification of the nanoparticles are assessed using a diffractometer (PANalytical XPERT PRO) operated at room temperature with a diffraction angle varied from 10° to 90°. Figure 11.1 demonstrates that the XRD analysis of the nanoparticles showing the representative peak corresponding to a diffraction angle at $2\theta = 28°$ validates the crystalline nature of the nanoparticles, confirming the purity of the particles, which is well matched with the published reports [75]. The surface morphological characteristics of the nanoparticles using SEM are illustrated in Figure 11.2. The EDX spectra of the synthesized TiO_2 nanoparticles showing the composition analysis are displayed in Figure 11.3. The average size of the nanoparticles is shown in Figure 11.4. Only one peak seen in Figure 11.4 characterizes the particles are of uniform size distribution with an average particle size of around 5 nm.

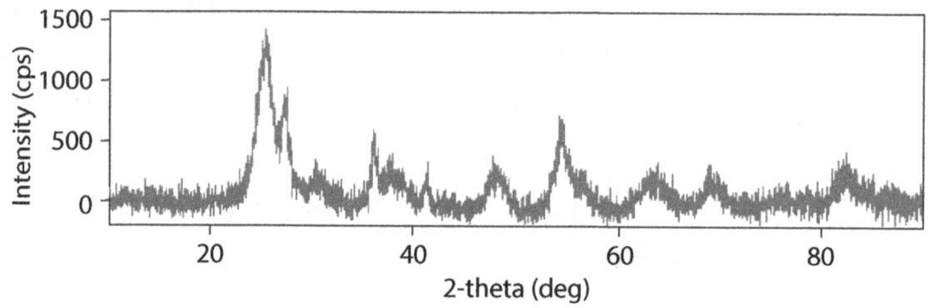

Figure 11.1 XRD spectra of TiO_2.

Figure 11.2 SEM characterization of TiO_2.

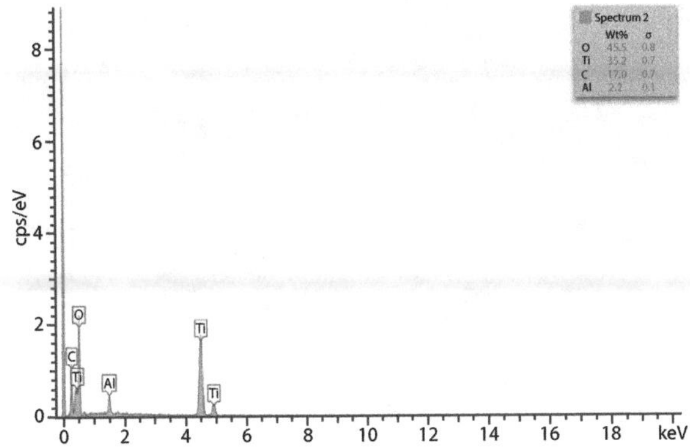

Figure 11.3 EDX analysis of TiO_2.

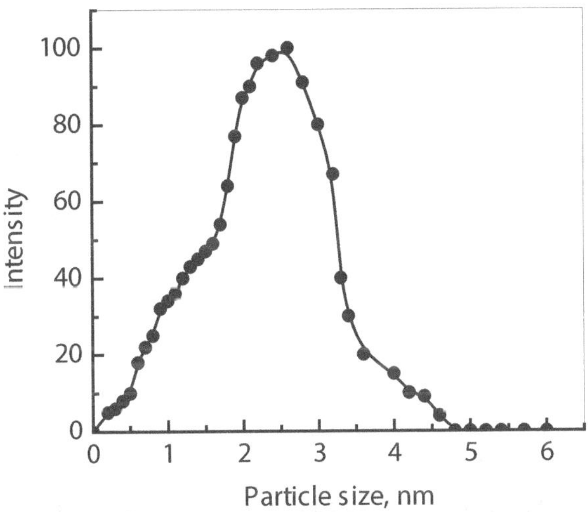

Figure 11.4 Particle size analysis of TiO_2.

The SEM image of the nanocomposite thin film is shown in Figure 11.5. As seen in Figure 11.5, the morphology of the coated surface looks smooth and evenly distributed with good coverage of thin film, which indicates the attainment of perfect coating. The three-dimensional surface topological features of the coated thin film at various dipping conditions are shown in Figure 11.6(a) to Figure 11.6(d). The images represent the AFM 3D features of PANI-TiO_2–coated thin film exposed to different environmental conditions of 0.1 N NaCl, seawater, and 0.1 N H_2SO_4, respectively.

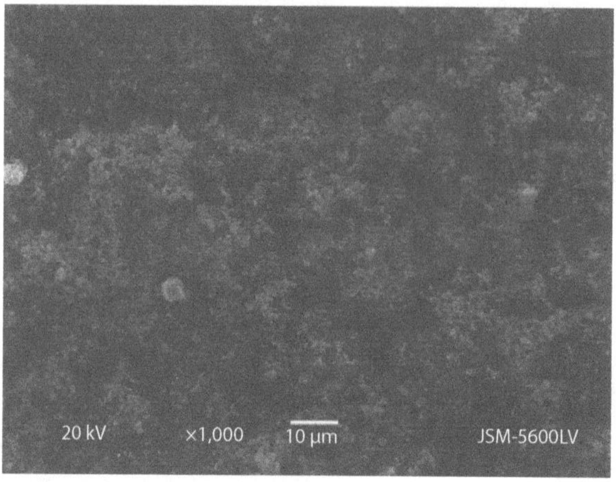

Figure 11.5 SEM image of coated mild steel specimen.

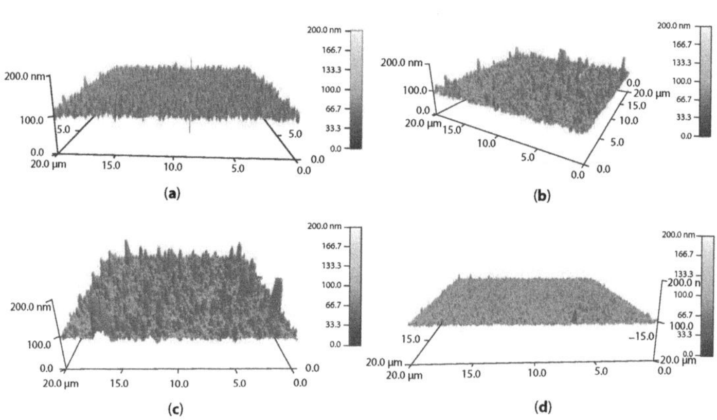

Figure 11.6 (a) AFM 3D image: 0.5 g of TiO_2-coated specimen. (b) AFM 3D image: Coated specimen dipped in 0.1 N NaCl. (c) AFM 3D image: Coated specimen dipped in seawater. (d) AFM 3D image: Coated specimen dipped in 0.1 N H_2SO_4.

The study demonstrates that the protective layer on the mild steel surface acts as a barrier to safeguard the specimen from direct interaction with the surrounding atmospheric oxygen, moisture content, and corrosive environment. The direct exposure to the corrosive medium will be affected by changing the proportion of nanoparticles in the PANI solution. The passivation layer was developed, and the resulting layer acted as an additional barrier to the mild steel specimen [76]. The surface characteristic of the secondary barrier was improved by the addition of TiO_2 in the polymer solution compared to the PANI solution alone. The increased loading of

TiO_2 will lead to the diffusive transfer to the coated specimen by improving the properties of the protective layer surface. The life of the stable thin film was increased due to the uniform deposition of the composite layer. It was observed that the composite coating layer was most stable at a dosage of 0.5 g of TiO_2 and there was a direct correlation between the rate of corrosion and the extent of TiO_2 ranging from 0.5 g to 1.0 g [77, 78]. Based on the experimental outcome, the best result was achieved when the mild steel specimen coated with 0.5 g of TiO_2 showed a reduced corrosion rate with respect to other coated coupons in varying compositions. The inhibitory behavior was studied by changing the proportion of PANI and TiO_2 and the study revealed that a surge in the dosage of TiO_2 in the coating medium demonstrated promising results, but there was a little variation in corrosion rate in all the other three corrosive environments.

The polarization curves of the bare specimen and PANI-coated mild steel specimen are studied by the potentiostatic measurement as displayed in Figure 11.7. From the outcome of the polarization curves, the mild steel specimen with PANI and 0.5 g of TiO_2 coating showed encouraging corrosion inhibition efficiency. The layer deposition and the surface characteristics were investigated through SEM imaging. The duration of the corrosion inhibition study was fixed as 60 days, and the loss in weight was assessed. The coupons coated with 0.5 g of TiO_2 nanoparticles exhibited a corrosion rate of 0.058 mm/year and the pure PANI-coated coupon showed a corrosion rate of 0.35 mm/year, which is a much higher value. The corrosion behaviors of coated and uncoated mild steel coupons at different proportions of TiO_2 are shown. The result indicates that the corrosion of mild steel specimens without coating was initiated from week 2 onward and

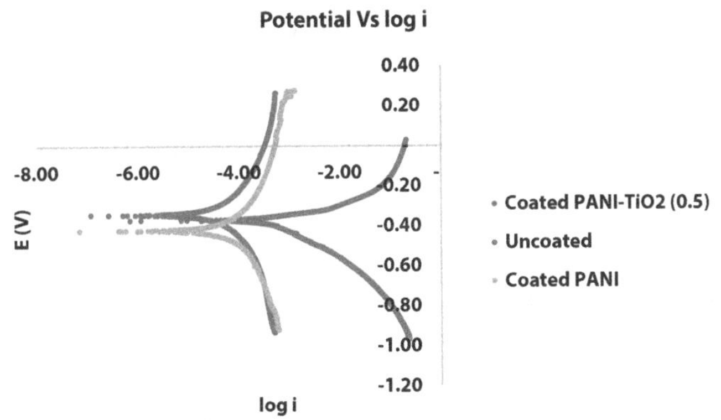

Figure 11.7 Polarization curves: (–) bare specimen; (–) PANI and 0.5 g of TiO_2-coated specimen; (–) PANI-coated specimen (scan rate of 0.2 mV/s).

gradually extended until the eighth week. Three different coating schemes with varying quantities of TiO_2 ranging from 0 to 0.5 g were prepared and the anticorrosion behavior of each sample was investigated for a duration of 8 weeks. The mild steel specimen coated with 0.5 g of TiO_2 showed the best result up to 7 weeks of study. The bare mild steel sample was corroded after 2 weeks, whereas the coated sample (0.1 g of TiO_2) was corroded after 4 weeks [78].

11.8 Conclusion

Corrosion is a serious issue in the oil and gas sectors and its control is done through the application of nanotechnology-mediated techniques, which are considered to be an attractive technique to protect the pipe surface from a corrosive environment. This chapter focused on the synthesis of TiO_2 nanoparticles and the characterizations using SEM, XRD, EDX, and DLS are highlighted. PANI-TiO_2 nano–thin films were deposited on mild steel specimens by a dip coating method. The stability of the thin film coating was carried out at different pH conditions and the best processing conditions were established. The resulting stable nanocoating was exposed to different environmental conditions. Atmospheric, potentiostatic, and wet/dry tests were performed to examine the corrosion performance of the coated specimen. The PANI-TiO_2–coated specimens were submerged into three different media, *viz*, seawater, 1.0 N H_2SO_4, and 1.0 N NaCl. The surface topology of the coated nanocomposite thin films and the films exposed to various environmental conditions were investigated by AFM. The surface topological structure of the composite thin layer with minimum thickness displayed exceptional corrosion inhibition efficiency. The effectiveness of corrosion inhibition potential was also scrutinized by calculating the corrosion rate after 2 months of exposure duration and the resulting weight difference was appraised. The study shows that the PANI-TiO_2 nanocoatings showed excellent corrosion resistance with good mechanical strength.

11.9 Future Scope

In the past few decades, the worldwide advancement of materials research has shown continuous progress to promote the development of novel coating technologies for the growing demand for corrosion control. Nanotechnology-oriented solutions are considered to be the most efficient, reliable, and cost-effective techniques to enhance the corrosion resistance of steel and improve surface protection. Also, the manipulation of the chemical composition of the material and the thickness of the coating to the nanometer scale will protect the metal from corrosion. The development of 2D materials, nanocomposite, hybrid, and eco-friendly materials can be investigated to gain more attention in the near future. Recent developments have shown significant consideration in corrosion control by the application of nanofabrication principles in material synthesis. Different techniques are being practiced to minimize corrosion. More research initiatives are directed towards the protection of material from corrosion, and there are ample scopes for extended research. A better understanding of the cause of corrosion and implementation of efficient control techniques using efficient technologies can suppress the rate of corrosion to a permissible limit range, and in certain cases, it can even be stopped. Future research will focus on the improvement of corrosion resistance by minimizing the negative effects of protection techniques and their impact on the environment, employing nontoxic ingredients, and improving the stability of coating with reuse options. Advanced nanostructured coatings for pipelines, tanks, and offshore structures will help the oil and gas sector by enhancing the life span of equipment and minimizing maintenance costs. In the future, novel coatings incorporating nanosized particles can be employed to significantly improve corrosion protection performance and for self-healing, antifouling characteristics, and the capacity to release corrosion inhibitors or antibacterial compounds as necessary.

11.10 Challenges

A major challenge still exists in safeguarding the long-term durability of nanostructured thin films. The extreme environmental conditions of high temperature, exposure to chemicals and UV radiation, pH, exposure time, etc., are the contributing factors that affect the stability of nanocoatings. It is a difficult procedure to move from small-scale production to large-scale development. Maintaining homogeneity, quality control, and cost-effectiveness are some scalability issues. For widespread application, it is crucial to evaluate how much more affordable nanostructured coatings are than conventional corrosion prevention techniques. The production, processing, and disposal of nanoparticles may raise health and safety issues that need to be addressed. For sustainable practices, it is essential to evaluate the long-term impact on the environmental effects of nanostructured coatings, particularly when they deteriorate or are discarded.

References

1. Hegde, M., Kavanagh, Y., Duffy, B., Tobin, E.F., Abrasion and Cavitation Erosion Resistance of Multi-Layer Dip Coated Sol-Gel Coatings on AA2024-T3. *Corros. Mater. Degrad.*, 3, 661, 2022.
2. Atrens, A., Chen, X., Shi, Z., Mg Corrosion—Recent Progress. *Corros. Mater. Degrad.*, 3, 566, 2022.
3. Song, G.L. and Feng, Z., Modification, degradation and evaluation of a few organic coatings for some marine applications. *Corros. Mater. Degrad.*, 1, 408, 2020.
4. Andrade, C. and Izquierdo, D., Statistical Treatments of Chloride Threshold and Corrosion Propagation Rate. *Corros. Mater. Degrad.*, 3, 598, 2022.
5. Chang, C.H., Huang, T.C., Peng, C.W., Novel anticorrosion coatings prepared from polyaniline/graphene composites. *Carbon.*, 50, 14, 5044e5, 2012.
6. Pourhashem, S., Rashidi, A., Vaezi, M.R., Excellent corrosion protection performance of epoxy composite coatings filled with amino-silane functionalized graphene oxide. *Surf. Coat. Tech.*, 317, 1e9, 2017.
7. Jafari, Y., Ghoreishi, S., Shabani-Nooshabadi, M.J.S.M., Polyaniline/ graphene nanocomposite coatings on copper: electropolymerization, characterization, and evaluation of corrosion protection performance. *Synth. Met.*, 217, 220e30, 2016.
8. Sambyal, P., Ruhi, G., Mishra, M., Conducting polymer/biomaterial composite coatings for corrosion protection. *Mater. Corros.*, 69, 3, 402e17, 2018.
9. Talbot, D.E. and Talbot, J.D., *Corrosion Science and Technology*, p. 596, CRC Press, Boca Raton, FL, USA, 2018.
10. Kausar, A., Ahmad, I., Eisa, M.H., Maaza, M., Avant-Garde Polymer/ Graphene Nanocompositesfor Corrosion Protection: Design, Features, and Performance. *Corros.Mater. Degrad.*, 4, 33, 2023. https://doi.org/10.3390/ cmd4010004.
11. Andrade, C. and Izquierdo, D., Statistical Treatments of Chloride Threshold and Corrosion Propagation Rate. *Corros. Mater. Degrad.*, 3, 598, 2022.
12. Kumar, A.M., Adesina, A.Y., Veeramani, J., Rahman, M.M., Nirmal Ram, J., Hybrid Polyurethane/Polypyrrole Composite Coatings on Passivated 316L SS for Surface Protective Action against Corrosion in Saline Medium. *Corros. Mater. Degrad.*, 3, 612, 2022.
13. Alamri, A.H., Localized Corrosion and Mitigation Approach of Steel Materials Used in Oil and Gas Pipelines -An overview, Eng. *Failure Analy.*, 116, 104735, 2020.
14. Aisha, H., Al-Moubaraki, Ime Bassey, O., Corrosion challenges in petroleum refinery operations: Sources, mechanisms, mitigation, and future outlook. *J. Saudi Chem. Soc.*, 25, 12, 101370, 2021.
15. Ehsan, S.A., Ahmad, A., Chang, H.Y., Mohammed, A.K., Gaurav, R., Corrosion protection through naturally occurring films: new insights from

iron carbonate. *ACS Appl. Mater. Interf.*, 11, 33435, 20190. https://doi.org/10.1021/ acsami.9b10221.

16. Sharland, S., A review of the theoretical modelling of crevice and pitting corrosion. *Corros. Sci.*, 27, 289, 1987.
17. Was, G.S., Corrosion and Stress Corrosion Cracking Fundamentals, Fundamentals of RadiationMaterials Science. *Springer*, 857, 2017.
18. Was, G.S., *Fundamentals of radiation materials science: metals and alloys*. Springer, 2016.
19. Popoola, L.T., Grema, A.S., Latinwo, G.K., Gutti, B., Balogun, A.S., Corrosion problems during oil and gas production and its mitigation. *Int. J. Ind. Chem.*, 4, 35, 2013.
20. Roberge, P.R., *Handbook of corrosion engineering*, McGraw-Hill, 2000.
21. Ossai, C.I., Advances in asset management techniques: An overview of corrosion mechanisms and mitigation strategies for oil and gas pipelines. *ISRN Corros.*, 2012, 2012.
22. Izionworu, V.O., Ukpaka, C.P., Ogwuzie, E., Oguzie, E.E., Green and eco-benign corrosion inhibition agents: Alternatives and options to chemical based toxic corrosion inhibitors. *Chem. Int.*, 6, 232, 2020.
23. Ahmad, Z., *Principles of corrosion engineering and corrosion control*, Elsevier, 2006.
24. Streicher, M. and Begum, S., *Corrosion, Intergranular*, 2016.
25. Stein-Brzozowska, G., Flórez, D.M., Maier, J., Scheffknecht, G., Nickel-base super alloys for ultra supercritical coal-fired power plants: Fireside corrosion Laboratory studies and power plant exposures. *Fuel*, 108, 521, 2013.
26. Combrade, P., Crevice corrosion of metallic materials. *Corros. Tech. New York Basel*, 17, 349, 2002.
27. Fatah, M.C., Ismail, M.C., Wahjoedi, B.A., Effects of sulphide ion on corrosion behaviour of X52 steel in simulated solution containing metabolic products species: a study pertaining to microbiologically influenced corrosion (MIC). *Corros. Eng. Sci. Technol.*, 482013, 211.
28. Skovhus, T.L., Eckert, R.B., Rodrigues, E., Management and control of microbiologically influenced corrosion (MIC) in the oil and gas industry—Overview and a North Sea case study. *J. Biotech.*, 256, 31, 2017.
29. Hashemi, S.J., Bak, N., Khan, F., Hawboldt, K., Lefsrud, L., Wolodko, J., Bibliometric Analysis of Microbiologically Influenced Corrosion (MIC) of Oil and Gas Engineering Systems. *Corrosion*, 74, 468, 2018.
30. Faes, W., Corrosion and corrosion prevention in heat exchangers, *J. Corros. Rev.* 22, 4, 2019.
31. Koch, G.H., Brongers, M.P., Thompson, N.G., Virmani, Y.P., Payer, J.H., Corrosion cost and preventive strategies in the United States. *Techn. Rep.*, 2002.
32. Health and Safety Executive. Catastrophic failure of shell and tube production cooler, 2006. URL: http://www.hse.gov.uk/ offshore/alerts/sa_01_06.htm. Accessed 23 April, 2018.

33. Xie, J., Yazdanfar, K., Anning, G., Rockwell, T., Ahmed, A., Corrosion of a vertical shell and tube heat exchanger. *Corrosion*, 2015.
34. Julian, R., Lichti, K., Mroczek, E., Mountain, B., Heavy metal scaling and corrosion in a geothermal heat exchanger, in: *Proceedings World Geothermal Congress*, vol. 2010, p. 2015.
35. Deen, K., Virk, M., Haque, C., Ahmad, R., Khan, I., Failure investigation of heat exchanger plates due to pitting corrosion. *Eng Fail Analy.*, 17, 886, 2010.
36. Kuppan, T., *Heat exchanger design handbook*, Marcel Dekker, Inc., New York, 2000.
37. Eich, G., Erosion corrosion on oil coolers for navy ships. *J. Fail. Analy. Prev.*, 12, 490, 2012.
38. Wang, L., Sundén, B., Manglik, R.M., *Plate heat exchangers: design, applications and performance*, vol. 11, Wit Press, Southampton, UK, 2007.
39. Tawancy, H., Failure of hydrocracker heat exchanger tubes in an oil refinery by polythionic acid-stress corrosion cracking. *Eng Fail Analy.*, 16, 2091, 2009.
40. Wang, L., Sundén, B., Manglik, R.M., *Plate heat exchangers: design, applications and performance*, vol. 11, Wit Press, Southampton, UK, 2007.
41. Walker, G., *Industrial heat exchangers: a basic guide*, Hemisphere Pub. Corp., Washington, 1990.
42. Roberge, P.R., *Handbook of corrosion engineering*, McGraw-Hill, New York, 2000.
43. Yang, Z.G., Gong, Y., Yuan, J.Z., Failure analysis of leakage on titanium tubes within heat exchangers in a nuclear power plant. Part I: Electrochemical corrosion. *Mat. Corros.*, 63, 7, 2012.
44. Rao, T. and Nair, K., Microbiologically influenced stress corrosion cracking failure of admiralty brass condenser tubes in a nuclear power plant cooled by freshwater. *Corros. Sci.*, 40, 1821, 1998.
45. Huttunen, S.E., Honkanen, M., Lepistö, T., Kuokkala, V.T., Koivisto, L., Berg, C.G., Microbiologically influenced corrosion (MIC) in stainless steel heat exchanger. *Appl. Surf. Sci.*, 258, 6512, 2012.
46. El-Yazgi, A.A. and Hardie, D., Stress corrosion cracking of duplex and super duplex stainless steels in sour environments. *Corro. Sci.*, 40, 6, 909, 1998.
47. Fayomi, O.S.I., Atayero, A.A., Mubiayi, M.P., Akande, I.G., Adewuyi, P.A., Fajobi, M.A., Ayara, W.A., Mechanical and opto-electrical response of embedded smart composite coating produced via electrodeposition technique for embedded system in defence application. *J. Alloys Compd.*, 773, 305, 2019.
48. Li, X., Cui, C., Sun, J., Enhanced product quality in lubricant type vacuum distillation unit by implementing dividing wall column. *Chem. Eng. Process.: Process Intensif.*, 123, 1, 2018.
49. Waheed, M.A. and Oni, A.O., Performance improvement of a crude oil distillation unit. *Appl. Therm. Eng.*, 75, 315, 2015.
50. Blumer, A.F., Crude Still Overhead System Corrosion. *Corrosion*, 5, 5, 135, 1949. https://doi.org/10.5006/0010-9312-5.5.135.

51. Blumer, A.F., Crude Still Overhead System Corrosion. *Corrosion*, 5, 5, 135, 1949. https://doi.org/10.5006/0010-9312-5.5.135.
52. Mostafaei, A., Peighambari, S.M., Nasirpouri, F., Failure analysis of monel packing in atmospheric distillation tower under the service in the presence of corrosive gases. *Eng. Fail. Anal.*, 28, 241, 2013.
53. Papavinasam, S., *Corrosion control in the oil and gas industry*, Elsevier, 2013.
54. Alec, G., *Corrosion Problems and Solution in Oil Refining and Petrochemical Industry*, vol. 356, SpringerDordrecht, 2016.
55. Heidersbach, R., *Metallurgy and Corrosion Control on Oil and Gas Production*, John Wiley & Sons, Inc, Hoboken, New Jersey, USA, 2011.
56. Groysman, A., *Corrosion for Everybody*, vol. 368, SpringerVerlag, Dordrecht, 2010.
57. Craig, B.D., Selection guidelines for corrosion resistant alloys in the oil gas industry, Technical Publication No. 10073. *Nickel Dev. Inst.*, 11, 1995. http://www.stainless-steel-world.net/pdf/10073.pdf.
58. Kane, R.D., Craig, B., Venkatesh, A., Titanium Alloys for Oil and Gas Service: A Review. Paper no. 09078. *NACE Corrosion 2009 Conference, NACE International*, Houston, TX, p. 21, 2009.
59. McKeen, L.W., Albert, R., Hofmans, J., Corrosion Testing of Fluorocoatings for Oil/Gas Production Tubing, Paper no. 08029. *NACE Corrosion 2008 Conference, NACE International*, Houston, TX, USA, p. 9, 2008.
60. Kau, C.J., Luc, V., Veith, C.A., Polyketone Polymers: A New Liner Material for Corrosion Control in Oil and Gas Industry, Paper no. 00787. *NACE Corrosion 2000 Conference, NACE International*, Houston, TX, USA, p. 16, 2000.
61. Groysman, A., *Corrosion in Systems for Transportation and Storage of Petroleum Products and Biofuels*, p. 297, Springer, Dordrecht, 2014.
62. NACE, Standard Practice SP0169-2013, Control of External Corrosion on Underground or Submerged Metallic Piping Systems. *NACE International*, Houston, TX, USA, p. 60, 2013.
63. NACE, Standard Test Method TM0101-2012, Measurement Techniques Related to Criteria for Cathodic Protection of Underground Storage Tank Systems. *NACE International*, Houston, TX, USA, p. 34, 2012.
64. Foss, M., Gulbrandsen, E., Sjoblom, J., Effect of Corrosion Inhibitors and Oil on Carbon Dioxide Corrosion and Wetting of Carbon Steel with Ferrous Carbonate. *Corrosion*, 653, 1, 2009.
65. Paul, J.S., Corrosion Inhibitors for Oil and Gas Production, in: *Corrosion, 13*, Ninth Edition, p. 478, ASM International, USA, 1987.
66. Alec, G., *Corrosion Problems and Solution in Oil Refining and Petrochemical Industry*, p. 356, Springer, Dordrecht, 2016.
67. Zhu, T.T., Zhou, C.H., Kabwe, F.B., Exfoliation of montmorillonite and related properties of clay/polymer nanocomposites. *Appl. Clay Sci.*, 169, 48, 2019. https://doi.org/10.1016/j.clay.2018.12.006.

68. Yeganeh, M., Nguyen, T.A., Rajendran, S., Kakooei, S., Corrosion protection at the nanoscale: an introduction. *Micro Nano Technol.*, 4, 3, 2020. https://doi.org/10.1016/B978-0-12-819359-4.00001-5.
69. Arian, M., Hamidreza, G.T., Mohammad, S.L., Preparation and comparative study of anticorrosion nanocomposites of polyaniline/graphene oxide/clay coating. *J. Mat. Res. Tech.*, 13, 2325, 2021. https://doi.org/10.1016/j.jmrt.2021.05.098.
70. Moiseeva, L.S., Carbon dioxide corrosion of oil and gas field equipment. *Prot. Met.*, 41, 1, 76–83, 2005. https://doi.org/10.1007/s11124-005-0011-6.
71. Kermani, M.B. and Morshed, A., Carbon dioxide corrosion in oil and gas production - a compendium. *Corrosion*, 59, 8, 659, 2003. https://doi.org/10.5006/ 1.3277596.
72. Duarte, T., Meyer, Y.A., Os´orio, W.R., The holes of Zn phosphate and hot dip galvanizing on electrochemical behaviors of multi-coatings on steel substrates. *Metals*, 12, 863, 2022. https://doi.org/10.3390/met12050863.
73. Konovalova, V., The effect of temperature on the corrosion rate of iron-carbon alloys. *Mater. Today Proc.*, 38, 4, 1326, 2021. https://doi.org/10.1016/j.matpr.2020.08.094.
74. Verma, V., Al-Dossari, M., Singh, J., Rawat, M., Kordy, M.G.M., Shaban, M., A Review on Green Synthesis of TiO_2 NPs: Photocatalysis and Antimicrobial Applications. *Polym. (Basel)*, 14, 7, 1444, Apr 12022, doi: 10.3390/polym14071444. PMID: 35406317; PMCID: PMC9002645.
75. Aravind, M., Amalanathan, M., Sony Michael, M.M., Synthesis of TiO_2 nanoparticles by chemical and green synthesis. *SN Appl. Sci.*, 3, 409, 2021. https://doi.org/10.1007/s42452-021-04281-5.
76. Aravind, M., Amalanathan, M., Sony Michael, M.M., Synthesis of TiO_2 nanoparticles by chemical and green synthesis. *SN Appl. Sci.*, 3, 409, 2021. https://doi.org/10.1007/s42452-021-04281-5.
77. Priyanka, K.P., Sukirtha, T.H., Balakrishna, K.M., Varghese, T., Microbicidal activity of TiO_2 nanoparticles synthesized by sol-gel method. *IET Nano Biotech.*, 10, 2, 81, 2016. https://doi.org/10.1049/iet-nbt.2015.0038.
78. Al Jabri, H.M., Geetha, D., Muhannad, A.A., Development of polyaniline – TiO_2 nano composite films and its application in corrosion inhibition of oil pipelines. *J. Indian Chem. Soc.*, 100, 100826, 2023. https://doi.org/10.1016/j.jics.2022.100826.

12

Anticorrosion and Antifouling Coating Materials

N. Haridharan* **and R. V. Shiva Kumar**

Department of Chemistry, Vel Tech Rangarajan Dr. Sagunthala R&D Institute of Science and Technology, Avadi, Chennai, Tamil Nadu, India

Abstract

Over the past few decades, metallic equipment has seen widespread use in marine infrastructure and applications. However, pristine metallic surfaces are highly susceptible to contamination from marine fouling and corrosion, brought about by interactions with marine water, bacteria, and other organisms. The presence of microorganisms on these surfaces triggers corrosion reactions due to changes in pH and the presence of oxygen. Fouling and corrosion pose significant challenges for maritime industries, leading to structural failures, increased fuel consumption, and high maintenance costs. To address these issues and enhance the efficiency of marine-related activities, the marine-shipping industry has focused on developing innovative metallic components and infrastructure.

One approach involves altering the metal surface with protective coatings that serve a dual function of preventing fouling and corrosion. This chapter discusses recent advancements and challenges in designing such materials in four distinct sections. The final section explores future developments and acknowledges critical limitations.

The first section outlines the requirements for new antifouling and anti-corrosion materials. It underscores the drawbacks of using bare metallic components in marine applications and emphasizes the need for novel materials. The section also highlights the detrimental effects of marine organisms and salty water on marine infrastructure and discusses the benefits of protective coatings.

**Corresponding author*: drnharidharan@veltech.edu.in

Hari Murthy, Vinay Jha Pillai, Kukatlapalli Pradeep Kumar and Matthew Cowan (eds.) Novel Anti-Corrosion and Anti-Fouling Coatings and Thin Films, (353–398) © 2024 Scrivener Publishing LLC

In the second section, key issues in formulating anti-corrosive materials are examined. Challenges in creating effective inhibitors are discussed, as many materials fail to demonstrate optimal inhibitory properties. Additionally, factors influencing the lifespan of protected metals, such as leaching, are explored.

The third section explores modeling and optimization techniques for antifouling and anti-corrosive coatings. By utilizing modeling, researchers can reduce the need for extensive experimental work and predict coating behavior under various conditions. The section also addresses strategies for controlling microorganism release rates.

In the fourth section, the focus shifts to the development of new nanomaterials and oxide thin films for enhanced antifouling and anti-corrosive properties. The discussion covers the non-toxicity, biocompatibility, and photochemical stability of nanomaterials, as well as their ability to combat bacteria and corrosion. Additionally, the section examines the influence of physical and chemical parameters on the efficacy of nanomaterials.

Finally, the fifth section provides a summary of current advancements and future prospects in the field. It explores ways to mitigate critical limitations through adaptive methodologies, ensuring continued progress in combating fouling and corrosion in marine environments.

Keywords: Corrosion, marine equipment, coatings, anti-corrosion, anti-fouling, nano-materials, modelling and optimization

12.1 Introduction

The functions of modern society are fundamentally dependent on metals in the form of vehicles, buildings, machinery, and infrastructure. In particular, the "major metals", which have a long history of human use, provide society with goods and services that are required for meeting basic human needs. Metals are considered to be the main components in the construction industry; especially, iron and steel play a vital role in this field. Since, in general, metals are good conductors of electricity, IT equipment, and gadgets are made out of specific metals.

12.1.1 Demerits of Using Bare Metallic Components as a Substrate Material in Marine Applications

While bare metallic components can be used as substrate materials in marine applications, they have several demerits or disadvantages that make them less ideal compared to other alternatives. Some of the main demerits are the following:

- **Corrosion:** Metals are highly susceptible to corrosion in marine environments due to the presence of saltwater. Exposure to seawater, moisture, and oxygen can lead to the formation of rust, which weakens the structure and reduces its life span. Regular maintenance and protective coatings are required to prevent corrosion, increasing maintenance costs.
- **Fouling**: Bare metallic surfaces tend to promote the growth of marine organisms such as barnacles, algae, and mussels. This biofouling can adversely affect the performance and efficiency of marine equipment, increase drag on surfaces, and decrease fuel efficiency. It also requires additional maintenance and cleaning efforts. **Galvanic Corrosion**: When dissimilar metals come into contact in a marine environment, they can undergo galvanic corrosion. The electrolytic action between these metals can accelerate corrosion, leading to rapid degradation and structural failures. **Mechanical Weakness**: Some metals used in marine applications may not possess the necessary mechanical strength and resistance to impacts, fatigue, and stress. In harsh marine conditions with waves, currents, and heavy loads, the structural integrity of bare metallic components can be compromised over time. **Electrical Conductivity**: Metals are good conductors of electricity, which can lead to galvanic corrosion and electrical hazards if proper insulation and grounding measures are not in place. Electrolytic action between dissimilar metals or stray electrical currents can cause accelerated corrosion and damage to the equipment. **Weight**: Some metallic materials can be relatively heavy, which can impact the overall weight and buoyancy of marine structures. This increased weight can affect the stability, fuel efficiency, and maneuverability of vessels and other marine equipment.

To mitigate these demerits, alternative materials such as corrosion-resistant alloys, composites, and protective coatings are often used in marine applications to enhance durability, reduce maintenance costs, and improve performance.

12.1.2 Essentiality in Designing and Developing Novel Materials for Anticorrosive and Antifouling Purposes

Designing and developing novel materials for anticorrosive and antifouling purposes are crucial in various industries, especially those dealing with infrastructure, marine applications, and chemical processing. The essentiality of such materials lies in their ability to mitigate the detrimental effects of corrosion and fouling, which can lead to significant economic losses, safety hazards, and environmental damage. Let us explore the importance of these materials in more detail:

- Corrosion Prevention:
  Corrosion is a natural process that can cause the degradation and deterioration of materials, leading to structural weaknesses, reduced life span, and increased maintenance costs. By designing anticorrosive materials, the following benefits can be achieved:
  a. Increased Durability: Anticorrosive materials can extend the life span of structures and components, reducing the need for frequent repairs or replacements.
  b. Cost Savings: By minimizing corrosion-induced damages and maintenance requirements, businesses can save significant costs associated with repairs, downtime, and material replacement.
  c. Enhanced Safety: Corrosion can compromise the structural integrity of infrastructure, machinery, and equipment, potentially resulting in accidents or failures. Anticorrosive materials help maintain safety standards and prevent catastrophic incidents.
  d. Environmental Protection: Corrosion often involves the release of pollutants into the environment. By minimizing corrosion, anticorrosive materials contribute to reducing the environmental impact and preserving natural resources.
- Fouling Mitigation: As was mentioned earlier, fouling refers to the accumulation of unwanted substances on surfaces, such as marine organisms (biofouling) or deposits from chemical reactions (e.g., scaling). Developing materials with antifouling properties offers several advantages:
  a. Improved Performance: Antifouling materials ensure optimal performance by preventing the accumulation of

contaminants on surfaces, which can interfere with fluid flow, heat transfer, or signal transmission.

b. Energy Efficiency: Fouling deposits can increase the energy consumption of equipment and systems, leading to higher operational costs. Antifouling materials help maintain efficiency and reduce energy waste.
c. Maintenance Reduction: The presence of fouling substances necessitates regular cleaning or maintenance, which can be time-consuming and costly. Antifouling materials minimize the need for such interventions, saving both time and resources.
d. Environmental Preservation: Biofouling can harm ecosystems when invasive species are transported through maritime activities. Antifouling materials help prevent the spread of harmful organisms and minimize the ecological impact.

In summary, the design and development of novel materials with anticorrosive and antifouling properties are essential for improving durability, reducing maintenance costs, ensuring safety, conserving resources, and protecting the environment. These materials play a vital role in various industries, contributing to sustainable and efficient operations.

12.1.3 The Serious Effects of Marine Organisms and Salty Water on the Surface of Marine Infrastructure and How It Affects Marine Construction and Transport

Marine organisms and the corrosive nature of saltwater can have significant effects on marine infrastructure, including marine construction and transport. Here are some serious effects:

- Biofouling: Marine organisms, such as barnacles, mussels, and algae, can attach themselves to the surface of marine structures, forming a layer of biofouling. This biofouling can increase drag and reduce the efficiency of ships, resulting in increased fuel consumption and decreased speed. It can also affect the performance of underwater structures, such as pipelines and offshore platforms.
- Corrosion: Saltwater is highly corrosive due to its high salt content and the presence of dissolved oxygen. Corrosion

can damage the structural integrity of marine infrastructure, including ships, piers, bridges, and offshore platforms. It weakens the materials, leading to decreased load-bearing capacity and potential structural failures. **Wear and Tear**: The constant movement of water, waves, and tides can cause mechanical stress on marine structures, leading to wear and tear. Over time, this can result in cracks, erosion, and degradation of the infrastructure, requiring costly repairs and maintenance. **Clogging**: Marine organisms can accumulate in pipes, valves, and intake systems, causing clogging and reducing the efficiency of water circulation systems. This can affect the cooling systems of ships and power plants, leading to decreased performance and increased maintenance requirements. **Environmental Impact**: The release of antifouling paints and chemicals used to combat biofouling can have negative environmental consequences. Some of these chemicals contain toxic substances that can harm marine organisms and disrupt the marine ecosystem.

To mitigate these effects, various measures are taken, including regular cleaning and maintenance of marine structures to remove biofouling, application of antifouling coatings and paints to prevent the attachment of marine organisms, use of corrosion-resistant materials and coatings, installation of sacrificial anodes to protect metal structures from corrosion, implementation of monitoring systems to detect early signs of degradation and corrosion, and compliance with environmental regulations and use of environmentally friendly antifouling solutions.

By understanding and addressing the serious effects of marine organisms and salty water on marine infrastructure, steps can be taken to minimize their impact, ensure the safety and efficiency of marine construction and transport, and reduce the environmental footprint of these activities.

12.1.4 The Advantages of Protective Coatings on the Surface of the Material to Overcome this Difficulty

Protective coatings play a vital role in enhancing the performance and durability of marine equipment.

Some advantages of using protective coatings on the surface of marine equipment are the following: **Corrosion Resistance**: Marine equipment is constantly exposed to harsh environmental conditions, including saltwater, moisture, and chemicals. Protective coatings, such as marine-grade paints or corrosion-resistant coatings, create a barrier between the equipment surface and the corrosive elements, preventing corrosion and extending the equipment's life span. **Antifouling Properties**: Marine equipment is susceptible to fouling, which refers to the accumulation of organisms, such as barnacles, algae, and marine organisms, on the surface. Fouling can increase drag, reduce fuel efficiency, and damage equipment. Antifouling coatings contain biocides or special surface properties that deter the attachment and growth of organisms, keeping the equipment's surface clean and efficient. **UV Protection**: Exposure to sunlight, especially in marine environments, can cause degradation and fading of equipment surfaces. UV-resistant coatings provide protection against harmful ultraviolet (UV) radiation, preventing color fading, cracking, and degradation of the underlying material. **Improved Durability**: Protective coatings provide an additional layer of toughness and abrasion resistance to the surface of marine equipment. They can withstand mechanical wear, impact, and abrasion, protecting the equipment from damage during operation, transportation, and maintenance activities. **Ease of Cleaning and Maintenance**: Coated surfaces are generally easier to clean and maintain. The smooth and nonstick properties of certain coatings prevent the adhesion of dirt, oil, and other contaminants. This simplifies cleaning processes, reduces maintenance time and costs, and ensures the equipment remains in optimal condition. **Thermal Insulation**: Some specialized coatings offer thermal insulation properties, helping to regulate the temperature of marine equipment. This is particularly beneficial for equipment exposed to extreme heat or cold, preventing thermal stress and enhancing operational performance. **Aesthetic Appeal**: Protective coatings can improve the aesthetic appearance of marine equipment. They come in various colors, finishes, and textures, allowing customization options to match the vessel's design or company branding. **Environmental Protection**: Certain coatings are designed to be environmentally friendly, reducing the impact on marine ecosystems. Low VOC (volatile organic compound) coatings and nontoxic antifouling solutions help minimize pollution and harm to marine life. Overall, protective coatings offer a range of advantages for marine equipment, including corrosion resistance, antifouling properties, UV protection, improved durability, ease of maintenance, thermal insulation, aesthetic appeal, and environmental protection. These benefits contribute to the longevity, performance, and operational efficiency of marine equipment, ensuring its optimal functionality in challenging marine environments.

12.1.5 Coating Thickness—A Good Alternative But Yet…

Thicker coatings do indeed offer better protection in terms of enhanced barrier properties and increased resistance to corrosion, but they can pose challenges in terms of application, curing, and maintenance. Some promising ways to achieve and manage thicker coatings while addressing the associated difficulties are the following: **High-Build Coatings:** High-build coatings are formulated to be applied in thicker layers while maintaining proper adhesion and performance. They are designed to self-level and cure without sagging or running. **Multilayer Systems:** Applying multiple layers of coatings in a controlled manner can achieve a thicker overall protective system. Each layer serves a specific purpose, such as adhesion, corrosion resistance, and topcoat. **Spraying Techniques:** Advanced spraying techniques, such as airless or plural-component spraying, allow for controlled application of thicker coatings with reduced risk of runs or sags. **Brush and Roller Application:** For smaller surfaces or complex geometries, applying thicker coatings with brushes or rollers allows for more precise control over the coating thickness. **Thixotropic Coatings:** Thixotropic coatings have a higher viscosity when at rest but become more fluid when agitated or applied. This property helps prevent sagging during application. **Slow-Curing Formulations:** Slow-curing coatings provide a longer open time, allowing the coating to self-level and flow evenly, resulting in a smoother and more uniform finish. **Controlled Drying and Curing Conditions:** Maintaining optimal drying and curing conditions, such as temperature and humidity, during application ensures proper adhesion and reduces the risk of defects. **Use of Structural Coatings:** Structural coatings are designed to be applied at thicker thicknesses, providing both protection and reinforcement to structures, such as bridge decks and pipelines. **Spray-and-Build Techniques:** Some coatings are designed to be sprayed in multiple thin layers, allowing each layer to partially dry before applying the next. This approach builds up the desired thickness gradually. **Application Training and Expertise:** Proper training of personnel in coating application techniques is crucial for achieving uniform and effective thicker coatings. **Thickening Agents:** Certain additives can be mixed with coatings to increase their viscosity and thixotropy, aiding in the application of thicker coatings. **Coating Flow Control:** Implementing flow control measures, such as tilting or rotating the substrate during application, can help evenly distribute the coating and prevent runs. **Testing and Verification:** Regular thickness measurement using tools like ultrasonic gauges ensures that the desired thickness is achieved and maintained. **Maintenance Planning:**

Thicker coatings might require more diligent inspection and maintenance. Planning for routine inspections and maintenance can prevent issues.

It is important to note that while thicker coatings offer benefits, they may not always be necessary for every application. A thorough understanding of the project's requirements, substrate condition, and environmental factors will help determine the optimal coating thickness and application approach. Additionally, working closely with coating manufacturers, experts, and experienced contractors can provide valuable insights into achieving and managing thicker coatings effectively.

12.1.6 Coating Integrity

The integrity of the coating film is crucial for effective corrosion protection. Defects in the coating film can compromise its ability to act as a barrier against corrosive agents, allowing them to penetrate and reach the underlying substrate. Some common coating film defects and their impact on corrosion protection are described here: **Cracking:** Cracks in the coating film can occur due to factors like poor adhesion, substrate movement, or thermal stresses. These cracks provide pathways for corrosive agents to reach the substrate, accelerating the corrosion process. **Blistering:** Blisters form when trapped moisture or gasses cause localized swelling and separation of the coating from the substrate. Blisters can create hollow spaces where corrosive agents accumulate and attack the substrate. **Pinholes and Holidays:** Pinholes are small voids or gaps in the coating, while holidays are areas where the coating is missing altogether. Both defects expose the substrate directly to the environment and promote localized corrosion. **Peeling and Delamination:** Poor adhesion between the coating layers or between the coating and substrate can lead to peeling or delamination. These defects create areas of reduced protection where corrosive agents can enter. **Cratering and Pitting:** Surface imperfections like craters and pits can be caused by contaminants, air bubbles, or improper application. These defects can trap corrosive agents, leading to localized corrosion. **Uneven Application:** Variations in coating thickness across the substrate can create areas of thinner coverage, which are more susceptible to corrosion. **Orange Peel Effect:** This textured surface defect reduces the effectiveness of the barrier and can trap moisture and contaminants, promoting corrosion. **Under-curing or Over-curing:** Improper curing can lead to reduced film integrity. Under-curing leaves the coating vulnerable to mechanical damage and chemical attack, while over-curing can make the coating brittle and prone to cracking.

To enable the integrity of the coating film and minimize defects, we need to ensure the following:

Proper Surface Preparation: Ensure that the substrate is properly cleaned, prepped, and primed before applying the coating to promote good adhesion. **Quality Application:** Use proper application techniques, including the right equipment, application rate, and curing conditions. **Quality Control and Inspection:** Regularly inspect coatings during and after application to detect defects early and take corrective actions. **Testing Methods:** Utilize nondestructive testing methods like ultrasonic, radiographic, or visual inspection to identify defects that might not be visible to the naked eye. **Maintenance and Repairs:** Address defects promptly through maintenance and repair actions. Smaller defects can be patched or repaired to prevent further deterioration. **Education and Training**: Ensure that personnel involved in coating application and inspection are properly trained and educated about the importance of film integrity. We need to remember that the effectiveness of the coating is directly linked to its ability to maintain its protective barrier over time. Proper application, inspection, and maintenance practices are essential to prevent coating film defects and maintain corrosion protection.

12.1.7 Choosing the Right Coating Material

Choosing the right or optimal coating material involves a systematic evaluation process that considers various factors to ensure the coating's effectiveness, durability, and compatibility with the substrate and environment. We give here a brief step-by-step guide to help make an informed decision: **Identify Requirements:** Define the specific requirements of the project, such as corrosion resistance, chemical resistance, UV resistance, mechanical durability, and service life expectancy. **Understand Substrate and Environment:** Analyze the substrate material (metal, concrete, etc.) and the environmental conditions the coating will be exposed to (temperature, humidity, chemicals, etc.). **Review Industry Standards and Regulations:** Check if there are any industry-specific standards, regulations, or certifications that the coating needs to meet. **Research Available Coating Types:** Research the different types of coatings available, such as epoxy, polyurethane, zinc-rich coatings, ceramic coatings, etc. Understand their properties, benefits, and limitations. **Evaluate Performance Characteristics:** Compare the performance characteristics of different coatings, including their corrosion resistance, adhesion, flexibility, impact resistance, UV stability, and chemical resistance. **Consider Application Method:** Determine the most suitable application method for the coating (spraying, brushing, rolling, etc.) based on the substrate, surface area, and accessibility. **Budget and Cost Analysis:** Consider the initial cost of the coating, as well as the potential cost savings in terms of maintenance and life-span extension. **Compatibility with Other Coatings:** If other coatings will be applied on top of the chosen coating in the future, ensure compatibility between coatings to prevent adhesion or chemical issues. **Consult Coating Manufacturers:** Reach out to coating manufacturers and suppliers to discuss your project's requirements. They can provide recommendations based on their expertise. **Sample Testing and Trials:** Conduct small-scale testing and trials using different coatings on sample substrates in conditions similar to the actual environment. Evaluate the performance of each coating. **Consider Long-Term Performance:** Assess the long-term performance and durability of coatings, including how well they resist degradation and maintain their protective properties over time. **Environmental Impact:** Evaluate the environmental impact of the coating material, including considerations of VOC emissions, toxicity, and disposal. **Consult Experts and Peers:** Seek advice from experts in the field, colleagues, or peers who have experience with similar projects. **Risk Assessment:** Consider the potential risks associated with choosing a particular coating, such as the risk of premature failure or maintenance challenges. **Make Informed Decision:** Based on the gathered information, analysis, and considerations, make an informed decision on the coating material that best meets your project's requirements. **Document and Plan:** Document the rationale for your choice and develop a detailed plan for coating application, inspection, and maintenance. **Monitor and Evaluate:** Continuously monitor the performance of the chosen coating over time and be prepared to make adjustments if issues arise.

Choosing the right coating material involves a balance between technical requirements, practical considerations, budget constraints, and long-term goals. A thorough evaluation process ensures that the selected coating material will effectively protect the substrate in its intended environment for an extended period.

12.1.8 Factors to be Considered While Choosing a Protective Coating

Choosing the right protective coating is crucial for ensuring the long-term performance and durability of structures, equipment, and surfaces in various industries, including marine environments. Selecting the right protective coating involves a balance between various technical, practical, and economic considerations. Consulting with coating manufacturers, industry experts, and professionals can help you make an informed decision based on your specific needs and requirements. Several factors should be considered when selecting a protective coating: **Substrate and Surface Conditions:** The type of substrate (metal, concrete, wood, etc.) and its condition (clean, rusted, and oily) influence the choice of coating and preparation needed for proper adhesion. **Corrosion and Chemical Resistance:** Consider the specific corrosive agents, chemicals, and environmental conditions the coating will be exposed to, such as saltwater, acids, alkalis, or pollutants. **Service Life Expectancy:** Determine the expected life span of the coating, as coatings vary in terms of durability and longevity. Consider the cost-effectiveness of longer-lasting coatings. **Application Method and Environment:** Some coatings are better suited for certain application methods (spraying, brushing, and dipping), and environmental conditions (temperature, humidity, and ventilation) can impact the application process. **Curing Time and Drying Properties:** Coatings have different curing times and drying properties. These factors affect the downtime during application and the speed at which the coated surface can be put back into service. **Ease of Application:** Consider the ease of applying the coating, especially for complex or hard-to-reach surfaces. Some coatings require specialized equipment or skills for proper application. **Compatibility with Other Coatings and Substrates:** If the substrate has multiple layers of coatings or will have additional coatings applied in the future, ensure compatibility between coatings to prevent adhesion or chemical compatibility issues. **UV Resistance and Color Retention:** Coatings exposed to sunlight need UV resistance to prevent degradation and color fading. Choose coatings that maintain their appearance over time. **Flexibility and Impact Resistance:** Consider whether the coated surface will experience flexing, vibration, or impact. Flexible coatings are essential for surfaces prone to movement. **Environmental and Health Considerations:** Check if the coating meets environmental regulations and health standards. Some coatings release harmful VOCs or toxic substances during application. **Maintenance and Repair:** Evaluate the ease of maintenance and repair. Some coatings allow for spot repairs without the need to recoat entire surfaces. **Budget and Cost-effectiveness:** Consider the initial cost of the coating, as well as the potential long-term cost savings due to reduced maintenance and longer life span. **Industry Standards and Certifications:** **We need to l**ook for coatings that meet industry-specific standards and certifications for performance and safety. **Previous Performance and Experience:** If applicable, we also consider the performance of similar coatings in similar conditions and learn from past experiences.

12.1.9 Regular Inspection and Maintenance

Regular inspection and maintenance are indeed essential to ensure that anticorrosive materials continue to provide effective protection and prevent potential issues. Proper maintenance practices help identify and address any signs of degradation, damage, or wear early on, minimizing the risk of corrosion-related problems. Regular inspection and maintenance should be part of a comprehensive corrosion management strategy, and any identified issues should be addressed promptly to prevent further damage. This proactive approach helps extend the life of anticorrosive materials and reduces the risk of unexpected failures. There are several methods available in this context and the choice of methods depends on the resources available to the organization. **Visual Inspection:** Regular visual inspections are one of the most straightforward methods. Trained personnel assess the condition of coatings, looking for signs of peeling, cracking, blistering, or other visible damage. **Thickness Measurement:** Nondestructive thickness measurement tools, such as ultrasonic gauges, can assess the thickness of coatings. This helps ensure that coatings remain within the specified thickness range for optimal protection. **Cathodic Protection Monitoring:** If cathodic protection systems are in place, monitoring the performance of sacrificial anodes or impressed current systems is crucial to ensure their effectiveness in preventing corrosion. **Electrochemical Techniques:** Electrochemical methods like electrochemical impedance spectroscopy (EIS) and polarization resistance can provide insights into the corrosion rate and condition of coated surfaces. **Destructive Testing:** In some cases, small sections of coatings or substrates might be removed for laboratory analysis. Destructive tests help assess the condition of coatings and the extent of corrosion underneath. **Environmental Monitoring:** Regular assessment of the environmental conditions in which the anticorrosive materials are exposed (humidity, temperature, and salt concentration) can provide insights into potential risks. **Thermal Imaging:** Thermal imaging can identify areas of temperature variation that might indicate hidden corrosion or damage beneath the coating. **Ultrasonic Testing:** Ultrasonic techniques can be used to detect hidden defects, such as delamination or voids, within coatings and substrates. **Corrosion Coupons and Probes:** Corrosion coupons or probes made from the same material as the coated substrate can be periodically inspected to assess the rate of corrosion and the condition of the coating. **Acoustic Emission Testing:** This technique involves monitoring for acoustic emissions caused by crack growth or other structural changes, helping to identify hidden defects. **Regular Cleaning and Washing:** Regular cleaning of coated surfaces removes accumulated dirt, salts, and pollutants that can degrade the protective coating. **Reapplication or Touch-up Coating:** Depending on the inspection findings, touch-up or reapplication of coatings may be necessary to restore or reinforce protection. **Documentation and Record-Keeping:** Maintaining detailed records of inspections, test results, maintenance actions, and repairs helps track the condition of the coatings over time. **Preventive Maintenance Schedules:** Establishing regular maintenance schedules based on the specific environmental conditions and the coating's expected life span helps ensure timely inspections and actions. **Training and Expertise:** Having trained personnel with expertise in corrosion protection and inspection techniques is crucial for accurate assessments and decision-making.

12.2 Key Issues in the Formulation of Anticorrosive Materials

The formulation of anticorrosive materials involves several key issues that need to be addressed to ensure effective corrosion protection. Some important considerations are as follows:

1. **Corrosion Mechanisms**: Understanding the corrosion mechanisms that occur in a particular environment is crucial. Different types of corrosion, such as uniform corrosion, pitting corrosion, or galvanic corrosion, require different approaches for effective protection. The formulation of anticorrosive materials should take into account the specific corrosion mechanisms that need to be addressed. There are several different types of corrosion that can occur, each with its own characteristics and mechanisms.
   Uniform Corrosion: Also known as general corrosion, this type of corrosion occurs uniformly over the entire surface of a material. It happens when the material is exposed to a corrosive environment, such as moisture or chemicals. Uniform corrosion leads to a gradual degradation of the material's surface. **Pitting Corrosion**: Pitting corrosion is localized corrosion characterized by the formation of small pits or cavities on the material's surface. It occurs when there are localized variations in the chemical composition or environmental conditions, causing accelerated corrosion in specific

areas. Pitting corrosion can be particularly dangerous as it can lead to rapid penetration and failure of the material. **Galvanic Corrosion**: Galvanic corrosion happens when two different metals or alloys with different electrochemical potentials are in contact with each other in the presence of an electrolyte. The more active (less noble) metal undergoes accelerated corrosion, while the less active (more noble) metal remains protected. This type of corrosion is common in situations where dissimilar metals are joined together, such as in plumbing systems or metal structures. **Crevice Corrosion**: Crevice corrosion occurs in narrow gaps or crevices between two surfaces, such as gaps between overlapping metal sheets or under gaskets and seals. The restricted access to oxygen and stagnant conditions within the crevices create an environment that promotes localized corrosion. Crevice corrosion can be challenging to detect and mitigate due to its hidden nature. **Stress Corrosion Cracking (SCC)**: Stress corrosion cracking is a type of corrosion that occurs under tensile stress in the presence of a corrosive environment. It can lead to the formation of cracks and ultimately catastrophic failure of the material, even at stress levels below its yield strength. SCC is commonly observed in materials such as stainless steel, aluminum alloys, and brass. **Intergranular Corrosion**: Intergranular corrosion occurs along the grain boundaries of a material, typically due to the presence of impurities or changes in the alloy's composition. This type of corrosion weakens the material along the grain boundaries, leading to cracking and structural failure. **Erosion Corrosion**: Erosion corrosion is the combined effect of corrosion and mechanical erosion. It occurs when a corrosive fluid, such as a liquid or gas, flows at high velocities over a material surface. The fluid's flow enhances the corrosive attack, leading to accelerated material degradation. **Microbiologically Influenced Corrosion (MIC)**: MIC is corrosion caused or influenced by the presence and activity of microorganisms, such as bacteria, algae, or fungi. These microorganisms can produce corrosive byproducts or create localized environments that accelerate corrosion. MIC is often encountered in industries involving water systems, pipelines, and marine environments.

2. **Material Selection**: Choosing the right materials for anticorrosive formulations is essential. The materials should possess inherent corrosion resistance or have properties that can be enhanced through additives. Commonly used materials for corrosion protection include coatings, polymers, alloys, and inhibitors. Choosing the right materials for anticorrosive formulations is crucial for developing effective corrosion protection. Some key considerations when selecting materials are as follows: **Inherent Corrosion Resistance**: Look for materials that possess inherent corrosion resistance. For example, stainless steel, aluminum, and certain alloys have a natural resistance to corrosion due to their composition and protective oxide layers. **Coatings**: Coatings are commonly used for corrosion protection. Consider materials such as epoxy, polyurethane, acrylics, or fluoropolymers that offer excellent barrier properties against moisture, chemicals, and environmental factors. **Polymers**: Polymers can provide corrosion resistance as well. Look for polymers that have good chemical resistance and can withstand exposure to harsh environments. Examples include polyethylene, polypropylene, and PVC (polyvinyl chloride).
   Corrosion Inhibitors: Incorporating corrosion inhibitors into formulations can help mitigate corrosion. These substances can either form a protective layer on the material surface or modify the electrochemical reactions that drive corrosion. Examples include organic and inorganic inhibitors like phosphates, chromates, amines, and benzotriazoles. **Composites**: Composite materials can offer enhanced corrosion resistance by combining different materials with complementary properties. For example, fiber-reinforced polymers (FRPs) or metal matrix composites (MMCs) can provide superior strength and corrosion resistance compared to individual materials. **Barrier Films**: Thin barrier films can be applied to protect metal surfaces. Look for materials like oxides, nitrides, or carbides that can form a protective layer on the surface, preventing direct contact with corrosive agents. **Sacrificial Anodes**: In some applications, sacrificial anodes are used to protect the main metal structure from corrosion. Materials like zinc, aluminum, or magnesium are chosen as sacrificial anodes because they are more reactive and corrode preferentially, diverting corrosion away from the

main structure. **Nanostructured Materials**: Nanostructured materials, such as nanocoatings or nanocomposites, offer improved corrosion resistance due to their unique properties at the nanoscale. These materials can provide enhanced barrier properties, increased surface hardness, or self-healing capabilities. **Environmental Considerations**: Consider the specific environmental conditions the material will be exposed to. Factors such as temperature, humidity, exposure to chemicals, UV radiation, or abrasion should be taken into account to ensure the selected materials can withstand the specific environment. **Compatibility**: Ensure compatibility between the anticorrosive material and the substrate it will protect. Consider factors such as adhesion, expansion/contraction coefficients, and the potential for galvanic corrosion between dissimilar materials. **Longevity and Durability**: Assess the expected service life and durability requirements of the anticorrosive material. Choose materials that can withstand long-term exposure without significant degradation or loss of protective properties. **Cost-effectiveness**: Balancing the cost of the anticorrosive materials with their performance is essential. Consider the overall lifecycle cost, including material acquisition, application, maintenance, and potential downtime due to corrosion-related issues.

By carefully considering these factors, it is possible to select materials that offer effective corrosion protection for specific applications and environments. Additionally, regular testing and evaluation should be conducted to validate the chosen materials' performance and optimize the formulation if needed.

3. **Environmental Factors**: The formulation should consider the environmental conditions in which the material will be used. Factors such as temperature, humidity, pH, presence of contaminants, and exposure to chemicals can significantly impact the corrosion process. Tailoring the material composition to withstand these conditions is crucial.
4. **Compatibility**: Anticorrosive materials must be compatible with the substrate they will protect. The formulation should consider the type of substrate, whether it is metal, concrete, or another material, and ensure that the material adheres well and provides effective protection without causing any adverse effects.

5. **Durability**: Long-term durability is an important factor in anticorrosive material formulation. The material should have the ability to withstand corrosion over extended periods, providing reliable protection for the intended service life. Factors such as resistance to UV radiation, mechanical stress, and aging should be considered.
6. **Application Method**: The formulation should be designed to be applied using appropriate methods such as spraying, brushing, dipping, or electroplating. The application process should be practical and efficient, allowing for uniform coverage and adherence to the substrate.
7. **Cost-Effectiveness**: Balancing the cost of the anticorrosive material with its performance is crucial. The formulation should aim to achieve the desired level of corrosion protection while considering the economic feasibility of production and application.
8. **Environmental Impact**: With increasing focus on sustainability, it is important to consider the environmental impact of anticorrosive materials. Formulations should aim for low toxicity, reduced VOC emissions, and minimal waste generation during production and application.
9. **Testing and Evaluation**: Thorough testing and evaluation of anticorrosive materials are essential to ensure their effectiveness. This includes laboratory testing, accelerated corrosion tests, and field trials to assess performance and validate the formulated material's corrosion protection capabilities.

Addressing these key issues during the formulation of anticorrosive materials can help in developing effective solutions to combat corrosion and extend the life span of various structures and components.

The Difficulties in Formulating a Good Anticorrosive and Antifouling Material

As most materials fail to prove their best inhibitor tendency, formulating the best anticorrosive and antifouling material can indeed be challenging, and not all materials may demonstrate optimal inhibitor tendencies. The difficulties encountered in the formulation process touch upon multiple engineering aspects as discussed next.

- **Complexity of Corrosion and Fouling Mechanisms**: Corrosion and fouling are complex processes influenced by various factors such as the specific environment, material composition, temperature, pH, and flow conditions. Understanding the intricate mechanisms involved in corrosion and fouling and identifying the most effective inhibition strategies can be a significant challenge. **Compatibility Issues**: Achieving compatibility between the anticorrosive or antifouling material and the substrate is crucial. Some inhibitors may have adverse effects on the material being protected or exhibit poor adhesion, leading to reduced performance. Finding materials that are both effective inhibitors and compatible with different substrates can be difficult. **Selectivity and Specificity**: Ideally, an anticorrosive or antifouling material should target the specific corrosion or fouling mechanisms without interfering with the normal functioning of the system. Developing materials that are selective and specific in their inhibitory actions, focusing on the problem areas while leaving other components unaffected, is a challenge. **Long-Term Stability**: Anticorrosive and antifouling materials should provide long-term stability to ensure continuous protection. Some inhibitors may degrade or lose their effectiveness over time due to environmental factors, UV exposure, chemical reactions, or mechanical stress. Achieving long-term stability and durability can be a difficult task. **Environmental Impact and Regulations**: Many traditional anticorrosive and antifouling materials may pose environmental risks due to their toxicity or the release of harmful substances during application, use, or disposal. Formulating materials that are effective inhibitors while being environmentally friendly and compliant with regulations can be a challenge. **Resistance Development**: Over time, some organisms or corrosion mechanisms can adapt and develop resistance to certain inhibitors. This necessitates the continuous development of new and improved materials to counteract evolving resistance and maintain effective protection. **Performance Testing and Evaluation**: Assessing the performance of anticorrosive and antifouling materials requires rigorous testing under various conditions, including laboratory simulations and real-world environments. **Obtaining Accurate and Reliable Data:** Obtaining accurate and reliable data on the effectiveness of the inhibitors can be challenging, particularly when dealing with long-term exposure or complex systems. **Cost and Availability**: Developing cost-effective anticorrosive and antifouling materials can be demanding, especially when considering the expense of raw materials, manufacturing processes, and application methods. Additionally, ensuring the availability of the required materials on a large scale can pose challenges.

Addressing these difficulties requires a multidisciplinary approach including material science, chemistry, engineering, and environmental considerations. Continuous research, innovation, and collaboration across different domains can help overcome these challenges and lead to the development of more effective anticorrosive and antifouling materials.

Corrosion and fouling mechanisms are complex processes influenced by various factors. We take a look at some recognized complexities involved in corrosion and fouling mechanisms:

- **Multiphase Systems**: Corrosion and fouling often occur in multiphase systems where multiple phases (e.g., solid, liquid, and gas) coexist. The interactions between different phases and their interfaces introduce complexities, such as the role of interfacial tension, mass transfer, and phase changes, which can significantly affect the corrosion and fouling behavior. **Electrochemical Reactions**: Corrosion is also an electrochemical process involving complex reactions at the metal-electrolyte interface. The simultaneous occurrence of anodic and cathodic reactions, coupled with mass transfer processes, governs the corrosion rate. **Understanding:** Understanding the kinetics and mechanisms of these electrochemical reactions is a challenging aspect of corrosion analysis. **Corrosion Products**: Corrosion of metals often leads to the formation of various corrosion products, such as oxides, hydroxides, or sulfides, which can influence the corrosion process. These corrosion products can have different morphologies, compositions, and protective properties, adding complexity to the overall corrosion mechanism and its impact on the material. **Localized Corrosion**: Localized corrosion phenomena, such as pitting corrosion or crevice corrosion, are highly complex and difficult to predict. Localized corrosion occurs at specific sites or regions on the metal surface, and it is influenced by factors such as microstructure, defects, composition gradients, and local environmental conditions. **Galvanic Corrosion**: Galvanic corrosion arises when two dissimilar metals or alloys come into electrical contact in the presence of an electrolyte. This process involves complex electrochemical interactions between the different metals, leading to accelerated corrosion of the less noble metal. The potential for galvanic corrosion depends on the specific metal combinations and their relative positions in the galvanic series. **Fouling Mechanisms**: Fouling refers to the accumulation of unwanted substances, such as scale, biofilms, or deposits, on surfaces. Fouling mechanisms can be influenced by biological, chemical, and physical factors, making them multifaceted. Factors such as fluid flow patterns, surface roughness, organic matter deposition, and microbial activities contribute to the complexity of fouling mechanisms. **Environmental Factors**: Corrosion and fouling are highly dependent on the surrounding environment. Environmental factors, including temperature, humidity, pH, dissolved oxygen, pollutants, and chemicals, can significantly influence the corrosion and fouling mechanisms. The variability of environmental conditions across different applications and settings adds complexity to the understanding and prediction of corrosion and fouling behavior. **Synergistic Effects**: In practical scenarios, corrosion and fouling mechanisms often interact and influence each other. For example, fouling layers can modify the local environment, creating differential aeration cells or crevice corrosion sites that accelerate corrosion. Understanding the synergistic effects between corrosion and fouling processes requires a comprehensive understanding of both phenomena.

To tackle these complexities, interdisciplinary approaches combining materials science, electrochemistry, fluid dynamics, microbiology, and environmental sciences are often necessary. Experimental investigations, theoretical models, and advanced analytical techniques are employed to unravel the intricacies of corrosion and fouling mechanisms and develop effective mitigation strategies.

The Factors Affecting the Lifetime of the Protected Metal due to Leaching

When considering the lifetime of a protected metal with regard to leaching, several factors come into play. Leaching refers to the process of material loss or dissolution from the protective coating or film, which can potentially expose the underlying metal to corrosion. **Coating Composition**: The composition of the protective coating or film plays a crucial role in its resistance to leaching. The choice of materials, binders, additives, and pigments used in the coating formulation can influence its stability and durability against leaching. Certain coatings may be more susceptible to leaching, while others may offer better resistance. **Environmental Exposure**: The environmental conditions to which the protected metal is exposed can impact the leaching process. Factors such as temperature, humidity, pH, exposure to chemicals, UV radiation, and moisture can accelerate leaching. Aggressive or corrosive environments can increase the likelihood and rate of leaching, thereby shortening the lifetime of the protected metal. **Coating Thickness**: The thickness of the protective coating can affect its resistance to leaching. Thicker coatings generally offer better protection against leaching as they provide a greater barrier between the metal and the environment. Thinner coatings may be more prone to leaching, especially if they have discontinuities or defects. **Adhesion and Film Integrity**: The adhesion of the protective coating to the metal surface and the integrity of the film are critical for preventing leaching. If the coating has poor adhesion or contains defects such as cracks, pinholes, or delamination, it can allow the ingress of corrosive agents, leading to accelerated leaching and corrosion of the underlying metal. **Leaching Medium**: The nature of the leaching medium, such as the presence of corrosive substances or solvents, can impact the rate of leaching. Some chemicals or environments may have a more aggressive leaching effect, causing faster degradation of the protective coating and increased exposure of the metal to corrosion. **Coating Porosity**: The porosity of the protective coating can influence leaching. Highly porous coatings may be more prone to leaching as corrosive agents can penetrate through the pores and reach the metal surface. Lower porosity or denser coatings can provide better resistance against leaching. **Coating Age and Degradation**: Over time, protective coatings may undergo degradation due to aging, exposure to UV radiation, chemical reactions, or mechanical stress. Degraded coatings are more susceptible to leaching, as their protective properties can diminish, leading to accelerated material loss and increased corrosion risk for the underlying metal. **Maintenance and Inspection**: Regular maintenance and inspection of the protective coating are crucial for detecting and addressing any signs of degradation or leaching. Timely repair or reapplication of the coating can help extend the lifetime of the protected metal and minimize the effects of leaching.

By considering these factors and selecting appropriate coating materials, ensuring proper application, and implementing effective maintenance practices, it is possible to mitigate the impact of leaching on the lifetime of protected metals and enhance their overall corrosion resistance.

12.3 Formulating a Good Model for Leaching

This process involves considering various factors and parameters that influence the leaching process. Some critical steps to help formulate a robust model: **Define the Objective**: Clearly define the objective of the leaching model. Determine what specific aspects of leaching we want to study or predict, such as the concentration profile of leached species over time, the rate of leaching, or the effect of different parameters on leaching behavior. **Identify Relevant Parameters**: Identify the parameters that have a significant influence on the leaching process. This may include factors such as coating composition, environmental conditions, coating thickness, adhesion, porosity, leaching medium properties, and metal properties. We need to consider both intrinsic properties and external factors that affect leaching. **Gather Experimental Data**: Collect experimental data on leaching from relevant studies or conduct your own experiments. This data will serve as the basis for model development, calibration, and validation. Ensure the data covers a range of conditions and provides sufficient information about the system being studied. **Select a Modeling Approach**: Choose an appropriate modeling approach based on the nature of the leaching process and the available data. Various modeling techniques can be employed, including mathematical models (e.g., mass transfer models, diffusion models, and kinetics models), empirical models, statistical models, or computational modeling using numerical methods. **Formulate Governing Equations**: Based on the selected modeling approach, develop the governing equations that describe the leaching process. These equations should represent the physical or chemical phenomena occurring during leaching, accounting for the parameters identified earlier. The equations may involve diffusion, reaction kinetics, mass transfer, or other relevant mechanisms. **Incorporate Boundary Conditions**: Define the appropriate boundary conditions for the leaching model. This includes specifying the initial conditions, environmental conditions, and any other boundary constraints that are relevant to the system being modeled. **Parameter Estimation and Calibration**: Calibrate the model by estimating the model parameters using the experimental data. This involves adjusting the model parameters to best fit the experimental results, typically through optimization techniques or regression analysis. The calibrated model should accurately represent the leaching behavior observed in the experimental data. **Model Validation**: Validate the model by comparing its predictions with additional experimental data or real-world observations that were not used in the calibration stage. Assess the model's ability to predict leaching under different conditions and determine its reliability and accuracy. **Sensitivity Analysis**: Conduct sensitivity analysis to identify the parameters or factors that have the most significant impact on leaching. This analysis helps understand the relative importance of different parameters and can guide future experimental or modeling efforts. **Model Refinement and Improvement**: Continuously refine and improve the model based on new data, insights, or advances in understanding. Incorporate feedback from experimental observations or further research to enhance the accuracy and applicability of the leaching model.

By following these steps and iteratively refining the model, a good model for leaching can be developed, that captures the essential factors and provides valuable insights into the leaching process, aiding in the design of effective protective coatings and materials.

The DLL Model

One well-known model for leaching is the diffusion-limited leaching (DLL) model. The DLL model is widely used to describe the leaching behavior of solid materials, particularly in the context of environmental impact assessments, waste management, and contaminant transport studies. This model assumes that leaching is primarily controlled by diffusion through a porous medium. The key features of the DLL model: **Fick's Law of Diffusion**: The DLL model is based on Fick's law of diffusion, which states that the mass flux of a solute through a medium is proportional to the concentration gradient. The concentration gradient drives the diffusion of the solute from the interior of the solid material to the surrounding leaching solution. **Porosity and Tortuosity**: The DLL model incorporates the concepts of porosity and tortuosity to account for the complex transport pathways within the porous medium. Porosity represents the fraction of void space in the material, while tortuosity represents the indirect or convoluted diffusion path resulting from the porous structure. **Concentration Profile**: The DLL model predicts the concentration profile of the leached species within the solid material over time. It accounts for the diffusion process, which gradually depletes the solute concentration in the material as leaching progresses. The model can provide insights into the temporal and spatial distribution of leached species. **Boundary Conditions**: The DLL model considers various boundary conditions, such as the initial concentration profile, the concentration of the leaching solution, and the interface between the solid material and the leaching solution. These boundary conditions play a critical role in determining the leaching behavior and can be adjusted to represent different scenarios. **Material Properties**: The DLL model takes into account material properties such as diffusivity and solubility. These properties influence the rate and extent of leaching and can vary depending on the specific material being studied. **Experimental Calibration**: The DLL model can be calibrated using experimental data to estimate model parameters such as diffusivity or tortuosity. By fitting the model predictions to experimental observations, the model's accuracy and reliability can be improved.

It is important to note that the DLL model represents a simplified representation of leaching processes and may not capture all factors involved in complex real-world scenarios. However, it provides a valuable framework for understanding and predicting the leaching behavior of solid materials under diffusion-limited conditions.

Density Functional Theory and Corrosion

The connection between density functional theory (DFT) and the corrosion behavior of metals lies in the ability of DFT to provide insights into the electronic structure and properties of materials at the atomic scale. Corrosion, which refers to the degradation of metals due to chemical reactions with their environment, is a complex process influenced by several factors, including the electronic structure of the metal.

DFT is a computational method used to calculate the electronic structure of materials. It is based on quantum mechanics and provides information about the distribution of electrons within a material, as well as various properties derived from the electronic structure, such as energy levels, bonding, and reactivity.

By applying DFT calculations, researchers can study the corrosion behavior of metals by investigating key factors that influence corrosion, such as follows: **Electronic Structure**: DFT allows the calculation of electronic properties, such as band structure, density of states, and work function, which can provide insights into the reactivity of the metal surface. By examining the electronic structure, researchers can understand the tendency of metals to donate or accept electrons, which is crucial in corrosion processes. **Adsorption and Reactions**: DFT can simulate the adsorption of species on the metal surface and provide information about the stability and energetics of adsorbed species. This helps in understanding the initial steps of corrosion, such as the adsorption of corrosive species (e.g., water, oxygen, or corrosive ions) on the metal surface and subsequent reaction pathways. **Passivation and Protective Films**: DFT can also shed light on the formation and stability of passivating oxide films that protect metals from further corrosion. By studying the electronic structure and properties of these oxide films, researchers can evaluate their effectiveness in preventing corrosion. **Alloying Effects**: DFT calculations can explore the influence of alloying elements on the corrosion behavior of metals. Alloying can significantly alter the electronic structure and surface reactivity of metals, leading to changes in corrosion resistance. DFT helps in understanding how different elements affect the corrosion behavior and designing alloys with improved corrosion resistance.

In summary, DFT provides a theoretical framework to investigate the electronic structure and properties of metals, enabling researchers to understand the underlying mechanisms of corrosion. By using DFT, scientists can gain insights into the reactivity of metal surfaces, adsorption processes, formation of protective films, and the impact of alloying on corrosion behavior, ultimately aiding in the development of corrosion-resistant materials and protective strategies.

DFT and Antifouling Modeling

DFT can be applied in antifouling modeling to understand and predict the interactions between antifouling coatings and foulants, as well as to design more effective antifouling materials. There are many ways DFT can be utilized in antifouling research: **Surface Interactions**: DFT can be used to study the interactions between antifouling coatings and foulants at the molecular level. By simulating the adsorption of foulants (e.g., proteins, bacteria, or marine organisms) on different coating surfaces, researchers can evaluate the strength and nature of the interactions. This information can guide the design of coatings with specific properties that deter fouling. **Surface Energy and Wettability**: DFT can calculate the surface energy and wettability of materials, which are important factors in antifouling performance. By analyzing the electronic structure and surface properties, DFT can help determine the hydrophobic or hydrophilic nature of a coating surface. Controlling surface wettability is crucial for preventing foulants from adhering to the surface or promoting their easy removal. **Electronic Properties and Charge Transfer**: DFT can provide insights into the electronic properties of antifouling coatings and their impact on fouling organisms. The electronic structure of a coating can affect its ability to repel or inhibit the attachment of foulants. Additionally, DFT can help understand charge transfer processes between the coating and foulants, which can influence adhesion and biofilm formation. **Coating**

Stability and Degradation: DFT can investigate the stability and degradation mechanisms of antifouling coatings in different environments. By studying the reaction pathways and energies involved in degradation processes, researchers can design more durable and long-lasting coatings. This knowledge can also help identify potential degradation products that may have adverse effects on the marine ecosystem. **Coating Optimization**: DFT calculations can aid in the optimization of antifouling coating formulations. By screening different molecular structures or compositions, researchers can predict the effectiveness of various coating materials in inhibiting fouling. DFT can assist in identifying promising candidates with desirable properties, such as low fouling affinity, high stability, and environmentally friendly characteristics.

Overall, DFT provides a powerful tool for understanding the fundamental interactions between antifouling coatings and foulants, as well as for designing and optimizing novel antifouling materials. By employing DFT in antifouling modeling, researchers can contribute to the development of more efficient and environmentally sustainable antifouling strategies.

DFT Applied to the Study of Surface Interactions—A Theoretical Perspective

DFT can be used to study surface interactions at the molecular level. Here is how DFT can be applied in investigating surface interactions: Adsorption studies: DFT can simulate the adsorption of molecules or atoms on a surface and provide information about the energetics, geometry, and electronic structure of the adsorbate-surface system. By calculating the adsorption energies and analyzing the charge redistribution, researchers can understand the strength and nature of the adsorption interactions. This is particularly useful in studying the adsorption of species on catalytic surfaces or in understanding the behavior of molecules on factionalized surfaces. Surface energy and surface tension: DFT can calculate the surface energy and surface tension of a material. Surface energy is a measure of the stability and reactivity of a surface, while surface tension affects wetting behavior and the spreading of liquids on a surface. By comparing the surface energies of different facets or materials, researchers can predict the preferred orientation or stability of surfaces and understand wetting phenomena. **Surface Reconstructions**: DFT can investigate surface reconstructions, which occur when a material's surface rearranges its atoms to achieve a lower energy configuration. By calculating the energies of different surface reconstructions and analyzing the resulting surface structures, researchers can gain insights into the stability and structural changes of surfaces under various conditions. This is crucial in understanding the behavior of materials at nanoscale dimensions or in catalytic processes where surface restructuring plays a role. **Surface Reactions**: DFT can model surface reactions and provide information about the reaction energetics, reaction pathways, and intermediates involved. By calculating the energy barriers for different reaction steps and analyzing the electronic structure of reaction intermediates, researchers can understand the mechanisms of surface reactions and identify the key factors influencing reaction rates and selectivity. This is particularly important in the field of heterogeneous catalysis, where surface reactions play a vital role. **Intermolecular Interactions**: DFT can also investigate intermolecular interactions between adsorbents on a surface or between molecules at a liquid-solid interface. By calculating the interaction energies and analyzing the electronic structure, researchers can gain insights into the nature and strength of these interactions. This information is useful in understanding the behavior of molecules on surfaces, such as self-assembly processes or the formation of molecular monolayers.

Using DFT for Studying Surface Energy and Wettability

DFT can be utilized to study the wettability of surfaces, which refers to the ability of a liquid to spread and form a uniform film on a solid surface. **Surface Energy**: DFT can calculate the surface energy of a material, which is a fundamental parameter influencing wettability. By comparing the surface energies of different materials or surface orientations, researchers can predict the relative wettability of these surfaces. A lower surface energy typically corresponds to higher hydrophobicity and lower wettability, while a higher surface energy indicates higher hydrophilicity and better wetting. **Contact Angle Calculations**: DFT can simulate the contact angle formed by a liquid droplet on a solid surface. The contact angle is a measure of wettability and is determined by the balance between adhesive forces (between the liquid and solid) and cohesive forces (within the liquid). By calculating the interaction energies between the liquid and solid, DFT can provide insights into the contact angle and help understand the wetting behavior of different liquids on specific surfaces. **Surface Reconstruction**: DFT can investigate surface reconstructions that occur on materials' surfaces, leading to changes in their wettability. By modeling the atomic rearrangement on the surface and calculating the resulting surface energies, researchers can predict the impact of surface reconstructions on wettability. This is particularly relevant when studying materials exposed to various environmental conditions or when designing surfaces with tailored wettability properties. **Surface Functionalization**: DFT can aid in the design of surface coatings or functionalization strategies to control wettability. By simulating the adsorption of molecules on a surface and calculating the resulting surface energies, researchers can identify molecules that can modify the wettability of the surface. This knowledge can guide the development of coatings or surface treatments that enhance or reduce wettability as desired. **Electrostatic Interactions**: DFT can also consider electrostatic interactions between the liquid and solid surfaces. By analyzing the charge distribution and electrostatic potential, researchers can gain insights into how electrostatic forces influence the wetting behavior. This is particularly relevant in systems where electrostatic interactions play a significant role, such as in polar liquids or charged surfaces.

By employing DFT calculations, researchers can investigate the underlying factors that influence wettability, including surface energy, contact angle, surface reconstructions, and electrostatic interactions. This information can guide the design and development of materials with tailored wettability properties for applications ranging from self-cleaning surfaces and anti-fog coatings to microfluidics and biomedical devices.

DFT, Electronic Properties, and Charge Transfer of Antifouling Coatings

DFT can be applied to study the electronic properties and charge transfer processes in antifouling coatings, providing insights into their performance and effectiveness. Here's how DFT can be used in this context: **Electronic Structure**: DFT can calculate the electronic structure of antifouling coatings, including the band structure, density of states, and energy levels. By examining the electronic properties, such as the HOMO (highest occupied molecular orbital) and LUMO (lowest unoccupied molecular orbital) energies, researchers can assess the reactivity and potential interactions with foulants. Understanding the electronic structure helps in determining the coating's ability to donate or accept charges and its electronic compatibility with the surrounding environment. **Charge Transfer at Interfaces**: DFT can simulate charge transfer processes occurring at the interface between the antifouling coating and foulants. This includes the evaluation of electron or hole injection from the coating into the foulant or vice versa. By calculating the energy barriers and analyzing the electronic states involved, researchers can gain insights into the charge transfer mechanisms and the factors influencing them. This knowledge is crucial for understanding the adhesion, biofilm formation, or repulsion of foulants on the coating surface. **Redox Reactions**: DFT can investigate redox reactions that may occur at the interface between the antifouling coating and foulants. By calculating the reaction energies and analyzing the electronic structure of the species involved, researchers can assess the likelihood and kinetics of these redox reactions. This information helps in understanding the effectiveness of the coating in preventing or inhibiting fouling by interfering with essential redox processes in foulants. **Reactive Sites and Functional Groups**: DFT can identify reactive sites and functional groups on the surface of antifouling coatings. By evaluating the local reactivity descriptors, such as Fukui functions or local softness, researchers can pinpoint regions with high electron density or susceptibility to nucleophilic or electrophilic attacks. This knowledge aids in understanding the fouling mechanisms and can guide the modification or optimization of the coating surface to enhance antifouling properties. **Intermolecular Interactions**: DFT can study the intermolecular interactions between the antifouling coating and foulants. By calculating the interaction energies, analyzing the charge distribution, and investigating the nature of bonding, researchers can assess the strength and nature of these interactions. This information is valuable in designing coatings that repel or inhibit the attachment of foulants through non-covalent interactions such as hydrogen bonding, van der Waals forces, or electrostatic repulsion.

DFT for Studying Stability and Degradation Mechanisms of Antifouling Coating Materials

DFT can be employed to study the stability and degradation mechanisms of anti-coating materials, providing insights into their performance and durability. **Reaction Energetics**: DFT can calculate the energy barriers and reaction energies associated with the degradation processes of anti-coating materials. By studying the potential reaction pathways and identifying the most favorable reaction steps, researchers can understand the degradation mechanisms and predict the stability of the coating under different environmental conditions. This information aids in the design of more stable and long-lasting coatings. **Surface Reactions**: DFT can model surface reactions that occur during the degradation of anti-coating materials. By calculating the energetics and analyzing the electronic structure of intermediates and transition states, researchers can elucidate the reaction mechanisms and identify key factors influencing the degradation process. This knowledge helps in understanding the specific degradation pathways and designing coatings that are more resistant to these reactions. **Oxidation and Corrosion Resistance**: DFT can assess the oxidation and corrosion resistance of anti-coating materials. By examining the electronic structure and adsorption energies of reactive species, such as oxygen or corrosive ions, researchers can predict the susceptibility of the coating to oxidation or corrosion. DFT calculations provide insights into the stability of the protective oxide or passivation layers formed on the coating surface, aiding in the design of materials with enhanced resistance to degradation. **Stability of Functional Groups**: DFT can study the stability of functional groups present in anti-coating materials. By calculating the energies associated with bond breaking or formation, researchers can assess the stability of specific chemical functionalities under different conditions. This knowledge helps in selecting or modifying functional groups to improve the stability of the coating and mitigate degradation processes. **Environmental Effects:** DFT can investigate the influence of environmental factors, such as temperature, humidity, or pH, on the stability of anti-coating materials. By considering the impact of these factors on reaction energies and kinetics, researchers can predict the stability of the coating under realistic conditions. This information is crucial in assessing the long-term performance and durability of the coating in real-world applications.

Computational Improvements Beyond DFT

While DFT has been highly successful in many applications, it does have limitations and approximations that can affect the accuracy of its results, especially in cases where strong correlations between electrons or van der Waals interactions are involved. Researchers are continually working on developing more advanced theories and methods to address these limitations and improve accuracy. There are some approaches that aim to provide more accurate results: **Beyond DFT Methods:** Various methods that go beyond traditional DFT have been developed to handle specific challenges. These include methods like hybrid functionals (mixing DFT and Hartree-Fock), meta-GGA (generalized gradient approximation), and range-separated functionals. **Wave Function Methods:** Wave function–based methods, such as coupled cluster theory and Møller-Plesset perturbation theory, offer high accuracy but are computationally demanding and may be limited in terms of system size. **Many-Body Perturbation Theory (MBPT):** MBPT approaches, like GW (Gross-Wilson) approximation and Bethe-Salpeter equation, provide accurate results for electronic excitations, but they are computationally intensive and may not be suitable for large systems. **Tight-Binding and Semi-Empirical Methods:** Tight-binding models and semi-empirical methods provide a compromise between accuracy and computational cost. They use simplified Hamiltonians or empirical parameters to describe electronic interactions. **Correlated Electron Methods:** Correlated electron methods, such as Quantum Monte Carlo (QMC), offer high accuracy for electronic correlation effects but can be computationally demanding and challenging to apply to complex systems. **Machine Learning and Data-Driven Approaches:** Machine learning techniques are being explored to accelerate calculations and improve accuracy. Neural networks and other models can learn from databases of accurate calculations to predict properties efficiently. **Advanced**

Exchange-Correlation Functionals: Developing improved exchange-correlation functionals is an ongoing effort. Non-local and van der Waals functionals aim to better describe dispersion forces that are challenging for standard DFT. **Embedding and QM/MM Methods:** Embedding methods combine quantum mechanics (QM) with classical mechanics (MM) to accurately describe regions of interest in larger systems. **Combining Different Approaches:** Hybrid methods that combine different theories, such as DFT with post-DFT corrections or coupling electronic structure methods with molecular dynamics, aim to capture a wider range of interactions.

While each of these methods addresses specific challenges, no single theory or method is universally applicable or accurate for all situations. The choice of method depends on the nature of the system, the accuracy required, and the available computational resources. Researchers continually seek to strike a balance between accuracy and computational efficiency, and the field of electronic structure theory is dynamic and evolving with ongoing developments and discoveries.

Modeling and Optimization of Antifouling and Anticorrosive Coatings

An antifouling coating is a type of specialized coating that is usually applied on ship or boat hulls to inhibit organism growth on the hull. The presence of organisms in this part of the vessel could greatly affect the vessel's durability and performance. Other than fighting the growth of microorganisms, this paint also serves as a corrosion barrier for hulls, which improves performance and water flow across the vessel hull. The antifouling coating is also known as bottom paint. [Ref: corrosionpedia.com]

The antifouling coating prevents fouling growth through the production of ingredients that prevent organisms from settling. Its use plays a huge role in decreasing not only the commercial costs of vessels, but the total operational expenses of maritime industries and others.

For instance, fuel constitutes more than half of the entire operating costs of vessels. If antifouling coatings are not used, the consumption of fuel as well as the emission of carbon dioxide could rise significantly. This can be the same in other industrial machinery.

Moreover, most high-quality antifouling coatings could deliver lifetime benefits like extended intervals of dry dock or decreased dry dock, which in turn lowers maintenance expenses as well as harmful emissions. Additionally, this type of coating prevents hulls and other metal structures from undergoing corrosion. The antifouling coating can improve the corrosion resistance of steel structures other than stopping the growth of microorganisms.

There are various types of antifouling coatings ranging from linear polishing polymers (LPPs) to more advanced self-polishing copolymers (SPCs). With this type of product, operators can effectively extend the life of vessels and other structures, reducing the entire operation's costs. [Ref: corrosionpedia.com]

Linear Polishing Polymers

LPPs are a type of additive used in antifouling coatings to enhance their performance. LPPs are designed to continuously release a low concentration of biocides over time, thereby inhibiting the attachment and growth of fouling organisms on the coated surfaces. The concept behind LPPs is to provide a controlled release of biocides while minimizing their leaching into the environment. Some key points about LPPs for antifouling coatings: **Mechanism of Action**: LPPs work through a process called controlled surface erosion. The polymer matrix containing the biocide gradually erodes over time due to hydrolysis or other environmental factors, exposing fresh biocide molecules to the coating surface. This continuous erosion and release process helps maintain the antifouling efficacy of the coating over an extended period. **Biocide Incorporation**: LPPs are typically formulated by dispersing biocide particles, such as copper or zinc compounds, within a polymer matrix. The choice of biocide depends on its efficacy against target fouling organisms and compatibility with the coating formulation. Copper-based biocides, such as copper oxide or copper pyrithione, are commonly used due to their broad-spectrum antifouling properties. **Release Rate Control**: The release rate of biocides from LPPs can be controlled by several factors, including the polymer matrix composition, molecular weight, cross-linking density, and environmental conditions. By adjusting these parameters, the release rate can be tailored to meet specific requirements, such as the expected fouling pressure and the desired coating life span. **Longevity and Effectiveness**: LPPs are designed to provide long-term antifouling protection by continuously releasing biocides. The gradual erosion of the polymer matrix ensures a sustained release of the biocides, maintaining their effectiveness over an extended period. This helps reduce the frequency of coating maintenance and extends the service life of the coated surfaces. **Environmental Considerations**: LPPs are developed with a focus on minimizing the environmental impact associated with biocide release. By controlling the release rate, the concentration of biocides in the surrounding water can be reduced compared to conventional antifouling coatings, thereby mitigating potential ecological harm. **Coating Compatibility**: LPPs can be incorporated into various types of antifouling coating systems, including both self-polishing and fouling-release coatings. They can be added during the formulation stage or applied as a separate topcoat to existing coatings. Compatibility with the base coating and overall coating performance should be considered during the selection and application of LPPs.

LPPs have gained attention in recent years as a promising approach to improving the performance and environmental sustainability of antifouling coatings. Research and development in this field continue to explore novel polymer compositions, biocide combinations, and optimization strategies to enhance the effectiveness and longevity of LPP-based antifouling coatings.

Self-Polishing Copolymers

Self-polishing copolymers (SPCs) are a class of antifouling coating agents that provide controlled and continuous release of biocides to inhibit the attachment and growth of fouling organisms on submerged surfaces. SPCs are designed to gradually erode, exposing fresh biocide molecules and maintaining the antifouling efficacy of the coating over an extended period. What follows is an overview of self-polishing copolymers as antifouling coating agents: **Mechanism of Action**: SPCs work through a process known as controlled surface erosion. The polymer matrix containing biocides undergoes controlled degradation in contact with seawater. As the coating erodes, new biocide molecules are exposed to the surface, providing continuous and controlled release over time. This sustained release helps prevent fouling organisms from colonizing the surface. **Copolymer Composition**: SPCs are typically composed of a combination of hydrolyzable and non-hydrolyzable monomers. The hydrolyzable monomers contribute to the controlled degradation of the polymer matrix, while the non-hydrolyzable monomers provide structural integrity and mechanical properties to the coating. The selection of monomers depends on their compatibility, desired release rate, and overall coating performance requirements. **Biocide Incorporation**: Biocides, such as copper or zinc compounds, are incorporated into the polymer matrix of SPCs. The choice of biocide depends on its effectiveness against specific fouling organisms and regulatory requirements. Copper-based biocides, such as copper oxide or copper acrylate, are commonly used due to their broad-spectrum antifouling properties. **Release Rate Control**: The release rate of biocides from SPCs is controlled by the rate of polymer erosion. Factors such as the composition of the copolymer, hydrolysis kinetics, and environmental conditions influence the degradation rate. By adjusting these factors, the release rate of biocides can be tailored to provide optimal antifouling performance and minimize environmental impact. **Environmental Considerations**:

SPCs are designed to optimize antifouling efficacy while minimizing the release of biocides into the environment. The controlled release mechanism ensures that biocides are released at a relatively low concentration, reducing their impact on non-target organisms. Furthermore, advancements in SPC technology aim to improve the biodegradability and eco-toxicological profile of these coatings. **Coating Application and Maintenance**: SPCs are typically applied as a topcoat over a compatible primer or tie coat. The coating should be applied with proper surface preparation and coating application techniques to ensure good adhesion and long-term performance. **Periodic Maintenance**, such as recoating or hull cleaning, may be necessary to maintain the coating's antifouling effectiveness.

SPCs have been widely used in the marine industry as effective antifouling coatings. Their ability to provide continuous biocide release and adapt to varying environmental conditions makes them suitable for protecting ship hulls, offshore structures, and other submerged surfaces. Ongoing research and development focus on optimizing SPC formulations, enhancing their performance, and improving their environmental sustainability.

LPPs vs. SPCs

LPPs and Self-Polishing Copolymers (SPCs) are both types of antifouling coating additives designed to release biocides gradually over time. While they share some similarities in terms of their controlled release mechanism, there are notable differences between LPPs and SPCs.

1. Mechanism of Action:
   LPPs: LPPs work through controlled surface erosion. The polymer matrix containing biocides erodes over time, exposing fresh biocide molecules and maintaining antifouling efficacy.
   SPCs: SPCs also rely on controlled degradation but involve a self-polishing mechanism. As the coating erodes, a bioactive layer is formed on the surface, continuously releasing biocides from this layer.
2. Polymer Composition:
   LPPs: LPPs are typically linear polymers formulated with biocides dispersed within a polymer matrix. The polymer matrix erodes to release biocides gradually.
   SPCs: SPCs are copolymers composed of both hydrolyzable and non-hydrolyzable monomers. The hydrolyzable monomers contribute to controlled degradation, while the non-hydrolyzable monomers provide structural integrity to the coating.

3. Release Rate Control:
   LPPs: The release rate of biocides from LPPs is primarily controlled by the erosion rate of the polymer matrix. Adjusting the polymer composition and environmental conditions can influence the release rate.
   SPCs: The release rate of biocides from SPCs is primarily controlled by the self-polishing mechanism. As the coating erodes, the bioactive layer continuously releases biocides.
4. Environmental Impact:
   LPPs: LPPs are designed to minimize environmental impact by controlling the release rate of biocides, reducing their concentration in the surrounding water compared to conventional coatings.
   SPCs: SPCs also aim to reduce the environmental impact by providing controlled biocide release. The self-polishing mechanism helps maintain low biocide concentrations in the environment.
5. Coating Application:
   LPPs: LPPs can be incorporated into various types of antifouling coatings, including both self-polishing and fouling-release coatings, either during the formulation stage or as a separate topcoat.
   SPCs: SPCs are typically applied as a topcoat over a compatible primer or tie coat. They form a self-polishing bioactive layer on the surface that continuously releases biocides.

Both LPPs and SPCs contribute to the development of effective antifouling coatings with controlled biocide release. The choice between LPPs and SPCs depends on factors such as coating system requirements, desired release rate control, and specific application needs.

The development of effective antifouling coatings involves a combination of modeling and formulation considerations. An overview of the process follows here: **Understanding Fouling Mechanisms**: To develop effective antifouling coatings, it is essential to have a good understanding of the fouling mechanisms involved. This includes studying the attachment and growth of different organisms, the surface properties influencing fouling, and the environmental conditions where the coating will be used. **Surface Modeling and Modification**: Modeling the surface properties and interactions is crucial for antifouling coating development. This involves studying the surface energy, roughness, wettability, and charge characteristics of the substrate. Different techniques like contact angle measurements, surface energy calculations, and atomic force microscopy can be used to analyze and modify the surface properties. **Selection of Active Ingredients**: Antifouling coatings typically contain active ingredients that deter or inhibit the attachment and growth of fouling organisms. These ingredients can be biocides, such as copper or zinc compounds, or nontoxic substances with fouling-release properties. The selection of active ingredients depends on their efficacy, environmental impact, regulatory compliance, and compatibility with the coating formulation. **Coating Formulation**: The formulation of antifouling coatings involves combining the active ingredients with binders, solvents, additives, and other components to create a stable and effective coating system. The formulation needs to consider factors such as compatibility of ingredients, film formation properties, durability, and ease of application. **Coating Performance Testing**: Once the antifouling coating is formulated, it needs to undergo extensive testing to evaluate its performance. This includes laboratory tests to assess its antifouling efficacy, durability, leaching rate, and toxicity. Field trials in real marine environments are also important to validate the coating's effectiveness under practical conditions. **Optimization and Improvement**: Based on the performance testing results, the coating formulation can be optimized and improved. This iterative process involves adjusting the concentration of active ingredients, modifying the binder system, fine-tuning the coating thickness, and incorporating additional additives to enhance specific properties. **Regulatory Compliance**: Antifouling coatings must comply with various regulations, such as those related to environmental impact and toxicity. It is important to consider these regulations during the formulation and development process to ensure the coating meets the required standards.

Overall, the modeling and formulation of antifouling coatings involve a multidisciplinary approach, combining knowledge from surface science, material science, chemistry, and biology. The goal is to create coatings that provide effective fouling protection while minimizing environmental impact.

Mathematical Models

Mathematical models play a crucial role in formulating antifouling coating materials by providing insights into the physical and chemical processes involved. These models help optimize the composition, performance, and properties of the coatings. Described next are some commonly used mathematical models in formulating antifouling coating materials: **Diffusion Models**: Diffusion models are used to describe the release of biocides from coatings and their transport through the surrounding environment.

These models consider factors such as diffusion coefficients, concentration gradients, and interaction with fouling organisms. They help predict the release rate and distribution of biocides from the coating. **Kinetic Models**: Kinetic models are used to describe the degradation or erosion of the coating matrix over time. These models incorporate degradation rates, reaction kinetics, and environmental factors to predict the evolution of the coating and the release of active components. Kinetic models are essential for understanding the long-term performance and life span of the coating. **Surface Interaction Models**: Surface interaction models describe the interactions between the coating surface and fouling organisms. These models consider parameters such as surface roughness, surface energy, and fouling organism behavior. They help understand the adhesion mechanisms and provide insights into how surface properties can be modified to prevent fouling attachment. **Coating Film Formation Models**: Film formation models are used to understand the behavior of coating materials during the application and curing processes. These models incorporate variables such as viscosity, drying kinetics, and film thickness. By simulating the coating process, these models help optimize the application parameters and ensure the desired coating quality. **Computational Fluid Dynamics (CFD) Models**: CFD models simulate fluid flow patterns and mass transport phenomena around coated surfaces. They help predict the movement and settlement of fouling organisms by considering factors such as flow velocity, turbulence, and shear stress. CFD models provide insights into the performance of coatings under different hydrodynamic conditions. **Optimization Models**: Optimization models use mathematical algorithms to optimize the formulation of coating materials based on defined objectives and constraints. These models consider variables such as ingredient concentrations, binder ratios, and performance targets. By solving optimization problems, these models help identify the optimal combination of components to achieve desired antifouling properties.

These mathematical models can be used individually or in combination, depending on the specific requirements and complexities of the antifouling coating formulation. They provide valuable tools for understanding the underlying processes, guiding experimental design, and optimizing the performance of antifouling coatings.

12.4 The Advent of Nanotechnology

Nanotechnology offers significant potential for the development of new and improved anticorrosive and antifouling materials in the maritime industry. By manipulating materials at the nanoscale level, researchers can create coatings and surfaces with enhanced properties that effectively combat corrosion and fouling. **Enhanced Barrier Properties**: Nanoparticles can be incorporated into coatings to create more effective barriers against corrosive agents and fouling organisms. Nanostructured coatings provide a higher surface area, allowing for better coverage and adhesion, which can lead to improved protection. **Self-Healing Coatings**: Nanotechnology enables the development of self-healing coatings that can repair minor damage and cracks caused by corrosion. Nanoparticles with encapsulated corrosion inhibitors or polymer precursors can be released when damage occurs, thereby restoring the protective layer. **Smart Release of Active Agents:** Nanoparticles can be designed to release anticorrosive or antifouling agents in a controlled manner. This ensures a sustained and targeted effect, maximizing the efficiency of these agents and minimizing their environmental impact. **Surface Modification:** Nanotechnology allows for precise surface modification to create hydrophobic (water-repellent) and oleophobic (oil-repellent) surfaces. These surfaces discourage the adhesion of both waterborne and oilborne fouling agents. **Photocatalytic Properties:** Nanoparticles with photocatalytic properties, such as titanium dioxide (TiO2), can break down organic matter on surfaces when exposed to light. This property can help prevent fouling by disrupting the attachment process of organisms. **Nanoscale Topography:** Creating specific nanoscale textures on surfaces can affect the attachment and adhesion of fouling organisms. These textures can make it difficult for organisms to establish a foothold, reducing fouling buildup. **Reduced Environmental Impact**: Nanotechnology allows for the development of eco-friendly antifouling solutions. By using nanoparticles with controlled release mechanisms, it is possible to minimize the use of harmful biocides while still achieving effective antifouling properties. **Advanced Monitoring**: Nanosensors can be integrated into coatings to monitor the corrosion and fouling status of surfaces in real time. This enables proactive maintenance and the timely application of corrective measures.

Even though nanotechnology offers promising solutions, there are also challenges to consider, such as regulatory approvals, potential toxicity of nanoparticles, and scalability of production processes. Safety and environmental concerns associated with the release of nanoparticles into the marine ecosystem need to be thoroughly evaluated.

In summary, nanotechnology has the potential to revolutionize the development of anticorrosive and antifouling materials by providing enhanced properties and novel mechanisms of action. As research and development continue in this field, we can expect to see more innovative and effective solutions to combat corrosion and fouling in the maritime industry.

Types of Nanomaterials

Several types of nanomaterials are being explored for marine corrosion prevention due to their unique properties at the nanoscale. These materials offer improved corrosion resistance and can be incorporated into coatings, paints, and surface treatments to enhance protection against the harsh marine environment. Some of the commonly used nanomaterials for marine corrosion prevention include the following:

Nanoparticles of Corrosion Inhibitors:

- **Zinc Oxide (ZnO) Nanoparticles:** ZnO nanoparticles release zinc ions, which act as sacrificial anodes to protect the underlying metal from corrosion.
- **Cerium Oxide (CeO2) Nanoparticles:** CeO2 nanoparticles provide corrosion protection by forming a protective layer on the metal surface, hindering the corrosion process.

Nanocomposite Coatings:

- **Polymer-Nanoparticle Composites:** Incorporating nanoparticles (e.g., clay nanoparticles) into polymer coatings improves barrier properties, durability, and resistance to corrosive agents.
- **Metal Matrix Nanocomposites:** These composites consist of metal matrices with nanoparticles dispersed within. They enhance mechanical properties and corrosion resistance.

Nanocoatings with Self-Healing Properties:

- **Nano-capsules:** Coatings containing nano-capsules filled with corrosion inhibitors can release the inhibitors in response to damage, healing micro-cracks and protecting the underlying metal.
- **Nano-gels:** These three-dimensional networks of cross-linked nanoparticles can hold and release corrosion inhibitors, providing a continuous source of protection.

Conversion Coatings:

- **Nanocrystalline Ceramics:** Thin conversion coatings formed by depositing nanocrystalline ceramics (e.g., aluminum oxide) on metal surfaces provide corrosion protection.

Nanocomposite Anodes:

- **Nanostructured Metal Oxides:** Nanostructured metal oxides (e.g., nano-TiO2) can be used as anodes in cathodic protection systems, releasing ions to prevent corrosion.

Nanotubes and Nanofibers:

- **Carbon Nanotubes (CNTs) and Nanofibers:** These materials can be incorporated into coatings to improve adhesion and mechanical properties, providing corrosion resistance.
- **Inorganic Nanotubes:** Materials like boron nitride nanotubes have shown promise in enhancing corrosion resistance due to their unique chemical stability.

Nano-Porous Materials:

- **Nano-Porous Silica:** These materials can be loaded with corrosion inhibitors and incorporated into coatings to release the inhibitors over time, providing ongoing protection.

Photocatalytic Nanomaterials:

- **Titanium Dioxide (TiO2) Nanoparticles:** TiO2 nanoparticles possess photocatalytic properties that can break down organic matter on surfaces, inhibiting corrosion and fouling.

Anticorrosive Nanoparticle Primers:

- **Layered Double Hydroxides (LDHs):** LDHs loaded with corrosion inhibitors can be used as primers to enhance the adhcsion of coatings and provide corrosion protection.

Research and development in the field of nanomaterials for marine corrosion prevention are ongoing, aiming to create effective and sustainable solutions to combat corrosion and extend the life span of marine structures and equipment.

12.5 Summary of the Current Developments

Summarizing whatever we have discussed so far, the status of current developments is as given below:

Antifouling Coatings:
Environmentally Friendly Formulations: Researchers and manufacturers were focusing on developing antifouling coatings that are effective

against marine growth while minimizing the release of harmful biocides into the environment. These coatings often utilize natural or less toxic compounds. **Bioinspired Approaches:** Some antifouling coatings were being developed based on the principles of bio-mimicry, taking inspiration from marine organisms that naturally resist fouling. Mimicking the textures, surface chemistry, or defense mechanisms of these organisms was explored. **Copper-Free Solutions:** Due to concerns about the environmental impact of copper-based coatings, alternative copper-free antifouling solutions were gaining attention. These coatings typically employ different mechanisms, such as surface texturing or the release of nontoxic inhibitors. **Hydrodynamic Surface Designs:** Coatings with specific textures or patterns were being researched to reduce the ability of fouling organisms to adhere. These coatings aim to minimize drag and improve fuel efficiency by reducing biofouling.

Anticorrosion Coatings:
Advanced Coating Technologies: The development of coatings with enhanced barrier properties and improved adhesion for extended corrosion protection was ongoing. This includes the incorporation of nanomaterials for better corrosion resistance. **Self-Healing Coatings:** Self-healing coatings that can repair minor damage and cracks in the coating film were being explored. These coatings release inhibitors or healing agents when damage occurs, maintaining the integrity of the coating. **Smart Coatings:** Smart coatings with embedded sensors were developed to monitor the condition of the coating and substrate in real time. This enables proactive maintenance and targeted repairs. **High-Performance Coatings for Harsh Environments:** Coatings designed to withstand extreme marine environments, including offshore structures and vessels operating in rough seas, were being optimized for better durability. **Hybrid Coating Systems:** Researchers were investigating the combination of different coating types to provide comprehensive protection against both corrosion and fouling. These hybrid systems aimed to leverage the strengths of each coating type. **Digitalization and Predictive Maintenance:** The integration of digital technologies and data analysis enhanced the ability to predict coating degradation and plan maintenance more efficiently. It is important to keep in mind that the development of marine antifouling and anticorrosion coatings is an ongoing process, driven by regulatory changes, environmental concerns, technological advancements, and the need for more durable and sustainable solutions. To stay up-to-date on the latest developments, I recommend consulting recent scientific literature, industry publications, and reports from coating manufacturers and research organizations.

Future Prospects in These Areas of Research

The field of marine antifouling and anticorrosion coatings is continually evolving to address environmental concerns, regulatory changes, and the demand for more effective and sustainable solutions. The following is a brief description of the future prospects and trends that could shape this area:

Antifouling Coatings:

1. **Eco-Friendly Solutions:** The push for more environmentally friendly antifouling coatings is likely to continue. Research and development will focus on coatings that effectively prevent fouling while minimizing harm to marine ecosystems. **Bioinspired Innovations:** Bioinspired coatings that mimic the properties of marine organisms resistant to fouling are likely to gain further traction. Developing surfaces with tailored textures and chemistries can deter fouling organisms. **Smart Release Mechanisms:** Antifouling coatings with controlled and smart release of active agents will become more sophisticated. These coatings will provide sustained protection by releasing inhibitors only when needed. **Alternative Strategies:** The exploration of nontoxic or non-biocidal strategies to prevent fouling, such as using physical barriers or manipulating surface properties, will continue to be an active area of research.

Anticorrosion Coatings:

1. **Nanotechnology Advancements:** Continued advancements in nanotechnology will lead to the development of more effective and durable anticorrosion coatings. Nanostructured coatings, self-healing nanocomposites, and corrosion-resistant nanomaterials will likely become more prevalent. **Smart Coatings and Sensors:** The integration of sensors within coatings for real-time corrosion monitoring will become more common. This will enable predictive maintenance and reduce downtime for repairs. **Hybrid Coating Systems:** The development of hybrid systems that combine antifouling and anticorrosion properties in a single coating or complementary coatings will be explored further for comprehensive protection. **Digitalization and Data Analytics:** Digital tools and data analytics will play a larger role in assessing coating performance, optimizing maintenance schedules,

and making informed decisions about coating selection. **Advances in Coating Application:** Innovations in coating application techniques, such as 3D printing and advanced spraying methods, will improve the uniformity and quality of coating application. **Corrosion-Resistant Materials:** Research into new corrosion-resistant materials and alloys, as well as surface treatments and coatings that can mitigate galvanic corrosion, will continue. **Sustainable Coating Practices:** As sustainability becomes a focal point in many industries, the development of coatings with reduced environmental impact during production, application, and disposal will be emphasized. **Climate Change Adaptation:** Coatings that are designed to withstand the effects of climate change, such as increased saltwater exposure, temperature variations, and extreme weather events, will become more relevant. **Regulatory Compliance:** As environmental regulations evolve, the coatings industry will continue to adapt to meet new standards and requirements. Overall, the future of marine antifouling and anticorrosion coatings is likely to involve a combination of advanced materials, innovative technologies, and sustainable practices to provide effective protection while minimizing environmental impact. Collaboration between researchers, manufacturers, and regulatory bodies will be essential to drive these advancements and meet the growing demand for durable and environmentally responsible solutions.

12.6 Conclusion

In the present book chapter, an attempt has been made to present data to emphasize the importance of smart anticorrosive materials to monitor environmental changes, effectively prevent corrosive deterioration in real time, avoid unnecessary waste of resources, and have great practical value, which researchers and companies are frequently concerned. This chapter also discusses the lack of understanding of the physical and chemical mechanisms underlying the failure of the anticorrosive coating during service requirements. In the development of new interesting products, and in order to ensure a correlation between rapid exposure and natural exposure tests, it is appropriate to thoroughly understand and quantify degradation mechanisms by mathematical models. Moreover, anticorrosive coating systems that provide excellent protection against corrosion might be achieved by further studies on the degradation mechanisms and progress made in developing binders and pigments. However, other characteristics of existing systems must not be compromised in order to achieve high anticorrosive performance.

Bibliography

1. Pemberton, R., Summerscales, J., Jones, J.G., *Marine Composites - Design and Performance*, 1st edition, Woodhead Publishing.
2. Cramer, S. *et al.*, *ASM Handbook*, vol. 13, ASM international, 2003.
3. Russel, P., Embleton, W., Jackson, L., *Applied Thermodynamics for Marine Engineers*, 5th edition, Adlard Coles Nautical Bloomsbury.
4. Sheir, Jarman, Burstein, *Corrosion*, 3rd edition, vol. 12, Butterworth Heinemann.
5. Lei, Y., *Corrosion and Protection of Marine Engineering materials*, 1st edition, Taylor and Francis.
6. Kraal, E., *Advanced Electrotechnology for marine engineers*, 3rd edition, Adlard Coles Nautical Bloomsbury.
7. Roberge, P., *Handbook of Corrosion engineering*, 1st edition, McGraw Hill.
8. Phillip Marcus, Florian Mansfeld. Analytical methods in corrosion science and engineering. 1st edition. Taylor and Francis.
9. Soylev, T.A. and Richardson, M.G., Corrosion Inhibitors for steel in concrete. State of the art report. *Constr. Build. Mater.*., 22, 4, 609–622, 2008.
10. Kendig, M., Hon, M., Warren, L., Smart Corrosion Inhibiting Coatings. *Prog. Org. Coat.*, 47, 3-4, 183–189, 2003.
11. Sehrish Habib, R.A. and Shakoor, R.K., A focused review on smart carriers tailored for corrosion protection: Developments, applications, and challenges. *Prog. Org. Coat.*, 154, 106218, 2021.
12. Grundmeier, G., Schmidt, W., Stratmann, M., Corrosion protection by organic coatings: electrochemical mechanism and novel methods of investigation. *Electrochim. Acta*, 45, 15–16, 2515–2533, 2000.
13. Eivaz Mohammadloo, H., Sarabi, A.A., Asemani, H.R., Ahmadi, P., A comparative study of eco-friendly hybrid thin films: With and without organic coating application. *Prog. Org. Coat.*, 125, 432–442, 2018.
14. Deyab, M.A., Anticorrosion Properties of Nanocomposite Coatings: A critical Review. *J. Mol. Liq.*, 313, 1, 113533, 2020.
15. Rajan, R., Venu, G., Jayadev, D., Kumar, S.S., Appukuttan, S., Functionalized Nanofiber-based Anticorrosive Coatings. *Functionalized Nanofibers. Synth. Ind. Appl.*, 307–328, 2023.
16. Kumar Das, A. and Islam, N., Applications of organic-based nanocomposites in corrosion protection. *Nanocomposites Adv. Mater. Energy Environ. Aspects*, 579–590, 2023.

17. Coelho, L.B., Zhang, D., Van Ingelgem, Y., Reviewing machine learning of corrosion prediction in a data-oriented perspective. *NPJ Mater Degrad*, 6, 8, 2022.
18. Sharun, V., Rajasekaran M, S.S., Tripathi, V., Sharma, R., Puthilibai, G., Sudhakar, M., Negash, K., Study on Developments in Protection Coating Techniques for Steel. *Adv. Mater. Sci. Eng.*, 2022. https://doi.org/10.1155/2022/2843043.
19. Pal, S., Chatterjee, T., Manoharan, P., Naskar, K., Polymer-Carbon Composites as Anti-corrosive Materials, in: *Carbon-Containing Polymer Composites. Springer Series on Polymer and Composite Materials*, M. Rahaman, D. Khastgir, A. Aldalbahi (Eds.), Springer, Singapore, 2019, https://doi.org/10.1007/978-981-13-2688-2_16.
20. Hashimoto, K., Asami, K., Naka, M., Masumoto, T., The role of alloying elements in improving the corrosion resistance of amorphous iron base alloys. *Corros. Sci.*, 19, 857–867, 1979.
21. Sørensen, P.A., Kiil, S., Dam-Johansen, K., Weinell, C.E., Anti-corrosive coatings: a review. *J. Coat Technol. Res.*, 6, 135–176, 2009.
22. Mandal, N., *Ship Construction and Welding*, 1st edition, vol. 2, Springer Publications, 2017.
23. Hellio, C. and Yebra, D., *Advances in marine antifouling coatings and technologies*, 1st edition, Woodhead Publishing, 2009.
24. Stansbury, and Buchanan, *Fundamentals of electrochemical corrosion*, 1st edition, ASM Publishing, 2000.
25. Bahadori, A., *Corrosion and Materials selection*, 1st edition, John Wiley & Sons, 2014.
26. Schweitzer, P., *Metallic materials – Physical, Mechanical and corrosion properties*, 1st edition, Marcel Dekker, 2003.
27. Perez, N., 1st edition, Kluwer Academic Publishers, 2004.
28. Campbell, and Walsh, *Developments in marine corrosion*, 1st edition, Royal Society of Chemistry, 1998.
29. Googan, C., *Marine Corrosion and Cathodic protection*, 1st edition, Taylor and Francis, 2022.
30. Feron, D., *Marine Corrosion of stainless steels*, 1st edition, IOM Communications, 2001.
31. Powel, C. and Francis, R., *The corrosion performance of metals for the marine environment: a basic guide*, 1st edition, Maney Publishing, 2012.
32. Fusetani, N. and Clare, A.S., *Antifouling compounds*, 1st edition, Springer, 2006.
33. Zhou, F., *Antifouling surfaces and materials from land to marine environment*, 1st edition, Springer, 2015.
34. Konstantinou, I., *Antifouling paint biocides*, 1st edition, Springer, 2006.
35. Arai, T., Harino, H., Ohji, M., Langston, W.J., *Ecotoxicology of antifouling biocides*, 1st edition, Springer, 2009.

13

Metal(II) Complexes as Potential Anticorrosion and Antifouling Agents—A Review

Asha M. S.[1*], Zabiulla[1], Othbert Pinto[1], Arjun S. R.[1], Alen Eldose[1] and Sangamesha M. A.[2]

[1]PG Studies and Research in Chemistry, St. Philomena's College Bannimantap, Mysuru, Karnataka, India
[2]Department of Chemistry, National Institute of Engineering College, Mysuru, Karnataka, India

Abstract
The reason for this view is to introduce a number of novel anticorrosion and antifouling agents that are friendlier to the environment in diverse situations. These days, the use of metal complexes plays an essential and vital role in anticorrosion activity due to free electron pairs. Inhibitor adsorbed on the surface of the metal in the form of a wall, which performs an imperative role in putting a stop to the corrosion activity and consequently inhibits the cathodic and anodic reactions. Advanced factors like electrochemical impedance study, weight loss, adsorption isotherms, and thermodynamic parameters were elaborated here. The current review addressed the presence of heteroatoms like oxygen, nitrogen, sulfur, and π electron clouds in conjugation and the inhibitory competence of metal complexes, which is greatly appreciable. We focused on the study of new complexes that are more potent in antifouling activity.

Keywords: Metal(II) complexes, Schiff base, corrosion, inhibitor and antifouling agent

**Corresponding author*: mysha12ms@gmail.com

Hari Murthy, Vinay Jha Pillai, Kukatlapalli Pradeep Kumar and Matthew Cowan (eds.) Novel Anti-Corrosion and Anti-Fouling Coatings and Thin Films, (399–424) © 2024 Scrivener Publishing LLC

13.1 Introduction

In the past two decades, there have been major developments in the chemical properties of metal complexes, with wide applications mainly in the fields of medicine, agriculture, paints, etc. The ability of imines to form stable chelates with transition metals makes them desirable chelating ligands in coordination chemistry [1–5]. Various studies have been done on coordination and heterocyclic metallic complexes because of their extraordinary chemical and physical characteristics [6–14]. The metal complexes got esteem recognition for their immense binding and various structural potentials [14–21]; hence, they present different appliances for electrochemical sensing [22], catalytic [23–26], electroluminescent properties [27–29], nonlinear optical activities [30], organic photovoltaic material [31], and fluorescence properties [32–34].

In this context, the researchers have been exploring applications of metal complexes for the inhibition of corrosion to prevent metal damage. Inhibitors are an effective tool and have the potential to reduce the degree of corrosion [35–39].

Corrosion damage is repaired using an inhibitor, which is the most effective and affordable technique for the maintenance of metal structures, especially defense-essential supplies [40]. Metals are extensively used as building materials in most major industries, including those that produce food, fuel, electricity, chemicals, electrochemical products, and other things, with steel being one of the most common [41, 42].

The enlarged sizes, along with the compactness of metallic complexes containing organic ligands, as well as potential synergistic behavior between metal-organic mixtures, have all contributed to their inhibitory efficiency. Because they block processes involving hydrogen evolution as well as adsorb on the surface of the material, metal complexes are employed in anticorrosion applications. The efficient employment of the coordination metallic complexes for the purpose of applying corrosion inhibition is made possible by the organic inhibitors since they have a great affinity for inorganic surfaces and have a minimal impact on the atmosphere [43]. By regulating the potential of media that are corrosive, inhibitors perform an essential role in lowering the chances of corrosion. The job of inhibiting agents is to adhere to the exterior surface of metal. The increased efficacy of inhibitors in the order phosphorus > sulfur > nitrogen > oxygen is influenced by the presence of atoms that have significant electron densities around them, the quantity and size of binding areas that facilitate adsorption within the receptor's structure, the adsorption system, and the production associated with metal chelates [44–48]. Due to the existence of the azomethine linkage, the analysis of Schiff base metal chelates revealed a variety of biological properties, including deoxyribonucleic acid binding and cleavage [49–51], anticancer [50, 52], antiulcer [53], antimicrobial [54–56], antioxidant [56, 57], antibacterial [57], antiviral [58, 59, 64, 65], anti-inflammatory [60], antitubercular [61–63], antidiabetic [64], herbicidal, clinical, and analytical activity.

In another context, a number of investigations have also been conducted to find innovative antifouling chemical compounds from the sources of marine ecosystems after the lifting of the restriction on tributyltin in 2008 [66]. Massive economic damage as well as environmental issues have been brought on by marine bacterial fouling. It was necessary to develop technology that efficiently resists biofouling while being environmentally beneficial. Despite the fact that numerous effective antifouling compounds have been found, they are rarely made available for marketing; hence, the development of new antifouling compounds has been hampered [67–70]. This leads to the creation of several environmentally friendly, novel metal complexes as antifouling agents.

13.2 Outline and Mechanism of Metal(II) Complexes as Corrosion Inhibitors

Complexes are able to be absorbed onto the surface of the metal by means of their Van der Waals actions, generating a shielding film that reduces the action centers upon the metal's exterior. The metal complexes adsorb on the metal surface by means of their Van der Waals actions (chemisorption). This is an essential requirement in the development of metallic-based inhibitors.

The toxicity of the ingredients in both varnish and paint materials has been receiving special attention due to the worsening of environmental issues. Toxic corrosion pigments in inhibiting-type primers need to be replaced as one of the most urgent issues demanding quick attention.

To overcome the above issues, the cobalt complexes are employed since they exhibit outstanding thermal stability, rigidity, and excellent corrosion resistance. These complexes are stable even at very high temperatures. Recent studies have demonstrated that switching chromium pigments for manganese compounds is one method for reducing the toxicity of primer corrosion [71].

Zinc is well known for its capacity to resist corrosion. Zinc is a chemically active metal that readily forms alloys with other metallic substances. Furthermore, when exposed to air, it forms a thin film of gray oxide called patina, which prevents the metal's surface from corroding significantly due to profound oxidation. Corrosion resistance is a crucial quality for using metal. Zinc-silicate coatings protected metal from corrosion when the surfaces were scraped, in addition to covering it to keep out sea spray. By considering these facts about transition metals, the formation of bivalent metals like manganese, zinc, and cobalt complexes has been investigated for anticorrosion properties.

In the recent work carried out by Benguerba *et al.* (2023), the degree to which the furfural semicarbazone azomethine bivalent complexes of Mn, Zn, and Co resisted corrosion over C steel grade XC_38 encased in a 1 M hydrogen chloride solution was evaluated. Researchers employed the approach of density functional theory along with Monte Carlo simulation as theoretical methodologies for analyzing new inhibitors' inhibitory capabilities. The inhibiting impact of these complexes has also been investigated using EIS (electrochemical impedance spectroscopy) and PDP (potentiodynamic polarization). The findings of electrochemical testing demonstrated the exceptional potency of these chemical compounds as inhibitors, with FSC-Mn(II) offering a remarkable inhibition of 91.48% at 500 ppm. The investigation's findings demonstrated that the inhibitory potency of all four inhibitors increased with inhibitor concentration, which shows the efficacy of these substances significantly adhering to the external layer of the steel. SEM measurements showed the FSC, as well as its chelates, adsorbed over the external layer of carbon steel XC38 by establishing an obstacle that separates the inhibitors from the empty d-orbital of metal. The analysis presented shows that the imine-based complexes have been determined to be efficient inhibitors, eliminating the component of H_2O from low-carbon steel and resulting in heterogeneous coverings over the C steel grade XC_38 surface that prevents corrosion.

Quantum studies supported the experimental results on the ability of the four preventatives to protect C steel grade XC_38 from being corroded in 1 M HCl solution. Polarization observations of the FSC ligand and its associated manganese(II), zinc(II), and cobalt(II) metallic complexes of different concentrations were used to conduct corrosion tests on C steel grade XC_38 immersed in 1 M aqueous solution of hydrogen chloride. In addition to the Tafel polarization plots, corrosion statistics, such as corrosion potential, Ecorr; corrosion current density, icorr; and percentage of inhibition efficiency, Ecorr (%), were also employed. Researchers were able to observe that the current density had significantly decreased by measuring polarization. The investigation reported that the efficiency of inhibition enhanced with increasing inhibitory concentrations and quoted that the 2-furaldehyde semicarbazone Schiff base Mn(II) complex comes first, followed by the 2-furaldehyde semicarbazone Schiff base Co(II), and then the 2-furaldehyde semicarbazone Schiff base Zn(II) complex. The researchers concluded that the 2-furaldehyde semicarbazone Schiff base Mn(II) complex showed the best inhibitory efficiency, with a value of about 91.48% [72]. Two of the most dependable and long-lasting metals that are frequently utilized in a variety of sectors are mild steel (or MS for short) and austenitic stainless steel, also known as 304 SS. The reason for this is partly because of their low cost, excellent mechanical qualities, and resistance to corrosion [73–75].

A review of studies showed that there is very little research concerning the use of Schiff bases, including their complexes of metals, as effective corrosion inhibitors against steel in acidic solutions. A novel >C=N ligand obtained from ofloxacin was developed just recently by researchers Sayed *et al.* (2023) using 2-aminophenol and a number of d-transition metallic complexes. Using EFM (electrochemical frequency modulation), EIS (electrochemical impedance spectroscopy), and PP (potentiodynamic polarization), the anticorrosive performance of ligand and Ni(II) chelates for mild steel was investigated and demonstrated that the molecular structure, which is electron-rich in nature, results in high inhibitory effectiveness [76].

Even so, these steels deteriorate and experience pitting corrosion when they become subject to a corrosive atmosphere comprised of chloride ions [77]. It is highly challenging to forecast and identify pitting corrosion since it does not show any changes in the metal's properties at an early stage. This is due to the fact that the pits that develop on metallic alloys are often small and confined and produce relatively little net metal loss. As a result, it may cause an infrastructure loss that is catastrophic and unplanned. Inhibitors are frequently used to deal with this issue. But, the majority of all of these inhibitors are extremely hazardous and dangerous for both humans and the environment [78, 79]. This has led to restrictions on their usage, and there is a determined effort to replace them with anticorrosion substances that have a less detrimental effect on the environment [79, 80]. Over time, researchers have discovered that several rare earth element (REE) complexes compare favorably to chromate compounds and have good inhibitory properties when used in acidic or alkaline conditions. In this way, chromates might be replaced through the use of REE inhibitors. As a result of this information, Potgieter *et al.* (2022) explored REE-diketone chelates-cerium acetylacetone, $Ce(acac)_3$; cerium hexafluoroacetylacetone, $Ce(hfac)_3$; lanthanum acetylacetone, La(acac)$_3$; and lanthanum hexafluoroacetylacetone, La(hfac)$_3$, as effective anticorrosion agents over 3.5% sodium chloride solution for 304 stainless steel in addition to mild steel. Weight loss measures and potentiodynamic polarization scans were utilized to examine the REE-diketone complexes' ability to suppress corrosion.

Several researchers investigated the electrochemical properties of alloys, mild steel, and 304 stainless steel using a prepared solution of 3.5% sodium chloride, both with and without inhibitors, at 20°C, 40°C, and 60°C from the PDP curves. The REE-diketone inhibitor's absorption onto the metal surface is the cause of this occurrence. Comparatively with that of mild steel, according to the same circumstances, with no inhibitor employed at all, the results below showed that none of the four corrosion inhibitors studied significantly changed the mild steel's corrosion potential or corrosion current density when exposed to a prepared solution of 3.5% sodium chloride. The finding was consistent with the low corrosion-preventing efficiency that was determined for each of the four anticorrosion agents examined. The results were compared to situations where there had been no inhibitor, and the typical trend was shown to be a reduction in the cathodic Tafel constants, demonstrating that a concentration of 0.5% was not sufficient for providing corrosion resistance for mild steel in 3.5% NaCl across all of the temperatures examined. Moreover, the probability became notably more negative prior to the La(hfac)3 inhibitor employed, compared with the samples that were made aware of the solution containing NaCl without any inhibitor. Considering these two details, the study concluded that the La(hfac)$_3$ inhibitor, or, at least possibly, the REE-diketone inhibitor, performs the role of an inhibitor of cathodic corrosion. Results of surface analysis using Raman spectra showed that Fe_2O_3/FeO(OH) and REE oxide had formed a protective film layer on MS and SS. As the temperature improved, corrosion progressed more quickly, as was supposed to be expected [81].

13.2.1 Comparison of Corrosion Rate and Inhibition Efficiency [73]

Figure 13.1 shows the evidence of the formation of complex between metal (cerium and lanthanum) and (a) acetylacetonate and (b) hexafluoroacetylacetone.

a. For Mild Steel
   i. With inhibitor (μm/year): cerium acetylacetone, 42; cerium hexafluoroacetylacetone, 41; lanthanum acetylacetone, 43; lanthanum hexafluoroacetylacetone, 45.

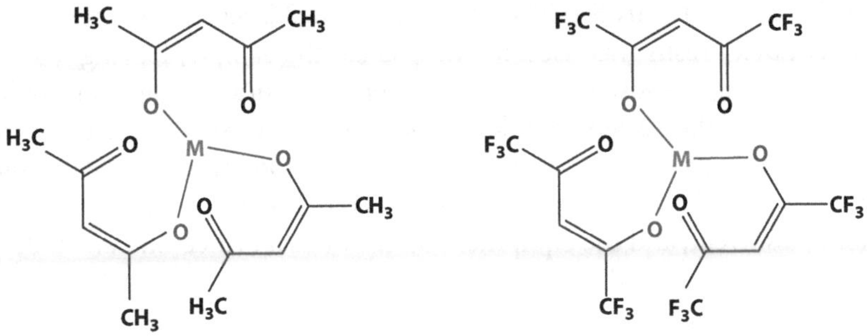

(a) Metal acetylacetonate **(b) Metal hexafluoroacetylacetone**

Figure 13.1 M = La or Ce [82].

   ii. Without inhibitor (μm/year): cerium acetylacetone, 156; cerium hexafluoroacetylacetone, 166; lanthanum acetylacetone, 180; lanthanum hexafluoroacetylacetone, 159.
   iii. IE (%): cerium acetylacetone, 74; cerium hexafluoroacetylacetone, 75; lanthanum acetylacetone, 76; lanthanum hexafluoroacetylacetone, 72.

b. For 304 SS
   i. With inhibitor (μm/year): cerium acetylacetone, 0.0; cerium hexafluoroacetylacetone, 0.0; lanthanum acetylacetone, 0.0; lanthanum hexafluoroacetylacetone, 0.0.
   ii. Without inhibitor (μm/year): cerium acetylacetone, 0.7498; cerium hexafluoroacetylacetone, 0.6722; lanthanum acetylacetone, 0.5430; lanthanum hexafluoroacetylacetone, 0.6378.
   iii. IE (%): cerium acetylacetone, 100; cerium hexafluoro acetylacetone, 100; lanthanum acetylacetone, 100; lanthanum hexafluoroacetylacetone, 100.

The effectiveness of the inhibitory is dependent on its kind of environment, the heteroatoms present (such as oxygen, nitrogen, and sulfur), the character of their substitute entities, electronic clouds, as well as the presence of these heteroatoms.

Wang *et al.* (2021) synthesized 2,2'-bipyridine-3,3'dicarboxylic acid (bpda) as well as its nickel(II), manganese(II), and zinc(II) complexes and evaluated them as anticorrosion agents for an element of carbon using steel (0.5 M aqueous solution of hydrogen chloride) with electrochemical testing. The theoretical chemical composition of the substance was 0.31 C, 0.27

Si, 0.55 Mn, 0.02 P, 0.01 S, 0.95 Cr, 0.2 Ni, 0.2 Cu, 0.2 Mo, and the remainder was Fe. Epoxy resin was used to encapsulate the sample, leaving a 1-cm^2 working surface exposed. After using a series of abrasive papers to polish the surface of the sealed sample to 800 #, it was washed with deionized water and anhydrous ethanol. The corrosive solution was 0.5 M HCl, which the synthesized ligand and its complexes can dissolve in without the assistance of a solvent, and the equipment utilized was a Princeton Parstat 2273. The OCP was observed for 30 min previous to the electrochemical testing to achieve a quasi-steady mode. Following that, studies involving EIS were carried out at OCP by employing a sinusoidal potential perturbation corresponding to 5 mV with frequencies that range from 100 kHz to 20 MHz. By screening the potential at a rate of 0.33 mV between 300 mV under OCP and 250 mV above, the potentiodynamic polarization curves were taken into consideration. The corrosive inhibition efficacy of the metallic complexes is more efficient than that of ligand, according to the researchers, who found that both the bpda and its metal complexes exhibit efficiency in preventing corrosion of a P110 steel through 0.5 M HCl solution, and corrosion resistance is enhanced by increasing the inhibitor concentration. A relationship among Ni(bpda)2, Zinc(bpda), and Manganes(bpda)2 affects the corrosion efficiency of the three complexes, with the greatest value of 95.9% occurring in the case of 400 ppm nickel(bpda). Figure 13.2 shows the formation of co-ordination complex between Ni, Mn, Zn and C12H8N2O4-2,2'-bipyridine-3,3' dicarboxylic acid complexes.

Furthermore, researchers demonstrated the evidence that the inclusion of the inhibitors greatly lowers the anodic and cathodic current densities. The subsequent inspection of the corrosion morphology demonstrated the significantly higher corrosion inhibition effectiveness of the three complexes. The corrosion morphological analysis supports the electrochemical findings and adsorption analysis showing the bpda and nickel$(bpda)_2$ complexes have the lowest and best corrosion inhibition performances, respectively.

The process of corrosion and the spontaneously occurring adsorption of BPDA along with its complexes over the stellar surface were confirmed using the Fourier transform infrared and XPS investigations, as well as an adsorption study that revealed the fact that physisorption and chemisorption play a role in the overall adsorption. Studies showed that the Gibbs free energy of adsorption, which was calculated using the adsorption analysis, has a similar correlation to the effectiveness of preventing corrosion. It proved that BPDA and its chelate's potential for adsorption impacts the extent to which they inhibit corrosion. In the concluding part, a justification for the variation in the four inhibitor adsorption capacities was put forth from the standpoint of the stability of a coordination bond, which

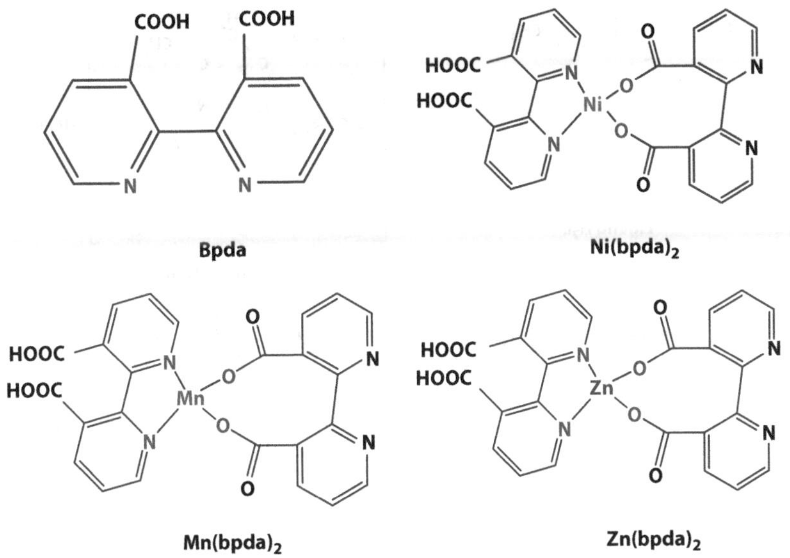

Figure 13.2 $C_{12}H_8N_2O_4$-2,2'-bipyridine-3,3'dicarboxylic acid complexes [82].

provided an explanation for why the Ni(bpda)$_2$ complex has the best corrosion inhibition performance [82].

Figure 13.3 shows the formation metal N-(2-hydroxyphenyl)-3-(-2-hydroxyphenylimino) complexes. Emran *et al.* (2020) synthesized the novel seven metallic complexes nickel(II), cobalt(II), copper(II), manganese(II), platinum(IV), silver(I), and iron(III) of imine base N-(2-hydroxyphenyl)-3-(-2-hydroxyphenylimino). In-depth experiments on electrochemical polarization as well as impedance demonstrated that the H_2L imine base acts as a mixed-type inhibitor with better H_2L azomethine molecule adsorption on C steel surfaces at a higher rate of concentrations, with 83.08% utmost effectiveness. The authors concluded that the existence of the Cl^- ion within the coordination sphere or complex, encompassing a number of lone pair electrons and strengthening its interaction with the steel, considerably increases the inhibitory effects of the newly prepared azomethine cobalt metallic complex compound on C steel. From the investigations, they concluded that the formation of a dense protective coating on the carbon steel surface as a result of physical and chemical adsorption prevents the corrosive attack of carbon steel by acidic solutions [83].

Steels have been protected from galvanic corrosion by cadmium coatings, which also act as a natural lubricant for interconnected applications. It is plated by using an aqueous solution comprising cyanide salts, though this serves as a poisonous metal and is a proven carcinogen. Since

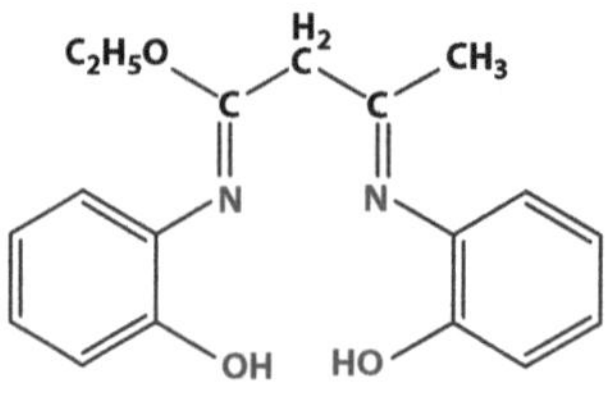

Ligand H_2L

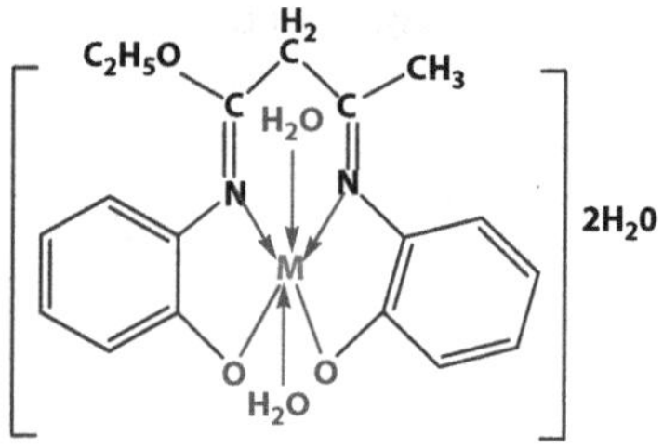

M = Cu(II) or Ni(II)

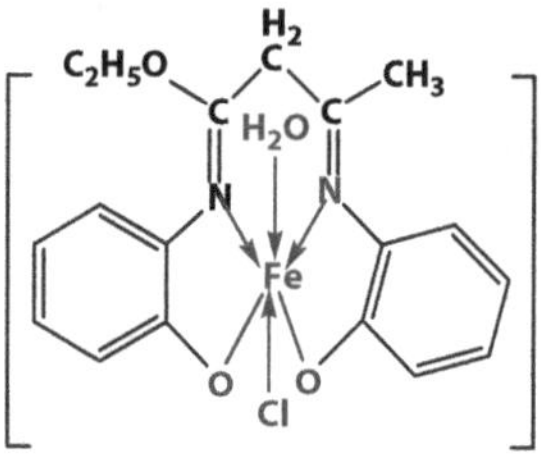

$[(HL)Fe(H_2O]Cl)$

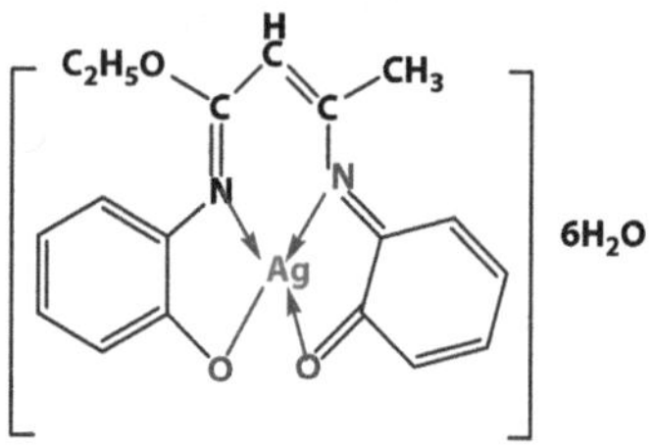

$[(HL)AgI]\ 6H_2O$

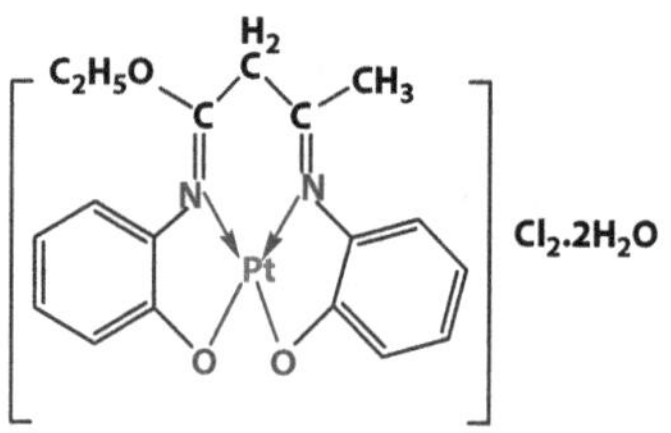

$[(L)Pt]Cl_2.2H_2O$

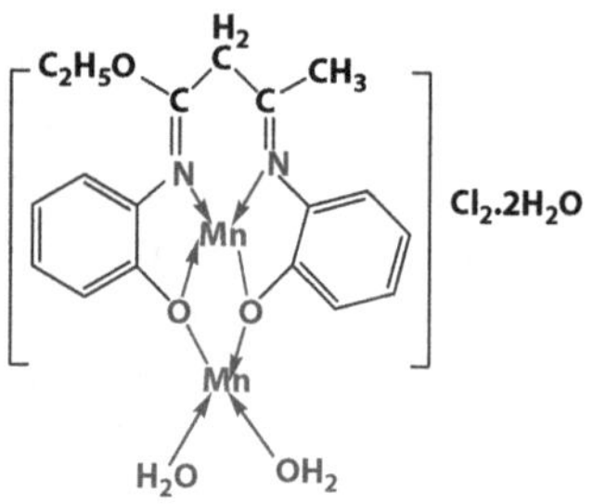

$[(LMn_2(H_2O)_2]Cl_2.4H_2O$

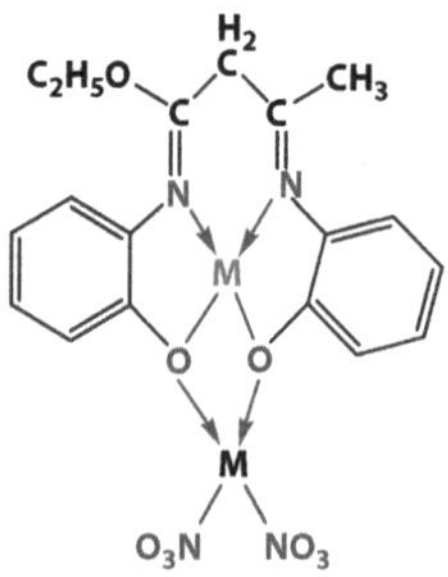

$[LM_2(NO_3)2]Cl_2.4H_2O$ M=Cu or Co

Figure 13.3 N-(2-hydroxyphenyl)-3-(-2-hydroxyphenylimino) complexes [83].

the fundamental metal has been exposed, the cadmium coating will corrode first if a part that is cadmium-plated includes the same damage that exposes a maximum of a millimeter of the base metal's surface. So even the adjoining base metal is protected by cadmium, making even the possibility of tiny pores and holes irrelevant. By considering this fact, the authors, Taha *et al.* (2019) condensed 4-amino-N(quinoxalin-2-yl)benzenesulfonamide and 4-chloro-benzaldehyde to aid in the formation of imine-based cadmium(II) and uranium dioxide(II) complexes. They demonstrated that the loss of weight measurement is an excellent tool for examining how well the two types of imine Cd(II) and UO_2(II) complexes resist corrosion. It is also a great way to confirm the outcomes of the electrochemical investigation. In order to assess how metal complexes affected the corrosion process of mild steel using a 1 M hydrogen chloride solution, researchers evaluated the weight loss data of mild steel at room temperature. The estimation of corrosion and the proportion of effective inhibition determined by measuring weight loss after 72 hours.

Investigation revealed that the two complexes are effective inhibitors, as shown by the findings of the inhibitory effectiveness percentage. This may be because coordination involving donor-acceptor relationships exists between the pairs of unpaired electron atoms among the organic ligand and transition metal. The authors concluded that the larger size as well as the structural planarity of the organic metal complexes, which have been demonstrated to be more effective as inhibitors in comparison to the ligand molecule, maybe the cause of this. Therefore, the UO_2(II) metal complex is more efficient than the Cd(II) complex. By tracking a variation in the rate of corrosion (Ecorr) over time, the electrochemical behavior of mild steel in a 1 M hydrogen chloride solution was examined. These outcomes demonstrated that mild steel corrosion was effectively inhibited due to the existence of UO_2(II) metal complexes in acidic solutions. The morphology of the surface of mild steel revealed the existence of an inhibitory film that shields the metal from dissolving in a solution of hydrogen chloride and delays hydrogen evolution, therefore favoring mixed-type inhibitory activity [84].

13.3 Outline of Metal Complexes as Antifouling Agents

Currently, antibiofouling coating is considered to be the simplest and most popular method of protecting boats and other marine equipment. Following the 2003 ban on the traditional organotin-based antifouling coating, the copper-based antifouling coating gained popularity.

Hamaida *et al.* (2008) synthesized the novel ligand, 3-(2-furylidine) hydrazine-5,6-diphenyl-1,2,4-triazine (L) and its seven cobalt(II) and copper(II) complexes. By testing the toxicity of the invented copper and cobalt complexes on nauplius larvae of the balanus amphitrite, they were evaluated as antifouling agents. According to the results listed below, researchers concluded that, compared to its metallic complexes, the organic ligand molecule is less hazardous to barnacle larvae [85].

13.3.1 The Outcome of Barnacle Larvae Bioassay for Ligand and Its Complexes

a. Ligand: Concentration in mg/L: 300, 700, and 1000; % of molarity control: 76, 70, 93, 30, and 100; LC_{50} mg/L: 240.
b. $LCo(H_2O)_3(ClO_4)_3$: Concentration in mg/L: 300, 700, and 1000; % of molarity control: 20.00, 26.70, and 26.70; LC_{50} mg/L: >1000.
c. $LCuCl_2$: Concentration in mg/L: 10, 30, 50, 70, and 100; % of molarity control: 10.00, 83.30, 86.70, 100.00, and 100.00; LC_{50} mg/L: 20.
d. $L[Cu(AC)_2]$ $2H_2O$: Concentration in mg/L: 100, 300, 700, and 1000; % of molarity control: 86.70, 90.00, and 100.00; LC_{50} mg/L: 200.
e. $L_2[Cu(ClO_4)_2]$: Concentration in mg/L: 300, 700, and 1000; % of molarity control: 66.70, 83.00, and 90.00; LC_{50} mg/L: 250.

Figure 13.4 shows the evidence of the formation metal complex with ligand. The above findings also revealed that when the ligand is complexed with $CuCl_2$, its toxicity increases significantly. The work reported that the most effective complexes in this natural analysis are $LCuCl_2$ $L[Cu(AC)_2]$ $2H_2O$ and $L2[Cu(ClO_4)_2]$ with LC_{50} values of 20, 200, and 240 mg/L, accordingly. The authors predicted that the capacity of the copper(II) complexes to pass through the tested larvae's cell membrane and suppress some critically important metabolic processes may be the cause of their significant toxicity. In contrast, cobalt(II) complexes presented decreased toxicity in the direction of barnacle larvae when compared to the imine-based ligand followed by its metallic complexes, according to measurements. The findings elaborated on the fact that the ligand's toxicity significantly increased after complexing with $CuCl_2$. This might be explained by copper complexes' capacity to enter barnacle larvae's cell membrane [85].

Copper is the ideal antifoulant because it is a substance that occurs naturally and a metal that all plants and animals require for proper growth.

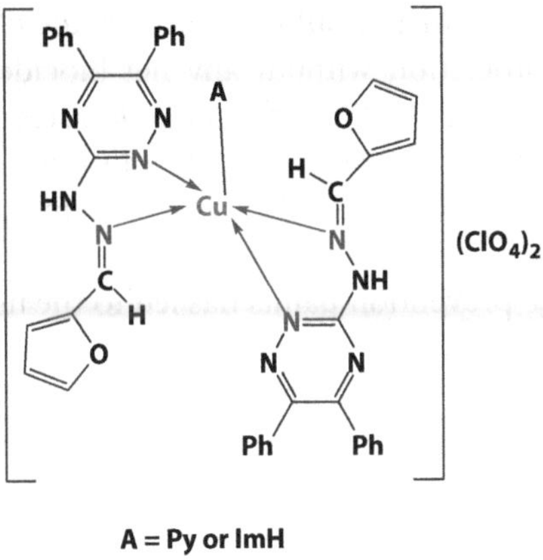

Figure 13.4 $LCu(ClO_4)_2$ [84].

Copper's bioavailability may also be affected by chemical changes brought on by the environment. The cupric ion, Cu(II), is regarded to be primarily responsible for the bioactivity of copper. But, numerous studies have shown that most of the copper present in marine environments cannot be found as cupric ions. Actually, compared to the quantity of "dissolved" as well as total copper, the quantity of cupric ions in the environment seems orders several times lower. An important factor in the environmental impact of copper-based antifoulants is the reality that the active form, cupric ion, only exists for a brief period of duration when it is on the external layer of the coating. During that time, it accomplishes its goal by preventing the emergence of biological life; after that, it is rendered inactive by its interaction with dissolved organic components in the water. Based on this information, investigations have been done on Cu(II) complexes as antifouling agents.

Trojer *et al.* (2013) described that azole chemistry for coordination exhibits a great deal of the potential of antifouling coatings because of their inhibition action and properties. Numerous antifouling substances have azole moieties, and imidazole, which contains the biocide medetomidine, is a promising antifouling substance in terms of barnacle fouling defenses. Immobilizing such biocide upon macromolecules as well as nanoparticles and controlling their release (which includes sustained and triggered release conceivable) can be conducted by making utilize of the azole moiety's function as a ligand. Researchers further elaborated that azoles' exceptionally high affinity for Cu^{2+}, an excellent antifouling presents the most promising antifouling strategy. The abundant amounts of naturally occurring copper can then be absorbed by an antifouling coating made of azole ligands, providing antifouling protection without any net biocide release. The most bioactive state of copper is Cu^{+}, which can be lowered from the oxidation state Cu^{2+}. Authors reported that triazole ligands offer promise for various uses; therefore, this can be accomplished by modifying the ligand chemistry [86]. The International Maritime Organization's (IMO) ban on the usage of tributyltin in fouling-prevention paints has led to the usage of booster biocides as a substitute. Numerous substitute biocides have been created after the removal of tributyltin from antifouling paints to stop the growth and colonization of ship hulls by aquatic organisms. Figure 13.5 shows the Zn pyrithione complex. A biocide antifouling booster, zinc pyrithione (ZnPT), and the biological outcome expected high research since it thought that this chemical substance would not persist in marine ecosystems.

Using biological sample materials, such as the gills and digestive gland from the bioindicator organisms *Mytilus galloprovincialis*, Konstantinou and Albanis (2004) established a methodology that has been successfully proven to be effective and highly sensitive in separating ZnPT as well as its chief secondary products, Zn and ionized pyrithione (PT), in both seawater as well as biological samples. In addition to investigating ZnPT's general chemical properties, toxicity traits, as well as environmental consequences, the authors also discussed the scientific difficulties associated with researching relevant processes.

ZnPT is often harmful at concentrations of microgram per liter to a wide variety of marine creatures, particularly algae, crustaceans, bivalves, polychaetes, sea urchins, and fish. If ZnPT is exposed to UV light, it photodegrades and may become transchelated to different compounds by metal ions. There are additional claims that ZnPT is biodegraded as well as hydrolyzed into a number of metabolites with different levels of stability and toxicity. If ZnPT does not decompose under particular environmental circumstances, it will nevertheless accumulate inside the water stream or sediment. Studies have examined the ecological spread of ZnPT by employing different chromatographic and voltammetric techniques in order to identify the potential hazards brought on by ZnPT in the marine environment. Unfortunately, the majority of the techniques used were hampered by quick transchelation as well as degradation by ZnPT in both the aquatic environment and the laboratory, making it challenging to determine its ecological distribution. Unfortunately, the majority of the techniques used were hampered by quick transchelation as well as degradation by ZnPT in both the aquatic environment and the laboratory, making it challenging to determine its ecological distribution. To accurately describe the dispersal and release of ZnPT into the environment, more reliable and sensitive analytical techniques must be created. Complete processes for degradation, including their possible byproducts, must also be carefully studied in order to fully comprehend the potential hazards associated with the introduction of ZnPT entering the marine ecosystem. The toxicology and bioaccumulation of ZnPT were studied using short-term trials. The authors examined the effects of sublethal doses of the biocidal for up to 7 days on adult mussels in naturally occurring populations that were collected in Porto Santo Stefano, Italy, and subjected to ZnPT and its secondary products.

The investigation identified that a baseline level of Zn pyrithione and ionized pyrithione was found in the mussels, suggesting that ZnPT availability within the sampling site has reached a level high enough to cause an identifiable growth in members of the native population. Zinc pyrithione was demonstrated to remain persistent within the experimental marine water for short periods of time. Considering the gills as well as the digestive gland as significant targets along the biological route of the pollutants, ZnPT gradually deposited within tissues in exposed mussels, corresponding to exposure concentration as well as duration.

The results showed that this biocide may pose a threat to marine life in coastal zones; therefore, more studies regarding its biological consequences and sublethal doses are required. The experiment revealed the fact that the 7-day median lethal quantity (LC50) = 8.27 mM implemented right here seems high in comparison to reported ZnPT concentrations in the environment. The authors came to the conclusion that tracking ZnPT as well as its secondary products in ecosystems, marine waters, and sediment has become necessary, particularly in fishery harbors and yacht marinas in which exchange of water is constrained and antifouling paints with pyrithione, like zinc omadine and copper omadine, are widely used on fishing and pleasure boats. In order to get the essential data on the sustainability of the biocide as a secure substitute for TBT in marine antifouling products, additional research on the ecotoxicological impacts of sublethal and nonlethal doses in said biocide is desirable [87].

Chen *et al.* (2016) used the grafting approach to create a covalent link between carboxymethyl chitosan (CmCs) and multiwall carbon nanotubes

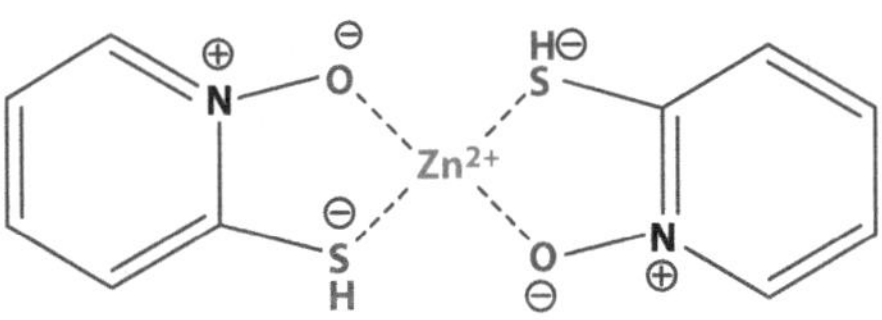

Figure 13.5 Zn pyrithione [87].

(MWCNTs). The MWCNT-CmCs composites were further coordinated with the Cu(II) and Zn(II) ions. The researchers demonstrated outstanding antibacterial action against *Staphylococcus aureus*, *Escherichia coli*, and *Vibrio anguillarum* by the coordinated MWCNT-CmCs. After 21 days, the MWCNT-CmCs complex still showed effective antibacterial properties against *S. aureus*. The investigation of biological evaluation concluded that composites had a considerable potential to replace other antifouling substances [88].

TOB (tobramycin) from *Streptomyces* and PR (protocatechualdehyde) from black fern were employed to prepare Schiff base compounds by Liu *et al.* (2023). Five catechol molecules were present in each monomer of the imines and formed a complex with Fe^{3+}. The coating's tannic acid-like composition allowed it to be applied to a range of material surfaces, and it was made entirely of natural ingredients. A particular thickness of the Schiff base metallic composite encapsulating has been reported to provide enhanced antibiofouling capacity due to the discharge of biological antifouling products from degradation and to exhibit controlled decomposition capability due to the hydrolysis properties of an imine bond.

Imine bonds have the special characteristics of hydrolysis. The coating of Schiff base metal composite to the optimum thickness is anticipated to exhibit control over degradation as well as intensified antibiofouling competence. It happens mainly because of the release of natural antifouling degradation products. The degradation, antibacterial properties, and antibioadhesion properties of Schiff base metal composite coating $(Fe/TOB\text{-}PF)_{20}$ were calculated, and the biofouling resistance mechanism was analyzed by many researchers.

As a result, a completely novel antibiofouling coating that is fully biodegradable was investigated in this work, potentially opening the way for the creation of innovative antifouling coatings. Testing for hydrolytic degradation revealed that the $(Fe/TOB\text{-}PF)_{20}$ coating degraded in a predictable and controlled manner. Figure 13.6 shows the formation of (Fe/TOB-PF)20 complex. Furthermore, after 24 hours, the antibacterial efficiency of $(Fe/TOB\text{-}PF)_{20}$ achieved 97% compared to the blank glass substrate, and the study further demonstrated that the thicker coating helps the antibacterial efficacy remain effective for longer. In addition, the coating subsequently degraded into a simple macromolecule monomer, preventing the formation of microplastics, and it presented new opportunities for creating eco-friendly antibiofouling coatings. Such coatings demonstrated outstanding antibacterial as well as antialgae adhesion characteristics, which contributed to the extension of the surface and the formation of antimicrobial substance (TOB) found in the coating [89].

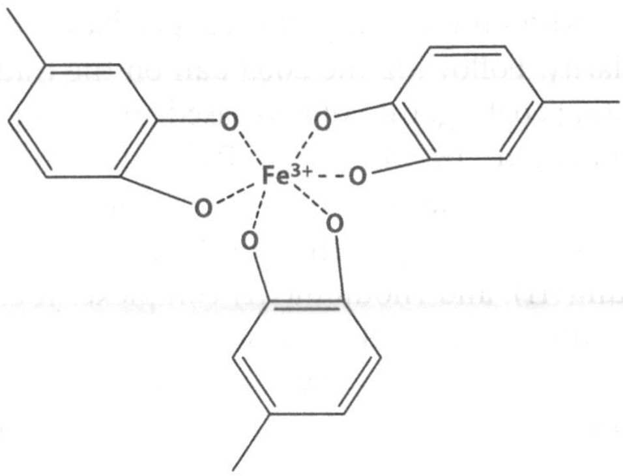

Figure 13.6 $(Fe/TOB\text{-}PF)_{20}$ [89].

Singh *et al.* (2023) synthesized novel Ir(III) and Rh(III) complexes of $[\eta^5\text{-}C_5Me_5)M(L)Cl_2]$ [M=Ir; L = atz, ctz; M = Rh, L = atz, catz]. When 10 marine fouling bacteria were evaluated, the *in vitro* antimicrobial properties of both iridium(III) and rhodium(III) complexes comprising 2-acetamidothiazole organic ligands exhibited moderate to good activity [90].

13.4 Conclusion

There is a need for inhibitors of corrosion in mild steels exposed to aggressive environments across a variety of industries, particularly in power plants, automobiles, petro-automotive, petrochemicals, and the pharmaceutical sector. Accordingly, metal complexes have proven to have effective protective activity in corrosive environments as a class of organic inhibitors. The presented overview addressed the mild steel corrosion inhibitory performance of various bivalent metals complexes like cobalt, nickel, zinc, copper, cadmium, uranium dioxide, and manganese resisted corrosion on XC_38 carbon alloy steel immersed in a 1 M hydrogen chloride solution and highlighted studies in which the protecting effectiveness and applicability of complexes have been impacted by the introduction of various reagents as well as taking the environment into account. Quantum chemical computations are employed for the prevention of corrosion analyses to present pertinent insights into the interactions between metals and inhibitors; as a result, these powerful tools aid in the prediction and design of more potent inhibitor compounds. Currently, antibiofouling coatings are considered to be the most affordable and effective way of protecting boats as well as other marine equipment. Following the 2003 ban on the traditional organotin-based antifouling coating, the copper-based antifouling coating gained popularity. Following the 2003 ban on the traditional organotin-based antifouling coating, the copper-based antifouling coating gained popularity. Recently, zinc pyrithione (ZnPT), an alternative antifouling booster biocide, has become one of the most often used booster biocides in antifouling paints. At present, novel antibiofouling coatings $(Fe/TOB\text{-}PF)_{20}$, iridium(III), and rhodium(III) complexes focus on the attention of researchers due to their fully biodegradable nature and provide new opportunities for creating eco-friendly antibiofouling coatings.

Conflicts of Interest: There is no conflict of interest stated by the researchers.

Acknowledgements

The entire team gratefully acknowledges the assistance of the Director and management of the PG Studies and Research in Chemistry (A recognized research center of the University of Mysore), St. Philomena's College Bannimantap, Mysuru, Karnataka, India.

Sangamesha M. A. expresses gratitude to the Department of Chemistry, The National Institute of Engineering College in Mysore.

References

1. Inada, Y., Mochizuki, K., Tsuchiya, T., Tsuji, H., Funahashi, S., Equilibrium and kinetics of the dinuclear complex formation between N, N′-ethylenebis (salicylideneiminato) copper (II) and metal (II, I) ions in acetonitrile. *Inorg. Chim. Acta*, 358, 3009, 2005.
2. Khalaf, M.M., El-Lateef, H.M.A., Alhadhrami, A., Sayed, F.N., Mohamed, G.G., Gouda, M., Shaaban, S., Abu-Dief, A.M., Synthesis, Spectroscopic, Structural and Molecular Docking Studies of Some New Nano-Sized Ferrocene-Based Imine Chelates as Antimicrobial and Anticancer Agents. *Materials*, 15, 3678, 2022.
3. Abosadiya, H.M., Hasbullah, S.A., Yamin, B.M., Synthesis, X-ray, NMR, FT-IR, UV/vis, DFT and TD-DFT studies of N-(4-chlorobutanoyl)-N′-(2-, 3-and 4-methylphenyl) thiourea derivatives. *Spectrochim. Acta, Part A*, 144, 115, 2015.
4. Abu-Dief, A.M., Abdel-Rahman, L.H., Abdel-Mawgoud, A.A.H., A robust in vitro anticancer, antioxidant and antimicrobial agents based on new metal-azomethine chelates incorporating Ag (I), Pd (II) and VO (II) cations: Probing the aspects of DNA interaction. *Appl. Organomet. Chem.*, 34, e5373, 2020.

5. Abu-Dief, A.M., El-Sagher, H.M., Shehata, M.R., Fabrication, spectroscopic characterization, calf thymus DNA binding investi-gation, antioxidant and anticancer activities of some antibiotic azomethine Cu (II), Pd (II), Zn (II) and Cr (III) complexes. *Appl. Organomet. Chem.*, 33, 4943, 2019.
6. Chandra, S. and Kumar, A., Spectral studies on Co (II), Ni (II) and Cu (II) complexes with thiosemicarbazone (L1) and semicarbazon (L2) derived from 2-acetyl furan. *Spectrochim. Acta, Part A*, 66, 1347, 2007.
7. Abu-Dief, A.M., El-Khatib, R.M., Aljohani, F.S., Alzahrani, S.O., Mahran, A., Khalifa, M.E., El-Metwaly, N.M., Synthesis and intensive characterization for novel Zn (II), Pd (II), Cr (III) and VO (II)-Schiff base complexes; DNA-interaction, DFT, drug-likeness and molecular docking studies. *J. Mol. Struct.*, 1242, 130693, 2021.
8. El-Lateef, H.M.A., Khalaf, M.M., Shehata, M.R., Abu-Dief, A.M., Fabrication, DFT Calculation and Molecular Docking of Two Fe(III) Imine Chelates as Anti-COVID-19 and Pharmaceutical Drug Candidate. *Int. J. Mol. Sci.*, 23, 3994, 2022.
9. Kızılcıklı, I., Ülküseven, B., Da ̦sdemir, Y., Akkurt, B., Zn (II) and Pd (II) Complexes of Thiosemicarbazone-S-alkyl Esters Derived from 2/3-Formylpyridine. *Synth. React. Inorg. Met. Org. Chem.*, 34, 653–665, 2004.
10. Abdel-Rahman, L.H., Abu-Dief, A.M., El-Khatib, R.M., Abdel-Fatah, S.M., Adam, A.M., Ibrahim, E.M.M., Sonochemical synthesis, structural inspection and semiconductor behavior of three new nano sized Cu (II), Co (II) and Ni (II) chelates based on tri-dentate NOO imine ligand as precursors for metal oxides. *Appl. Organomet. Chem.*, 32, e4174, 2018.
11. Abdel-Rahman, L.H., Abu-Dief, A.M., El-Khatib, R.M., Abdel-Fatah, S.M., New Cd (II), Mn (II) and Ag (I) Schiff base complexes: Synthesis, characterization, DNA binding and antimicrobial activity. *Inter. J. Nano Chem.*, 2, 83, 2016.
12. Liberta, A.E. and West, D.X., Antifungal and antitumor activity of heterocyclic thiosemicarbazones and their metal complexes. *Curr. status. Biomet.*, 5, 121, 1992.
13. Al-Saeedi, S.I., Abdel-Rahman, L.H., Abu-Dief, A.M., Abdel-Fatah, S.M., Alotaibi, T.M., Alsalme, A.M., Nafady, A., Catalytic Oxidation of Benzyl Alcohol Using Nanosized Cu/Ni Schiff-Base Complexes and Their Metal Oxide Nanoparticles. *Catalyst*, 8, 452, 2018.
14. Al-Abdulkarim, H.A., El-Khatib, R.M., Aljohani, F.S., Mahran, A., Alharbi, A., Mersal, G.A.M., El-Metwaly, N.M., Abu-Dief, A.M. Optimization for synthesized quinoline-based Cr3+, VO2+, Zn2+ and Pd2+ complexes: DNA interaction, bio-logical assay and in-silico treatments for verification. *J. Mol. Liq.*, 339, 116797, 2021.
15. Zangrando, E., Casanova, M., Alessio, E., Trinuclear metallacycles: Metallatriangles and much more. *Chem. Rev.*, 108, 4979, 2008.
16. Aljohani, E.T., Shehata, M.R., Abu-Dief, A.M., Design, synthesis, structural inspection of Pd2+, VO2+, Mn2+ and Zn2+ chelates incorporating

ferrocenyl thiophenol ligand: DNA interaction and pharmaceutical studies. *Appl. Organomet. Chem.*, 35, e6169, 2021.

17. Qasem, H.A., Aouad, M.R., Al-Abdulkarim, H.A., Al-Farraj, E.S., Attar, R.M.S., El-Metwaly, N.M., Abu-Dief, A.M., Tailoring of some novel bis-hydrazone metal chelates, spectral based characterization and DFT calculations for pharmaceutical applications and in-silico treatments for verification. *J. Mol. Struct.*, 1264, 133263, 2022.
18. Abu-Dief, A.M., Abdel-Rahman, L.H., Shehata, M.R., Abdel-Mawgoud, A.A.H., Novel azomethine Pd (II)-and VO (II)-based metallo-pharmaceuticals as anticancer, antimicrobial, and antioxidant agents: Design, structural inspection, DFT investigation, and DNA interaction. *J. Phys. Org. Chem.*, 32, e4009, 2019.
19. Dalgarno, S.J., Power, N.P., Atwood, J.L., Metallo-supramolecular capsules. *Coord. Chem. Rev.*, 252, 825, 2008.
20. Ibrahim, E.M.M., Abdel-Rahman, L.H., Abu-Dief, A.M., Elshafaie, A., Hamdan, S.K., Ahmed, A.M., The synthesis of CuO and NiO nanoparticles by facile thermal decomposition of metal-Schiff base complexes and an examination of their electric, thermoelectric and magnetic Properties. *Mater. Res. Bull.*, 107, 492, 2018.
21. Abu-Dief, A.M., El-Khatib, R.M., Aljohani, F.S., Al-Abdulkarim, H.A., Alzahrani, S., El-Sarrag, G., Ismael, M., Synthesis, structuralelucidation, DFT calculation, biological studies and DNA inter-action of some aryl hydrazone Cr3+, Fe3+, and Cu2+ chelates. *Comput. Biol. Chem.*, 97, 107643, 2022.
22. Zoubi, W.A., Kandil, F., Chebani, M.K., Acyclic and cyclic imines and their metal complexes: recent progress in biomaterials and corrosion applications. *Arab. J. Chem.*, 9, 626, 2016.
23. Lehwess-Litzmann, A., Neumann, P., Parthier, C., Ludtke, S., Golbik, R., Ficner, R., Tittmann, K., Twisted Schiff base intermediates and substrate locale revise transaldolase mechanism. *Nat. Chem. Biol.*, 7, 678, 2011.
24. Adaeo, P., Kuznetsov, M.L., Barroso, S., Martins, A.M., Avecilla, F., Pessoa, J.C., Amino alcohol-derived reduced Schiff base V(IV)O and V(V) compounds as catalysts for asymmetric sulfoxidation of thioanisole with hydrogen peroxide. *Inorg. Chem.*, 51, 11430, 2012.
25. Gong, D., Wang, B., Jia, X., Zhang, X., The enhanced catalytic performance of cobalt catalysts towards butadiene polymerization by introducing a labile donor in a salen ligand. *Dalton Trans.*, 43, 4169, 2014.
26. Das, P. and Linert, W., Schiff base-derived homogeneous and heterogeneous palladium catalysts for the Suzuki–Miyaura reaction. *Coord. Chem. Rev.*, 311, 1, 2016.
27. Che, C.M., Chan, S.C., Xiang, H.F., Chan, M.C., Liu, Y., Wang, Y., Tetradentate Schiff base platinum(ii) complexes as new class of phosphorescent materials for high-efficiency and white-light electroluminescent devices. *Chem. Commun.*, 1, 1484, 2004.

28. Zhou, L., Kwok, C.C., Cheng, G., Zhang, H., Che, C.M., Efficient red organic electroluminescent devices by doping platinum(II) Schiff base emitter into two host materials with stepwise energy levels. *Opt. Lett.*, 38, 2373, 2013.
29. Sano, T., Nishio, Y., Hamada, Y., Takahashi, H., Usuki, T., Shibata, K., Design of conjugated molecular materials for optoelectronics. *J. Mater. Chem.*, 10, 157, 2000.
30. Nayar, C.R. and Ravikumar, R., Review: Second order nonlinearities of Schiff bases derived from salicylaldehyde and their metal complexes. *J. Coord. Chem.*, 67, 1, 2014.
31. Jeevadason, A.W., Murugavel, K.K., Neelakantan, M.A., *Renew.* Review on Schiff bases and their metal complexes as organic photovoltaic materials. *Sustain. Energy. Rev.*, 36, 220, 2014.
32. Li, S.H., Chen, F.R., Zhou, Y.F., Wang, J.N., Zhang, H., Xu, J.G., Enhanced fluorescence sensing of hydroxylated organotins by a boronic acid-linked Schiff base. *Chem. Commun.*, 1, 4179, 2009.
33. Lee, S.A., You, G.R., Choi, Y.W., Jo, H.Y., Kim, A.R., Noh, I., Kim, S.J., Kim, Y., Kim, C., A new multifunctional Schiff base as a fluorescence sensor for $Al^{3+}$and a colorimetric sensor for CN^- in aqueous media: an application to bioimaging. *Dalton Trans.*, 2014, 43, 6650.
34. Ganguly, A., Paul, B.K., Ghosh, S., Kar, S., Guchhait, N., *Analyst*, 138, 6532, 2013.
35. Abd El-Lateef, H.M., Synergistic effect of polyethylene glycols and rare earth Ce4+ on the corrosion inhibition of carbon steel in sulfuric acid solution: Electrochemical, computational, and surface morphology studies. *Res. Chem. Intermed*, 42, 3219, 2016.
36. Abd El-Lateef, H.M., Ismael, M., Mohamed, I., Novel Schiff base amino acid as corrosion inhibitors for carbon steel in CO_2-saturated 3.5% NaCl solution: Experimental and computational study. *Corros. Rev.*, 33, 77, 2015.
37. Wu, H., Kou, F., Jia, F., Liu, B., Yuan, J., Bai, Y., A V-shaped ligand 1, 3-bis (1-methylbenzimidazol-2-yl)-2-oxapropane and its Cu(II) complex: Synthesis, crystal structure, antioxidation and DNA-binding properties. *J. Photochem. Photobiol. B Biol.*, 105, 190, 2011.
38. Sun, W. and Nesic, S., A mechanistic model of H2S corrosion of mild steel. Paper No. 07655, in: *Proceedings of the NACE International Corrosion Conference & Expo*, Houstan, TX, USA, vol. 1, 2007.
39. Marshall, K.C., Stout, R., Mitchell, R., Mechanism of the initial events in the sorption of marine bacteria to surfaces. *J. Gen. Microbiol*, 68, 337, 1971.
40. Gupta, N.K., Verma, C., Quraishi, M.A., Mukhrajee, A.K., Schiff's Bases Derived from L-lysine and Aromatic Aldehydes as GreenCorrosion Inhibitors for Mild Steel: Experimental and theoretical studies. *J. Mol. Liq.*, 215, 47, 2016.
41. Abd El-Lateef, H.M. and Tantawy, A.H., Synthesis and evaluation of novel series of Schiff base cationic surfactants as corrosion Inhibitors for Carbon

Steel in Acidic/Chloride Media: Experimental and theoretical investigations. *RSC Adv.*, 6, 8681, 2016.

42. Abd El-Lateef, H.M., Soliman, K.A., Tantawy, A.H., Novel synthesized Schiff Base-based cationic Gemini surfactants: Electro-chemical investigation, theoretical modeling and applicability as biodegradable inhibitors for mild steel against acidic corrosion. *J. Mol. Liq.*, 232, 478, 2017.
43. Asha, M.S., Sangamesha, M.A., Pinto, O., Sandra, T.O., Shaji, R.P., Synthesis and characterization of hetero cyclic imine and its metal complexes for anti-corrosion application. *Mater. Today: Proc.*, 46, 2436, 2021.
44. Nassar, A.M., Hassan, A.M., Shoeib, M.A., Synthesis, characterization and anticorrosion studies of new homobimetallic Co (II), Ni(II), Cu (II), and Zn (II) Schiff base complexes. *J. Bio. Tribo Corros.*, 1, 1, 2015.
45. Jawad, Q.A., Zinad, S.D., Dawood Salim, R., A. Al-Amiery, A., Sumer Gaaz, T., Takriff, M.S., H. Kadhum, A.A., Synthesis, characterization, and corrosion inhibition potential of novel thiosemicarbazone on mild steel in sulfuric acid environment. *Coatings*, 9, 729, 2019.
46. Danaee, I., RameshKumar, S., RashvandAvei, M., Vijayan, M., Electrochemical and Quantum Chemical Studies on Corrosion Inhibition Performance of 2,2′-(2-Hydroxyethylimino)bis[N-(alphaalphadimethylphenethyl)-N-methylacetamide] on Mild Steel Corrosion in 1M HCl Solution. *Mater. Res.*, 23, 1, 2020.
47. Mohammed, M.A. and Kubba, R.M., Experimental Evaluation for the Inhibition of Carbon Steel Corrosion in Salt and Acid Media by New Derivative of Quinolin-2-One. *Iraqi J. Sci.*, 61, 1861, 2020.
48. Inada, Y., Mochizuki, K., Tsuchiya, T., Tsuji, H., Funahashi, S., Equilibrium and kinetics of the dinuclear complex formation between N, N′ethylenebis(salicylideneiminato)copper(II) and metal(II,I) ions in acetonitrile. *Inorg. Chimica. Acta*, 358, 3009, 2005.
49. Asha, M.S., Zabiulla, H.A.K., Al-Ostoot, F.H., Pinto, O., Gopichand, T.V., Shivappaa, M., In-silico docking, synthesis, structure analysis, DFT calculations and energy frameworks of metal complexes to regress angiogenesis activity. *J. Mol. Struc.*, 1253, 132272, 2022.
50. Arora, M., Saravanan, J., Bhattacharjee, S.S., Synthesis, characterization and antimicrobial activity of some schiff bases of 2-amino-n-(p-acetamidophenyl carboxa. *Int. J. Pharm. Pharm. Sci.*, 1, 5, 2013.
51. Subbaraj, P., Ramu, A., Raman, N., Dharamraja, J., Synthesis, charac-terization, DNA interaction and pharmacological studies of substituted benzophenone derived Schiff base metal (II) complexes. *J. Saudi. Chem. Soc*, 19, 207, 2014.
52. Shokohi-pour, Z., Chiniforoshan, H., Momtazi-borojeni, A.A., Notash, B., A novel Schiff base derived from the gabapentin drug and copper (II) complex: synthesis, characterization, interaction with DNA/protein and cytotoxic activity. *J. Photochem Photobiol.*, 162, 34, 2016.

53. Sheikh, R.A., Wani, M.Y., Shreaz, S., Hashmi, A.A., Synthesis, characterization and biological screening of some Schiff base macrocyclic ligand based transition metal complexes as antifungal agents. *Arab. J. Chem.*, 9, 743, 2011.
54. Kumar, C.T.K., Keshavayya, J., Rajesh, T., Peethambar, S.K., Synthesis, characterization and biological activity of heterocyclic azo dyes derived from 2-Aminobenzothiozole. *Int. J. Pharm. Pharm. Sci.*, 1, 1, 2013.
55. Ramadan, A.M., Ibrahim, M.M., Shaban, S.Y., Synthesis, characterization, and tyrosinase biomimetic catalytic activity of copper(II) complexes with schiff base ligands derived from a-diketones with 2- methyl-3-amino-(3H) quinazolin-4-one. *J. Mol. Struct.*, 1006, 348, 2011.
56. Zabiulla, S.K., Joythi, M., Bushra Begum, A., Asha, M.S., Al-Ostoot, F.H., Lakshmeesha, D.P., Ramu, R., Khanum, S.A., Molecular docking, synthesis and antimicrobial evaluation of metal complexes with Schiff base. *Res. Chem.*, 5, 100650, 2023.
57. Viswanathamurthi, P. and Natarajan, K., Ruthenium(II) Carbonyl complexes containing tetradentate schiff bases and their catalytic activities. *Synth. React. Inorg. Met. Org. Chem.*, 36, 415, 2006.
58. Abdel-Rahman, H.L., El-Khatib, R.M., Nassr, L.A.E., Abu-Dief, A.M., DNA binding ability mode, spectroscopic studies, hydrophobicity and in vitro antibacterial evaluation of some new Fe(II) complexes bearing ONO donors amino acid Schiff bases. *Arab. J. Chem.*, 10, 1835, 2013.
59. Chang, E.L., Simmers, C., Knight, D.A., Cobalt complexes as antiviral and antibacterial agents. *Pharmaceuticals*, 3, 1711, 2010.
60. Asha, M.S., Pinto, O., Nagendra Murthy, V.N., Raj, V., Delarasoltanesmaeili, Synthesis, characterization, anti-inflammatory, antimicrobial and DNA binding properties of Co(II). Ni(II), Cd(II) and Cu(II) Complexes with Schiff Base. *Int. J. Adv. Res.*, 4, 8, 1065, 2016.
61. Patel, R.N., Structural, magnetic and spectroscopic characterization of two unusual end on bis(l-acetato/l-nitrato) bridged copper(II) complexes with N-[phenyl (pyridin-2-yl) methylidene] furan-2- carbohydrazide and (2E,4Z)-N,2dimethylhepta-2,4,6- trienamide-1-phenyl-1-pyridin-2ylmethanimine (1:1) as capping ligands. *Inorg. Chim. Acta*, 363, 3838, 2010.
62. Budhani, P., Iqbal, S.A., Malik, S., Bhattacharya, M., Mitu, L., Synthesis, characterization and spectroscopic studies of pyrazinamide metal complexes. *J. Saudi. Chem. Soc*, 14, 281, 2010.
63. Ibrahim, M.M., Ali, H.M., Abdullah, M.A., Hassandarvish, P., Acute toxicity and gastroprotective effect of the Schiff base ligand 1H-indole-3-ethylene-5-nitro- salicylaldimine and its Nickel (II) complex on ethanol induced gastric lesions in rats. *Molecules*, 17, 12449, 2012.
64. Miyazaki, R., Yasui, H., Yoshikawa, Y., α-Glucosidase inhibition by new Schiff base complexes of Zn (II). *Open J. Inorgan. Chem.*, 6, 114, 2016.
65. S.E., Bhandarkar, P.P., Pathare , B.P., Khobragade, Nickel(II), copper(II) and cobalt(II) complexes based salicyaldehyde Schiff Base: synthesis, characterisation, and antiviral activity, *Mater. Today: Proc.*, 92, 807–816, 2023.

66. Gipperth, L., The legal design of the international and European Union ban on tributyltin antifouling paint: Direct and indirect effects. *J. Environ. Manage.*, 90, 86, 2009.
67. Qian, P.Y., Xu, Y., Fusetani, N., Natural products as antifouling compounds: Recent progress and future perspectives. *Biofouling*, 26, 223, 2010.
68. Fusetani, N., Antifouling marine natural products. *Nat. Prod. Rep.*, 28, 400, 2011.
69. Dobretsov, S., Dahms, H.U., Qian, P.Y., Inhibition of biofouling by marine microorganisms and their metabolites. *Biofouling*, 22, 43, 2006.
70. Xu, Y., He, H., Schulz, S., Liu, X., Fusetani, N., Potent antifouling compounds produced by marine Streptomyces. *Bioresour. Technol.*, 101, 1331, 2010.
71. Ziganshina, M. and Nurislamova, E., Anti-corrosion properties of coatings with manganese compounds pigmentation. *Mater. Sci. Eng.*, 311, 012017, 2018, doi: 10.1088/1757-899X/311/1/012017.
72. Boulechfar, C., Ferkous, H., Delimi, A., Berredjem, M., Kahlouche, A., Madaci, A., Djellali, S., Boufas, S., Djedouani, A., Errachid, A., AliKhan, A., Boublia, A., Lemaoui, T., Benguerba, Y., Corrosion inhibition of Schiff base and their metal complexes with [Mn (II), Co (II) and Zn (II)]: Experimental and quantum chemical studies. *J. Mol. Liq.*, 378, 121637, 2023.
73. Fouda, A.S., Abd El-Wahab, S.M., Attia, M.S., Youssef, A.O., Elmoher, H.O., Rare earth metals as eco-friendly corrosion inhibitors for mild steel in produced water. *Der Pharma Chem.*, 7, 74–87, 2015.
74. Al-Amiery, A.A., Binti Kassim, F.A., Kadhum, A.A.H., Mohamad, A.B., Synthesis and characterization of a novel eco-friendly corrosion inhibition for mild steel in 1 M hydrochloric acid. *Sci. Rep.*, 6, 19890, 2016.
75. Pandey, A., Singh, B., Verma, C., Ebenso, E.E., Synthesis, characterization and corrosion inhibition potential of two novel Schiff bases on mild steel in acidic medium. *RSC Adv.*, 7, 47148–47163, 2017.
76. Sayed, F.N., Ashmawy, A.M., Saad, S.M., Omar, M.M., Mohamed, G.G., Design, spectroscopic characterization, DFT, molecular docking, and different applications: Anti-corrosion and antioxidant of novel metal complexes derived from ofloxacin-based Schiff base. *J. Organomet. Chem.*, 993, 1, 122698, 2023.
77. Potgieter, J.H., Lawal, O.J., Billing, C., Whitefield, D.J., The Influence of REE β-Diketone Complexes on the Corrosion Behaviour of Mild Steel and 304 SS in 3.5% NaCl Solution. *Minerals*, 12, 416, 2022.
78. Esmailzadeh, S., Aliofkhazraei, M., Sarlak, H., Interpretation of Cyclic Potentiodynamic Polarization Test Results for Study of Corrosion Behavior of Metals: A Review. *Prot. Met. Phys. Chem. Surf.*, 54, 976–989, 2018.
79. Volari ̌c, B. and Milošev, I., Rare earth chloride and nitrate salts as individual and mixed inhibitors for aluminium alloy 7075-T6 in chloride solution. *Corros. Eng. Sci. Technol.*, 52, 201–211, 2017.
80. Gharbi, O., Thomas, S., Smith, C., Birbilis, N., Chromate replacement: What does the future hold? *NPJ Mater. Degrad.*, 2, 23–25, 2018.

81. Blin, F., Leary, S.G., Wilson, K., Deacon, G.B., Junk, P.C., Forsyth, M., Corrosion mitigation of mild steel by new rare earth cinnamate compounds. *J. Appl. Electrochem*, 34, 591–599, 2004.
82. Liu, X., Sun, Y., Lu, M., Pan, X., Wang, Z., Electrochemical and surface analytical studies of transition metal bipyridine dicarboxylic acid complexes as corrosion inhibitors for a mild steel in HCl solution. *J. Adhesion Sci. Technol.*, 2021. https://doi.org/10.1080/01694243.2021.1929017.
83. Kashar, T., Emran, K.M.,., Mo'ala, A., Novel metal complexes as antimicrobial and antcorrosion in acidic nedia. *Arab. J. Chem. Environ. Res.*, 7, 1, 29–49, 2020.
84. Taha, R.H., Gaber, G.A., Mohamed, L.Z., Ghanem, W.A., corrosion inhibition of two Schiff base complexes on the mild steel in 1M HCl solution. *Egypt. J. Chem.*, 62, 367–381, 2019.
85. Hamaida, H.A.E., Ali, E.-D., Sadek, S.M.M., Potential anti-fouling agents: metal complexes of 3-(2-furylidine)hydrazine-5,6-diphenyl-1,2,4-triazine. *Pigm. Resin Technol.*, 37, 4, 243–249, 2008.
86. Trojer, A., Movahedi, A., Blanck, H., Nyden, M., Imidazole and Triazole Coordination Chemistry for Antifouling Coatings. *Markus J. Chem.*, 23, 2013.
87. Konstantinou, I.K. and Albanis, T.A., Worldwide occurrence and effects of antifouling paint booster biocides in the aquatic environment: *A review. Environ. Int.*, 30, 235–248, 2004.
88. Hao, X., Chen, S., Yu, H., Liu, D., Sum, W., Metal ion-coodrdinated carboxymethylated chitosan grafted carbon nanotubes with enhanced antibacterial properties. *RSC Adv.*, 1, 2016.
89. Liu, W., Yan, M., Zhao, W., Antibacterial-renew dual-function antifoulingn-strategy: Self-assembled Schiff-base metal complex coatings built from natural products. *J. Colloid Interface Sci.*, 629, 496–507, 2023.
90. Singh, K.S., Devi, P., Gaonkar, R.R., Kaminsky, W., Half-sandwich iridium(III) and rhodium(III) complexes: synthesis, characterization and antifouling properties. *J. Coord. Chem.*, 2023. https://doi.org/10.1080/00958972.2023.2217517.

81. Rila, L., [illegible], V., [illegible], G.B., [illegible] ... marine coatings [illegible]

82. [illegible] ... steel [illegible]

83. [illegible]

84. [illegible], L.Z., Obanijesu, V.A., [illegible] ... of two Schiff base complexes on the mild steel [illegible] ... 2019.

85. [illegible], Sadek, S.M.M., [illegible]

86. [illegible], Rana, [illegible]

87. [illegible]

88. [illegible]

89. [illegible]

90. [illegible]

14

Thermodynamic Modeling of Carbonaceous Coating of Oxides and Sulfides Thin Films Grown by CVD for Enhancing Surface Quality

Kranthi Kumar Vaidyula*, Sukanya Dhar, Anjali Lalithambika and S. A. Shivashankar

Centre for Nanoscience and Engineering (CeNSE), Indian Institute of Science, Bangalore, Karnataka, India

Abstract

Over the years, to achieve high quality large-area thin films, chemical vapor deposition (CVD) technique is used. However, achieving high quality is a challenge due to controllability of process parameters like carrier and precursor gas flows, reactor pressure, sublimation temperature and reactor temperature. To solve this problem, a generic approach involving thermodynamic modelling is used. In this work, using thermodynamic modelling an appropriate CVD process window is attained. Firstly, thermodynamic analysis of the solid and gas phases for the Al-C-O-H-Ar system to grow alumina in an CVD chamber predicts large amount of carbon in Ar ambient. As oxygen is introduced as the reacting gas, carbon content in the film reduces by significant extent. X-ray diffraction and Raman spectroscopy used to characterise the carbon less films. Secondly, a detailed thermodynamic modelling of MoS_2 in argon and hydrogen ambient is presented. Raman Spectroscopy and X-ray diffraction are used to characterize the films grown. Using the model, at 700°C and higher pressures in hydrogen ambient, pure MoS_2 is grown.

Keywords: Chemical vapor deposition, thermodynamic modelling

**Corresponding author*: kranthikumar081986@gmail.com

Hari Murthy, Vinay Jha Pillai, Kukatlapalli Pradeep Kumar and Matthew Cowan (eds.) Novel Anti-Corrosion and Anti-Fouling Coatings and Thin Films, (425–434) © 2024 Scrivener Publishing LLC

14.1 Introduction

The major challenge for many industry sectors today is dealing with metal corrosion as it raises operational costs. Over the past few decades, the industry has implemented various transition metal oxide coatings as anti-corrosion protective layers using different approaches such as sol-gel [1, 2], spray pyrolysis [3, 4], and chemical vapor deposition (CVD) [5, 6].

Films coated with CVD have gained attention as the approach is bottom-up, robust, and widely used to grow uniform films with excellent mechanical properties. It also offers superior performance in terms of conformality and is therefore particularly adapted to coat porous substrates and complicated-shaped surfaces. Numerous studies have shown that Al_2O_3 films show anticorrosion properties, with less tool wear and improved productivity [7, 8]. Unlike traditional anticorrosion coatings, CVD coatings are thin and dense while serving as a highly effective barrier between the surface and corrosive media [9].

In this work, we develop a robust approach to grow continuous large-area alumina films on industry-relevant stainless-steel substrate at temperatures as low as 700°C using the metal-organic chemical vapor deposition (MOCVD) method, in oxygen ambient using precursor tris(2,4-pentanedionato)aluminum(III). For growing contamination-free Al_2O_3 films, thermodynamic modeling was used to predict the growth window of pure solids. Owing to the unique layered structure, transition metal dichalcogenides (TMDs) are also explored as protective layers for anticorrosion coatings. MoS_2 being popular in the 2D materials family exhibits excellent chemical inertness and thermal stability. These traits make MoS_2 as an ideal candidate for anticorrosion applications [8]. Recently, MoS_2 growth using the CVD technique has been widely studied [9–11]. In this work, a growth window for pure MoS_2 is demonstrated with phase diagrams in both argon and hydrogen ambient at temperatures 350°C to 1050°C from 1 Torr to atmospheric pressures.

14.2 Experimental

14.2.1 Thermodynamic Modeling

HSC Chemistry, software version 6.1, developed by Outokumpu, Finland, was used to perform thermodynamic calculations of phase diagrams.

14.2.1.1 Al_2O_3 Growth

TiN/TiC/WC substrates (1 cm × 1 cm) were used for Al_2O_3 film growth and are supplied by Kennametal Ltd, Bangalore, India. The experiments were conducted in a custom-built in-house, horizontal, hot-wall MOCVD reactor. The reactor uses a fused quartz tube of 3.5-cm inner diameter and 50-cm length and is equipped with mass flow control systems, pressure control units, and temperature controls through which the total chamber pressure (P_c), flow rates of both carrier gases, growth temperature (T_s), and the precursor $Al(acac)_3$ vaporizer temperature (T_V) could be adjusted. The growth conditions used for alumina films in this work were the following: T_s is 700°C, P_c is 10 Torr, T_v is 165°C, oxygen flow used is 5 standard cubic centimeters per minute (sccm), and the argon flow used is 20 sccm. To obtain thicker films, the temperature of the vaporizer was kept higher to increase the concentration of the precursor. The deposited films were evaluated by various techniques. For film phase identification, X-ray diffraction (XRD) is used. For investigating surface morphology and thickness, an optical microscope (OM) and scanning electron microscope (SEM) are used.

14.2.1.2 MoS_2 Growth

The molybdenum precursor used for this study is $Mo(CO)_6$ and the sulfur source is H_2S gas (from Bhoruka, 99.5%). The experiment was conducted in an in-house–built horizontal CVD reactor, as displayed in Figure 14.1. SiO_2/Si of 1 cm × 1 cm are used as substrates, they were cleaned successfully in acetone followed by isopropyl alcohol and deionized water in an ultrasonicator for 300 s each and dried in an ultrahigh pure nitrogen stream. $Mo(CO)_6$ precursor was put in a bubbler setup and heated to 100°C to get the desired Mo flux. The vapor transport lines were heated and kept at 130°C. The carrier gasses used for this study were Ar and H_2, and H_2S

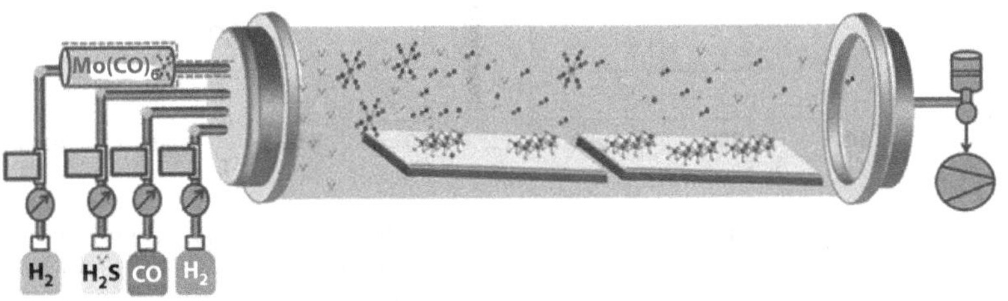

Figure 14.1 MOCVD setup for growing MoS_2 films.

as the reacting gas. The chamber temperature (T_c) varied from 350°C to 850°C at a ramp rate of 25°C/min using argon or hydrogen or a mixer of both. Before the growth, the chamber was pumped down to the desired reactor pressure. Carrier gas flow (15 sccm) was allowed through a vaporizer to carry the $Mo(CO)_6$ powder. The gas flow rates used in this setup are in the ratios 1:10:100 or 1:10:1000, Mo:H_2S:(Ar or H_2).

14.2.2 Characterization

The deposited Al_2O_3 films were investigated using Rigaku Smart Lab HR-XRD for the identification of crystalline phases. The Al_2O_3 and MoS_2 films were examined by Leica optical microscopy and Zeiss, Ultra-55, (SEM) to confirm the uniformity and surface morphology of the grown films. Horiba, LabRam with 532-nm laser was used to characterize the MoS_2 films and carbon contamination too.

14.3 Results and Discussion

14.3.1 Thermodynamic Calculations for Al_2O_3 and Growth

For the Al-C-O-H-Ar system, the thermodynamic modeling data predict a large amount of carbon deposited with Al_2O_3 in Ar ambient. As oxygen is introduced as the reacting gas, carbon contamination in the film is reduced to a significant extent. Al_2O_3-C composition is also a function of temperature and total pressure. Figure 14.2(a) presents Al_2O_3-C composition as a function of temperature at a pressure of 10 Torr, when 10 mol of oxygen

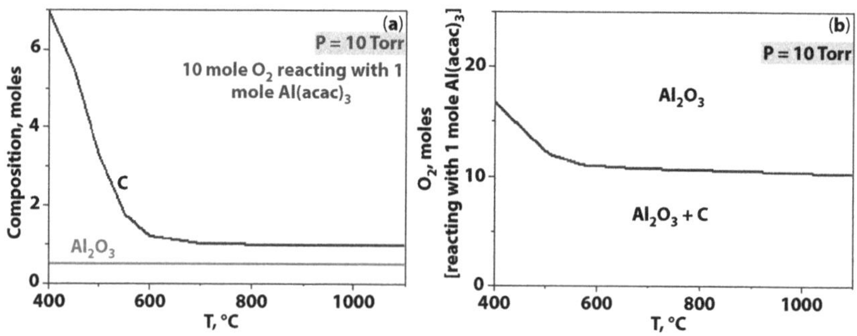

Figure 14.2 Al-C-O-H-Ar system phase stability diagrams. (a) Calculated stability windows for various condensed phases at a pressure of 10 Torr at 10 mol of O_2. (b) Stability diagram with varying oxygen concentration showing carbon contamination increases at low temperatures.

reacts with 1 mol of $Al(acac)_3$. Thus, by controlling the moles of reacting oxygen, P, and T, the carbon content in the MOCVD-grown alumina films can be optimized to the requirement of cutting tool applications. A qualitative "MOCVD phase stability diagram" depicted in Figure 14.2(b) shows the regions of pure and carbon-containing Al_2O_3 films as a function of CVD process parameters. The results of thermodynamic modeling have been verified quantitatively.

Alumina growth by MOCVD, conducted in an oxygen environment, resulted in glossy blackish coatings. The XRD pattern, displayed in Figure 14.3(a), shows the film constituents of Al_2O_3 of α-, κ-, and γ-crystal phases. Alumina crystallites are embedded in the ellipsoid-shaped features seen in the optical images shown in Figure 14.3(b). These ellipsoids, measuring around a few micrometers in diameter, primarily include amorphous carbon, which forms the matrix for the growth of crystallites of alumina films. The coated film was found to be approximately 7 μm by cross-sectional SEM shown in Figure 14.3(c).

14.3.2 Thermodynamic Modeling and Growth of MoS_2

In thermodynamic modeling, it mainly adjusts the moles of individual species and gets the minimum Gibbs free energy with mass balance conditions, assuming that equilibrium holds in the entire process, which is applicable in CVD only at low deposition rates [10, 11]. CVD growth of MoS_2 using H_2S, $MoCl_5$, and MoF_6 has been tried successfully previously along with thermodynamic calculations [10, 11]. Earlier, thermodynamic modeling was also performed for the deposition of molybdenum and other molybdenum compounds by using plasma-enhanced CVD from molybdenum hexacarbonyl and molybdenum hexafluoride as precursors

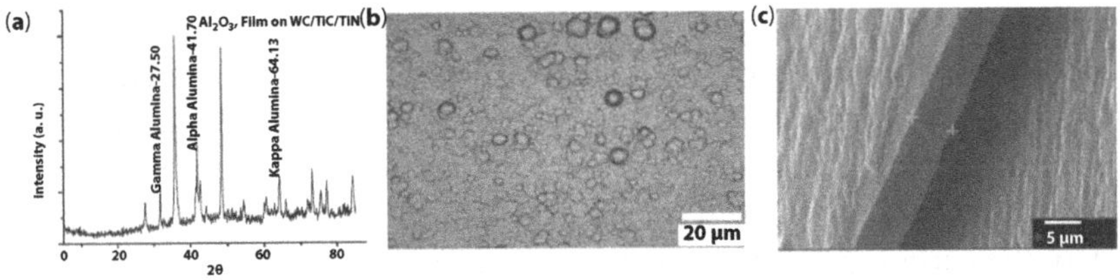

Figure 14.3 Alumina film characterization. (a) X-ray diffraction of the film deposited on TiN/TiC/WC in an O_2 environment at T_s of 700°C and pressure of 10 Torr at vaporizer temperature T_v of 165°C and argon flow of 45 sccm and O_2 flow of 10 sccm. (b) Optical microscopy images of the grown films on TiN/TiC/WC showing a "droplet"-like shape. (c) Cross-sectional SEM images of the alumina film grown.

and hydrogen or methane as reductants [12, 13]. However, to the best of our knowledge, there have been no reports exploring stability windows for CVD MoS_2 from $Mo(CO)_6$ and H_2S by thermodynamic investigations, covering a wide range of gas ambient in particular.

By assuming that the system is under equilibrium, using thermodynamic modeling for any CVD process, "phase diagrams" as a function of temperature, pressure, and gas flows can be found, which can predict "process windows" for the deposition of the desired film. All the films grown in this study for verifying the analytical results were carried out in the reactor shown in Figure 14.1. Due to kinetic limitations most of the growth runs are carried at T > 350°C. Carrier gas argon mainly controls the partial pressures in the chamber, thereby changing the concentrations at equilibrium. Figure 14.4(a) shows a solid phase stability diagram in the CVD chamber, measured across a range of higher to lower $Mo(CO)_6$ flows (x) than that of H_2S (y) as a function of chamber temperature at reactor pressure of 20 Torr. The phase diagram clearly shows that the pure MoS_2 phase at H_2S flows greater than 1 can supply enough sulfur for the formation of contamination-free MoS_2. To the left of this phase diagram at lower temperatures, different molybdenum carbides, MoO_2, and carbon are expected along with MoS_2. At higher temperatures, metallic molybdenum, Mo_2S_3, and Mo_2C are expected. For contamination-free MoS_2, elevated temperatures and the ratios of y/x > 2 are needed. The phase diagram regions were validated using XRD and Raman spectroscopy shown in Figures 14.4(b to d), respectively.

In Figure 14.5(a), the phase stability diagram for $Mo(CO)_6$:H_2S:H_2:Ar ratios of 1:y:50:50 is shown. This diagram reveals that, for the pure MoS_2 phase, it mainly depends on chamber temperature and reactor pressure, and not much on y. A portion of the diagram with Ar:H_2 = 50:50, at y of 10 and pressure of 20 Torr, is shown in Figure 14.5(b), indicating the solids and gaseous species formed [14]. It also shows that a steady increase in CO formed and a depletion in CH_4 concentration with temperature, with the proportion of carbon being the highest when CH_4 and CO are equal, and CO_2 is close to its maximum value. Portions of Figure 14.5(a) for y = 10 at different reactor pressures show maximum carbon. This is confirmed using Raman spectroscopy shown in Figure 14.5(c), the increase in the intensity of the carbon peak (G band at 1586 cm^{-1}) with chamber temperatures (350 to 500°C) and its absence at higher temperatures (700°C) confirms the phase diagram in Figure 14.5(a).

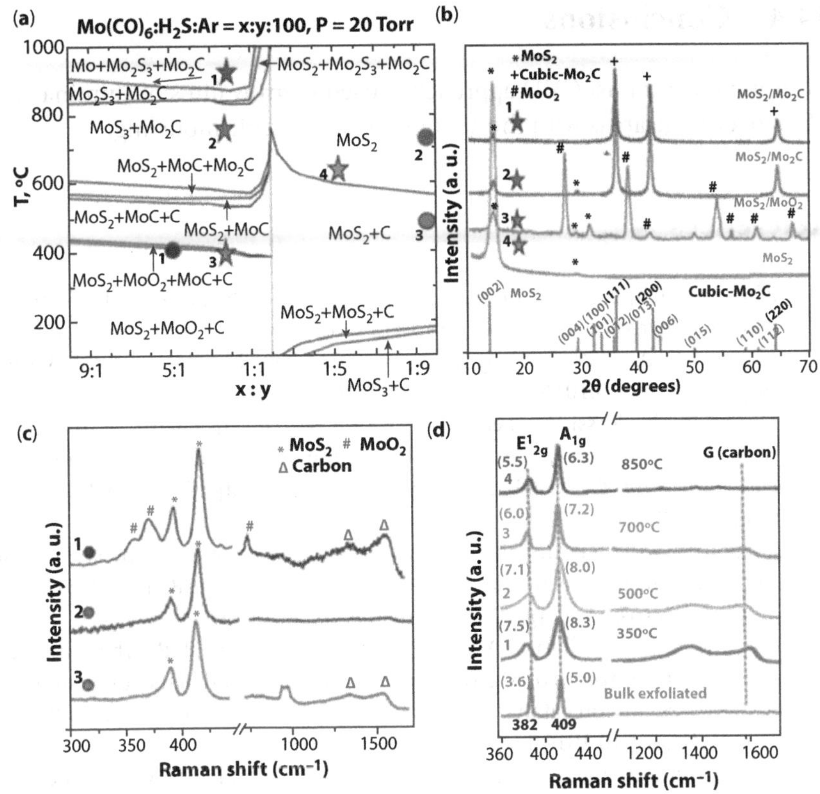

Figure 14.4 (a) Phase stability diagram of Mo–C–O–H–S system in an argon environment is shown. (b) Confirmation of the phase stability diagram using XRD. (c) Confirmation of films grown using Raman spectroscopy. (d) Raman peaks of MoS_2 across various temperatures for x:y =1:5. Symbols next to the patterns correspond to locations in the phase diagram where the symbols appear.

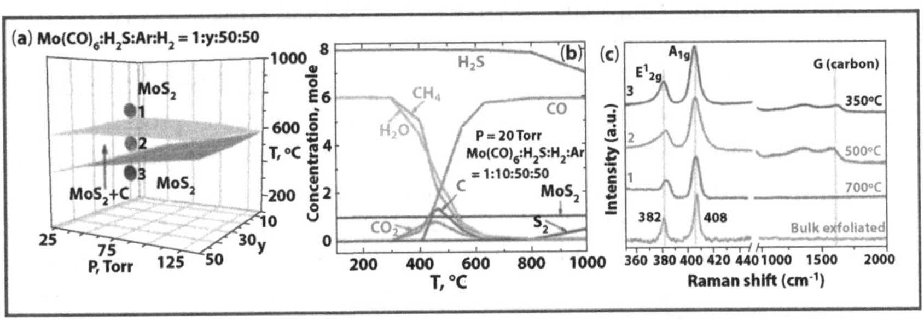

Figure 14.5 (a) The solid phase stability diagram depicts the growth window of contamination-free MoS_2 in argon and hydrogen mixture of 50 and 50. (b) Equilibrium molar concentration of gasses and solids formed at pressure of 20 torr. (c) Raman spectra of MoS_2 showing a "lump" around 1536 cm^{-1} indicate mole% of carbon.

14.4 Conclusions

In summary, a robust CVD approach is used to grow films of alumina and MoS_2 over large areas with the aid of thermodynamic modeling.

References

1. Zheludkevich, M.L., Serra, R., Montemor, M.F., Salvado, I.M.M., Ferreira, M.G.S., Corrosion protective properties of nanostructured sol–gel hybrid coatings to AA2024-T3. *Surf. Coat. Technol.*, 200, 9, 3084–3094, 2006, doi: https://doi.org/10.1016/j.surfcoat.2004.09.007.
2. Deflorian, F., Rossi, S., Fedel, M., Motte, C., Electrochemical investigation of high-performance silane sol–gel films containing clay nanoparticles. *Prog. Org. Coat.*, 69, 2, 158–166, 2010, doi: https://doi.org/10.1016/j.porgcoat.2010.04.007.
3. Angappan, S., Berchmans, L.J., Augustin, C.O., Sintering behaviour of MgAl2O4—a prospective anode material. *Mater Lett.*, 58, 17, 2283–2289, 2004, doi: https://doi.org/10.1016/j.matlet.2004.01.033.
4. Bocanegra, S.A., Ballarini, A.D., Scelza, O.A., de Miguel, S.R., The influence of the synthesis routes of MgAl2O4 on its properties and behavior as support of dehydrogenation catalysts. *Mater Chem. Phys.*, 111, 2, 534–541, 2008, doi: https://doi.org/10.1016/j.matchemphys.2008.05.002.
5. Paseuth, A., Fukui, H., Okuno, S., Kanaoka, H., Okada, Y., Microstructure, mechanical properties, and cutting performance of TiCxN1−x coatings with various x values fabricated by moderate temperature chemical vapor deposition. *Surf. Coat. Technol.*, 260, 139–147, 2014, doi: https://doi.org/10.1016/j.surfcoat.2014.09.068.
6. Larsson, A. and Ruppi, S., Microstructure and properties of Ti(C,N) coatings produced by moderate temperature chemical vapour deposition. *Thin Solid Films*, 402, 1, 203–210, 2002, doi: https://doi.org/10.1016/S0040-6090(01)01712-6.
7. Gunby, N.R., Krumdieck, S., Murthy, H., Masters, S.L., Miya, S.S., Study of precursor chemistry and solvent systems in pp-MOCVD processing with alumina case study. *Phys. Status Solidi*, 212, 7, 1519–1526, Jul. 2015, doi: https://doi.org/10.1002/pssa.201532309.
8. Jiang, A., Cao, X., Wang, Z., Ma, J., Xiao, J., Ma, S., Friction performance and corrosion resistance of MoS2/DLC composite films deposited by magnetron sputtering. *Results Phys.*, 25, 104278, 2021, doi: https://doi.org/10.1016/j.rinp.2021.104278.
9. Chai, Z., Liu, Y., Li, J., Lu, X., He, D., Ultra-thin Al2O3 films grown by atomic layer deposition for corrosion protection of copper. *RSC Adv.*, 4, 92, 50503–50509, 2014, doi: 10.1039/C4RA09179E.

10. Wang, X., Feng, H., Wu, Y., Jiao, L., Controlled Synthesis of Highly Crystalline MoS2 Flakes by Chemical Vapor Deposition. *J. Am. Chem. Soc.*, 135, 14, 5304–5307, Apr. 2013, doi: 10.1021/ja4013485.
11. Kranthi Kumar, V., Dhar, S., Choudhury, T.H., S.A., Raghavan, S., Shivashankar A predictive approach to CVD of crystalline layers of TMDs: the case of MoS2. *Nanoscale*, 7, 17, 7802–7810, 2015, doi: 10.1039/C4NR07080A.
12. Najmaei, S. *et al.*, Vapour phase growth and grain boundary structure of molybdenum disulphide atomic layers. *Nat. Mater*, 12, 8, 754–759, 2013, doi: 10.1038/nmat3673.
13. Kim, K.-W. *et al.*, Fabrication and Characterization of Ru Thin Films Prepared by Liquid Delivery Metal-Organic Chemical Vapor Deposition. *Jpn J. Appl. Phys.*, 41, 2, 820, 2002, doi: 10.1143/JJAP.41.820.
14. Dhar, S., Kranthi Kumar, V., Choudhury, T.H., Shivashankar, S.A., Raghavan, S., Chemical vapor deposition of MoS2 layers from Mo–S–C–O–H system: thermodynamic modeling and validation. *Phys. Chem. Chem. Phys.*, 18, 22, 14918–14926, 2016, doi: 10.1039/C6CP01617K.

15

Metal Nanoparticles: Biosynthesis Approach and Bio-Packaging Application

Priti Chaware[1], Amol Nande[2*], J. D. Punde[3], K. G. Rewatkar[4] and S. J. Dhoble[5]

[1]Dr. Ambedkar College, Nagpur, India
[2]Guru Nanak College of Science, Ballarpur, India
[3]S. S. Girls' College, Gondia, India
[4]Vidya Vikas College of Art, Commerce and Science, Samudrapur, India
[5]Department of Physics, RTM Nagpur University, Nagpur, India

Abstract

Nanoparticles' (NPs') widespread adoption and several ranges of promising new uses have astounded experts and laypeople alike. These are favored over their respective macrostructure analogs due to their distinctive chemical properties, physical properties, and superior performance. Nanotechnology has changed the ways in which foods are preserved, processed, tested, and packaged in a number of industries, including the food business. Modern packaging based on nanotechnology allows for the safe transport of perishable goods without compromising their flavor, nutritional value, or quality during transit. In this chapter, a concise picture of the fundamental processes and mechanisms underlying "green" synthesis techniques is discussed, in which the discussion mainly involves the synthesis of NPs using plant extracts. Furthermore, the necessary characterization (such as transmission electron microscopy, scanning electron microscopy, and x-ray diffraction analysis) in order to confirm the structure and size and the surface analysis of the green synthesized NPs have been discussed here. We concluded by discussing the uses of these synthetic goods in packaging and food engineering.

Keywords: Nanoparticles, bio-synthesis, bio-packaging, X-ray diffraction, SEM, TEM, green synthesis, FTIR

**Corresponding author*: nande.av@gmail.com

Hari Murthy, Vinay Jha Pillai, Kukatlapalli Pradeep Kumar and Matthew Cowan (eds.) Novel Anti-Corrosion and Anti-Fouling Coatings and Thin Films, (435–458) © 2024 Scrivener Publishing LLC

15.1 Introduction

Nanoparticles (NPs) are quite interesting topics for the current research which has very wide varieties of applications in the fields of science, technology robotics, magnetism, electronics, medical sciences, encapsulations, etc. [1, 2]. The varieties of applications are due to the distinguished properties that appeared after reducing the materials to their NPs. NPs provide surface effects and quantum effects which makes them a suitable candidate for the unconventional thermal, electrical, superconducting, mechanical optical, and chemical properties [1, 3–5]. However, after all these years of long research, the major problem is with the synthesis approach of the nanomaterials. It is very difficult to reproduce nanoparticles with the same properties and functions and it is quite a big challenge to produce toxicity-free NPs.

The NPs are prepared in two novel ways—top-down and bottom-up ways. These two fundamental approaches of synthesis have been explored in recent years to prepare NPs of required shapes, sizes, and applications [6, 7]. Using these two fundamental approaches, NPs are synthesized using a wide variety of chemical, physical, biological, or hybrid processes. Traditional chemical and physical methods for synthesizing metal NPs like gold, silver, cadmium, and iron include pyrolysis, chemical vapor deposition, microemulsion method, aerosol method, lithography, physical vapor depositions, and laser ablation [8–11]. These synthesis methods involve consumption energy, high temperature, pH, and high pressure which make them very expensive methods for NP preparation. Furthermore, chemical methods involve the use of organic solvents and reducing and stabilizing agents which release toxic byproducts and increase toxicity along with the nanoparticles. Therefore, their use in clinical, food, agricultural, and environmental settings has been thwarted due to the contamination of hazardous substances with the produced NPs. Recently, biosynthesis (also referred to as green synthesis) has emerged to circumvent these constraints by making use of readily biodegradable natural materials for its production. Moreover, the process does not involve any stabilizing agents.

In the biosynthesis process, nanoparticles are synthesized using microorganisms, including viruses, bacteria, fungi, algae, yeast, and plants (through their respective enzymes, proteins, DNA, lipids, carbohydrates, etc.) [12, 13]. Metal-reducing bacteria have been identified as eco-friendly catalysts for bioremediation and material production. It is possible that respiration mechanisms aided by bacteria aid in the creation of certain metal oxides. The potential for these organisms to be used in the preparation of materials at the nano- and microscale has propelled the establishment of a novel and underexplored field of study. However, the selection of medium and involvement risk of handling bacteria is very difficult. Similar risks are involved in the green preparation of NPs using viruses, fungi, algae, and yeast. Compared to others, utilizing plant extracts provides a relatively quick and easy approach to synthesizing metal or metal oxide NPs. It also makes the large-scale production of metal/metal oxide nanoparticles [14–16]. Furthermore, plant extracts and tissues in them make the process of preparation of NPs far simpler and easier with respective other biological approaches. It has been observed that a wide variety of plant species as well as parts of plants such as flowers, fruits, seeds, stems, and leaves can be used to synthesize metal or metal oxide NPs. The biomolecules in the extracts have the potential to carry out the reduction reaction and convert the reactant into nanosized materials in a one-step green manufacturing process. This method has also made use of plant extracts for their stabilizing or reducing properties of synthesized nanoparticles. For example, *Jasminum sambac* leaf extracts were used to prepare stable NPs which involved the preparation of silver, gold, and alloy NPs [17–19]. In contrast to NPs prepared by chemical procedures, NPs were prepared using olive leaf extract in hot water. The reaction was carried out at high temperatures to produce different shapes of Au NPs. Iron-polyphenol (Fe-P) NPs are prepared using plant leaf extracts of Australian natural plants such as *Melaleuca nesophila*, *Eucalyptus tereticornis*, and *Rosmarinus officinalis*. The NPs showed good photocatalytic activity against Acid Black 194 dye. The polyphenols in all three plants were found to react with ferric chloride ($FeCl_3$) solution in distinct ways, resulting in differently shaped Fe-P NPs and chelated ferric-polyphenols NPs [20, 21].

As discussed earlier, one of the promising applications of nanoparticles is bio-packaging. In bio-packaging, protecting, preserving, and increasing the shelf life of eatable materials are all important functions for which NPs are used in food technology and industry. In particular, NPs have improved antibacterial and antifungal activities as well as enhanced mechanical properties, and vapor permeability in food packaging, allowing the packed food for longer product shelf lives and better preservation of freshness, as well as the creation of safer and more environmentally friendly (degradable) food packaging.

This chapter provides detailed information with appropriate examples of the synthesis of NPs using the biosynthesis approach. Later, it provides information about suitable characterization techniques used to analyze the NPs. We provide a summarized table from the recent work which depicts the information about the preparation conditions and characterization of several metal and metal oxide NPs using plant extracts. In the end, it also gives information about the use of biosynthesized nanoparticles with examples from recent research works. Our overarching objective is to provide a comprehensive description of "biosynthesis" processes and their related components that are both informative to researchers in this new field and accessible to others with a more casual interest in the subject. Here, we take a look at the many strategies for producing nanoparticles from plant extracts, potential mechanisms of nanoparticle production, and their uses in bio-packaging.

15.2 Synthesis of NPs

Green synthesis or biosynthesis of NPs has acquired attraction in recent years as a result of their distinctive characteristics, applied applications in various fields, and low environmental impact. Although NPs can be prepared via chemical and physical routes, the former are generally considered less desirable than biological approaches because of their lower environmental impact and higher cost. By avoiding the use of poisonous and hazardous ingredients, biosynthesis involves inherent reducing and stabilizing agents. Furthermore, in this approach, the need for high pressure or temperature is not required to carry out the process [22, 23]. Recent years have seen a proliferation of methods for increasing the output of NPs of varying sizes, different shapes, and chemical stability [24–26].

Both top-down and bottom-up approaches are widely used to manufacture NPs. The powers of different materials are systematically grounded down to nanosized structures using mechanical processes in the top-down method. On the other hand, the bottom-up techniques are assembling of atomic or molecular building blocks into nanoscale structures [10, 27]. Top-down techniques in nanoparticle synthesis might result in imperfections in the synthesized NPs. This affects the chemical and physical properties of NPs significantly. However, the bottom-up strategy provides good crystal structure and perfection in NPs [28, 29]. In this context, constituent elements of the required material NPs are initially created and afterward arranged to generate the ultimate particle. The utilization of metal nanoparticles in both chemical and biological contexts relies on bottom-up methodologies.

The green synthesis method can be further categorized into two main types: intracellular synthesis and extracellular synthesis. Bacteria, yeast, fungi, algae, and plant extracts are only a few of many biological entities capable of achieving intra- or extracellular production of NPs [12, 30]. Although it is well known that biosynthesis using micro-organism is easily scalable and can be used for pharmaceutical products, it is often more expensive than plant-based material production. Also, the plant-based biosynthesis process does not involve high pressure and high temperature, as well as harmful chemicals. Thus, this is also an economical and environmentally friendly way to synthesize NPs. Furthermore, no toxicity in byproducts makes the plant extract–based biosynthesis approach very famous and has applications in the field of food safety and medical sciences [31–34]. Plant extracts are recognized for their diverse range of reducing and stabilizing agents, which are crucial in the process of synthesizing NPs. NP synthesis is highly dependent on the composition of the plant extract used. It is interesting to know that the NP's size and shape are influenced by the concentration and type of the plant extract source. Furthermore, this phenomenon might be attributed to varying quantities of biochemical reducing agents found in different plant extracts. Regional and seasonal variations in the content of biochemical reducing agents are seen in the majority of plant extracts [35–40]. Because of this, every batch of produced nanoparticles will be somewhat different. Biosynthesis of NPs is possible in many different plant components, including flowers, stems, leaves, fruits, seeds, roots, and complete plants. Plants can store varying levels of heavy metals in their many tissues. Biomolecules (such as carbohydrates, proteins, and coenzyme) that are present in plants have excellent potential for reducing metal salt into NPs [41]. Among all metals, silver and gold NPs synthesized using the biosynthesis process were widely studied. For instance, if a sample was synthesized from a plant leaf extract, fresh leaves would need to be obtained from a pesticide- and pollution-free location. Twice with running water and once with the extract's mixture, the leaves are given a thorough washing. Our target phytochemical can be effectively extracted using different solvents throughout the extraction processes. NPs are synthesized using these extracts. Plant extracts are subjected to metal ion solutions of varying concentrations in a 370°C, 150 rpm, agitated

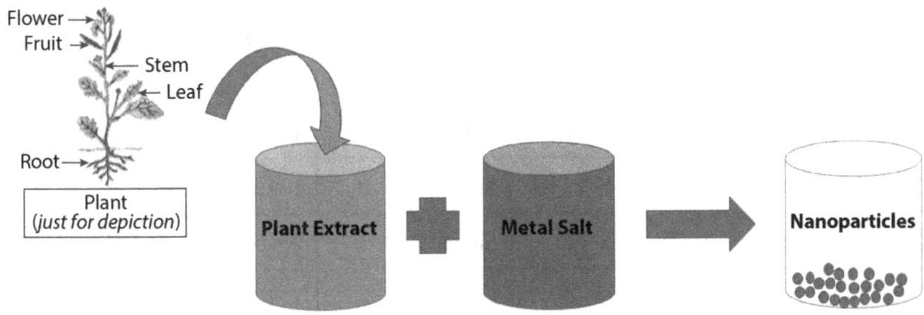

Figure 15.1 A schematic for the synthesis of nanoparticles using plant extract in the biosynthesis method.

environment for 24 to 48 hours. A change in color denotes the formation of nanoparticles [21, 42–44]. Figure 15.1 depicts a schematic of the synthesis of NPs using plant extract.

Table 15.1 lists works from the recent literature in which NPs were synthesized using plant extract. The table clearly reflects that leaf, flower, bark, roots, fruit, and seed extracts are utilized to synthesize metal and metal oxide NPs.

Table 15.1 Plant species and their different part extracts for the preparation of NPs. It contains the particle size of NPs.

Sr. no.	Plant name	Extract	Nanoparticle	Particle size	References
1	*Argemone mexicana*	Leaf	Silver	30	[45]
2	*Acalypha indica*	Leaf	Silver	20–30	[46]
3	*Cassia fistula*	Leaf	Silver	50–60	[47]
4	*Eucalyptus camaldulensis*	Leaf	Gold	5.5–7.5	[48]
5	*Diospyros kaki*	Leaf	Platinum	2–12	[49]
6	*Aloe vera*	Leaf	Silver and Gold	15.2	[50]
7	*Emblica officinalis*	Fruit	Silver and Gold	10–20 and 15–25	[19]
8	*Carica papaya*	Fruit	Silver	15	[51]
9	*Musa paradisica*	Fruit	Palladium	50	[52]
10	*Nyctanthes arbortristis*	Flower	Gold	19.8	[53]
11	*Cinnamon zeylanicum*	Bark	Silver	31–40	[54]
12	*Pinus resinosa*	Bark	Palladium	16–20	[55]
13	*Jatropha curcas*	Seed	Silver	15–50	[56]
14	*Syzygiumcumini*	Seed	Silver	73–92	[57]
15	*Cuminum cyminum*	Seed	Gold	1–10	[58]

15.3 Characterization of Nanoparticles

Prior to using nanoparticles, it is crucial to characterize their physico-chemical properties. Investigating the properties of NPs, including their size, shape, surface area, homogeneity, and stability, can help us understand nanoscale systems and manage their production for commercial uses. Characterization methods include the color change test, powder X-ray diffraction (XRD), Fourier transformation infrared spectroscopy (FTIR), ultraviolet-visible (UV-Vis) spectrometry, field emission scanning electron microscopy (FE-SEM), high-resolution transmission electron microscopy (HR-TEM), and energy-dispersive X-ray spectroscopy mapping (EDX).

15.3.1 X-Ray Diffraction (XRD)

Powder X-ray diffraction (XRD) is a commonly used tool for nanoscale material to investigate its crystal state and crystalline size. Powder X-ray diffraction analysis, when combined with other microscopic and spectroscopic techniques, can provide useful details about a sample, such as its phase, purity, crystallite size, and even morphology in some situations. When combined with microscopy data, this bulk approach can determine whether or not microscopic observations of a select number of particles are typical of the entire sample. Powder XRD data for nanoscale materials are widely used, but the information it contains is not always properly utilized and is sometimes misconstrued. Focusing on inorganic nanoparticles of varying sizes, shapes, and dimensions, this article discusses the essential features of powder XRD data that are commonly encountered during the analysis of samples of nanoscale materials [59, 60].

When going from a bulk size to a nanoscale size, the XRD peaks will expand. The Scherrer equation $D=\dfrac{k\lambda}{\beta\cos\theta}$ connects the crystalline domain size (D) to the breadth of the peak at half its height (β), providing a quantitative description of the widening of a peak at a specific diffraction angle (θ), although it is generally accepted that the Scherrer constant is 0.91, and it varies with crystalline domains. The wavelength of an X-ray (λ) is a fixed quantity that is determined by the X-ray source. If the sample can be thought of as a collection of relatively uniform, spherical particles, then evaluating each peak separately should result in consistent crystalline domain size. The Scherrer equation uses radians (not degrees) for the diffraction angle; hence, the normal XRD plot of 2θ is incorrect. It is important to remember that particles can be polycrystalline, meaning they include numerous crystallographic domains, and that domain size is not always proportional to particle size. When the average diameter of particles measured by TEM or other particle sizing methods is found to be the same as the crystalline domain size estimated using the Scherrer equation, this is indicative of the presence of single crystals rather than polycrystalline particles [61–65].

15.3.2 UV-Visible Absorption Spectroscopy

UV-Vis spectroscopy provides information about the interactions of ultraviolet (UV) or visible electromagnetic radiation with the NPs. UV-Vis spectroscopy provides accurate information on the size and band gap information about NPs. The UV-Vis spectra can be obtained in between 200- and 1100-nm range, based on the operational wavelengths of conventional commercial spectrophotometers. Conventional UV-Vis spectrometers had a limited wavelength limit of 180 nm because atmospheric gasses absorbed all UV light below that. The limit is increased to 175 nm when a spectrometer is purged with nitrogen gas. However, if the required data are beyond 175 nm, then a vacuum chamber and adequate UV light source are required in the spectrometer. In most cases, the long-wavelength limit was calculated by analyzing the spectrometer's detector's wavelength response. Modern, high-end UV-Vis spectrophotometers on the market allow detection in the near-infrared (NIR) spectrum up to 3300 nm [66, 67].

A tungsten-halogen lamp was utilized for visible and near-infrared measurements, while a deuterium discharge lamp was employed for UV observations. The gadgets are designed to automatically switch on and off lights while scanning between visible and UV zones. A holographic grating would typically scatter the wavelengths of these continuous light sources in a spectrograph or monochromator, whether it was a single or double unit. The width of the array-detector spectrometer's elements or the monochromator's slit was then used to determine the spectral bandpass. One of the limitations of quantitative absorbance measurements was stray light rejection, which was addressed in spectrometer designs and optical components. The detector in instruments with a single detector was a photomultiplier tube (PMT), phototube, or photodiode. A PMT and an infrared detector made of lead sulfide (PbS) that is cooled by a Peltier are used in UV-Vis-NIR spectrometers. The light beam was mechanically redirected to the correct detector whenever the scanning range was changed from visible to near-infrared.

15.3.3 FE-Scanning Electron Microscopy (FESEM)

An electron microscope, as opposed to a light microscope, is called a FESEM. The electrons in FESEM are generated using a field emission source. A zigzag pattern of electrons scans NPS. Surface topography characteristics of complete or fractioned objects may be seen using FESEM, a method. Structures as small as 1 nm may be studied using this technique by physicists, biologists, and chemists.

After leaving a FE source, electrons accelerate inside a large electrical field gradient. The arrangement is kept in a high-vacuum column, in which with the help of electronic lenses source of electrons focuses and deflects in order to generate a narrow scan beam that attacks the target. Each part of the thing releases secondary electrons. Secondary electron angle and velocity depend on the object's structure on the surface. A detector detects secondary electrons to form an electrical signal. The resulting signal is amplified and turned into a video scan image for display or a digital image for storage and processing [68, 69].

The FESEM uses a high-energy focused beam of electrons to generate signals on the surface of NPs. The information contains the chemical composition, shape, and orientation of NPs. It also may be uncovered through the signals produced by electron sample interactions. A two-dimensional image is obtained by collecting data throughout a specified area of the sample's surface. A variety of scanning modes with different magnifications along with conventional FESEM techniques are used to examine regions with widths ranging from about 1 cm to 5 micrometer. Research on individual points on the material is also possible with the FESEM; this technique is useful to get information on the chemical composition present, crystalline structure, shape, and orientations (by electron backscatter diffraction) of nanoparticles [70, 71].

15.3.4 EDS

EDS is a powerful yet easy method for identifying elements and chemical compounds in a material. To use EDS, an electron microscope is used to bombard a sample with high-energy electrons. Then, the specific X-rays produced by each element are detected. The elemental makeup of a sample may be superimposed on an expanded image using X-ray mapping.

During EDAX examination, the specimen is subjected to an intense burst of electrons within the SEM. How much energy an electron releases during a transfer depends on both the shell it is heading toward and the shell it is transferring to. In addition, different elements release X-rays with different energies as they undergo the transfer process. To find out what kind of atom released the X-ray, scientists measured how much energy was in the X-rays produced by the specimen when they bombarded it with electron beams. Energy levels that were exposed to the greatest number of X-rays tend to be the ones that show up as peaks in an EDAX spectrum. Every one of these peaks corresponds to a different element since they are all atom-specific. The size of the spectral peak depends on the amount of the element present in the specimen [72, 73].

Each peak in an EDAX spectrum map corresponds to one element, and the graphic also shows the X-ray type it belongs to. A K-β peak is the one that corresponds to the X-rays produced when electrons from the M-shell transition to the K-shell. Electron probe microanalysis and EDAX have the same fundamental idea. Atoms in a specimen are excited when an electron beam hits them and when the electrons come back to the ground state, leading to the emission of X-rays. Detecting these X-rays is the first step in elemental analysis using an electron microscope as their energy is directly related to the number of excited elements' atomic numbers [74, 75].

15.3.5 HR-TEM

High-resolution transmission electron microscopy involves subjecting NPs to a focussed beam of electrons in a highly vacuumed environment in order to examine the thin solid or diluted material [76]. Images and information on the material's crystallography and morphology may be obtained using the focussed electron beam in transmission electron microscopy (TEM). In theory, topography data in the nanoscale range may be adequately provided by TEM due to its great resolution [77, 78]. Due to material contrasts, TEM can view the sample surface and inside, such as layer organization, in addition to its high resolution. A filament of tungsten at the top of a 2-m cylindrical column emits a short-wavelength electron beam that illuminates TEM. A vacuum surrounds the microscope's optical system. A vacuum must be created in the column to prevent electrons from colliding with air molecules and scattering. Magnetic coils are spaced along the column. These coils of magnets in the electron microscope concentrate the electron beam like the glass lenses in the optical microscope. Magnetic coils spaced at regular intervals across the column create an electromagnetic condenser lens. Stained with an electron-dense substance, the specimen is vacuumed. The interior structures scatter electron beams as they move through the material.

Specialist transmission electron microscopes may directly image the NP atomic structure using HR-TEM. It is useful for researching semiconductors, metals, and nanoparticles at the atomic level. High-resolution TEM is possible with a large-diameter objective diaphragm which allows both transmitted and diffracted beams. The image-forming process recombines every beam that passes via the objective aperture, maintaining their amplitudes and phases. High-magnification images show periodic contrast fringes. Phase difference occurs in these fringes and Bragg diffracting planes. The high-resolution image shows fringes from Bragg reflecting planes with interplanar spacings greater than the device's lateral spatial resolution [79–82].

Using selected area electron diffraction (SAED), one may learn the crystal structure of certain NPs or even different parts of a sample. A high-resolution transmission image and SAED pattern on the screen of the microscope may be achieved by arranging TEM lenses properly, allowing a small portion of the material to be selected. One useful tool for species identification is the SAED pattern, which allows one to determine the lattice constant of the crystalline material. Spot patterns resulting from ring patterns are mainly recognized from one another in diffraction patterns due to the fact that polycrystallites, which are aggregations of randomly oriented crystals, form these patterns. Nanocrystallites will exhibit diffraction patterns that resemble diffused ring patterns. For materials with a polycrystalline or nanocrystalline structure, the electron diffraction pattern may be described by the following equation [83].

$$Rd = \lambda L$$

where, R, L, λ, and d are the radius of the diffraction ring, length of camera, wavelength, and the interplanar spacing, respectively.

15.3.6 FT-IR Spectroscopy

An IR spectrum is a sample's fingerprint. The absorption peaks in the spectrum identify the frequencies at which the surface and particle-based chemical bonds vibrate. As it is aware that material has a distinguished atom combination, different infrared spectrum is observed for different compounds. Therefore, infrared spectroscopy is used for qualitative analysis of any type of substance.

Another way to get the same data is via Fourier spectroscopy. This technique assesses the sample's absorption rate by simultaneously illuminating a beam with several light frequencies instead of flashing a monochromatic beam into it. Then, to get another data point, the beam is changed to cover a different range of frequencies. This phenomenon occurs at a high frequency. Subsequently, the computer processes the collected data, enabling the execution of computations to determine the absorption levels corresponding to each individual wavelength. FTIR is a highly efficient analytical method that enables rapid and precise determination of the chemical composition and classification of a material based on its chemical family. Typically, the fingerprint infrared spectrum is utilized to generate organic and polymeric compounds, enabling a comparison with the EAG complete reference database. This process facilitates the identification of the chemical family or real identity of the unknown component. The infrared spectrum of a fingerprint may be analyzed and compared to the complete reference database of the EAG. By comparing the two, we can determine the unknown component's true identity or the chemical family to which it belongs. A spectrum that depends on the functional groups in the material is produced by using a Fourier infrared spectrometer to measure the sample's infrared radiation absorption. Direct analysis of intractable or multilayer materials is made possible by EAG's employment of multiple attachments total reflection (ATR). To determine the absorption of the sample at each wavelength, an FTIR absorption spectrometer is used. Scatter spectroscopy is the most straightforward method, and it involves shining a single color of light onto a material, measuring how much light is absorbed, and doing this for each wavelength. Spectroscopy like this, which looks at electronic transitions between the ground and excited states, is as useful as fluorescent spectroscopy, which looks at the fluorescence that results from the opposite event. Under UV-Vis spectroscopy, NPs are subjected to electromagnetic radiation, which excites valence electrons, allowing them to gain energy and participate in the movement of electrons within the energy levels of NPs. The energy required for this change must be proportional to the magnitude of the energy gap [84–86].

15.3.7 Application of Biosynthesized Nanoparticles for Bio-Packaging

The food industry, along with technical and medicinal applications, made extensive use of nanostructured materials. Hence, it is imperative to develop economical and ecologically friendly methodologies for the rapid production of NPs. The utilization of NPs in food packaging has led to the incorporation of nanoparticles and nanocomposites, which provide improved, dynamic, intelligent, and environmentally friendly packaging solutions [34, 87–89]. Nanotechnology enables the utilization of a wide range of nanomaterials such as nanofillers, nanometals, and nanosensors, as well as in modern packaging. In addition to extending the food's shelf life and preserving its quality, this packaging also shields its contents from infection. Among the many uses for nanomaterials, metals and metal oxides (Au, CaO, Ag_2O, CuO, Ag, Au-Ag, ZnO, TiO_2, and MgO_2) are widely studied for bio-packaging applications [90–97]. The particles' characteristics at the nanoscale scale are distinct from those at the bulk amount level. Metal/metal oxide NPs provide a range of distinctive characteristics, including catalytic qualities, UV light absorption ability, physicochemical features, exceptional flexibility under elevated temperatures, as well as notable mechanical and biological properties. The primary concern associated with methodologies is to the use of hazardous compounds, such as protective agents employed for stability preservation. This reliance on toxic substances restricts their compatibility with biological systems and gives rise to several environmental challenges. Therefore, in recent times, there has been a significant focus from both researchers and producers "biosynthesis" process of metal/metal oxide NPs. These NPs have a wide range of applications, including antibacterial properties, target recognition capabilities, biomedical uses, biosensor development, and applications in food packaging [41, 98–103].

Here, we take a look at the latest research done on the topic, so that we may have a better grasp on how green-produced metal/metal oxide nanoparticles can be included in food packaging. Using particular packaging materials and procedures to minimize food toxicity and losses and guarantee the delivery of safe, nutritious food has always been the major emphasis of the food packaging industry. Because nanocomposite packaging materials may have their properties tailored to specific needs, this technology has piqued the interest of many in the food packaging industry.

Kumar *et al.* [104] created a green-manufactured zinc oxide (ZnO) NP nanocomposite film based on chitosan and gelatin. Cassia fistula fruit extract served as the reducing and capping materials for the preparation of green ZnO NPs (in the 20- to 40-nm size range). Results demonstrated that nanocomposite films doped with biosynthesized ZnO NPs have enhanced thermal and chemical stability, electron-hole pair (EB), and compactness. Two distinct clay-based packaging were tested for the amount of aluminum (Al) they leaked into food upon opening (the migration value was 51.65 and 24.14 ng/cm2, respectively) [105]. In addition, it was shown that clay nanoparticle migration was related to both temperature and contact duration [106]. Insoluble NPs made of metal oxides like ZnO, TiO2, and SiO2 activate blood mononuclear cells and phagocytes upon absorption of calcium and lipopolysaccharides in the gastrointestinal system [107]. Finally, it can be concluded that genotoxicity is caused by both direct and indirect exposure to NPs of metals and their oxides such as Co/Cr alloy, Fe/Pt alloy, SiO2, ZnO, and TiO2 [108]. Consequently, the toxicity of nanomaterials and their final destiny in food packaging may be controlled by their physicochemical qualities and dosage [108].

Marrez *et al.* [109] have produced a nanocomposite film made of cellulose acetate and biosynthesized manufactured silver NPs for use as antibacterial food packaging, which can kill some types of foodborne harmful bacteria without harming the environment. Biosynthesized silver nanoparticles (Ag NPs) were created by synthesizing silver with polyphenols present in extracts. The nanocomposite film exhibited potent antibacterial action, and all of the produced Ag NPs were effective against the tested pathogens. Using the nanocomposite film as an active food packaging system is more likely after the scientists claimed it had strong antibacterial activity without any harmful side effects. The researchers, Basumatary *et al.* [110], fabricated a nanocomposite film using agar as the matrix material and incorporating Ag NPs as the filler. Lagerstroemia speciosa, a common medicinal plant in northeastern India, has its aqueous fruit extract utilized to create Ag NPs. The findings of the study show that the inclusion of Ag NPs had a positive impact on the elongation at break (EB), antibacterial characteristics, thermal stability, and appearance of the nanocomposite films. However, the tensile strength (TS) was shown to decrease as a result. Furthermore, the orange-brown coloration shown by the composite sheets serves as a protective barrier against UV radiation, so safeguarding the packed food. TiO2 nanoparticles photocatalyze, especially under UV light with water and oxygen. These circumstances create ROS, which produces hydroxyl- and superoxide-free radicals. These radicals promoted a reaction with cell membrane phospholipids, damaging DNA as well as cell membrane integrity. A thorough TNP response mechanism has yet to be found. Due to TiO2 nanoparticles, TiO2 and chitosan nanocomposites inhibited more than pure chitosan films. TiO2 nanoparticles block UV better than white light or dark storage. TiO2 nanocomposites show more Gram-positive antibacterial activity than Gram-negative [111, 112]. As indicated before, lower pH boosts chitosan-based nanocomposites' microorganism suppression. Nanomaterials such as ZnO, TiO2, and silver (Ag) work particularly well when combined with chitosan to improve the films' antibacterial properties. However, as was previously indicated, these compounds lack sufficient data on their cytotoxicity to be used in the food business without more research.

15.4 Concluding Remarks

The chapter suggests that the biosynthesized nanoparticles using plant extracts are easy to make as well as economical. Furthermore, during the synthesis, there is no requirement for expensive and toxic chemicals, in fact, the compound present in the extracts acts as reducing and capping agents. These materials are mostly human-friendly and biodegradable. Thus, these nanoparticles are very useful in food processing, food packaging, cosmetics, and medical industries. The chapter provides information about the easy approach to synthesizing the nanoparticles; also, useful techniques generally are used for characterizing them. Furthermore, the summary from the table suggests that any part of the plant can be used in the synthesis process. The nanoparticles have potential applications in food packaging due to their less toxicity in nature but more research is required. This suggests that there is a huge scope of research for researchers in this field.

References

1. Joudeh, N. and Linke, D., Nanoparticle classification, physicochemical properties, characterization, and applications: a comprehensive review for biologists. *J. Nanobiotechnol.*, 20, 1–29, 2022, https://doi.org/10.1186/s12951-022-01477-8.
2. Chaware, P., Nande, A., Dhoble, S.J., Rewatkar, K.G., Structural, photoluminescence and Judd-Ofelt analysis of red-emitting Eu^{3+} doped strontium hexa-aluminate nanophosphors for lighting application. *Opt Mater (Amst)*, 121, 111542, 2021, https://doi.org/10.1016/j.optmat.2021.111542.
3. Roduner, E., Size matters: Why nanomaterials are different. *Chem. Soc. Rev.*, 35, 583–592, 2006, https://doi.org/10.1039/b502142c.
4. Lines, M.G., Nanomaterials for practical functional uses. *J. Alloys Compd.*, 449, 242–245, 2008, https://doi.org/10.1016/j.jallcom.2006.02.082.
5. Nande, A., Fostner, S., Grigg, J., Smith, A., Temst, K., Bael, M.J.V., Brown, S.A., Quantum fluctuations in percolating superconductors: An evolution with effective dimensionality. *Nanotechnology*, 28, 2017, https://doi.org/10.1088/1361-6528/aa5e88.
6. Segal, D., Chemical synthesis of ceramic materials. *J. Mater Chem.*, 7, 1297–1305, 1997, https://doi.org/10.1039/a700881c.
7. Srivastava, R., Synthesis and characterization techniques of nanomaterials. *Int. J. Green Nanotechnol Biomed.*, 4, 17–27, 2012, https://doi.org/10.1080/19430892.2012.654738.

8. Lakshmanan, A., The Role of Sintering in the Synthesis of Luminescence Phosphors, in: *Sintering of Ceramics - New Emerging Techniques*, 2012, https://doi.org/10.5772/33208.
9. Prasad Yadav, T., Manohar Yadav, R., Pratap Singh, D., Mechanical Milling: a Top Down Approach for the Synthesis of Nanomaterials and Nanocomposites. *Nanosci Nanotechnol.*, 2, 22–48, 2012, https://doi.org/10.5923/j.nn.20120203.01.
10. Prof, B., Singh, B.P., Drexler, K.E., TOP-DOWN AND BOTTOM-UP APPROACHES FOR SYNTHESIS OF NANOMATERIALS, in: *Introduction : Key issues in the fabrication of Nanomaterials*, 1986.
11. Fostner, S., Nande, A., Smith, A., Martinez Gazoni, R., Grigg, J., Temst, K., Van Bael, M.J., Brown, S.A., Percolating transport in superconducting nanoparticle films. *J. Appl. Phys.*, 122, 2017, https://doi.org/10.1063/1.5006694.
12. Li, X., Xu, H., Chen, Z.S., Chen, G., Biosynthesis of nanoparticles by microorganisms and their applications. *J. Nanomater.*, 2011, 2011, https://doi.org/10.1155/2011/270974.
13. Sharma, D., Kanchi, S., Bisetty, K., Biogenic synthesis of nanoparticles: A review, Arabian Journal of Chemistry. 12, 3576–3600, 2019, https://doi.org/10.1016/j.arabjc.2015.11.002.
14. Alayli, A., Nadaroğlu, H., Alayli Güngör, A., İnce, S., Synthesis of Nanoparticles by Green Synthesis Method. *Int. J. Innovative Res. Rev.*, 1, 6–9, 2017, http://www.injirr.com/article/view/4.
15. Purohit, J., Chattopadhyay, A., Singh, N.K., Green Synthesis of Microbial Nanoparticle: Approaches to Application. *Nanotechnol. Life Sci.*, 35–60, 2019, https://doi.org/10.1007/978-3-030-16534-5_3.
16. Salem, S.S. and Fouda, A., Green Synthesis of Metallic Nanoparticles and Their Prospective Biotechnological Applications: an Overview. *Biol. Trace Elem Res.*, 199, 344–370, 2021, https://doi.org/10.1007/s12011-020-02138-3.
17. Shah, S., Shah, S.A., Faisal, S., Khan, A., Ullah, R., Ali, N., Bilal, M., Engineering novel gold nanoparticles using Sageretiathea leaf extract and evaluation of their biological activities. *J. Nanostruct. Chem.*, 12, 129–140, 2022, https://doi.org/10.1007/s40097-021-00407-8.
18. Gudimalla, A., Jose, J., Varghese, R.J., Thomas, S., Green Synthesis of Silver Nanoparticles Using Nymphae odorata Extract Incorporated Films and Antimicrobial Activity. *J. Polym Environ.*, 29, 1412–1423, 2021. https://doi.org/10.1007/s10924-020-01959-6.
19. Ankamwar, B., Damle, C., Ahmad, A., Sastry, M., Biosynthesis of gold and silver nanoparticles using Emblica Officinalis fruit extract, their phase transfer and transmetallation in an organic solution. *J. Nanosci. Nanotechnol.*, 5, 1665–1671, 2005, https://doi.org/10.1166/jnn.2005.184.
20. Tan, H.L., Lim, Y.C., Ng, L.Y., Lim, Y.P., Plant-mediated synthesis of iron nanoparticles for environmental application: Mini review. *Mater Today Proc.*, 87, 64–69, 2023, https://doi.org/10.1016/J.MATPR.2023.02.101.

21. Saif, S., Tahir, A., Chen, Y., Green synthesis of iron nanoparticles and their environmental applications and implications. *Nanomaterials*, 6, 1–26, 2016, https://doi.org/10.3390/nano6110209.
22. Sengul, A.B. and Asmatulu, E., Toxicity of metal and metal oxide nanoparticles: a review. *Environ. Chem. Lett.*, 18, 1659–1683, 2020, https://doi.org/10.1007/s10311-020-01033-6.
23. Murthy, S., Effiong, P., Fei, C.C., 11 - Metal oxide nanoparticles in biomedical applications, in: *Metal Oxide Powder Technologies*, Y. Al-Douri (Ed.), pp. 233–251, Elsevier, 2020, https://doi.org/https://doi.org/10.1016/B978-0-12-817505-7.00011-7.
24. Prasad, S., Kumar, V., Kirubanandam, S., Barhoum, A., *Engineered nanomaterials: Nanofabrication and surface functionalization*, Elsevier Inc, 2018, https://doi.org/10.1016/B978-0-323-51254-1.00011-7.
25. Tahir, M.B., Iqbal, T., Rafique, M., Rafique, M.S., Nawaz, T., Sagir, M., *Nanomaterials for photocatalysis*, Elsevier Inc, 2020, https://doi.org/10.1016/b978-0-12-821192-2.00005-x.
26. Hadef, F., An Introduction to Nanomaterials, in: *Environmental Nanotechnology*, vol. 1, N. Dasgupta, S. Ranjan, E. Lichtfouse (Eds.), pp. 1–58, Springer International Publishing, Cham, 2018, https://doi.org/10.1007/978-3-319-76090-2_1.
27. Fu, X., Cai, J., Zhang, X., Di Li, W., Ge, H., Hu, Y., Top-down fabrication of shape-controlled, monodisperse nanoparticles for biomedical applications. *Adv. Drug Delivery Rev.*, 132, 169–187, 2018, https://doi.org/10.1016/j.addr.2018.07.006.
28. Stojanovic, B.D., Dzunuzovic, A.S., Ilic, N.I., Review of methods for the preparation of magnetic metal oxides, in: *Magnetic, Ferroelectric, and Multiferroic Metal Oxides*, pp. 333–359, 2018, https://doi.org/10.1016/B978-0-12-811180-2.00017-7.
29. Navaf, M., Sunooj, K.V., Aaliya, B., Akhila, P.P., Sudheesh, C., Mir, S.A., George, J., Impact of metal and metal oxide nanoparticles on functional and antimicrobial activity of starch nanocomposite film; A review. *Meas. Food*, 11, 100099, 2023, https://doi.org/10.1016/j.meafoo.2023.100099.
30. Manivasagan, P. and Kim, S.K., Biosynthesis of Nanoparticles Using Marine Algae: A Review. *Mar. Algae Extracts Process., Prod., Appl.*, 1–2, 295–304, 2015, https://doi.org/10.1002/9783527679577.ch17.
31. Hernández-Muñoz, P., Cerisuelo, J.P., Domínguez, I., López-Carballo, G., Catalá, R., Gavara, R., Nanotechnology in Food Packaging, Nanomaterials for Food Applications. 205–232, 2018, https://doi.org/10.1016/B978-0-12-814130-4.00008-7.
32. Duncan, T.V., Applications of nanotechnology in food packaging and food safety: Barrier materials, antimicrobials and sensors. *J. Colloid Interface Sci.*, 363, 1–24, 2011, https://doi.org/10.1016/j.jcis.2011.07.017.
33. Primožič, M., Knez, Ž., Leitgeb, M., (Bio)nanotechnology in food science—food packaging. *Nanomaterials*, 11, 1–31, 2021, https://doi.org/10.3390/NANO11020292.

34. Mahmud, J., Sarmast, E., Shankar, S., Lacroix, M., Advantages of nanotechnology developments in active food packaging. *Food Res. Int.*, 154, 2022. https://doi.org/10.1016/j.foodres.2022.111023.
35. Iravani, S., Green synthesis of metal nanoparticles using plants. *Green Chem.*, 13, 2638–2650, 2011, https://doi.org/10.1039/C1GC15386B.
36. Ghaffari-Moghaddam, M., Hadi-Dabanlou, R., Khajeh, M., Rakhshanipour, M., Shameli, K., Green synthesis of silver nanoparticles using plant extracts. *Korean J. Chem. Eng.*, 31, 548–557, 2014, https://doi.org/10.1007/S11814-014-0014-6.
37. N. Chaudhry *et al.*, Bio-inspired nanomaterials in agriculture and food: Current status, foreseen applications and challenges, *Microb. Pathogen.*, 123, 196–200, 2018/10/01/ 2018, doi: https://doi.org/10.1016/j.micpath.2018.07.013
38. Rao, K.E., Devi, D.R., Varaprasad, D., Divya, A., Aparna, N., Babu, M.R., Samsonu, D., Basavaiah, K., Green Synthesis of Palladium (Pd) Nanomaterials using Plant Extracts for Catalysis and Biological Applications. 15, 30–36, 2022, https://doi.org/10.9790/5736-1506023036.
39. Singh, A., Jain, D., Upadhyay, M.K., Khandelwal, N., Verma, H.N., Green synthesis of silver nanoparticles using Argemone Mexicana leaf extract and evaluation of their antimicrobial activities. *Dig J. Nanomater Biostruct*, 5, 483–489, 2010.
40. Sharma, K., Guleria, S., Salaria, K.H., Majeed, A., Sharma, N., Pawar, K.D., Thakur, V.K., Gupta, V.K., Photocatalytic and biological properties of silver nanoparticles synthesized using Callistemon lanceolatus leaf extract. *Ind. Crops Prod.*, 202, 2023, https://doi.org/10.1016/j.indcrop.2023.116951.
41. Parveen, K., Banse, V., Ledwani, L., Green synthesis of nanoparticles: Their advantages and disadvantages. *AIP Conf Proc.*, 17242016, https://doi.org/10.1063/1.4945168.
42. Yallappa, S., Manjanna, J., Dhananjaya, B.L., Phytosynthesis of stable Au, Ag and Au–Ag alloy nanoparticles using J. Sambac leaves extract, and their enhanced antimicrobial activity in presence of organic antimicrobials. *Spectrochim. Acta, Part A*, 137, 236–243, 2015, https://doi.org/10.1016/J.SAA.2014.08.030.
43. Poonkodi, B., Suguna Lakshmi, M., Tamilselvi, A., Sumathi Jones, C., Deepa, K., Pattabi, S., Bava Bakrudeen, H., Prabhu, K., Kim, S.C., Ranjith, M.S., Bionanocomposite films loaded with lemon leaf extract for bio packaging application. *J. King Saud Univ Sci.*, 34, 102333, 2022, https://doi.org/10.1016/j.jksus.2022.102333.
44. Das, M. and Smita, S.S., Biosynthesis of silver nanoparticles using bark extracts of Butea monosperma (Lam.) Taub. and study of their antimicrobial activity. *Appl. Nanosc. (Switzerland)*, 8, 1059–1067, 2018, https://doi.org/10.1007/s13204-018-0721-0.
45. D.G. Téllez-de-Jesús, N. Flores-Lopez, J. Cervantes-Chávez, and A. Hernández-Martínez, Antibacterial and antifungal activities of encapsulated

Au and Ag nanoparticles synthesized using Argemone mexicana L extract, against antibiotic-resistant bacteria and Candida albicans, *Surf. Interfaces*, 27, 101456, 2021
46. Krishnaraj, C., Jagan, E.G., Rajasekar, S., Selvakumar, P., Kalaichelvan, P.T., Mohan, N., Synthesis of silver nanoparticles using Acalypha indica leaf extracts and its antibacterial activity against water borne pathogens. *Colloids Surf. B.*, 76, 50–56, 2010, https://doi.org/10.1016/J.COLSURFB.2009.10.008.
47. Lin, L., Wang, W., Huang, J., Li, Q., Sun, D., Yang, X., Wang, H., He, N., Wang, Y., Nature factory of silver nanowires: Plant-mediated synthesis using broth of Cassia fistula leaf. *Chem. Eng. J.*, 162, 852–858, 2010, https://doi.org/10.1016/J.CEJ.2010.06.023.
48. Lintern, M., Anand, R., Ryan, C., Paterson, D., Natural gold particles in Eucalyptus leaves and their relevance to exploration for buried gold deposits. *Nat. Commun.*, 4, 1–9, 2013, https://doi.org/10.1038/ncomms3614.
49. Song, J.Y., Kwon, E.-Y., Kim, B.S., Biological synthesis of platinum nanoparticles using Diopyros kaki leaf extract. *Bioprocess Biosyst. Eng.*, 33, 159–164, 2010, https://doi.org/10.1007/s00449-009-0373-2.
50. Samrot, A.V., Saigeetha, S., Shobana, N., Chandrasekaran, K., Green-synthesized nanoparticle-based polymer nanocomposites: synthesis, characterizations, and applications. *Biodegradable Biocompatible Polym. Nanocomposites*, 241–270, 2023, https://doi.org/10.1016/B978-0-323-91696-7.00012-X.
51. El Mouzahim, M., Eddarai, E.M., Eladaoui, S., Guenbour, A., Bellaouchou, A., Zarrouk, A., Boussen, R., Effect of Kaolin clay and Ficus carica mediated silver nanoparticles on chitosan food packaging film for fresh apple slice preservation. *Food Chem.*, 410, 2023, https://doi.org/10.1016/j.foodchem.2023.135470.
52. Bankar, A., Joshi, B., Ravi Kumar, A., Zinjarde, S., Banana peel extract mediated synthesis of gold nanoparticles. *Colloids Surf B Biointerfaces*, 80, 45–50, 2010, https://doi.org/10.1016/j.colsurfb.2010.05.029.
53. Das, R.K., Gogoi, N., Bora, U., Green synthesis of gold nanoparticles using Nyctanthesarbortristis flower extract. *Bioprocess Biosyst. Eng.*, 34, 615–619, 2011, https://doi.org/10.1007/s00449-010-0510-y.
54. Sathishkumar, M., Sneha, K., Won, S.W., Cho, C.W., Kim, S., Yun, Y.S., Cinnamon zeylanicum bark extract and powder mediated green synthesis of nano-crystalline silver particles and its bactericidal activity. *Colloids Surf. B.*, 73, 332–338, 2009, https://doi.org/10.1016/J.COLSURFB.2009.06.005.
55. Shenton, W., Douglas, T., Young, M., Stubbs, G., Mann, S., Inorganic–Organic Nanotube Composites from Template Mineralization of Tobacco Mosaic Virus. *Adv. Mater.*, 11, 253–256, 1999, https://doi.org/10.1002/(sici)1521-4095(199903)11:3<253::aid-adma253>3.3.co;2-z.
56. Bar, H., Bhui, D.K., Sahoo, G.P., Sarkar, P., Pyne, S., Misra, A., Green synthesis of silver nanoparticles using seed extract of Jatropha curcas. *Colloids Surf., A*, 348, 212–216, 2009, https://doi.org/10.1016/J.COLSURFA.2009.07.021.
57. Kumar, V., Yadav, S.C., Yadav, S.K., Syzygiumcumini leaf and seed extract mediated biosynthesis of silver nanoparticles and their characterization. *J.*

Chem. Technol. Biotechnol., 85, 1301–1309, 2010, https://doi.org/10.1002/jctb.2427.

58. Gohari, A.R. and Saeidnia, S., A Review on Phytochemistry of Cuminum cyminum seeds and its Standards from Field to Market. *Pharmacogn. J.*, 3, 1–5, 2011, https://doi.org/10.5530/PJ.2011.25.1.
59. Gupta, S.K. and Gartley, M.G., *Denver X-ray Conference (DXC) on Applications of X-ray Analysis*, vol. 48, 1999, Sponsored by the International Centre for Diffraction Data (ICDD). This document is provided by ICDD in cooperation with the authors and presenters of.
60. Gawas, U.B., Mandrekar, V.K., Majik, M.S., Structural analysis of proteins using X-ray diffraction technique. *Adv. Biol. Sci. Res. A Pract. Approach*, 69–84, 2019, https://doi.org/10.1016/B978-0-12-817497-5.00005-7.
61. Raja, P.B., Munusamy, K.R., Perumal, V., Ibrahim, M.N.M., Characterization of nanomaterial used in nanobioremediation. *Nano-Biorem. Fundam. Appl.*, 57–83, 2022, https://doi.org/10.1016/B978-0-12-823962-9.00037-4.
62. Kauffman, G.B., B.B., He, *Two-dimensional X-ray diffraction, Found Chem*, vol. 14, pp. 187–188, 2012, https://doi.org/10.1007/s10698-011-9135-8.
63. Bykkam, S., Ahmadipour, M., Narisngam, S., Kalagadda, V.R., Chidurala, S.C., RETRACTED: Extensive Studies on X-Ray Diffraction of Green Synthesized Silver Nanoparticles. *Adv. Nanopart.*, 04, 1–10, 2015, https://doi.org/10.4236/anp.2015.41001.
64. Chaware, P. and Rewatkar, K.G., Structural and photoluminescence study of $SrAl_2O_4$:Eu^{3+} phosphors synthesized by combustion method. *Int. J. Chem. Math. Phys. (IJCMP)*, 5, 1–6, 2021, https://doi.org/10.22161/ijcmp.5.6.1.
65. Gahane, N.M., Chaware, P.J., Rewatkar, K.G., Structural, morphological, and photoluminescence study of Europium doped spinel ZnAl2O4 phosphors. *Int. J. Chem. Math. Phys.*, 7, 01–06, 2023, https://doi.org/10.22161/ijcmp.7.1.1.
66. Weckhuysen, B.M., CHAPTER 12 Ultraviolet-Visible Spectroscopy, 2014.
67. Passos, M.L.C. and Saraiva, M.L. M.F.S., Detection in UV-visible spectrophotometry: Detectors, detection systems, and detection strategies. *Measurement*, 135, 896–904, 2019, https://doi.org/10.1016/J.MEASUREMENT.2018.12.045.
68. Akhtar, K., Khan, S.A., Khan, S.B., Asiri, A.M., Scanning Electron Microscopy: Principle and Applications in Nanomaterials Characterization, in: *Handbook of Materials Characterization*, S.K. Sharma (Ed.), pp. 113–145, Springer International Publishing, Cham, 2018, https://doi.org/10.1007/978-3-319-92955-2_4.
69. Scanning Electron Microscopy (SEM), in: *Electron Microscopy of Polymers*, pp. 87–120, Springer, Berlin, Heidelberg, 2008, https://doi.org/10.1007/978-3-540-36352-1_5.
70. Han, W., Zheng, M., Banerjee, A., Luo, Y.Z., Shen, L., Khursheed, A., Quantitative material analysis using secondary electron energy spectromicroscopy. *Sci. Rep.*, 10, 1–14, 2020, https://doi.org/10.1038/s41598-020-78973-0.
71. Mohammed, A. and Abdullah, A., Scanning Electron Microscopy (Sem): a Review. *Proceedings of 2018 International Conference on Hydraulics and*

Pneumatics - HERVEX, vol. 1, pp. 77–85, 2018, https://fluidas.ro/hervex/proceedings2018/77-85.pdf.

72. Scimeca, M., Bischetti, S., Lamsira, H.K., Bonfiglio, R., Bonanno, E., Energy dispersive X-ray (EDX) microanalysis: A powerful tool in biomedical research and diagnosis. *Eur. J. Histochem.*, 62, 89–99, 2018, https://doi.org/10.4081/ejh.2018.2841.
73. Chaware, P.J., Choudhari, Y.D., Borikar, D.M., Rewatkar, K.G., Photoluminescence and Judd-Ofelt analysis of Eu3+ doped akermanite silicate phosphors for solid state lighting. *Opt Mater (Amst)*, 133, 2022, https://doi.org/10.1016/j.optmat.2022.112945.
74. Tummala, S.K., Bobba, P.B., Satyanarayana, K., SEM & EDAX analysis of super capacitor. *Adv. Mater. Process. Technol.*, 8, 2398–2409, 2022. https://doi.org/10.1080/2374068X.2022.2040175.
75. Titus, D., James Jebaseelan Samuel, E., Roopan, S.M., *Nanoparticle characterization techniques, Green Synthesis, Characterization and Applications of Nanoparticles*, pp. 303–319, 2019, https://doi.org/10.1016/B978-0-08-102579-6.00012-5.
76. Kulkarni, S.K., *Nanotechnology - Principles and Practices*, 3rd ed, Springer, 2015.
77. Kisielowski, C., Erni, R., Meyer, J., Electron beam induced damage: An atom-by-atom investigation with TEAM0.5, in: *EMC 2008 14th European Microscopy Congress*, Aachen, Germany, 1–5 September 2008 Springer, Berlin, Heidelberg, pp. 39–40.
78. Childres, I., Jauregui, L.A., Foxe, M., Tian, J., Jalilian, R., Jovanovic, I., Chen, Y.P., Effect of electron-beam irradiation on graphene field effect devices. *Appl. Phys. Lett.*, 97, 2010, https://doi.org/10.1063/1.3502610.
79. T. G, R. and B. R, A., Review Article TRANSMISSION ELECTRON MICROSCOPY-AN OVERVIEW. *Int. Res. J. Invent.*, 2321–7855, 2019, ISSN Online.
80. Williams, D.B. and Carter, C.B., The Transmission Electron Microscope, in: *Transmission Electron Microscopy: A Textbook for Materials Science*, pp. 3–17, Springer, US, Boston, MA, 1996, https://doi.org/10.1007/978-1-4757-2519-3_1.
81. Lin, Y., Zhou, M., Tai, X., Li, H., Han, X., Yu, J., Analytical transmission electron microscopy for emerging advanced materials. *Matter*, 4, 2309–2339, 2021, https://doi.org/10.1016/J.MATT.2021.05.005.
82. Inkson, B.J., Scanning electron microscopy (SEM) and transmission electron microscopy (TEM) for materials characterization, in: *Materials Characterization Using Nondestructive Evaluation (NDE) Methods*, pp. 17–43, 2016, https://doi.org/10.1016/B978-0-08-100040-3.00002-X.
83. Flewitt, P.E. and Wild, R.K., *Physical methods for materials characterisation*, CRC Press, 2017.
84. Fadlelmoula, A., Pinho, D., Carvalho, V.H., Catarino, S.O., Minas, G., Fourier Transform Infrared (FTIR) Spectroscopy to Analyse Human Blood over the Last 20 Years: A Review towards Lab-on-a-Chip Devices. *Micromachines (Basel)*, 13, 2022, https://doi.org/10.3390/mi13020187.

85. Eid, M.M., Characterization of Nanoparticles by FTIR and FTIR-Microscopy, in: *Handbook of Consumer Nanoproducts*, pp. 1–30, Springer, Singapore, 2021, https://doi.org/10.1007/978-981-15-6453-6_89-1.
86. Berthomieu, C. and Hienerwadel, R., Fourier transform infrared (FTIR) spectroscopy. *Photosynth Res.*, 101, 157–170, 2009, https://doi.org/10.1007/s11120-009-9439-x.
87. Fortunati, E., Luzi, F., Yang, W., Kenny, J.M., Torre, L., Puglia, D., Bio-based nanocomposites in food packaging. *Nanomater. Food Packag. Mater. Process. Technol. Saf. Issues*, 71–110, 2018, https://doi.org/10.1016/B978-0-323-51271-8.00004-8.
88. Cerqueira, M.A., Vicente, A.A., Pastrana, L.M., Nanotechnology in Food Packaging: Opportunities and Challenges. *Nanomater. Food Packag. Mater. Process. Technol. Saf. Issues*, 1–11, 2018, https://doi.org/10.1016/B978-0-323-51271-8.00001-2.
89. Chausali, N., Saxena, J., Prasad, R., Recent trends in nanotechnology applications of bio-based packaging. *J. Agric. Food Res.*, 7, 100257, 2022, https://doi.org/10.1016/j.jafr.2021.100257.
90. Lin, J., Nguyen, N.Y.T., Zhang, C., Ha, A., Liu, H.H., Antimicrobial Properties of MgO Nanostructures on Magnesium Substrates. *ACS Omega*, 5, 24613–24627, 2020, https://doi.org/10.1021/acsomega.0c03151.
91. Hassan, S.E.-D., Salem, S.S., Fouda, A., Awad, M.A., El-Gamal, M.S., Abdo, A.M., New approach for antimicrobial activity and bio-control of various pathogens by biosynthesized copper nanoparticles using endophytic actinomycetes. *J. Radiat. Res. Appl. Sci.*, 11, 262–270, 2018, https://doi.org/10.1016/j.jrras.2018.05.003.
92. Hameed, A.S.H., Karthikeyan, C., Ahamed, A.P., Thajuddin, N., Alharbi, N.S., Alharbi, S.A., Ravi, G., In vitro antibacterial activity of ZnO and Nd doped ZnO nanoparticles against ESBL producing Escherichia coli and Klebsiella pneumoniae. *Sci. Rep.*, 6, 1–11, 2016, https://doi.org/10.1038/srep24312.
93. Mohamad, N.A.N., Arham, N.A., Junaidah, J., Hadi, A., Idris, S.A., Green Synthesis of Ag, Cu and AgCu Nanoparticles using Palm Leaves Extract as the Reducing and Stabilizing Agents. *IOP Conf Ser. Mater Sci. Eng.*, 358, 2018, https://doi.org/10.1088/1757-899X/358/1/012063.
94. Kumar, S., Basumatary, I.B., Sudhani, H.P.K., Bajpai, V.K., Chen, L., Shukla, S., Mukherjee, A., Plant extract mediated silver nanoparticles and their applications as antimicrobials and in sustainable food packaging: A state-of-the-art review. *Trends Food Sci. Technol.*, 112, 651–666, 2021, https://doi.org/10.1016/j.tifs.2021.04.031.
95. Sneha, K., Sathishkumar, M., Lee, S.Y., Bae, M.A., Yun, Y.S., Biosynthesis of Au nanoparticles using cumin seed powder extract. *J. Nanosci Nanotechnol.*, 11, 1811–1814, 2011, https://doi.org/10.1166/jnn.2011.3414.
96. Urnukhsaikhan, E., Bold, B.E., Gunbileg, A., Sukhbaatar, N., Mishig-Ochir, T., Antibacterial activity and characteristics of silver nanoparticles biosynthesized from Carduus crispus. *Sci. Rep.*, 11, 1–12, 2021, https://doi.org/10.1038/s41598-021-00520-2.

97. Jafarzadeh, S., Salehabadi, A., Jafari, S.M., Metal nanoparticles as antimicrobial agents in food packaging, in: *Handbook of Food Nanotechnology: Applications and Approaches*, pp. 379–414, 2020, https://doi.org/10.1016/B978-0-12-815866-1.00010-8.
98. Jafarzadeh, S., Nooshkam, M., Zargar, M., Garavand, F., Ghosh, S., Hadidi, M., Forough, M., Green synthesis of nanomaterials for smart biopolymer packaging: challenges and outlooks. *J. Nanostruct. Chem.*, 2023, https://doi.org/10.1007/s40097-023-00527-3.
99. Nirmal, R.M. and Sudha, K., Green Synthesis of Silver Nanoparticles Using Plant (Azadirachtaindica) Leaf Extract. 477–478, 2015.
100. Goutam, S.P., Saxena, G., Roy, D., Yadav, A.K., Bharagava, R.N., Green Synthesis of Nanoparticles and Their Applications in Water and Wastewater Treatment, 2020, https://doi.org/10.1007/978-981-13-1891-7_16.
101. Ghosh, S., Ahmad, R., Zeyaullah, M., Khare, S.K., Microbial Nano-Factories: Synthesis and Biomedical Applications. *Front. Chem.*, 9, 2021, https://doi.org/10.3389/fchem.2021.626834.
102. Al-khayri, J.M., Ansari, M.I., Singh, A.K., *Nanobiotechnology*.
103. Rónavári, A., Igaz, N., Adamecz, D.I., Szerencsés, B., Molnar, C., Kónya, Z., Pfeiffer, I., Kiricsi, M., Green silver and gold nanoparticles: Biological synthesis approaches and potentials for biomedical applications. *Molecules*, 26, 1–39, 2021, https://doi.org/10.3390/molecules26040844.
104. Kumar, S., Mudai, A., Roy, B., Basumatary, I.B., Mukherjee, A., Dutta, J., Biodegradable hybrid nanocomposite of chitosan/gelatin and green synthesized zinc oxide nanoparticles for food packaging. *Foods*, 9, 2020, https://doi.org/10.3390/foods9091143.
105. Echegoyen, Y., Rodríguez, S., Nerín, C., Nanoclay migration from food packaging materials. *Food Addit Contam Part A Chem. Anal. Control Expo Risk Assess.*, 33, 530–539, 2016, https://doi.org/10.1080/19440049.2015.1136844.
106. Huang, J.-Y., Chieng, Y.Y., Li, X., Zhou, W., Experimental and Mathematical Assessment of Migration from Multilayer Food Packaging Containing a Novel Clay/Polymer Nanocomposite. *Food Bioproc Tech.*, 8, 382–393, 2015, https://doi.org/10.1007/s11947-014-1408-5.
107. Powell, J.J., Harvey, R.S.J., Ashwood, P., Wolstencroft, R., Gershwin, M.E., Thompson, R.P.H., Immune Potentiation of Ultrafine Dietary Particles in Normal Subjects and Patients with Inflammatory Bowel Disease. *J. Autoimmun*, 14, 99–105, 2000, https://doi.org/10.1006/JAUT.1999.0342.
108. Gurr, J.R., Wang, A.S.S., Chen, C.H., Jan, K.Y., Ultrafine titanium dioxide particles in the absence of photoactivation can induce oxidative damage to human bronchial epithelial cells. *Toxicology*, 213, 66–73, 2005, https://doi.org/10.1016/J.TOX.2005.05.007.
109. Marrez, D.A., Abdelhamid, A.E., Darwesh, O.M., Eco-friendly cellulose acetate green synthesized silver nano-composite as antibacterial packaging system for food safety. *Food Packag Shelf Life*, 20, 100302, 2019, https://doi.org/10.1016/J.FPSL.2019.100302.

110. Basumatary, K., Daimary, P., Das, S.K., Thapa, M., Singh, M., Mukherjee, A., Kumar, S., Lagerstroemia speciosa fruit-mediated synthesis of silver nanoparticles and its application as filler in agar based nanocomposite films for antimicrobial food packaging. *Food Packag. Shelf Life*, 17, 99–106, 2018, https://doi.org/10.1016/J.FPSL.2018.06.003.
111. Qu, L., Chen, G., Dong, S., Huo, Y., Yin, Z., Li, S., Chen, Y., Improved mechanical and antimicrobial properties of zein/chitosan films by adding highly dispersed nano-TiO2. *Ind. Crops Prod.*, 130, 450–458, 2019, https://doi.org/10.1016/j.indcrop.2018.12.093.
112. Zhang, X., Liu, Y., Yong, H., Qin, Y., Liu, J., Liu, J., Development of multi-functional food packaging films based on chitosan, TiO2 nanoparticles and anthocyanin-rich black plum peel extract. *Food Hydrocoll.*, 94, 80–92, 2019. https://doi.org/https://doi.org/10.1016/j.foodhyd.2019.03.009.

16

Statistical Data Analysis of Anticorrosion and Antifouling: Unveiling Insights from Performance and Trends

Preetham Noel P.[1], Kukatlapalli Pradeep Kumar[1*], Mani Joseph P.[2] and Vinay Jha Pillai[3]

[1]Department of Computer Science and Engineering, CHRIST (Deemed to be University), Bangalore, Karnataka, India
[2]Department of Mathematics and Computer Science, Modern College of Business and Science, Muscat, PC, Sultanate of Oman
[3]Department of Electronics and Communication Engineering, CHRIST (Deemed to be University), Bangalore, Karnataka, India

Abstract
This chapter provides intended insights about the current antifouling and anticorrosion conventional coatings, each with a distinct quality, that fall short of fully satisfying the contemporary requirements of material preservation in a marine environment. To solve these issues and accomplish the required goals, a number of novel terminologies and procedures have been investigated and several recently published research works were referred to and briefly examined in order to provide appropriate conclusions. In order to explore new methodologies that provide total protection for materials from biofouling and corrosion in maritime environments, this study looks into the complex realm of antifouling and anticorrosion methods.

Keywords: Polymers, MIC (microbiologically influenced corrosion), WCA (water contact angle), SLIPS (slippery lubricant-infused porous surfaces), CAH (contact angle hysteresis), CA (contact angle)

**Corresponding author*: kukatlapalli.kumar@christuniversity.in

Hari Murthy, Vinay Jha Pillai, Kukatlapalli Pradeep Kumar and Matthew Cowan (eds.) Novel Anti-Corrosion and Anti-Fouling Coatings and Thin Films, (459–474) © 2024 Scrivener Publishing LLC

16.1 Introduction

The ocean is an enormous reservoir of saline water that makes up about 71% of the surface of the planet. Ninety percent of all the goods transported globally use waterways. Therefore, marine resources and the marine shipping industry have resulted as one of the fundamental pillars of economic growth in almost all countries. The sea climate is very acidic and harsh for materials used in ocean business including metals. Limitations on metals in marine settings are primarily due to biofouling and corrosion. The initial cause of biofouling is bacterial colonization and infection and the subsequent production of biofilms exacerbates this process. Corrosion is the irreversible deterioration of a metal surface which undergoes various chemical processes that lead to the transformation of metals into their chemically more stable form such as sulfides, oxides, and hydroxides. To prevent and minimize the despoliation of metals and other materials, the concept of antifouling and anticorrosion has been brought into the picture. Current demands of maintaining the characteristics of the materials or metals cannot be fulfilled by conventional antifouling or anticorrosion coatings with a single attribute. Therefore, various terminologies and techniques have been experimented with and tested under various circumstances to foster and meet the required objectives and goals.

16.2 Literature

16.2.1 Tetrabromobisphenol-A Epoxy and Polyaniline Nanowires Coating

Tetrabromobisphenol-A epoxy (TEBP) is an epoxy bisphenol-A with benzene rings that have bromine substituents on them. Its ability to resist the environment's penetration and deterioration of corrosion confirms that it is a strong contender for anticorrosion and antifouling properties. Moreover, conductive polymers (CPs), polyaniline (PANI), and metal corrosion inhibitors were briefly studied and analyzed. Epoxy coatings are incorporated into PANI, and as compared to virgin coatings, they perform significantly against corrosion in harsh conditions. EP (bisphenol-A epoxy) coatings are incorporated into PANI, and compared to other coatings, their performance is feasible against corrosion in hostile conditions. Three types of coatings were prepared in the process:

- Covering 1: merely EP covering
- Covering 2: the mixture covering free of PANI

- Covering 3: combined coating with two percent by weight PANI nanowires

At the very end, covering 3 was completed, employing EP layer as the base covering and 2 wt% PANI.

TBEP covering along with 2 wt% PANI was implemented for the final layer. Furthermore, the three kinds of the coating were polished on materials made of polyethylene terephthalate, also known as (PET), and the coatings were removed from PET following the water absorption inspection. Finally, in order to evaluate the moisture retention of the coverings, water penetration and angle-of-contact experiments were evaluated. This information was used as a starting point to assess the coverings' hydrophobic nature and anti-adhesion properties [1]. After undergoing several experimental testing and analysis, the following outcomes are shown below:

- Antimicrobial properties of coatings: The immersion test for a period of 4 weeks was performed on *E. coli* and *B. subtilis* which inferred that covering 2 and covering 3 retained a substantial number of microbes.
- Anticontamination properties of coatings: After 4 days of exposure to an inoculated medium in three sterilization flasks. Covering 2 and Covering 3 effectively prevented advancement along with invasion by microbes. The antimicrobial attributes of covering 1 diminished as its submersion duration extended.
- Adhesion strength: Coating 1 and coating 2 showed great bonding power due to the EP coatings being used for their basic coverings.
- EIS measurement: The technique by which all three coverings deteriorated over the course of the 70-day submerged examination at 95°C in a 12 wt% solution of sodium chloride. The anticorrosive efficiency of covering 1 and covering 2 decreased as the destructive surroundings were encountered within the coverings' openings. The experimental measurement demonstrated that coating 3 improves the anticorrosion productivity compared with coating 2 under preventive measures of PANI nanowires.
- Modified EEC outcomes for measurement: The results stated that the bromine substituents have an optimistic effect on metal fabrication.

16.2.2 AZ31 Magnesium Alloy Fabrication by Establishing Superhydrophobic Needle-Like Ca-P Layering

The Ca-P layering preparation mechanism involves the following:

1. Grounding of AZ31 substrate with SiC sandpapers and ultrasonic cleansing with anhydrous ethanol.
2. The coating was fabricated employing a straightforward chemical modification procedure, and the material was immersed in a $CaCl_2$ and NaH_2PO_4 mixture at a stable temperature, after the immersion process, the sample was subjected to drying utilizing heated air.
3. Long-chain fatty acid (STA) devoid of fluorine was employed to lower the surface energy of the coating [2].

The coating was subjected to various characterization evaluations which involved a dispersive power spectrometer (EDS), scanning electron microscope examination (SEM), the X-ray scattering system (XRD), and Fourier transform infrared radiation spectrometer and derivative thermogravimetric (DTG).

The electrochemical assessments were carried out by the CS2350H electrochemical workstation. The test was conducted utilizing a conventional three-electrode setup and the working electrode used was Ag/AgCl and platinum plate.

Outcomes:

- The effect on morphology and wettability by various experimental analyses: As the temperature increased, the bond between the structures strengthened and the structures were turned and bent into shapes resembling nests. The coverings' water contact angle (WCA) increased and then decreased as it reached the peak value of 158.7°. Finally, the Cassie-Baxter equation was used for estimating the involvement of the needle-like framework, which increased the contact area between the covering and the water droplet size [12].
- Analytical study of chemical composition and structure: The XRD outcome showed the following Ca-P coating acquired using various intervals and varying settings. An increase in the temperature decreased the magnitude of the magnesium diffraction sequence.

- Antifouling property: It was found that the superhydrophobic surface's poor adherence favors the antimicrobial characteristic—preserving the specimen from getting contaminated.
- Stability factor: The degrees at which the highly hydrophobic specimens were exposed to heat differed for about 2 hours and the samples were cooled down to room temperature. WCA increased when the temperature was less than 180°C. However, WCA diminished to 61.5°C, while the environmental temperature was raised to 200°C. The WCA decreased, with an increase in peeling times (tape-peeling tests) —demonstrating the structure's destruction gradually.
- Anticorrosion property: The rate of deterioration is directly correlated with the deterioration intensity. Corrosion potential is inversely proportional to the corrosion tendency. Better corrosion inhibition performance is demonstrated by the extremely hydrophobic coating's competence cycle, which is significantly larger than that of the material.

16.2.3 Cerium Stearate Surfaces – Fluorine-Free

The 5083 Al was introduced as a substrate in this research experimental procedure. SHS (superhydrophobic surfaces) and SLIPS (slippery lubricant-infused porous surfaces) upon 5083 Al materials were developed using a fluorine-free, natural electrode placement approach. Cerium coating is used for anticorrosion.

The polished 5083 Al is made possible with the help of alcohol and filtered water. To conduct and proceed with the experiment, SHS along with various morphologies were submerged in the mixture as the anode and synthesized at 50 V for a definite amount of duration. At the cathode, an electrode made of platinum was used. SLIPS was developed by immersing SHS in silicone oil. CAH (contact angle hysteresis) was evaluated based on the variation between the advancing and retreading contact angles. After 40 days of immersion in simulated saline water (ASW), the variations in WCA (water contact angle) and CAH were used to assess the accuracy of SLIPS.

Outcomes:

- Structure and content of the surface: The colorless electrolyte close to the platinum electrode changes to a pale yellow color. The light-yellow region grew larger and the color darkened as the reaction proceeded. The coating formed on 5083 Al is cerium stearate [3].

- Wettability and resilience of the surface: The electrodeposition items and SA-prolonged carbon link are attributed to the lack of energy density at the surface and surface roughness. SHS shows good water-repelling properties. The SLIPS shows excellent mechanical durability because the lubricant layer has the ability to regenerate itself [11].
- Antifouling ability: The primary cause of the development of biofilms is due to the adherence of diatoms, bacteria, and other microorganisms to the substrate's outer layer. Therefore, the massive damage brought on by the later emergence of macrofouling biological species can be diminished by lowering the production of biofilms. The gas confined in the microscopic sphere and the covering on the SHS with the least energy density both minimize the forces created by van der Waals among the microbes and the rigid substrate. The increase in the likelihood of deterioration and a drop in the density of rusting forces indicated improvement in the anticorrosion performance of 5083 Al.

Finally, the results infer that the H-SLIPS's constant friction-reducing layer on the topmost layer allowed it to exhibit the best antifouling qualities [10].

16.2.4 Corrosion and Biofouling – Marine Microorganisms

Microbiologically influenced corrosion (MIC) is the term used to describe the type of deterioration of metallic materials that is either directly or indirectly stimulated by the life functioning of microorganisms through their metabolic byproducts. The anaerobic sulfate-physical iron-oxidizing bacteria (IOB) and reducing bacteria (SRB) are the major factors to favor MIC. The concept of oxygen quantity organisms was implemented to estimate and calculate the corrosion caused at localized areas by the heterogeneous biofilms. In this research article reviewed, a new theory was proposed, namely, BCSR, or biocatalytic cathodic sulfate reaction, which provides an overview of the MIC technique with angles of bioenergetics and bioelectrochemistry. BCSR also explains the detailed procedure of the incidence of microbiological [9]. Marine biofouling is caused by the uncontrolled settlement of plants and animals on the materials and apparatuses used in the marine industries for their betterment [4].

Table 16.1 Distinguishing MIC and marine biofouling: A comparative overview.

MIC	Marine biofouling
Corrosion occurring at the micro level.	Techniques of retention at the larger stage
Damage to structures and loss of integrity caused by localized corrosion.	Structural damage as a result of the weight and drag of fouling organisms, decreased vessel and equipment performance, higher fuel consumption, and other effects.
Potentially changes the microbial populations in the marine environment.	Disrupts food chains and encourages invasive species, which changes the local marine ecosystems.
Only microorganisms stimulate this type of corrosion.	This type of fouling is stimulated by various microorganisms, plants, and animals.
The damage caused is direct.	The damage caused is more problematic and larger.

Table 16.1 above provides a fundamental distinction between marine biofouling and MIC, though they are two distinct phenomena, they are interconnected phenomena in the world of marine engineering and infrastructure management.

To overcome the problems, chemically based antifoulants and bacteria are combined along with certain recent components, chemicals, and technologies through a hybrid procedure to minimize their intake as well as acquire an efficient bacterial effect simultaneously [8]. This article also focuses to examine and estimate the performance of the antifoulant and its functionality in marine biodiversity. During the experiment, diblock copolymers of methyl methacrylate (MMA) and tert-butyldimethylsilyl methacrylate (TBDMSiMA) were created by recoverable addition-fragmentation chain transfer (RAFT) polymerization, which, when contrasted to analytical copolymer mixture coverings, shows better deterioration rate monitoring and antifouling effectiveness.

16.2.5 Fluorinated Silica Is Classified as Superamphiphobic and Exhibits a Core-Shell Framework

Sol-gel process was used to synthesize pristine silica at a TEOS molar concentration designed: CH3CH2OH: H2O = 1.0:20.0:2.0 with NH3.H2O as the catalyst. Using a contact angle detector at ambient temperature, contact angles (CAs) and slide angles (SAs) were measured. X-ray photoelectron spectroscopy, energy-dispersive X-ray spectroscopic analysis, and infrared spectroscopy with the Fourier transform were used to analyze the covering's chemical structure. The electron propagation microscope was implemented to examine the characteristics of individual pristine and fluorinated silica nanoparticles. The mean density of corrosion charges was determined by taking the average of all the corrosion current densities in the contours of divergence. The samples mean corrosion charge mass will decrease if the anticorrosion properties are higher. After experimental analysis, the fluorinated silica colloidal solution exhibits great potential adaptability and it may be used on a range of flexible or rigid regions. The bonding of the epoxy/F-SiO2 mixture covering is more powerful than that of the fluorinated silica covering [5–7].

16.3 Results and Discussion

This segment offers a concise overview of the data analysis conducted on a specific dataset within the realm of anticorrosion and antifouling. It also deals with a number of variables related to anticorrosion and antifouling strategies, such as the environment, the materials employed, and the application techniques. We will evaluate the efficacy of these measurements, spot trends, and highlight the primary factors that either contribute to success or failure by utilizing statistical tools and procedures. The ecological effects and regulatory implications of these activities will also be discussed in this section.

Figure 16.1 shows the distribution of concentrations. Concentrations correspond to the horizontal axis, and frequency is situated on the vertical axis. The peak around the mean concentrations of 35.657 with a standard deviation of 31.714 indicates that most environments have concentrations close to the mean.

Figure 16.2 represents a simple line chart with "Temperature_In_Degree_Celsius" corresponding to the horizontal axis and "Mean Concentration" situated on the vertical axis. The chart demonstrates the significance of mean concentration varies with temperature because there is a peak and a dip. The peak is around 1000°C and 150,000 units of mean concentration, and the trough is approximately 500°C and 50,000 units of mean concentration. According to the graph, there appears to be a nonlinear and periodic pattern between the two variables.

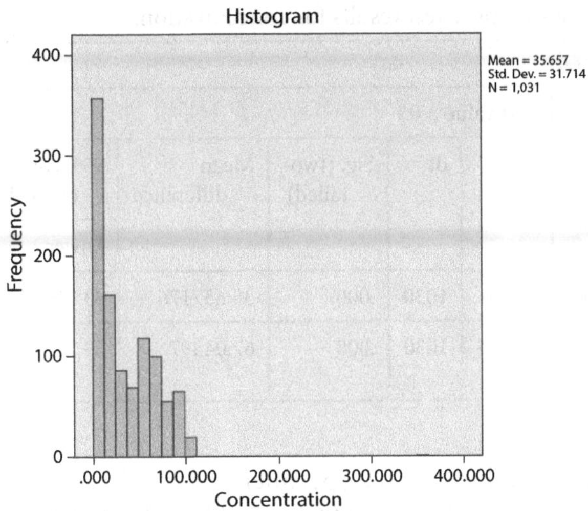

Figure 16.1 Relationship between concentration and the environment count using bar graph analysis.

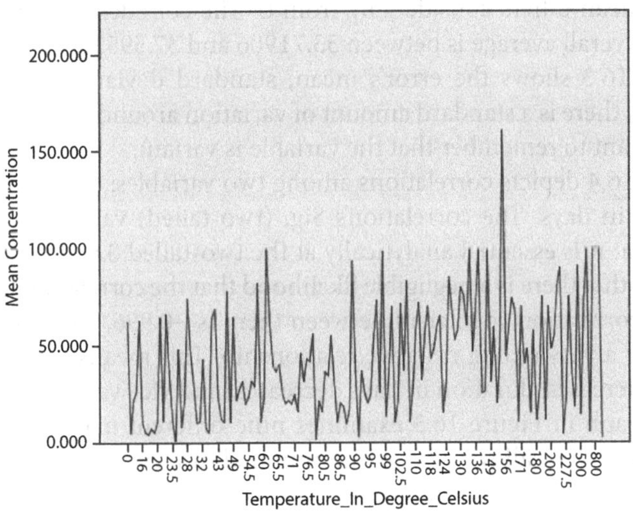

Figure 16.2 Relationship between temperature and the concentration mean count using simple line graph analysis.

Table 16.2 shows the outcomes of a one-sample t-test. The test contrasts the average of the sample with a predicted population mean. Two rows make up the table, one for each of the two variables: concentration and time in days. The mean difference is 35.657176, which is the variation

Table 16.2 One-sample t-test results for concentration.

One-sample test						
	Test value = 0					
	t	df	Sig. (two-tailed)	Mean difference	95% confidence interval of the difference	
					Lower	Upper
Concentration	36.101	1030	.000	35.657176	33.71906	37.59529
Duration_In_Days	18.065	1030	.000	67.04497	59.7623	74.3277

between the test value and the specimen mean. The average difference's 95% confidence interval is 33.71906 to 37.59529. The t-statistic is 67.04497 and the p-value is 0.000.

We dismiss the null hypothesis because the p-value is less than the statistically significant level of 0.05 and arrive at the conclusion that the population mean differs considerably from 0. The confidence interval shows that the overall average is between 33.71906 and 37.59529.

Table 16.3 shows the error's mean, standard deviation, and average. However, there is a standard amount of variation around these means, so it is important to remember that the variable is variant.

Table 16.4 depicts correlations among two variables: concentration and duration in days. The correlation's Sig. (two-tailed) value is 0.000, indicating that it is essential analytically at the two-tailed 0.01 threshold. This indicates that there is a negligible likelihood that the correlation is accidental. The correlation coefficient between them is –0.996, which means that they have a very strong negative relationship. This means that as concentration increases, duration in days decreases, and vice versa.

The graph in Figure 16.3 examines nine different material categories, including copper alloys, silicon aluminum alloys, and carbon steel alloys.

Table 16.3 One-sample t-test results for concentration and Duration_In_Days.

One-sample statistics				
	N	Mean	Std. deviation	Std. error mean
Concentration	1031	35.65718	31.714002	0.987692
Duration_In_Days	1031	67.0450	119.16872	3.71136

Table 16.4 Correlation output table for concentration and Duration_In_Days.

Correlations			
		Concentration	Duration_In_Days
Concentration	Pearson's correlation	1	0.168**
	Sig. (two-tailed)		0.000
	N	1031	1031
Duration_In_Days	Pearson's correlation	0.168**	1
	Sig. (two-tailed)	0.000	
	N	1031	1031

**Correlation is significant at the 0.01 level (two-tailed).

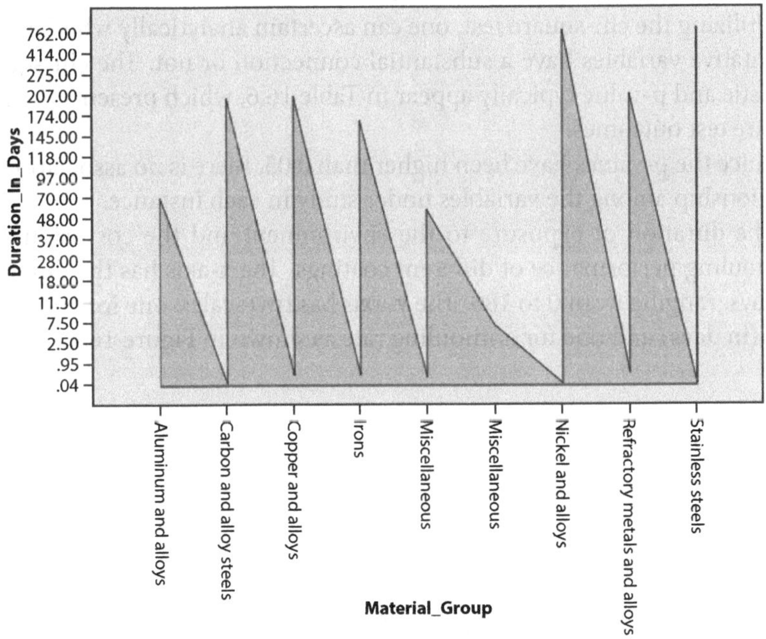

Figure 16.3 Simple area plot to illustrate the relationship between material group and duration in days.

Refractory metals and alloys have the longest life span (762 days), while silicon aluminum alloys have the shortest life span (94 days), as indicated by the graph. The graph also illustrates how the durations of the other material categories vary.

The one-way ANOVA test, which examined differences among group means, yielded an F-statistic of 3.125 and a corresponding significance value (p-value) of 0.000 is observed in Table 16.5. Based on the one-way ANOVA test results, we dismiss the null hypothesis that the group average differs from each other. There is a statistically substantial distinction in the group means, as indicated by the p-value of 0.000 is less than the statistically significant level of 0.05.

Another way to interpret the results is that the F-statistic of 0.957 is less than the critical value of the F-distribution. This indicates that there is no apparent distinction among the variances within and between the groups. The differences observed in the data are not statistically significant, and any variations are likely attributed to chance. This conclusion is specific to the data and analysis performed in this context.

Utilizing the chi-square test, one can ascertain analytically whether two qualitative variables have a substantial connection or not. The chi-square statistic and p-value typically appear in Table 16.6, which presents the chi-square test outcomes.

Since the p-values have been higher than 0.05, there is no association or relationship among the variables under study in each instance.

The duration of exposure to the environment and the corrosion and antifouling performance of different coatings. The x-axis has the duration in days, ranging from 0 to 180. The y-axis has two scales: one for corrosion rate (in days) and one for antifouling rate as shown in Figure 16.4.

Table 16.5 One-way ANOVA test.

ANOVA					
Concentration					
	Sum of squares	df	Mean square	F	Sig.
Between groups	399038.002	172	2319.988	3.125	0.000
Within groups	636913.232	858	742.323		
Total	1035951.234	1030			

Table 16.6 Chi-square test results.

Chi-square tests			
	Value	df	Asymp. Sig. (two-sided)
Pearson chi-square	17068.657[a]	18300	1.000
Likelihood ratio	3512.654	18300	1.000
N of valid cases	1031		

a. 18607 cells (99.9%) have an expected count of less than 5. The minimum expected count is 0.00.

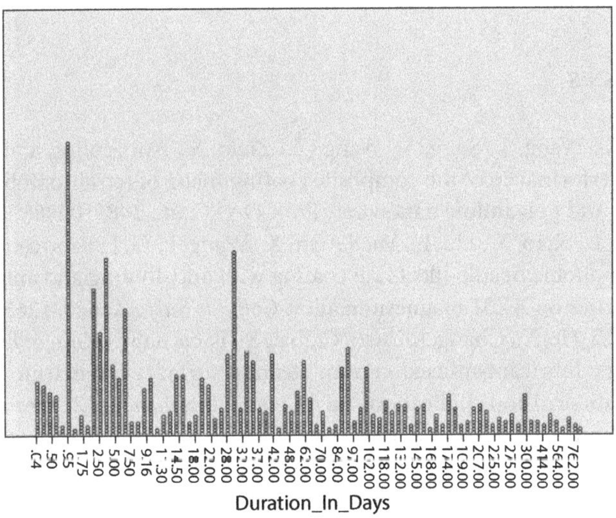

Figure 16.4 Simple dot plot illustrating the duration of days.

16.4 Conclusion

This chapter includes various analytical and statistical studies based on the findings and datasets. The datasets were downloaded from Kaggle and were structured in a formatted manner. Our research aimed at exploring the potential of unveiling the findings and results of antifouling and anticorrosion. Our findings revealed several key points. Coating 3 which consists of TBEP coating with 2 wt% PANI, EP coating with 2 wt% PANI, and PANI nanowire exhibited better antifouling and anticorrosion properties compared to the other two coatings. The prepared superhydrophobic covering demonstrated strong stability and outstanding antifouling characteristics. Moreover, increased corrosion resistance was exhibited by the extremely hydrophobic Ca-P covering. The hyacinth-like structure for SHS (superhydrophobic surfaces) demonstrated greater antifouling properties than the bayberry-like surface. Bacterial synergists that would disperse biofilms were established, the self-polishing copolymers were modified, the application of external electric field and conducting polymers was proposed, and the schema of controlling MIC (microbiologically influenced corrosion) and influence of microorganisms in biofilms. Great antiadhesion attributes are demonstrated by the superamphiphobic fluorinated silica covering. Suitable durability against corrosion has been demonstrated by the epoxy/F-SiO2 mixture covering.

References

1. Wu, C., Wang, J., Song, S., Wang, Z., Zhao, S., Antifouling and anticorrosion performance of the composite coating made of tetrabromobisphenol-A epoxy and polyaniline nanowires. *Prog. Org. Coat.*, *148*, 105888, 2020.
2. Wang, L., Xiao, X., Liu, E., Yu, S., Yin, X., Wang, J., Li, J., Fabrication of superhydrophobic needle-like Ca-P coating with anti-fouling and anti-corrosion properties on AZ31 magnesium alloy. *Colloids Surf., A*, *620*, 126568, 2021.
3. Yang, Z., He, X., Chang, J., Yuan, C., Bai, X., Facile fabrication of fluorine-free slippery lubricant-infused cerium stearate surfaces for marine antifouling and anticorrosion application. *Surf. Coat. Technol.*, *415*, 127136, 2021.
4. Li, Y. and Ning, C., Latest research progress of marine microbiological corrosion and bio-fouling, and new approaches of marine anti-corrosion and anti-fouling. *Bioact. Mater.*, *4*, 189–195, 2019.
5. Guo, C., Ding, H., Xie, M., Zhang, H., Hong, X., Sun, L., Ding, F., Multifunctional superamphiphobic fluorinated silica with a core-shell structure for anti-fouling and anti-corrosion applications. *Colloids Surf., A*, *615*, 126155, 2021.
6. Wang, P., He, B., Du, Y., Wang, B., Gao, J., Liu, S., Zhou, F., Functionalized Ti3C2Tx-based nanocomposite coatings for anticorrosion and antifouling applications. *Chem. Eng. J.*, *448*, 137668, 2022.
7. Jin, H., Wang, J., Tian, L., Gao, M., Zhao, J., Ren, L., Recent advances in emerging integrated antifouling and anticorrosion coatings. *Mater. Des.*, *213*, 110307, 2022.
8. Kumar, S.A., Balakrishnan, T., Alagar, M., Denchev, Z., Development and characterization of silicone/phosphorus modified epoxy materials and their application as anticorrosion and antifouling coatings. *Prog. Org. Coat.*, *55*, 3, 207–217, 2006.

9. Zhang, H., Liang, T., Liu, Y., Misra, R.D.K., Zhao, Y., Low-surface-free-energy GO/FSiAC coating with self-healing function for anticorrosion and antifouling applications. *Surf. Coat. Technol.*, *425*, 127690, 2021.
10. Kumar, K.P., Unal, A., Pillai, V.J., Murthy, H., Niranjanamurthy, M. (Eds.), *Data Engineering and Data Science: Concepts and Applications*, 2023.
11. Kumar, K. P., Murthy, H., Pillai, V. J., Prathap, B. R., Moses, M., & Urs, Y., Data Analysis and Machine Learning Observation on Production Losses in the Food Processing Industry, in: *2023 IEEE International Conference on Contemporary Computing and Communications (InC4)*, vol. 1, pp. 1–4, 2023, April.
12. Shibu, M., Kumar, K.P., Pillai, V.J., Murthy, H., Chandra, S., Structural health monitoring using AI and ML based multimodal sensors data. *Meas. Sens.*, *27*, 100762, 2023.

17

Comprehensive Data Analysis of Anticorrosion, Antifouling Agents, and the Efficiency of Corrosion Inhibitors in CO_2 Pipelines

Vineeth Simhadri[1], Kukatlapalli Pradeep Kumar[1]*, Vijaya P.[2] and Vinai George Biju[3]

[1]*Department of Computer Science and Engineering, CHRIST (Deemed to be University) Kengeri Campus, Bangalore, Karnataka, India*
[2]*Department of Mathematics and Computer Science, Modern College of Business and Science, Muscat, PC, Sultanate of Oman*
[3]*Faculty of Computer Science and Business Informatics, Technical University of Applied Sciences, Wurzburg-Schweinfurt, Germany*

Abstract

This study explores the various methods that are being proposed for their anticorrosion and antifouling capabilities and also reviews the unique properties that make them suitable for such applications. Special attention has also been given to the problem of corrosion in CO_2 pipelines, considering the corrosion inhibitors currently being used and performing statistical analysis about if and how various factors such as temperature, flow velocity, pH, and CO_2 pressure affect the rate of corrosion of the CO_2 pipelines. Tests including ANOVA, correlation, and graph analyses were conducted to explore their relationships, and suitable conclusions were drawn for the data collected.

Keywords: Anticorrosion, antifouling, coatings, S-FP, CPANOA, PANI, antibacterial, CO_2 pipelines

**Corresponding author*: kukatlapalli.kumar@christuniversity.in

Hari Murthy, Vinay Jha Pillai, Kukatlapalli Pradeep Kumar and Matthew Cowan (eds.) Novel Anti-Corrosion and Anti-Fouling Coatings and Thin Films, (475–496) © 2024 Scrivener Publishing LLC

17.1 Introduction

Oxidation is the natural process in which metal forms a more stable metal oxide, under certain atmospheric conditions. This is not ideal in many applications and can compromise the safety and effectiveness of the product. Ships, also made of metals, operate in the ocean, which comprises salt water—the ideal condition for corrosion. In addition to the corrosion, biofouling occurs due to living organisms accumulating on the ship's bottom surface, weakening the ship's structural integrity. Also, it was reported that micro-thick biofilm [1] on the ship's bottom surfaces increases the surface drag by 70% and that heavy fouling can decrease the ship's speed by 80%, meaning there will be a necessity to burn more fuel to maintain optimal speed and that leads to pollution. Thus, the prevention of corrosion of the ship's hull that is immersed in this seawater is of critical importance. Antifouling is the process of protecting a ship's hull to prevent the accumulation of unwanted material on the surface.

With water transport being the predominant mode of international trade in today's world, the safety and longevity of the ships used for such purposes are of utmost importance. The World Corrosion Organization (WCO) predicts that the economic loss from corrosion-related problems amounts to roughly 3.8% of the world's GDP [2]. On the other hand, the National Association of Corrosion Engineers (NACE) estimated that the economic loss due to corrosion-related problems was 3.4% of the world's GDP [3]. Due to this, extensive research is being done to discover new antifouling materials and develop new ways to synthesize these materials.

CO_2 pipelines are vital and are seen in many industrial applications. They are seen in the enhanced oil recovery (EOR) processes to transport the carbon dioxide released as a result of oil extraction, to transport carbon dioxide released as a result of processes carried out in power plants, and so on. Corrosion in the pipelines carrying CO_2 can happen due to many reasons, including the reaction of CO_2 with the metal, the formation of carbonic acid when CO_2 reacts with water, and local inaccuracies in the lattice structure of the metal with welds and cracks. The leakage of CO_2 as a result of cracks arising from increased corrosion is detrimental as it not only hampers the process for which it is being used but also has substantial environmental impacts, as CO_2 is a significant greenhouse gas contributing greatly to climate change. It is thus of utmost importance to study methods to mitigate such corrosion and also provide reliable analysis of the efficiency of the currently used anticorrosion materials being used for the purpose, and that is what this study has attempted to do.

17.2 Literature Review

Through the years, a lot of research has been done on the aspects related to anticorrosion and antifouling techniques, i.e., to study these processes and also to come up with more efficient and sustainable ways to mitigate the losses that occur due to corrosion and fouling. Some recent research into this field has been studied and included in this section.

17.2.1 Superoleophobic Composite Coating

Superoleophobic materials have a unique structure that makes them highly repellent to oils and other liquids having low surface tension. It is essential to understand why these materials have significant importance in the field of antifouling techniques.

Marine biofouling refers to the accumulation of organic material and marine living organisms or the deposits left by these organisms on materials immersed in water. Biofouling poses significant challenges to the global shipping business on which trade largely depends. Toxic biocides like tributyltin and zinc pyrithione were used, but the accumulation of these toxic substances over a long period of time poses a threat to marine life.

It is believed that the marine fouling process begins with the accumulation of oily molecules and microorganisms, which develop a conditioning film, thus contaminating the surface of the material. This is where the superhydrophilic and underwater superoleophobic (SUS) coatings have been gaining popularity in marine antifouling research. This can prevent the contamination of hull surfaces by controlling the attachment of oil particles on the surface.

The research focused on a new method to synthesize an antifouling material using nanotechnology. The "Glue + Powder" strategy was employed to fabricate a layer containing a superhydrophobic surface (SHS), a super hydrophilic and underwater superoleophobic surface layer.

Specimens like glass, aluminum, steel, and tetrafluoroethylene were used as substrates in the preparation. Superhydrophobic HD-SiO_2 nanoparticles were prepared. Generally, SiO_2 contains a mass of hydroxyl. It exhibits hydrophilicity, dispersed in a hybrid suspension of EtOH, deionized water, and HDTMS in appropriate proportions, and exposed to favorable conditions like temperature and pressure. The mixture is centrifugally rinsed with ethanol. The obtained HD-SiO_2 was sprayed onto substrates after treating it with PDMS appropriately, with a spray gun of nozzle diameter of 0.4 mm to fabricate a superhydrophobic PDMS/HD-SiO_2 coating surface. The fabrication of a superoleophobic PDMS/PAA-ZnO surface was done by the spraying method, where a carefully prepared mixture of ZnO particles dispersed in ethanol, deionized water, and PDMS was sprayed onto the PDMS/ HD-SiO_2 layer and the surface was obtained after curing at high temperature.

The materials prepared were subjected to various tests, including characterization, electrochemical, antibacterial, and antifouling tests. The surface chemistry of the materials was tested with various microscopic techniques:

- The morphologies of the surface were analyzed with the help of field-emission microscopy
- Transmission electron microscopy to assess the micromorphology and chemical composition of the materials
- Atomic force microscopy to measure the surface roughness of the coatings
- Thermal gravimetric analyzer to evaluate the thermal stability of PDMS/HSS and PDMS/PZS coatings.

A system containing three electrodes was used to know and compare the corrosion behavior of uncoated and coated aluminum samples (the metal ship hulls are mostly made of). CS Studio software was used to conduct and analyze the Tafel polarization curve and electrochemical impedance spectroscopy.

The antibacterial performance was assessed by using *Pseudomonas* sp. as an experimental strain. After inserting the sample into a sessile bacterial environment, their antibacterial effects were compared with the control specimen. The antifouling properties of the materials were assessed by using *Chlorella vulgaris* into which the samples were placed, and their antifouling properties were studied.

Many results were derived from the tests mentioned above. On the Fourier transform infrared spectroscopy (FTIR) conduction, the chemical compositions of the diverse nanoparticles like SiO_2, HD-SiO_2, and ZnO were confirmed. Spectrometry yields characteristic graphs for the compounds containing peak values on the graphs in which the wave number is usually taken as the abscissa. These peaks correspond to specific bonds' bending/stretching vibrations due to the functional groups on the molecule being analyzed—like -OH and -COO. Some graph peaks, like Si-O-Si's, correspond to molecular bonds' absorption and emission bands. These peaks found in the graphs are observed to correspond to the molecular compositions of the materials synthesized like HDTMS [4].

An analysis of the surface wettability and morphologies of materials like PDMS and PAA-ZnO was conducted, which revealed that the surface roughness, which is ideal for underwater superoleophobicity, increases with decreasing mass ratio of PDMS to PAA-ZnO. The results of the extensive analysis proved that an optimal mass ratio of PDMS to PAA-ZnO nanoparticles (2:1), a small and stable PDMS/PZS surface could be obtained.

So as to achieve the necessary antifouling and anticorrosion properties, a method for synthesizing superoleophobic and superhydrophobic composite coating was reported in this research. The mass ratio can be used to modify the coating's surface morphology and wettability, and the method was found to be sufficiently feasible. Even after 25 days of immersion, the results of many experiments show that the application of PDMS/HSS-PZS composite coating slows the rate of corrosion of aluminum alloy by seven times. Because of its simplicity, affordability, and universal adaptability to a variety of substrates, the spraying approach might be used to create the composite material on a wide scale [4].

17.2.2 Recent Advancements in Nanotechnological Research Related to Nanocomposite Coatings for Applications in Anticorrosion, Antifouling, and Self-Healing

Recent advancements in nanotechnology enable nanomaterials to explore the prospects of minimizing the degree of corrosion on metallic surfaces *via* surficial modification using nanocrystalline structural coatings. Nanocoatings have effectively reduced the harmful effects of corrosion on metal surfaces. PNC coating has been used chiefly for inhibiting, hindering, and reducing or mitigating the effects of corrosion due to various factors like environmental parameters and coating techniques.

The economic case for the use of nanocoating technique employment was also provided, as though the initial cost of the coating procedures can be high, it is cheaper in the long run than when the corrosion has already occurred. The paper delves into the latest developments in nanotechnology for protective coatings in polymer nanocomposites, providing comprehensive perspectives on the potential of hybrid substrates comprising both inorganic and organic materials.

The market for PNC coatings has also been widely expanding, with a projection of US$ 10.57 billion in revenue toward the end of 2027. A significant market for the materials has been Latin America, majorly due to the islander economies in the region and the vast expanse of the shipping industry therein, with this region alone forecasted to contribute upwards of US$ 2.3 billion by 2024.

Properties of PNC coatings and classification: The process of applying coatings on surfaces can be done by using different approaches—the most common among them is to deposit the coating as a multiple layering system composed of primer layering and an upper coat, with the number of coating layers varying subject to where it is being done and for what purpose the coated metal will be used.

Furthermore, the deposition techniques are divided into three broad categories: mechanical, physical, and chemical.

- Mechanical deposition: It is a low-cost technique achieved by spraying, painting, dip-coating, or spin coating.
- Physical deposition: It is conducted by sputtering, bonding, or condensation. Mild pressure and temperature are utilized for this type of coating.

Nanocoatings typically categorize corrosion-inhibiting substances into organic and inorganic types. In water-based PNC coatings, the primary utilization involves inorganic inhibiting entities that function through adsorption or oxidation, whereas solvent-based PNC coatings predominantly employ organic corrosion-inhibiting entities [5].

PNC coatings have a certain makeup of tightly packed grains, dislocations, and interphase boundaries which influence their unique properties. The qualities shown by PNC coatings rest solely on the attributes of the grains. The advancement of nanotechnology in recent times has resulted in the emergence of significantly better PNC coatings with additions such as self-cleaning functions and resistance to wear and scratches among other features. With the infusion of clever nanocoatings, the capacity of PNCs to resist corrosion and fouling has been tremendously boosted. Emerging are functional PNC coatings that offer both decorative and protective properties while harboring the potential for additional functionalities. To meet specific conditions, these coatings must exhibit adequate behavior, scratch resistance, and abrasion resistance to perform. Mechanical, chemical, physical, and thermal attributes serve as key determinants of these functional PNCs' efficiency.

Polymeric materials have excellent anticorrosion properties, and including ceramic nano-fillers in polymeric matrices enhances the resistance to corrosion. Organic and inorganic synergetic coatings, PNC, and hydrophobic coatings increase the shelf-life in corrosion-prone environments. The polymeric coating mainly protects metallic structures from corroding entities like O_2–, H+, and other atmospheric conditions that aid in corroding the metal.

Recent research studied the applicability of developing an advanced PMMA/silica composite (PSC) coating for anticorrosive properties. In the development of most anticorrosive materials, it is noted that the requirement of organic substances is hard to avoid. Most polymeric resins used to attain anticorrosive properties use organic solvents in preparation. However, the regulations on the emissions of volatile organic compounds (VOCs) necessitate the use of other materials in their place.

The inclusion of nanomaterials in PNC coating systems and polymeric matrices has increased their anticorroding properties, durability, and integrity. These materials imbibed critical efficiency at repairing micro-deterioration that occurs in anticorrosion coatings. The advancements in the applications of nanotechnology in anticorrosion and antifouling arenas make further enhancements in architectural properties and production parameters to develop viable and high-performance PNC protective coatings.

17.2.3 Rapid Curing and Self-Stratifying Lacquer Coatings

Lacquer refers to a liquid made of shellac dissolved in alcohol that dries to form a hard protective coating. Polymeric natural lacquer has been found to be very stable, durable, impermeable, and to have excellent anticorrosion properties. Thus, there has been a critical interest in researching increasing the drying speed of lacquer and improving its anticorrosion properties.

According to a recent publication on polymerization drying of Chinese lacquer, it was reported that adding a layer of titanium dioxide nanoparticles as a coating to natural lacquer can increase the rate of inhibition rate of the material against some bacteria and thus increase the material's antifouling capabilities. However, it has also been pointed out that TiO_2 nanoparticles have less resistance to certain kinds of bacteria and may be prone to agglomeration.

In the study conducted, the application of certain self-stratifying amphiphilic hybrid lacquer as a viable anticorrosion and antifouling coating has been proposed, and the preparation techniques, processing, and tests for the end product have all been extensively covered. The focus is on adding suitable substances to natural lacquer so that the resultant material will have excellent anticorrosion and antifouling capabilities. For this purpose, natural lacquer was coupled by the sol-gel process with hyperbranched polysiloxane terminated by multi-amine (HPSi) and silane-terminated amphiphilic telomer (S-FP). HPSi, due to its highly branched structure, is highly reactive at the terminal amines and has a lower viscosity, due to which it was added to the natural lacquer to improve the drying process and the mechanical properties of the lacquer [6].

Similarly, S-FP has some properties that are beneficial for the antifouling application they are being used for. Due to its hydrophobic nature, it has the ability to repel water molecules and thus prevent the attachment of aquatic organic matter, thereby reducing the fouling process on immersed surfaces. The chemical stability of S-FP also makes it resistant to environmental conditions like UV radiation and salt water, increasing its longevity and suitability for antifouling applications. The adhesive property of S-FP enables the substance to remain attached to the target metal surface for whose coating the material is being made. It will increase the duration for which the antifouling material remains attached to the metal surface. The low surface energy of the silane-terminated telomers lowers the chances of foulants to attach onto the metal surface. These properties of S-FP make it an excellent component to be added to lacquer to improve its antifouling capabilities.

Different tests were conducted on the hybrid lacquer to measure its efficiency in being a viable antifouling material:

- Impact resistance test: The drop weight, fracture roughness, etc., were measured.
- Flexibility test
- Gloss test: Measure by the use of a gloss meter by using triplicate samples to calculate the mean.
- Durability test: To test the durability of the coatings, a tinplate was coated and attached to the instrument's steel plate using a QUV accelerated weathering tester. The coatings' shine was evaluated again after coating.
- Abrasion-resistance test: It is used to determine the abrasion resistance properties with the help of the abrasion-resistance tester, and each coating was examined.
- Moisture resistance test: This is done by monitoring the relative mass of the coating after a set period of time in a controlled environment of 35°C and 90% relative humidity.
- Adhesion test: Four data points were averaged to determine the value of each sample, and a pull-off adhesion tester was used to help. The pull rate was set to 0.2 MPa s^{-1}.
- Measurements of surface wettability: Using the Theta Auto 113, the static water contact angle of each hybrid lacquer coating sample was obtained.
- XPS: To examine the surface compositions, X-ray photoelectron spectroscopy (XPS) was performed at 90° takeoff angles using an Axis Ultra DLD and a monochromate Al Kα source.

Some conclusions were drawn by analyzing various parameters regarding the antifouling properties of the hybrid lacquer. The increase in the concentration of HPSi in the sample led to a remarkable shortening of the drying time. This indicated that the hybrid lacquer has a much faster rate of polymerization than natural lacquer. As the concentration of HPSi increased, the hybrid material exhibited a corresponding rise in pencil hardness. Nevertheless, the flexibility of the material demonstrated an initial increase followed by a decline. The patterns observed seem to have been caused by the increased presence and addition of HPSi, which could elevate the cross-linking density of the natural lacquer covering to an extent that compromises flexibility. Impact resistance was superior in the absence of HPSi. Analysis of the XPS spectra revealed distinct peaks, confirming the self-stratifying property of amphiphilic oligomers and providing insights into the surface composition of the hybrid lacquer coatings. Consequently, during coating formation, the amphiphilic telomer S-FP can undergo self-enrichment, contributing to an enhanced antifouling effect.

The paper, thus, focuses on developing a rapid-curing hybrid lacquer that has less drying time by using various processes like a sol-gel reaction and an alcoholysis reaction. In addition to having a shorter drying time, the resulting hybrid lacquer coating possesses beneficial qualities such as strong adhesion, superior mechanical qualities, exceptional flexibility, and high hardness, which makes it a good antifouling material, and this was achieved with the addition of various substances that helped in the optimization of the factors involved in the process of increasing the antifouling capabilities of lacquer, like HPSi and S-FP.

17.2.4 Poly(Aniline-Co-Nitroaniline) Nanoparticle Coatings

It is observed that corrosion can be mitigated by slowing down the transmission of corrosive medium and controlling the rate of the corrosion reaction. The anticorrosion techniques thus used include the reasonable material selection and optimization of metal structures and coatings for protection against corrosion. In providing protective coatings for materials against corrosion, organic coatings have gained much attention. Polymer antifouling agents such as polyaniline (PANI), polypyridine, and polythiazole have become prominent for their numerous favorable characteristics and their contribution to enhancing the antifouling attributes of organic coatings. The modification of PANI with functional groups or its combination with chemical compounds is undertaken to enhance both antifouling properties and processability. The nitro group is one such functional group which, when introduced onto the benzene ring of the PANI molecule, due to its strong electronegativity and positive charge density on the N atom, improves the defense of the molecule against microorganisms, and can thus reinforce the corrosion protection and antifouling application performance of PANI.

In the preparation of PANI in which nanowires and other materials were used, the ratio of aniline monomer and nitroaniline monomers was varied as 3:1, 2:1, and 1:1, and the samples were labeled as CPANOA1, -2, and -3, respectively. Then, epoxy coatings were prepared with CTO nanosheets with each CPANOA sample [7].

Characterization of PANI and CPANOA included the following:

- The study of their morphologies by using field emission SEM. It was observed that PANI exhibits a wire shape with a similar length; that of CPANOA1 wires was shorter and thinner, while CPANOA2 and -3 had an irregular nanoparticle morphology.
- Surface composition analysis by XPS revealed the presence of carbon and nitrogen.
- Chemical properties' characterization by XRD.
- The thermal stability on a Netzsch STA449F5 instrument under an N2 atmosphere.
- Their optical homogeneity in cured coatings by Carl Zeiss Microscopy GmbH.
- The assessment of their antibacterial efficacy employed the plate count measurement technique. The sterilization number R was computed using the formula: R = (number of bacterial colonies in the suspension without CPANOA and PANI) divided by (number of bacterial colonies in the suspension with a particular concentration of CPANOA and PANI). Additionally, the coatings' resistance to fouling was evaluated by immersing them in a bacterial *E. coli* suspension with an initial inoculation level of colony-forming units per milliliter for a duration of 5 days.

17.2.5 Emerging Integrated Anticorrosion and Antifouling Coatings

This article discusses the recent advances and developments in integrated anticorrosion and antifouling coatings (IAAC). In recent years, superhydrophobic and slipper liquid-infused porous surfaces (SLIPS) have intrinsic anticorrosion and antifouling properties.

Polyaniline (PANI), a conductive polymer, has been gaining much attention in anticorrosion applications because of its unique electrochemical properties and stability. PANI has a limitation in its ability to be easily dedoped in an acidic environment of pH < 3, where it loses much of its antibacterial properties. To circumvent this, PANI combines with other antifouling units like organic biocides and nanomaterials or uses its derivatives to improve its antifouling capability.

- Organic bromine, tetrabromobisphenol-A epoxy, nanoparticles like graphene, etc., are discussed as suitable additives to PANI to enhance its antifouling properties due to their suitable and beneficial characteristics like hydrophobicity or antibacterial properties.
- Aniline derivatives of PANI, like poly(m-aminophenol) (PmAP) synthesized by oxidation polymerization of m-aminophenol in an alkaline medium, have also been found to enhance the antifouling properties of PANI. They also showed excellent antibacterial properties against *E. coli* and *B. subtilis*.
- Amorphous carbon has been significant in increasing the antifouling capabilities. The addition of copper, owing to its antibacterial properties, can enhance the antifouling capability of amorphous carbon. This has been found to be especially helpful when added to the allotropes of carbon like diamond-like carbon (DLC) and laser cladding (LC). When corrosion takes place, copper particles provide a barrier-like mechanism; thus, the addition of copper enhances the anticorrosion properties of the coatings.
- Polydopamine (PDA), due to its catechol-OH groups and the resulting increase in the string metal coordination ability, can act as an anticorrosion barrier to protect substrates.
- Polymer brushes are high-density materials consisting of polymers tethered to a surface and given as a surface coating. Grafting these brushes to surfaces has been used to inhibit the adhesion of proteins, bacteria, and microorganisms, thereby reducing the fouling effects on the material.
- Lacquers have also been discussed as effective coatings to reduce corrosion and fouling on metal surfaces. However, due to the long drying time of natural lacquer, it has been mixed with other, primarily organic compounds like S-FP and HPSi. Organic-inorganic hybrid sol-gel coatings are also

promising due to their compatibility and good barrier performance, and their modified versions can be more chemically stable, have excellent adhesion strength, and exhibit hydrophobicity [8].

Thus, the article focussed on the techniques to increase the antifouling and anticorrosion capabilities, the synthesis of such coatings, and the development of IAACs, as IAACs have huge advantages over traditional antifouling and anticorrosion coatings.

17.3 Results and Discussion

As mentioned previously, the corrosion in CO_2 pipelines is of particular concern due to the long-lasting procedural and environmental concerns that arise with leaky pipes. A dataset was considered that looked at the rate of corrosion of a CO_2 pipeline in the presence of a corrosion inhibitor. Various external factors were considered. In this section, some statistical tests were conducted on this dataset and the results obtained thereon were recorded.

(i) Dataset

The dataset contains information about the efficiency of corrosion inhibitors in CO_2 pipelines, taking into consideration some factors like temperature, the velocity of flow, pH, pressure of CO_2, internal pressure, and shear stress. The data were collected at varying external and internal conditions, and the efficiency of the corrosion inhibitor and the rate of corrosion were recorded.

(ii) Statistical Tests

- **pH and rate of corrosion**

  A one-way ANOVA test was conducted for the pH and corrosion rate to visualize the relationship between them.

  We consider the null hypothesis 'There is no significant relationship between the pH of CO_2 in the pipeline and the rate of corrosion'. The significant value obtained was 0.000 as evident from Table 17.1, which is less than the benchmark value of 0.05. Therefore, the alternate hypothesis is accepted, concluding that there is a significant relationship between the pH of CO_2 in the pipeline and the rate of corrosion.

  The value for Karl Pearson's coefficient of correlation from Table 17.2 was obtained as −0.429, which indicates a negative

Table 17.1 One-way ANOVA test for pH and the rate of corrosion.

ANOVA					
Corr_rate					
	Sum of squares	df	Mean square	F	Sig.
Between groups	116.850	2	58.425	52.493	0.000
Within groups	267.122	240	1.113		
Total	383.971	242			

Table 17.2 Correlation analysis for pH and rate of corrosion.

Correlations			
		Corr_rate	pH
Corr_rate	Pearson's correlation	1	−0.429**
	Sig. (two-tailed)		0.000
	N	243	243
pH	Pearson's correlation	−0.429**	1
	Sig. (two-tailed)	0.000	
	N	243	243

**Correlation is significant at the 0.01 level (2-tailed).

relationship between both variables. From the value, it can be said that as the value of pH increases, the rate of corrosion decreases. This conclusion is substantiated by the line plot obtained in Figure 17.1, as the rate of corrosion shows a decreasing trend as the pH value increases and progresses toward the basic side of the pH scale. This is true, as corrosion occurs at a faster rate in the presence of H^+ ions, which are present more in acidic solutions where pH is less. The corrosion rate decreased, and the passivation capability strengthened with increasing pH value [9].

- **Corrosion inhibitor efficiency and rate of corrosion**
  We consider the null hypothesis as 'there is no relationship between the corrosion inhibitor efficiency and the rate of corrosion. The significant value obtained after performing the

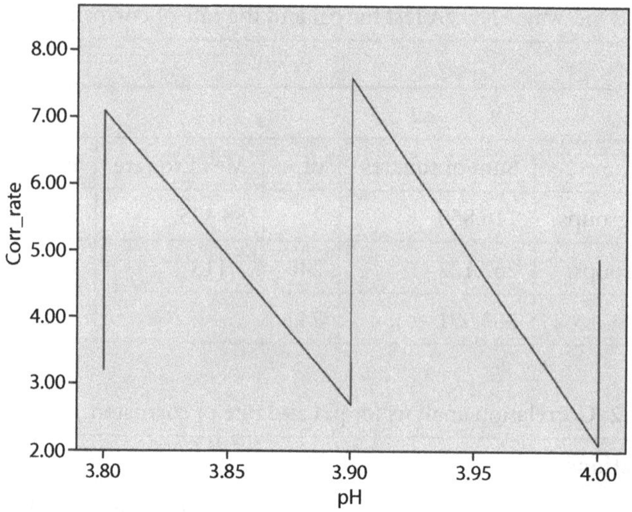

Figure 17.1 Line plot of pH vs. rate of corrosion.

ANOVA test for the variables was 0.000 (from Table 17.3), which is less than the benchmark value of 0.05. Therefore, the null hypothesis is rejected, and it is concluded that there is a relationship between the corrosion inhibitor efficiency and the rate of corrosion.

The obtained correlation value from Table 17.4 is –0.569, indicating a negative correlation. That is, a decrease in the efficiency of the corrosion inhibitor leads to an increase in the rate of corrosion and vice versa. The area plot shows that there is a relation between the rate of corrosion and the efficiency of the corrosion inhibitor (Figure 17.2).

Table 17.3 ANOVA test for the corrosion inhibitor efficiency and the rate of corrosion.

ANOVA					
Corr_rate					
	Sum of squares	df	Mean square	F	Sig.
Between groups	124.119	2	62.059	57.318	0.000
Within groups	259.852	240	1.083		
Total	383.971	242			

Table 17.4 Correlation analysis of corrosion inhibitor efficiency and the rate of corrosion.

Correlations			
		Corr_rate	Corr_inhibitor_ efficiency
Corr_rate	Pearson's correlation	1	−0.569**
	Sig. (two-tailed)		0.000
	N	243	243
Corr_inhibitor_ efficiency	Pearson's correlation	−0.569**	1
	Sig. (two-tailed)	0.000	
	N	243	243

**Correlation is significant at the 0.01 level (two-tailed).

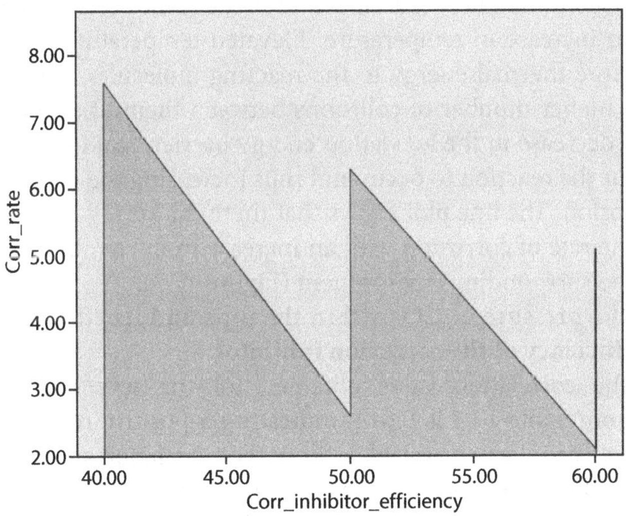

Figure 17.2 Area plot of corrosion inhibitor and the rate of corrosion.

- **Relation between the ambient temperature and the rate of corrosion**
  The rate of corrosion, being an electrochemical process, undoubtedly is dependent even on the level of the ambient temperature [10]. From the graph obtained above, it is

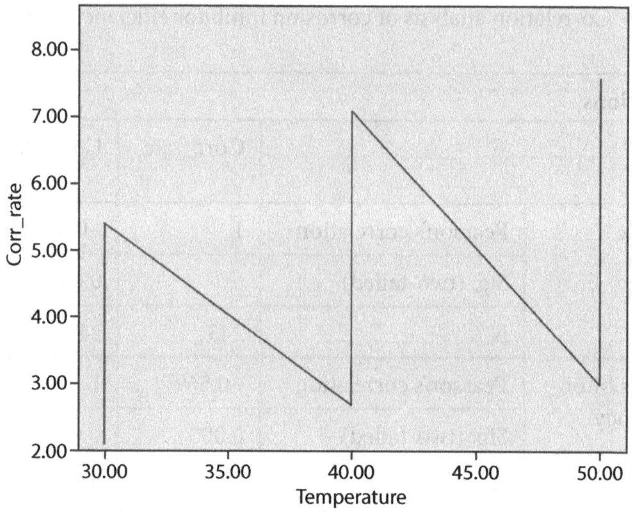

Figure 17.3 Line plot of ambient temperature and rate of corrosion.

evident that there is an increase in the rate of corrosion with an increase in temperature. Elevated temperatures provide more thermal energy to the reacting molecules, leading to a higher number of collisions between them. It also causes a decrease in the activation energy barrier, making it easier for the reaction to occur and thus increasing the rate of corrosion. The line plot shows that there is a steady increase in the rate of corrosion with an increase in the temperature of the surrounding environment (Figure 17.3).

- **The pressure of CO_2 within the pipe and its effect on the efficiency of the corrosion inhibitor**
  The correlation value obtained for the set of variables from Table 17.5 is 0.614, indicating a positive relationship between the two variables. From the correlation analysis and the line plot given in Figure 17.4, it can be concluded that the increase in the pressure of CO_2 in the pipeline causes a moderate increase in the rate of corrosion.
- **Relation between the rate of corrosion and other factors—velocity of flow, shear stress, and internal pressure**
  The Karl Pearson's correlation value obtained for the variables, from Table 17.6 was 0.107, which shows a very weak level of positive correlation between them. As also observed

Table 17.5 Correlation analysis of pressure of CO2 in the pipeline and the rate of corrosion.

Correlations			
		Corr_rate	CO2_Pressure
Corr_rate	Pearson's correlation	1	0.614**
	Sig. (two-tailed)		0.000
	N	243	243
CO2_Pressure	Pearson's correlation	0.614**	1
	Sig. (two-tailed)	0.000	
	N	243	243

**Correlation is significant at the 0.01 level (two-tailed).

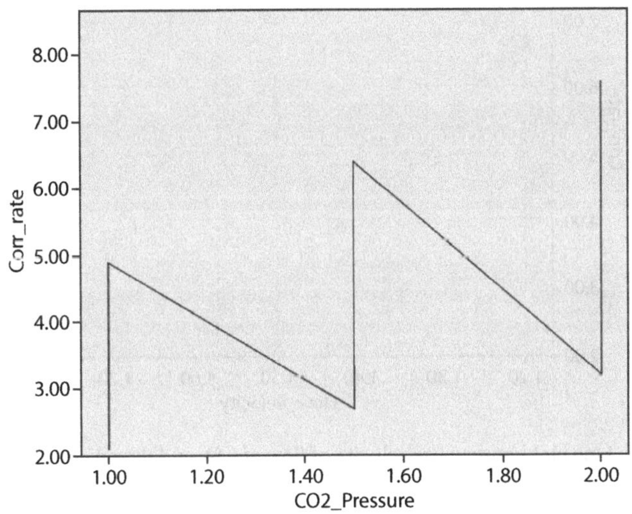

Figure 17.4 Line plot between the pressure of CO_2 in the pipeline and the rate of corrosion.

from Figure 17.5, there is a small (almost negligible) increase in the rate of corrosion as the velocity of flow of CO_2 in the pipe increases [11, 12].

Table 17.6 Correlation analysis between the velocity of flow of CO2 and the rate of corrosion.

Correlations			
		Corr_rate	Flow_velocity
Corr_rate	Pearson's correlation	1	0.107
	Sig. (two-tailed)		0.096
	N	243	243
Flow_velocity	Pearson's correlation	0.107	1
	Sig. (two-tailed)	0.096	
	N	243	243

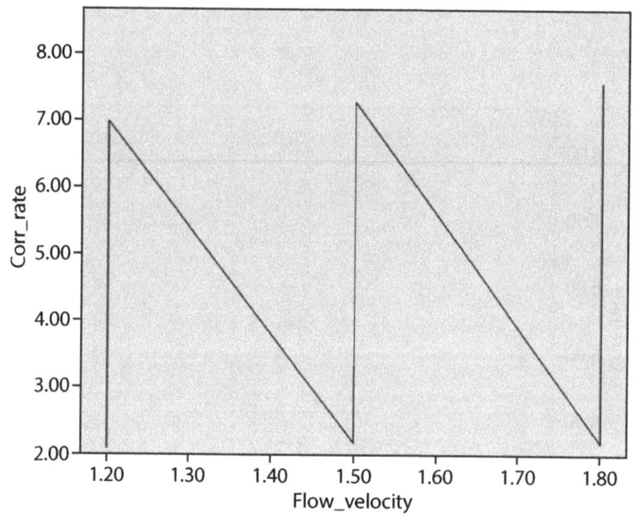

Figure 17.5 Line plot between the velocity of flow of CO_2 and the rate of corrosion.

The correlation value obtained for the variables from Table 17.7 was −0.031, which is a very weak level of correlation. Thus, we can say that the rate of corrosion is not very much dependent on the internal pressure of the CO_2 flowing in the pipe. The plot shows that the variation of the corrosion rate with the internal pressure is periodic and shows no increasing or decreasing trends (Figure 17.6).

The correlation value obtained for the set of variables, as seen in Table 17.8, was 0.107, which indicates a very weak level of positive correlation between them, i.e., there is no significant relationship between the shear stress and the

Table 17.7 Correlation analysis of the internal pressure and the rate of corrosion.

Correlations			
		Corr_rate	Internal_pressure
Corr_rate	Pearson's correlation	1	−0.031
	Sig. (two-tailed)		0.628
	N	243	243
Internal_pressure	Pearson's correlation	−0.031	1
	Sig. (two-tailed)	0.628	
	N	243	243

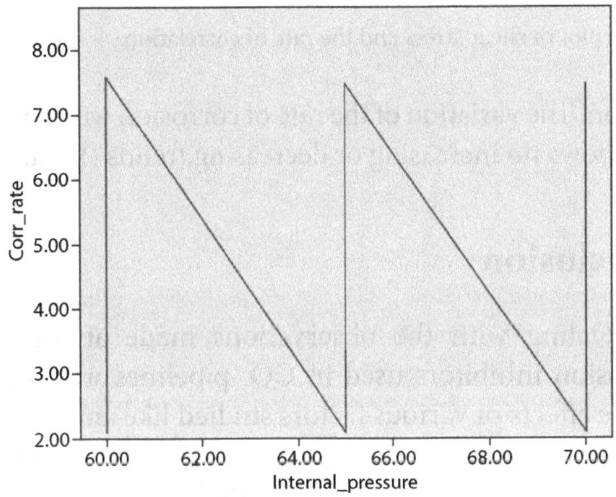

Figure 17.6 Line plot of internal pressure and the rate of corrosion.

Table 17.8 Correlation analysis of shear stress and the rate of corrosion.

Correlations			
		Corr_rate	**Shear_stress**
Corr_rate	Pearson's correlation	1	0.107
	Sig. (two-tailed)		0.096
	N	243	243
Shear_stress	Pearson's correlation	0.107	1
	Sig. (two-tailed)	0.096	
	N	243	243

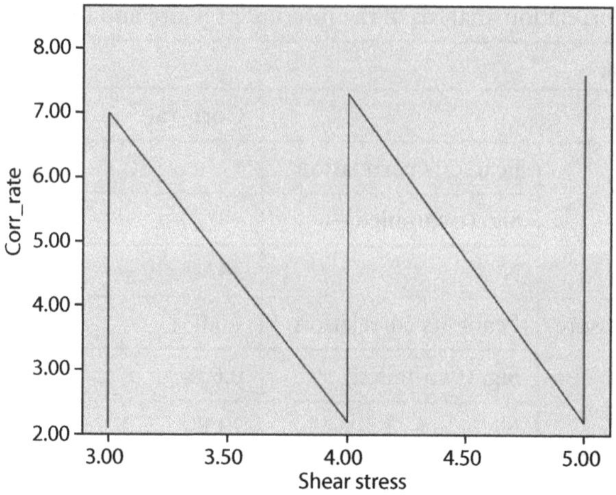

Figure 17.7 Line plot of shear stress and the rate of correlation.

rate of corrosion. The variation of the rate of corrosion with the shear stress is periodic and shows no increasing or decreasing trends (Figure 17.7).

17.4 Conclusion

The dataset dealing with the observations made about the efficiency of some corrosion inhibitors used in CO_2 pipelines was used to try and understand the effects of various factors studied like ambient temperature, shear stress, pH, and so on. Statistical analysis was carried out on different sets of variables, and the results were recorded.

The statistical tests like ANOVA, correlation, and the visual representation of them using line plots have solidified the previously researched topics like how an increase in pH and increase in the ambient temperature causes an increase in the rate of corrosion but also gave insights about the dependence of corrosion on other factors as well. As observed before, the rate of corrosion is seen to be increasing with an increase in the pressure of CO_2 in the pipeline. On the other hand, the factors of internal pressure within the pipe, the shear stress, and the velocity of flow of CO_2 within the pipe have not been found to be affecting the rate of corrosion significantly in this dataset.

As explained, the anticorrosion elements in CO_2 pipelines are of utmost importance. They must be analyzed to provide better and more anticorrosion materials that can be used for the purpose. The many tests with the variables provided in this dataset attempted to visualize the trends that the corrosion rate shows when other external as well as internal factors are varied, and the results were recorded.

References

1. Georgiades, E., Scianni, C., Tamburri, M.N., Biofilms associated with ship submerged surfaces: implications for ship biofouling management and the environment. *Front. Marine Sci.*, *10*, 1197366, 2023.
2. Schmitt, G., Schütze, M., Hays, G.F., Burns, W., Global needs for knowledge dissemination, research, and development in materials deterioration and corrosion control. *World Corros. Organ.*, *38*, 14, 2009.
3. National Association of Corrosion Engineers (n.d.). Assessment of the global cost of corrosion. NACE International. http://impact.nace.org/economic-impact.aspx#:~:text=The%20global%20cost%20of%20corrosion%20is%20estimated%20to,equivalent%20to%203.4%25%20of%20the%20global%20GDP%20%282013%29.
4. Xia, X.C., Cao, X.K., Cai, G.Y., Jiang, D., Zhang, F., Dong, Z.H., Underwater superoleophobic composite coating characteristic of durable anti-fouling and anti-corrosion properties in marine environments. *Colloids Surf., A*, *628*, 127323, 2021.
5. Idumah, C., II, Obele, C.M., Emmanuel, E.O., Hassan, A., Recently emerging nanotechnological advancements in polymer nanocomposite coatings for anti-corrosion, anti-fouling and self-healing. *Surf. Interfaces*, *21*, 100734, 2020.
6. Chen, Y., Zhang, G., Zhang, G., Ma, C., Rapid curing and self-stratifying lacquer coating with anti-fouling and anti-corrosive properties. *Chem. Eng. J.*, *421*, 129755, 2021.
7. Gao, M., Quan, X., Wang, J., Wang, Z., Preparation and characterisation of coatings incorporated with poly (aniline-co-nitroaniline) nanoparticles having anti-fouling and anti-corrosion behaviour. *Ind. Eng. Chem. Res.*, *59*, 51, 22173–22186, 2020.
8. Jin, H., Wang, J., Tian, L., Gao, M., Zhao, J., Ren, L., Recent advances in emerging integrated anti-fouling and anti-corrosion coatings. *Mater. Des.* *213*, 2022.
9. Li, X., Zhang, P., Huang, H., Hu, X., Zhou, Y., Yan, F., An electrochemical study of pH influences on corrosion and passivation for a Q235 carbon steel in HNO3–NaNO2, HAc–NaNO2 and HCl–NaNO2 solutions. *RSC Adv.*, 2019, https://doi.org/10.1039/C9RA08482G.
10. Živica, V., Significance and influence of the ambient temperature as a rate factor of steel reinforcement corrosion. *Bull. Mater. Sci.*, *25*, 375–379, 2002.
11. Kumar, K.P., Unal, A., Pillai, V.J., Murthy, H., Niranjanamurthy, M. (Eds.), *Data Engineering and Data Science: Concepts and Applications*, 2023.
12. Shibu, M., Kumar, K.P., Pillai, V.J., Murthy, H., Chandra, S., Structural health monitoring using AI and ML based multimodal sensors data. *Meas. Sens.*, *27*, 100762, 2023.

Index

Printed in the USA/Agawam, MA
September 16, 2026

910661.158